AF731394

Handbook of Number Theory I

Handbook of Number Theory I

by

József Sándor

Babes-Bolyai University of Cluj,
Cluj-Napoca,
Romania

Dragoslav S. Mitrinović

formerly of the University of Belgrade,
Servia

and

Borislav Crstici

formerly of the Technical University of Timisoara,
Romania

Springer

A C.I.P. Catalogue record for this book is available from the Library of Congress.

ISBN-10 1-4020-4215-9 (HB)
ISBN-13 978-1-4020-4215-7 (HB)
ISBN-10 1-4020-3658-2 (e-book)
ISBN-13 978-1-4020-3658-3 (e-book)

Published by Springer,
P.O. Box 17, 3300 AA Dordrecht, The Netherlands.

www.springeronline.com

Printed on acid-free paper

1st ed. 1995. 2nd printing

All Rights Reserved
© 2006 Springer
No part of this work may be reproduced, stored in a retrieval system, or transmitted
in any form or by any means, electronic, mechanical, photocopying, microfilming, recording
or otherwise, without written permission from the Publisher, with the exception
of any material supplied specifically for the purpose of being entered
and executed on a computer system, for exclusive use by the purchaser of the work.

Printed in the Netherlands.

TABLE OF CONTENTS

CHAPTER XII

PREFACE

It is the aim of this book to systematize and to present in an easily accessible framework the most important results from some parts of Number Theory, which are expressed by inequalities or by estimates.

This book focuses on the most important arithmetic functions in Number Theory, such as $\varphi(n)$, $\sigma(n)$, $d(n)$, $\omega(n)$, $\Omega(n)$, $\mu(n)$, $\pi(n)$, $P(n)$, $\psi(x, y)$, $e(\alpha)$, $\chi(n)$, $\binom{n}{k}$, $P(n, k)$ and so on, together with various generalizations, analogues and extensions of these functions, and also properties of some functions related to the distribution of the primes and of the quadratic residues and to partitions, etc. It is sufficient to take a look at the contents in order to realize the variety of the approached subjects in each chapter. The chapters are divided in consecutive "themes." Each theme expresses properties which are similar or contiguous by their nature.

We have attempted to make a selection which reflects the current situation in the domain regarded. On the other hand, as a basic characteristic of this book, we have included the results of the pioneers in the domains regarded, as well as some results reflecting the evolution from the pioneer works up to recent ones. Our aim was to give the most precise references, i.e. original ones, even when the results are standard and can be found in textbooks. To this purpose we have used a wealth of literature, consisting of books, monographs, journals, separates, reviews from *Mathematical Reviews* and from *Zentralblatt für Mathematik*, etc. Consequently, we hope that our book will also be useful for the nonspecialist, who – if need be – can find the result or the reference he needs. First of all, we consider the professional mathematician who works in a certain domain of Number Theory and who wishes to use material outside his own field in Number Theory. In this way, we hope to contribute to the unity of Number Theory despite of its great variety.

Of course, the choice of subjects reflects the personality of the authors. Therefore, we do not exclude the possibility that some important themes and aspects even with respect to our proclaimed goal – are missing. We will be grateful to all readers who will honour us with their remarks. Their opinions will be considered with the greatest attention by the authors.

Our book is not the first of this kind. The *Handbook of Estimates in the Theory of Numbers* by B. Spairman and K.S. Williams (Carleton University, Ottawa) appeared in 1975. The book by D.S. Mitrinović and M.S. Popadić, *Inequalities in Number Theory* (Naučni Podmladak, Univerzitet u Nišu) appeared in 1978.

The latter monograph served as impulse for the present book, as Prof. D.S. Mitrinović had the intention to publish a second edition – revised and enlarged – of the monograph written together with the late M.S. Popadić. Because of M.S. Popadić's death, this project could not be accomplished. Prof. D.S. Mitrinović then addressed the invitation for cooperation to Prof. J. Sándor. This circumstance led to an essentially new book, in concept, as well as in material. Prof. D.S. Mitrinović wishes to thank all mathematicians who have made remarks concerning his previous book. These remarks have been taken into account if they refer to the material included in the present book. Prof. J. Sándor wishes to thank the mathematicians all over the world who have had the kindness to offer him their papers. The gratefulness of J. Sándor is especially addressed to the colleagues from the Mathematics Institute of Budapest (Hungary) as well as from *Institutul Matematic al Academiei Române* – Bucharest (Romania). The authors hope that the mathematicians who have been in touch with them, in matters concerning the material of this book, will recognise themselves in the above acknowledgements. The list would be too long to mention them all.

The gratefulness of the authors is addressed to the staff of *Kluwer Academic Publishers*, especially to Dr. Paul Roos, Ms. Angelique Hempel and Ms. Anneke Pot for support while typesetting the manuscript.

The camera-ready manuscript for the present book was prepared by Mr. Antonius Stanciu (Timişoara, Romania) to whom the authors express their gratitude. The authors also acknowledge the assistance of Mr. Dan Magiaru in the final elaboration of the text.

The Authors

Unfortunately, after the manuscript was finished and during its preparation for printing, Professor D.S. Mitrinović died (the 2nd of April, 1995), not having the chance to see his last work in libraries.

June 1995

B.C.
J.S.

BASIC SYMBOLS

Below appear the most important symbols. The other ones are explained in the text.

$f(x) = O(g(x))$ or $f(x) \ll g(x)$	For a range of x-values, there is a constant A such that the inequality $$\lvert f(x)\rvert \le A\, g(x)$$ holds over the range
$f(x) \gg g(x)$	$g(x) \ll f(x)$, (or $g(x) = O(f(x))$)
$f(x) = o(g(x))$	as $x \to \infty$, means $$\lim_{x\to\infty} \frac{f(x)}{g(x)} = 0$$ ($g(x) \neq 0$ for x large.) The same meaning is used when $x \to \infty$ is replaced by $x \to \alpha$, for any fixed α.
$f(x) \sim g(x)$	as $x \to \infty$, means $$\lim_{x\to\infty} \frac{f(x)}{g(x)} = 1$$ ($g(x) \neq 0$ for x large.) The same is true when $x \to \infty$ is replaced with $x \to \alpha$.
$f(x) = \Omega(g(x))$	$f(x) = o(g(x))$ does not hold.
$f(x) = \Omega_+(g(x))$	There exists a positive constant K such that $f(x) > Kg(x)$ is satisfied by values of x surpassing all limit.
$f(x) = \Omega_-(g(x))$	$f(x) < -Kg(x)$ is satisfied by values of x surpassing all limit.
$f(x) = \Omega_\pm(g(x))$	we have both $f(x) = \Omega_+(g(x))$ and $f(x) = \Omega_-(g(x))$

BASIC NOTATIONS

All notations (excepting the most familiar ones) are specificated in the text. The following appear through all chapters of the work.

$\varphi(n)$	Euler's totient function
$\sigma(n)$	sum of divisors function
$d(n)$	number of distinct divisors of n
$\omega(n)$	number of distinct prime factors of n
$\Omega(n)$	total number of prime factors of n
$J_k(n)$	Jordan's arithmetical function
$\sigma_k(n)$	sum of kth powers of divisors of n
$\pi(n)$	number of all primes $\leq n$
$J_k^*(n), \varphi^*(n), d^*(n), \sigma_k^*(n)$	unitary analogues of the arithmetical functions $J_k, \varphi, d, \sigma_k$
$\zeta(s) = \sum_{n=1}^{\infty} \frac{1}{n^s}$	for Re $s > 1$ (Riemann's zeta function)
$\psi_k(n) = n^k \cdot \prod_{p\mid n} \left(1 + \frac{1}{p^k}\right)$, $\psi_k(1) = 1$	Dedekind's arithmetical function
$\psi(n)$	Dedekind's arithmetical function, or $\psi(n) = \sum_{m \leq n} \Lambda(m)$ Chebyshev's function
γ	Euler's constant, or an arbitrary constant, as specificated in the text
(a, b)	g.c.d. of a and b, or an ordered pair

$[a, b]$	l.c.m. of a and b
$\gamma(n)$	greatest squarefree divisor of n (the "core" of n)
$\varphi_F(n)$	Steven's generalization of φ
$\varphi_S(n)$	Cohen's generalization of φ
$\varphi(x, n)$	Legendre's totient function
$\Lambda(n) = \begin{cases} \log p, & n = p^m \\ 0, & \text{otherwise} \end{cases}$	von Mangoldt's function
$d_k(n)$	Piltz's divisor function
$f * g$	Dirichlet convolution
$P(n)$	greatest prime factor of n, or the number of unrestricted partitions of n
$p(n)$	least prime factor of n
$\left(\frac{a}{b}\right)$ or $(a \mid b)$	Legendre's symbol
$\theta(n)$	number of squarefree divisors of n
$d^e(n)$	number of exponential divisors of n
$\sigma^e(n)$	sum of exponential divisors of n
$d^{**}(n)$	number of bi-unitary divisors of n
$\sigma^{**}(n)$	sum of bi-unitary divisors of n
$a(n)$	number of nonisomorphic abelian groups of order n
$\psi(x, y)$	number of positive integers $\leq x$ and free of prime divisors $> y$

$\Phi(x, y)$	number of positive integers $\leq x$ with no prime divisors $< y$
$\psi(x, y; a, q)$	number of positive integers $\leq x$, free of prime factors $> y$, and satisfying $n = a \pmod q$
$s(n) = \sigma(n) - n$	number of aliquot divisors of n
$F_n = 2^{2^n} + 1$	Fermat's numbers
$M_p = 2^p - 1$	Mersenne's numbers
$B_k(n) = \sum a_i p_i^k$	if $n = p_1^{a_1} \cdots p_r^{a_r}$ (prime factorization)
$\beta_k(n) = \sum p_i^k$	
$B(n) = B_1(n)$	
$\beta(n) = \beta_1(n)$	
$B^1(n) = \sum_{p^\alpha \| n} p^\alpha$	
$\lambda(n) = (-1)^{\omega(n)}$	Liouville's function
$a \mid b$	a divides b
$\pi_i(x)$	number of integers $n \leq x$ satisfying $\Omega(n) = i$
$\rho(x)$	number of integers $n \leq x$ satisfying $\omega(n) = i$
$\omega(n; E)$	number of distinct primes in the set (of primes) E that divide n
$\mu(n)$	Möbius' function
$M(x) = \sum_{n \leq x} \mu(x)$	
$\mu_k(n)$	Möbius function of order k
$[x]$	integer part of x
$\{x\} = x - [x]$	fractional part of x

$q(k, l)$	smallest squarefree integer in the arithmetic progression $km + l\,(m = 0, 1, \ldots)$
$\|A\|$	cardinality of set A
$Q_k(x)$	number of k-free integers $\leq x$ ($k \geq 2$, integer)
$Q_r(x, k, l)$	number of r-free integers $\leq x$ in the arithmetic progression $kt + l\,(t = 0, 1, \ldots)$
$Q_{k,r}(x)$	number of (k, r)-integers $\leq x$
$N_k(x)$	number of k-full integers $\leq x$
$\mathrm{li}\, x = \int_0^x \frac{1}{\log t} \mathrm{d}t$	
$\theta(x) = \sum_{p \leq x} \log p$	Chebyshev's function
p_n	the nth prime
$d_n = p_{n+1} - p_n$	
$\|z\|$	modulus of a complex number z
$\arg z$	argument of the complex number z
$e(\alpha) = \exp(2\pi i \alpha)$	
$\exp(z) = e^z$	
$\tau(n)$	Ramanujan's arithmetic function
$d(A)$	asymptotic density of the set A
$\underline{d}(A)$	lower asymptotic density of the set A
$\\|x\\| = \min(x - [x], [x] + 1 - x)$	distance of x to the nearest integer
$\pi(x; k, l)$	number of primes $\leq x$ which are $\equiv l \pmod{k}$ $(k > 0)$

$p_n(k, l)$	the nth prime $= l \pmod k$
$p(k, l)$	the least prime $= l \pmod k$
$\theta(x; q, a) = \sum_{p \le x, p \equiv a (\mathrm{mod}\, q)} \log p$	
$\psi(x; q, a) = \sum_{n \le x, n \equiv a (\mathrm{mod}\, q)} \Lambda(n)$	
$\pi(x; \underline{a}, \underline{b})$	number of integers such that $1 < \underline{a}\, n + \underline{b} \le x$, $\underline{a}\, n + \underline{b}$ are primes ($\underline{a}$ and $\underline{b}$ are k-dimensional integer vectors)
$\pi_c(x)$	number of primes in the sequence $[n^c]$ not exceeding x
$\Pi_c(x)$	number of primes $p \le x$ for which $[p^c]$ is prime
χ	a character
$\binom{n}{k} = C_n^k$	binomial coefficient
$n! = 1 \cdot 2 \cdot \cdots \cdot n$	factorial of n
$A \setminus B$	the difference of sets A and B
$\Gamma(x)$	Euler's gamma function
$!(z) = \int_0^{+\infty} \frac{e^{-z} \cdot (t^z - 1)}{(t-1)} \mathrm{d}t$ ($\mathrm{Re}\, z > 0$)	the left factorial function
$P(n; k)$	number of partitions of n into exactly k positive integer parts
$P^*(n; k)$	number of partitions of n into at most k summands
$q(n; k)$	number of partitions of n into k distinct parts

$\underline{P}(n, A)$	number of partitions of n into numbers of the set A
$r(n, A)$	number of partitions of n with no parts belonging to A
$r(n, m)$	number of partitions whose parts are $\geq m$
$\rho(n, m)$	number of partitions into unequal parts $\geq m$
$R(n, a)$	number of partitions of n such that $n = n_1 + \cdots + n_t$ whose subsums $n_{i_1} + \cdots + n_{i_j}$ are all different from a
$Q(n, a)$	number of partitions of n such that $n = n_1 + \cdots + n_t$ whose subsums $n_{i_1} + \cdots + n_{i_j}$ are all different from a, and each part is allowed to occur at most once
$a \equiv b \pmod{m} \Leftrightarrow m \mid (a - b)$	
$n(p)$	smallest positive quadratic nonresidue (mod p)
$r_k(p)$	least prime kth power residue (mod p)
$n_k(p)$	least kth power nonresidue mod p
$g(p)$	least primitive root (mod p)
$N(f, m)$	number of solutions to the congruence $f(x) \equiv 0 \pmod{m}$ ($m > 1$, integer), counted mod m, including multiplicities ($f(x)$ a polynomial)

$$\Delta f(n) = f(n+1) - f(n)$$
$$\Delta^k f(n) = \Delta(\Delta^{k-1} f(n))$$

Chapter I

EULER'S φ-FUNCTION

§ I. 1 Elementary inequalities for φ

1) $\varphi(n) \geq \sqrt{n}$
for $n \neq 2$ and $n \neq 6$

A.M. Vaidya. *An inequality for Euler's totient function.* Math. Student 35 (1967), 79–80.

2) $\varphi(n) > n^{2/3}$
for $n > 30$

D.G. Kendall and R. Osborn. *Two simple lower bounds for Euler's function.* Texas J. Sci. 17 (1965), No. 3.

3) If $a > 6$ and $n > 2$, then

$$a^{\varphi(n)} > an$$

R.L. Goldstein. *An inequality for Euler's function* $\varphi(n)$. Math. Mag. 40 (1956), 131.

4) $\varphi(n) \leq n - \sqrt{n}$
if n is composite

W. Sierpiński. *Elementary theory of numbers.* Warsawa, 1964.

5) $\varphi(n) > \dfrac{\log 2}{2} \dfrac{n}{\log n}$
for $n \geq 3$

H. Hatalová and T. Šalát. *Remarks on two results in the elementary theory of numbers.* Acta Fac. Rer. Natur Univ. Comenian. Math. 20 (1969), 113–117.

§ I. 2 Inequalities for $\varphi(mn)$

1) $\varphi(m)\,\varphi(n) \leq \varphi(mn) \leq n \cdot \varphi(m); \quad m, n = 1, 2, 3, \ldots$
(Simple consequence of the formula expressing φ)

2) $(\varphi(mn))^2 \leq \varphi(m^2) \cdot \varphi(n^2); \quad m, n = 1, 2, 3, \ldots$

T. Popoviciu. Gaz. Mat. (Bucureşti), 46 (1940), p. 334.

§ I. 3 Relations connecting φ, σ, d

1) a) $\tau(n) \geq \varphi(n) + d(n), \quad n = 2, 3, \ldots$

H.d. Bagchi and M. Gupta. *Problem 343*. Jber. Dt. Math. Verein. 57 (1954), 8–9.

b) $\sigma_k(n) \geq (\varphi(n))^k + (d(n))^k, \quad n = 2, 3, \ldots$

E. Trost. *Problem 202*. Elem. Math. 9 (1954), 21.

c) $\sigma(n) \leq \varphi(n) + d(n) \cdot (n - \varphi(n)), \quad n = 1, 2, \ldots$

J. Sándor. *Some diophantine equations for particular arithmetic functions* (Romanian.) Seminarul de teoria structurilor. No. 53, Univ. Timişoara, 1989, pp. 1–10 (see p. 8.)

2) $\sigma(n) + \varphi(n) \leq n \cdot 2^{\omega(n)} \leq n \cdot d(n)$

where $\omega(n)$ denotes the number of distinct prime factors of n $(n \geq 2)$

3) If $\varphi(n) + \sigma(n) = k \cdot n$ $(k > 1$, integer), then

$$\omega(n) > \frac{\log(k-1)}{\log 2}$$

C.A. Nicol. *Some diophantine equations involving arithmetic functions*. J. Math. Analysis Appl. 15 (1966), pp. 154–161.

4) a) $\varphi(n) \cdot d(n) \geq n$

for all $n = 1, 2, 3, \ldots$

R. Sivaramakrishnan. *Problem E 1962*. Amer. Math. Monthly 74 (1967), p. 198.

b) $\varphi(n)\, d(n) \geq \sigma(n)$

for n odd

J. Sándor. *On Dedekind's Arithmetical Function*. Seminarul de teoria structurilor. No. 51, Univ. Timişoara, 1988, pp. 1–15 (see p. 11.)

c) $\varphi(n) d(n) \geq \varphi(n) + n - 1$

$n = 1, 2, 3, \ldots$

J. Sándor. As in 1) c), (p. 5.)

Remark. For other inequalities of this type, see also:

J. Sándor and R. Sivaramakrishnan. *The many facets of Euler's totient*. III. Nieuw Arch. Wiskunde 11 (1993), 97–130.

5) $\dfrac{6}{\pi^2} < \dfrac{\sigma(n)\, \varphi(n)}{n^2} = x_n < 1$

for $n \geq 2$ and $\limsup\limits_{n\to\infty} x_n = 1$, $\liminf\limits_{n\to\infty} x_n = 6/\pi^2$

G.H. Hardy and E.M. Wright. *An Introduction to the Theory of Numbers.* 4th ed. Oxford, New York, 1965 (See Theorem 329.)

6) a) $\liminf_{n\to\infty} \frac{\sigma(n)}{\varphi(n)} = 1$

b) $\limsup_{n\to\infty} \frac{\sigma(n)}{\varphi(n)} = +\infty$

c) The sequence

$$\left(\frac{\sigma(n)}{\varphi(n)}\right)$$

is everywhere dense in $(1, +\infty)$

B.S.K.R. Somayajulu. *The sequence* $\sigma(n)/\varphi(n)$. Math. Student 45 (1977), 52–54.

7) a) $\varphi(n)\, d^2(n) \leq n^2$
for $n \neq 4$

S. Porubsky. *Problem E 2351.* Amer. Math. Monthly 79 (1972), 394.

b) $\varphi(n)\, d^2(n) \geq \sigma(n)$
for all $n = 1, 2, 3, \ldots$

A. Mąkowski. *Problem 538.* Math. Mag. 37 (1974), 55.

8) $\varphi(n) > d(n)$
for $n > 30$ and $\varphi(n) = d(n)$ only if $n \in \{1, 3, 8, 10, 24, 30\}$

G. Pólya and G. Szegö. *Problems and theorems in analysis.* II. Springer V. 1976, Part VIII, Problem 45.

9) $\varphi(n) \geq \pi(n)$
for $n \geq 90$

L. Moser. *On the equation* $\varphi(n) = \pi(n)$. Pi Mu Epsilon J. 1951, 101–110.

§ I. 4 Inequalities for J_k, σ_k, ψ_k

1) $n^k d(n) \geq J_k(n) + \sigma_k(n) \geq 2n^k$
for all $n, k = 1, 2, 3, \ldots$

A. Mąkowski. *Problem 339.* Elem. Math. 15 (1960), 39–40.

2) $J_k(n)\, d^2(n) \geq J_k(n) \sum_{i|n} d(i) \geq \sigma_k(n)$

J. Sándor. *On Jordan's arithmetical function.* Math. Student. 52 (1984), 91–96 (1988.)

3) a) $J_k(n) \cdot d(n) \geq n^k$

b) $J_k(n) \sum_{i|n} \sigma_k(i) \geq n^k \cdot d(n)$

J. Sándor. *Ibid.*

4) a) $(\sigma_k(n))^{J_k(n)} < n^{k \cdot n^k}$
for $n \geq 2, k \geq 1$ natural numbers.

b) If for every prime divisor p of n we have $p^k \geq 5$, then

$$(J_k(n))^{\sigma_k(n)} > n^{k \cdot n^k}$$

J. Sándor. *On the arithmetical functions $\sigma_k(n)$ and $\varphi_k(n)$*. Math. Student. 58 (1990), 49–54.

5) a) $J_k(n) + \psi_k(n) \geq 2n^k$

b) $\psi_k(n) - (\varphi(n))^k \geq 2^{k\omega(n)}$
where ψ_k is the generalized Dedekind function,

i.e. $\psi_k(n) = n^k \cdot \prod_{p|n} \left(1 + \frac{1}{p^k}\right)$

J. Sándor. *On Dedekind's arithmetical function*. Seminarul de teoria structurilor. No. 51, Univ. Timişoara, 1988, pp. 1–15 (see p. 3.)

Remark. Result 5) a) in case $k = 1$ is attributed to
Ch.R. Wall. *Problem B-510*. Fib. Quart. 22 (1984), 371.

6) If $\omega(n) \geq 2$, then

$$\sigma_k(n) \cdot J_k(n) \leq n^{2k} - \sum_{p|n} p^k + 1$$

J. Sándor. *Note on the function σ and φ*. Bull. Number Theory Rel. Topics 12 (1988), 78–80.

§ I. 5 Unitary analogues of J_k, σ_k, d

Let J_k^* denote the unitary analogue of the Jordan totient function. Then:

1) $J_k^*(n) + d^*(n) \leq \sigma_k^*(n)$

2) $J_k^*(n) + \sigma_k^*(n) \leq n^k \cdot d^*(n)$

3) $\frac{1}{\zeta(2k)} < \frac{\sigma_k^*(n) J_k^*(n)}{n^{2k}} < 1$

4) $d^*(n) \cdot n^k \leq J_k^*(n)\, (d^*(n))^2 \leq n^{2k}$

where d^* and σ_k^* are the unitary analogues of d and σ_k

J. Sándor and L. Tóth. *On certain number–theoretic inequalities*. Fib. Quart. 28 (1990), 255–258.

§ I. 6 Composition of φ, σ, ψ

1) $\varphi\left(n \cdot \left[\dfrac{\sigma(n)}{n}\right]\right) \leq n$

A. Oppenheim. *Problem 5591*. Amer. Math. Monthly 75 (1968), 552.

2) $\varphi\left(n \cdot \left[\dfrac{n}{d(n)}\right]\right) \leq (\varphi(n))^2$

(Here $[x]$ denotes, as usual, the integer part of x)

J. Sándor. *Some arithmetic inequalities*. Bulletin Number Theory Rel. Topics, 11 (1987), 149–161.

3) $(\sigma(n))^{\varphi(n)} < n^n$
for all $n \geq 2$

J. Sándor. *Ibid.*

4) a) $\varphi\left(n \cdot \left[\dfrac{\psi(n)}{n}\right]\right) \leq n$

b) $(\psi(n))^{\varphi(n)} < n^n$
for all $n \geq 2$

$(\varphi(n))^{\psi(n)} > n^n$
if all prime factors of n are ≥ 5

J. Sándor. *On Dedekind's arithmetical function*. Seminarul de teoria structurilor. No. 51, Univ. Timişoara, 1988, pp. 1–15 (see pp. 6–7.)

Here $\psi(n)$ denotes Dedekind's arithmetical function, i.e. $\psi(n) = n \cdot \prod_{p|n}\left(1 + \dfrac{1}{p}\right)$

§ I. 7 Composition of σ, φ

1) $\liminf_{n\to\infty} \dfrac{\sigma(\varphi(n))}{n} \leq \inf_{4|n} \dfrac{\sigma(\varphi(n))}{n} \leq \dfrac{1}{2} + \dfrac{1}{2^{34} - 4}$

A. Mąkowski and A. Schinzel. *On the functions $\sigma(n)$ and $\varphi(n)$*. Colloq. Math. 13 (1965–65), 95–99.

2) $\inf \frac{\sigma(\varphi(n))}{n} > 0$

C. Pomerance. *On the composition of the arithmetic functions σ and φ*. Colloq. Math. 58 (1989), 11–15.

3) $\limsup_{n\to\infty} \frac{\varphi(\sigma(n))}{n} = \infty$

$$\limsup_{n\to\infty} \frac{\varphi(\varphi(n))}{n} = \frac{1}{2}$$

A. Mąkowski and A. Schinzel. *See 1)*.

4) $\liminf_{n\to\infty} \frac{\varphi(\sigma(n))}{n} = 0$

$$\limsup_{n\to\infty} \frac{\sigma(\varphi(n))}{n} = \infty$$

L. Alaoglu and P. Erdős. *A conjecture in elementary number theory*. Bull. Amer. Math. Soc. 50 (1944), 881–882.

5) $\limsup_{n\to\infty} \frac{\sigma_k(\varphi(n))}{n^k} \geq \frac{\zeta(k)}{\zeta(2k)}$

for $k > 1$

$$\liminf_{n\to\infty} \frac{\varphi(\sigma_k(n))}{n} = 0$$

for k odd.

J. Sándor. *Note on Jordan's arithmetical function*. Seminar Arghiriade, Univ. Timişoara, No. 19, 1989.

Remark. For other results on $\sigma_k \circ \varphi_s$ or $\varphi_k \circ \varphi_s$ (with "∘" denoting the composition of functions), see

J. Sándor. *A note on the functions $\sigma_k(n)$ and $\varphi_k(n)$*. Studia Univ. Babeş-Bolyai Math. 35 (1990), 3–6.

§ I. 8 On the function $n/\varphi(n)$

1) a) $\liminf_{n\to\infty} \frac{\varphi(n)\log\log n}{n} = e^{-\gamma}$

E. Landau. *Über den Verlauf der zahlentheoretischen Funktion φ(x)*. Archiv der Mathematik und Physik, 5 (1903), 86–91.

b)
$$\liminf_{n\to\infty} (\log\log n)^{1/k} \cdot \max\left(\frac{\varphi(n+1)}{n+1}, \ldots, \frac{\varphi(n+k)}{n+k}\right) =$$
$$= e^{-\gamma/k} \cdot \prod_{n<k}\left(1-\frac{1}{p}\right)^{1/p} \cdot f(k)$$

where $f(k) = \prod_{p|k,p<k}\left(1-\frac{1}{p}\right)^{\frac{1}{p}} \cdot \prod_{p\nmid k,p<k}\left(1-\frac{1}{p}\right)^{\frac{1}{k}\cdot\left[\frac{k}{p}\right]+\frac{1}{k}}$

M. Hausman. *Generalization of a theorem of Landau*. Pac. J. Math. 84 (1979), 91–95.

Remark. For $k = 1$ one reobtains Landau's theorem a).

2) For infinitely many positive integers n we have
$$\frac{n}{\varphi(n)} > e^{\gamma} \cdot \log\log n$$
(where γ is Euler's constant)

J.-L. Nicolas. *Petites valeurs de la fonction d'Euler*. J. Number Theory 17 (1983), 375–388.

3) a) If $n \geq 3$, then
$$\frac{n}{\varphi(n)} < e^{\gamma}\log\log n + \frac{2.50637}{\log\log n}$$

J.B. Rosser and L. Schoenfeld. *Approximate formulas for some functions of prime numbers*. Illinois J. Math. 6 (1962), 64–94.

b) There is a positive constant $C > 0$ such that
$$\frac{n}{\varphi(n)} < C \cdot \log\log\varphi(n)$$
for all $n \geq 3$

J. Sándor. *Remarks on the functions $\varphi(n)$ and $\sigma(n)$*. Seminar on Math. Analysis, Babeş–Bolyai Univ., Preprint Nr. 7, 1989, pp. 7–12.

§ I. 9 Minimum of $\varphi(n)/n$ for consecutive values of n

1) For every $k \geq k_0$, for all but $O(x/k_0^2)$ of the $n \leq x$ we have
$$\frac{c_1}{\log\log k} \leq \min\left\{\frac{\varphi(n+1)}{n+1}, \ldots, \frac{\varphi(n+k)}{n+k}\right\} \leq \frac{c_2}{\log\log k}$$
(c_1, c_2 — suitable absolute positive constants)

I. Kátai, *Maximum of number-theoretical functions in short intervals*. Ann. Univ. Sci. Budapest 18 (1975), 69–74.

2) For integral $k > 1$,

$$\limsup_{n\to\infty} \min\left\{\frac{\varphi(n+1)}{n+1},\ldots,\frac{\varphi(n+k)}{n+k}\right\} = \min\left\{\frac{\varphi(1)}{1},\ldots,\frac{\varphi(k)}{k}\right\}$$

M. Hausman. *An upper limit property of the Euler function.* Canad. Math. Bull. 23 (1980), 375–377.

§ I.10 On $\varphi(n+1)/\varphi(n)$

1) $$\limsup_{n\to\infty} \frac{\varphi(n+1)}{\varphi(n)} = +\infty$$

$$\liminf_{n\to\infty} \frac{\varphi(n+1)}{\varphi(n)} = 0$$

B.S.K.R. Somayajulu. *On Euler's totient function* $\varphi(n)$. Math. Student 18 (1950), 31–32 (1951.)

2) The sequence $\left(\frac{\varphi(n+1)}{\varphi(n)}\right)$ is dense in $(0, +\infty)$

A. Schinzel. *Generalization of a theorem of B.S.K.R. Somayajulu on the Euler's function* $\varphi(n)$. Ganita 5 (1954), 123–128.

3) For each positive integer m there exist positive integers n, h such that

$$\frac{\varphi(n-1)}{\varphi(n)} > m$$
$$\frac{\varphi(n+1)}{\varphi(n)} > m$$
$$\frac{\varphi(h)}{\varphi(h-1)} > m$$
$$\frac{\varphi(h)}{\varphi(h+1)} > m$$

A. Schinzel et W. Sierpiński. *Sur quelques propriétés des fonctions* $\varphi(n)$ *et* $\sigma(n)$. Bull. Acad. Polon. Sci. Cl. III. 2 (1954), 463–466 (1955.)

4) For each positive integers m, k there exist positive integers n, h such that

$$\frac{\varphi(n+i)}{\varphi(n+i-1)} > m$$

and

$$\frac{\varphi(h+i-1)}{\varphi(h+i)} > m$$

for $i = 1, 2, \ldots, k$

A. Schinzel. *Quelques théorèmes sur les fonction* $\varphi(n)$ *et* $\sigma(n)$. Bull. Acad. Polon. Sci. Cl. III. 2 (1954), 467–469 (1955.)

5) Let $a_1, \ldots, a_h$ be any sequence of nonnegative numbers or infinity. Then there exists an infinitive sequence of natural numbers $n_1 < n_2 < \cdots$ such that

$$\lim_{k\to\infty} \frac{\varphi(n_k+i)}{\varphi(n_k+i-1)} = a_i$$
$$i = 1, 2, 3, \ldots, h$$

A. Schinzel. *On functions* $\varphi(n)$ *and* $\sigma(n)$. Bull. Acad. Polon. Sci. Cl. III. 3 (1955), 415–419.

6) Let $\lim\limits_{n\to\infty} g(n)/\log\log\log n = 0$. Then there exists an infinite sequence n_k such that for all $1 \le i \le g(n_k)$

$$1 - \varepsilon \le \frac{\varphi(n_k+i)}{\varphi(n_k+i-1)} < 1 + \varepsilon_k$$

where $\varepsilon_k \to 0$ $(k \to \infty)$

P. Erdős. *Some remarks on Euler's* φ *function*. Acta Arith. 4 (1955), 10–19.

7) Let $a_1, \ldots, a_k, b_1, \ldots, b_k$ be positive numbers. Then the necessary and sufficient condition for the existence of an infinite sequence (n_l) of natural numbers such that

$$\lim_{l\to\infty} \frac{\varphi(n_l+i)}{\varphi(n_l+i+1)} = a_i$$

$$\lim_{l\to\infty} \frac{\sigma(n_l+i)}{\sigma(n_l+i+1)} = b_i$$

$(i = 1, 2, \ldots, k)$ is the existence of a sequence $\left(n_l'\right)$ of natural numbers with

$$\lim_{l\to\infty} \frac{h(n_l'+i)}{h(n_l'+i+1)} = a_i b_i$$

$(i = 1, 2, \ldots, k)$, where $h(n) = \varphi(n)\,\sigma(n)/n^2$

P. Erdős, K. Győry, and Z. Papp. *On some new properties of functions* $\sigma(n), \varphi(n), d(n),$ *and* $\gamma(n)$ (Hungarian.) Mat. Lapok 28 (1980), 125–131.

8) a) For any given sequence of h non-negative numbers $a_1, a_2, \ldots, a_h$ and $\varepsilon > 0$, there exist positive constants $C = C(a, \varepsilon)$ and $x_0 = x_0(a, \varepsilon)$ such that the number of positive integers $n \le x$ satisfying

$$\left| \frac{\varphi(n+i)}{\varphi(n+i-1)} - a_i \right| < \varepsilon$$

$(1 \le i \le h)$ is greater than $Cx/\log^{h+1} x$, whenever $x > x_0$

A. Schinzel and Y. Wang. *A note on some properties of the functions* $\varphi(n)$, $\sigma(n)$ *and* $\theta(n)$. Bull. Acad. Polon. Sci. Cl. III 4 (1956), 207–209 and Ann. Pol. Math. 4 (1958), 201–213.

b) Let $N =$ card $\left\{ p < x, p \text{ prime} : \left| \frac{\varphi(p+\nu+1)}{\varphi(p+\nu)} - a_\nu \right| < \varepsilon, 1 \le \nu \le k \right\}$.

Then

$$N > C(a, \varepsilon) \cdot x/((\log x)^{k+2} \cdot \log\log x)$$

Y. Wang. *A note on some properties of the arithmetical functions* $\varphi(n), \sigma(n)$ *and* $d(n)$. Acta Math. Sinica 8 (1958), 1–11.

c) $N > c_1(a, \varepsilon)\dfrac{x}{\log x}$

where N is defined in b).

P. Erdős and A. Schinzel. *Distributions of the value of some arithmetical functions*. Acta Arith. 6 (1961), 473–485.

d) There exists the distribution function

$$\lim_{x\to\infty} \frac{1}{\pi(x)} \cdot \operatorname{card}\left\{p < x : \frac{\varphi(p+\nu)}{p+\nu} \geq C, \nu = 1, 2, \dots, k\right\}$$

P. Erdős and A. Schinzel. *Ibid.*

§ I.11 On $(\varphi(n+1), \varphi(n))$

Let $N_\varphi(x) = \operatorname{card}\{n \leq x : p\nmid(\varphi(n), \varphi(n+1), p \text{ odd prime})\}$. Then

$$\frac{x}{\log x} \cdot (\log\log\log x)^{-1} \ll N_\varphi(x) \ll \frac{x}{\log x} \cdot \exp(A \log\log x (\log\log\log x)^{-1/2})$$

where A is a constant.

E.J. Scourfield. *On the coprimality of values of Euler's function at consecutive integers*. J. Reine Angew. Math. 336 (1982), 91–109.

§ I.12 On $(n, \varphi(n))$

1) The sequence of positive integers n with $(n, \varphi(n)) \leq k$ (k—fixed positive integer) has density zero.

I. Niven. *The asymptotic density of sequences*. Bull. Amer. Math. Soc. 57 (1951), 420–434.

2) The sequence of positive integers n with

$$(\varphi(n), \sigma_j(n)) \leq k$$

(k, j—fixed positive integers) has density zero.

R.E. Dressler. *On a theorem of Niven*. Canad. Math. Bull. 17 (1974), 109–110.

Remark. For a generalization for general sequences, see

Š. Porubski. *On theorems of Niven and Dressler*. Math. Slovaca 28 (1978), 243–246.

3) If $A(m) = \operatorname{card}\{n : n \leq m \text{ and } (n, \varphi(n)) = 1\}$, then

$$A(m) = (1 + o(1))\frac{m \cdot e^{-\gamma}}{\log\log\log m}$$

P. Erdős. *Some asymptotic formulas in number theory*. J. Indian Math. Soc. (N.S.) 12 (1948), 75–78.

§ I.13 The difference of consecutive totients

If $a_{1,n} < \cdots < a_{\varphi(n),n}$ are the "totients" of n (i.e. the integers relatively prime to and smaller than n), then for some $c > 0$,

$$\sum (a_{i+1,n} - a_{i,n})^2 < \frac{cn^2}{\varphi(n)}$$

H.L. Montgomery and R.C. Vaughan. Ann. of Math. (2) 123 (1986), 311–333.

Remark. The above result was a famous conjecture of P. Erdős.
P. Erdős. *The difference of consecutive primes*. Duke Math. J. 6 (1940), 438–441.

§ I.14 Nonmonotonicity of φ. (A measure)

1) Let $F_\varphi(n) = \text{card}\,\{j < n : \varphi(j) \geq \varphi(n)\} + \text{card}\,\{j > n : \varphi(j) \leq \varphi(n)\}$
Then:

a) $\frac{1}{n} F_\varphi(n) = f\left(\frac{\varphi(n)}{n}\right) + O(\exp(-\sqrt{\log n}))$
Where f is a certain convex function.

b) $\frac{1}{n} F_\varphi(n)$
has a continuous distribution function.

H.G. Diamond and P. Erdős. *A measure of the nonmonotonicity of the Euler phi function*. Pacific J. Math. 77 (1978), 83–101.

2) The number n is named *sparsely totient* if $\varphi(m) > \varphi(n)$ whenever $m > n$. Let $P_j(n)$ and $Q_j(n)$ be the jth largest prime factor of n and the jth smallest, respectively. Then:

a) $P_j(n) \leq \left(\frac{j}{j-1} + o(1)\right) \log n$

$j \geq 2$

b) $Q_j(n) \geq \left(\frac{j}{j+1} + o(1)\right) \log n$

$j \geq 1$

c) $P_1(n) \ll (\log n)^\gamma$
for any $\gamma > 2 - \frac{8}{65}$

G. Harman. *On sparsely totient numbers*. Glasgow Math. J. 33 (1991), 349–358.

Remark. The first results (and the terminology) on sparsely totient numbers are due to Masser and Shiu, who showed that

$$P_2(n) \leq (1 + \sqrt{2} + o(1)) \log n$$

and

$$Q_1(n) \geq (\sqrt{2} - 1 + o(1)) \log n$$

D.W. Masser and P. Shiu. Pac. J. Math. 121 (1986), 407–426.

§ I.15 Nonmonotonicity of J_k

Let $F_t = \{n \in N^* : J_t(n) < J_t(m) \forall m > n\}$ and $F_t(x) = \text{card}\,\{n \in F_t : n \leq x\}$ Then:

1) $\log F_t(x) \ll \log^{1/2} x$

2) If n and n' are consecutive members of F_t, then

$$\frac{n}{n'} \to 1$$

as $n \to \infty$

J. Chidambaraswamy and P.V. Krishnaiah. *On integers n with $J_t(n) < J_t(m)$ for $m > n$*. Internat. J. Math. Math. Sci. 12 (1989), 123–130.

§ I.16 Number of solutions of $\varphi(x) = n!$

Let $S_k(m)$ denote the number of solutions of $\varphi(x) = m$, where x has exactly k prime factors which appear to the first power.

1) $S_1(n!) \geq 1$
for every n and $S_1(n!) \to \infty$ $(n \to \infty)$

H. Gupta. *On a problem of Erdős*. Amer. Math. Monthly 57 (1950), 326–329.

2) $S_k(n!) > c \cdot n^k/(\log n)^k$
for every k and sufficiently large n (c is a positive constant).

P. Erdős. *On a conjecture of Klee*. Amer. Math. Monthly 58 (1951), 98–101.

§ I.17 Number of solutions of $\varphi(x) = m$

Let $b_m = \text{card}\,\{n \in N^* : \varphi(n) = m\}$, $m = 1, 2, 3, \ldots$ Then:

1) There exists $\delta > 0$ such that $b_m > m^\delta$ for infinitely many m.

P. Erdős. *On the normal number of prime factors of p − 1 and some related problems concerning Eulers's φ function*. Quart. J. Math. Oxford Ser. 6 (1935), 205–213.

2) a) $b_m > m^\delta$
for infinitely many m if $0 < \delta < 3 - 2\sqrt{2}$ ($\approx 0.17\,157$)

K. Woolridge. *Values taken many times by Euler's phi-function*. Proc. Amer. Math. Soc. 76 (1979), 229–234 (by using sieve methods.)

b) The same holds for

$$0 < \delta < 1 - 625/512e$$

($\approx 0.55\,655$)

C. Pomerance. *Popular values of Euler's function*. Mathematika 27 (1980), 84–89.

Remark. The above results are in connection with Carmichael's conjecture, which states that for every n it is possible to find an $m \neq n$ such that

$$\varphi(m) = \varphi(n)$$

This conjecture is still open.

R.D. Carmichael. *Note on Euler's φ-function*. Bull. Amer. Math. Soc. 28 (1922), 109–110.

3) $b_m \leq m \cdot \exp(-(1 + o(1)) \log m \cdot \log\log\log m / \log\log m)$
for all m

C. Pomerance. *Popular values of Euler's function*. Mathematika 27 (1980), 84–89.

4) Let $X = \{x \in N : \{x\} = \varphi^{-1}(\varphi(x))\}$. Then, if $x \in X$, then

$$\varphi(x) > 10^{10\,000}$$

P. Masai and A. Valette. *A lower bound for a counterexample to Carmichael's conjecture*. Boll. Un. Mat. Ital. A (6) 1 (1982), 313–316.

5) For infinitely many m there are more than

$$m^{c/(\log\log m)^4}$$

pairwise relatively prime integers $i_1, \ldots, i_l$ for which $\varphi(i_k) = m$ for $1 \leq k \leq l$

P. Erdős. *Some remarks on the functions φ and σ*. Bull. Acad. Polon. Sci. 10 (1962), 617–619.

§ I.18 Number of values of φ less than or equal to x

1) Let $F(x)$ be the number of values of the φ-function less than or equal to x; i.e. $F(x) = \text{card}\{n : \varphi(n) \le x\}$. Then

$$F(x) = \frac{\zeta(2)\zeta(3)}{\zeta(6)} \cdot x + R(x)$$

where:

a) $R(x) \ll x \cdot \exp\left(-(1-\varepsilon)\left(\frac{1}{2}\log x \log\log x\right)^{1/2}\right)$

for any $\varepsilon > 0$

P.T. Bateman. *The distribution of values of the Euler function.* Acta Arith. 21 (1972), 329–345 (by methods of complex analysis.)

b) $R(x) \ll x/\log^2 x$

A. Smati. *Répartition des valeurs de la fonction d'Euler.* Enseign. Math., II. Sér. 35. No. 1/2 (1989), 61–76 (using only elementary methods.)

c) $R(x) \ll x/\log^k x$

for every $k > 0$

M. Balazard and A. Smati. *Approche élémentaire d'un théorème de Bateman.* Preprint; Univ. Limoges, Limoges, 1988.

Remark. In this preprint an improvement of c) is also proved.

d) $\left|F(x) - \frac{\zeta(2)\zeta(3)}{\zeta(6)}x\right| \le 1.4x \cdot \log x \cdot \log\log x \cdot$
$\cdot \exp\{-(1 - T(x))\sqrt{1/2\log x \cdot \log\log x}\}$

for $x \ge 3$, where $T(x) = (\log\log\log x + 4 - \log 2)/\log\log x$

A. Smati. *Evaluation effective du nombre d'entiers n tels que $\varphi(n) \le x$.* Acta Arith. 61 (1992), 143–159.

2) Let $G_n(x) = \text{card}\left\{m : 1 \le m \le n, \frac{\varphi(m)}{m} \le x\right\}$. Then

$$\frac{1}{n} \cdot G_n(x) = \phi(x) + O\left(\frac{1}{\log n} \cdot (\log\log n)^2 \cdot (\log\log\log n)^{-2}\right)$$

where $\phi(x)$ is a distribution function.

A.S. Faĭnleĭb. *A generalization of Essen's inequality and its applications in probabilistic number theory* (Russian.) Izv. Akad. Nauk SSSR Ser. Mat. 32 (1968), 859–879.

3) Let $V(x)$ denote the number of distinct values of $\varphi(n)$ for $1 \le n \le x$. Then:

a) $V(x) = O\left(\frac{x}{\log x} \cdot \exp(B \cdot (\log\log x)^{1/2})\right)$

for every $B > 2(2/\log 2)^{1/2}$

P. Erdős and R.R. Hall. *On the values of the Euler φ-function.* Acta Arith. 22 (1973), 201–206.

b) $V(x) \geq c \cdot \pi(x) \cdot \exp(A(\log\log\log x)^2)$
where $A, C > 0$ are absolute constants.

P. Erdős and R.R. Hall. *Distinct values of Euler's φ-function.* Mathematika 23 (1976), 1–3.

c) $V(x) = O\left(\frac{x}{\log x} \cdot \exp(C(\log\log\log x)^2)\right)$
for $C > (\log 4 - 2\log(2 - \log 2))^{-1} \approx 1.17501\ldots$

C. Pomerance. *On the distribution of the values of Euler's function.* Acta Arith. 47 (1986), 63–70.

d)
$$\log V(x) = \log x - \log\log x + \\ + C(\log\log\log x)^2 + o((\log\log\log x)^2)$$
with an explicit value of C

H. Maier and C. Pomerance. *On the number of distinct values of Euler's φ-function.* Acta Arith. 49 (1988), 263–275.

§ I.19 On composite n with $\varphi(n) \mid (n-1)$ (Lehmer's conjecture)

1) If n is composite and $\varphi(n) \mid (n-1)$, then
$$\omega(n) \geq 14$$

2) If n is composite, $(n, 15) = 1$ and $\varphi(n) \mid (n-1)$, then
$$\omega(n) \geq 17$$

G.L. Cohen and P. Hagis Jr. *On the number of prime factors of n if φ(n) | (n − 1).* Nieuw Arch. Wiskunde (3) 28 (1980), 177–185.

3) If n is composite, $3 \mid n$ and $\varphi(n) \mid (n-1)$, then
$$\omega(n) \geq 212$$

E. Lieuwens. *Do there exist composite numbers M for which k φ(M) = M − 1 holds?* Nieuw Arch. Wiskunde (3) 18 (1970), 165–169.

4) If n is composite, $\varphi(n) \mid (n-1)$ and $\frac{n-1}{\varphi(n)} \geq 3$, then
$$\omega(n) \geq 33$$

M. Kishore. *On the number of distinct prime factors of n for which φ(n) | (n − 1).* Nieuw Arch. Wiskunde (3) 25 (1977), 48–53.

5) If n is composite, $\varphi(n) \mid (n-1)$, then:

a) n is odd and squarefree

b) If $pq \mid n$ (p, q primes), then

$$q \not\equiv 1 (\mathrm{mod}\, p)$$

D.H. Lehmer. *On Euler's totient function.* Bull. Amer. Math. Soc. 38 (1932), 745–751.

Remark. In the above paper, Lehmer conjectured that $\varphi(n) \mid (n-1)$ implies $n =$ prime. This conjecture is still open.

6) Let $A = \{n \in \mathbb{N} : n = 2 \text{ or } p, \text{ or } 2p, \text{ for } p \neq 2 \text{ prime}\}$ and $B = \{n \in \mathbb{N} : (\varphi(n)+1) \mid n\}$. Then
card $\{n \in \mathbb{N} : n \le x, n \in B \setminus A\} = O(x^{1/2}(\log x)^{3/4}(\log\log x)^{-5/6})$

G.L. Cohen and S.L. Segal. *A note concerning those n for which $\varphi(n)+1$ divides n.* Fib. Quart. 27 (1989), 285–286.

§ I.20 Number of composite $n \le x$ with $\varphi(n) \mid (n-1)$

1) Let $N(x)$ denote the number of composite $n \le x$ for which $\varphi(n) \mid (n-1)$ Then

$$N(x) = O(x^{1/2}(\log x)^{3/4})$$

C. Pomerance. *On composite n for which $\varphi(n) \mid (n-1)$.* II. Pacific J. Math. 69 (1977), 177–186.

2) For $a \in \mathbb{I}$ let $F(a) = \{n : n = a (\mathrm{mod}\, \varphi(n))\}$ and $F'(a) = \{n \in F(a) : n \neq pa \text{ for } p \text{ prime}, p \nmid a\}$. Then:

a) $\mathrm{F}'(a)(x) = O(x^{1/2}(\log x)^{3/4}) \quad \forall a \in \mathbb{I}$

C. Pomerance. *Ibid.*

b) $\mathrm{F}'(a)(x) = O(x^{1/2}(\log x)^{1/2}(\log\log x)^{-1/2})$
where $F'(a)(x)$ denotes the counting function of the set $F'(a)$.

Z. Shan. *On composite n for which $\varphi(n) \mid (n-1)$.* J. China Univ. Sci. Tech. 15 (1985), 109–112.

§ I.21 $\sum_{n \le x} \varphi(n)$

$$\sum_{n \le x} \varphi(n) = \frac{3}{\pi^2}x^2 + R(x), \text{ where:}$$

1) $R(x) = O(x^{\delta})$

for some $\delta \in (1, 2)$

G.L. Dirichlet. *Über die Bestimmung der mittleren Werthe in der Zahlentheorie*. Werke, G. Riemer, 1897, II, pp. 49–66 (original 1849.)

2) $R(x) = O(x \log x)$

F. Mertens. *Über einige asymptotische Gesetze der Zahlentheorie*. Crelle's Journal 77 (1874), 289–338.

3) $R(x) = O(x \log^{2/3} x (\log \log x)^{4/3})$

A. Walfisz. *Weylsche Exponentialsummen in der neueren Zahlentheorie*. Leipzig: B.G. Teubner, 1963.

§ I.22 $\sum_{k \le n} f\left(\frac{k}{n}\right) \cdot \varphi(k)$

$$\lim_{n\to\infty} \frac{1}{n^2} \sum_{k=1}^{n} f\left(\frac{k}{n}\right) \cdot \varphi(k) = \frac{6}{\pi^2} \int_0^1 x f(x) \mathrm{d}x$$

for all functions f such that $x \cdot f(x)$ is continuous on $[0, 1]$

Ch. Radoux. *Note sur le comportement asymptotique de l'indicateur d'Euler*. Ann. Soc. Sci. Bruxelles Sér. I 91 (1977), 13–18.

§ I.23 On $\sum_{n \le x} \varphi(n) - \frac{3}{\pi^2} x^2$

Let $R(x) = \sum_{n \le x} \varphi(n) - \frac{3}{\pi^2} x^2$. Then:

1) There exists a positive constant c and infinitely many integers x such that

$$R(x) > c\, x \log \log \log \log x$$

and infinitely many integers x such that

$$R(x) < -c\, x \log \log \log \log x$$

P. Erdős and H.N. Shapiro. *On the changes of sign of a certain error function*. Canad. J. Math. 3 (1951), 375–384.

2) $R(x) \ne o(x \log \log \log x)$

S. Chowla and S.S. Pillai. *On the error terms in some asymptotic formulae in the theory of numbers*. I.J. London Math. Soc. 5 (1930), 95–101.

3) $R(x) = O(x \log^{2/3} x (\log \log x)^{1+\varepsilon})$

for all $\varepsilon > 0$

A.I. Szaltiikov. *On Euler's function* (Russian.) Mat. Sb. 6 (1960), 34–50,

and

A. Walfisz. *Weylsche Exponentialsummen in der neueren Zahlentheorie*. Berlin 1963.

4) a) $\sum_{n\le x} R(n) \sim \frac{3}{2\pi^2}\cdot x^2$

b) If $H(x) = \sum_{n\le x} \frac{\varphi(n)}{n} - \frac{6}{\pi^2}x$, then

$$\sum_{n\le x} H(n) \sim \frac{3}{\pi^2}\cdot x$$

S. Chowla and S.S. Pillai. *See 2).*

5) a) $\sum_{n\le x} R(n) = \frac{3}{2\pi^2}x^2 + O(x^2\delta(x))$

where $\delta(x) = \exp(-A\log^{3/5} x(\log\log x)^{-3/5})$, $A > 0$, constant.

D. Suryanarayana, R. Sitaramachandrarao. *On the average order of the function* $E(x) = \sum_{n\le x}\varphi(n) - 3x^2/\pi^2$ Ark. Mat. 10 (1972), 99–106.

b) $\sum_{n\le x} H(n) = \frac{3}{\pi^2}\cdot x + o(x\delta_1(x))$

where $\delta_1(x) = \exp(-A_1\log^{3/5} x(\log\log x)^{-1/5})$, $A_1 > 0$, constant.

D. Suryanarayana. *On the average order of the function* $E(x) = \sum_{n\le x}\varphi(n) - 3x^2/\pi^2$. J. Indian Math. Soc. (N.S.) 42 (1978), 179–195.

c) $\sum_{n\le x} \frac{R(n)}{n} = \frac{3}{\pi^2}\cdot x + O(x\exp(-A_2\log^{4/7-\varepsilon} x))$

where $A_2 > 0$ and ε is arbitrarily small, $\varepsilon > 0$

C.T. Pan. *On $\sigma(n)$ and $\varphi(n)$*. Bull. Acad. Polon. Sci. CI. III 4 (1956), 637–638.

6) a) $R(x) = \Omega_\pm(x\sqrt{\log\log x})$

b) $H(x) = \Omega_\pm(\sqrt{\log\log x})$

H.L. Montgomery. *Fluctuations in the mean of Euler's phi function*. Proc. Indian. Acad. Sci. Math. Sci. 97 (1987), 239–245.

7) $\lim_{x\to\infty} \frac{1}{x}\int_1^x H^k(t)\mathrm{d}t$

exists for all $k \in \mathbb{N}$ and

$\lim_{x\to\infty} \frac{1}{x}\int_1^x |H|^\lambda(t)\mathrm{d}t$

exists for all $\lambda > 0$

Y.-F.S. Pétermann. *Existence of all the asymptotic λth means for certain arithmetical convolutions*. Tsukuba J. Math. 12 (1988), 241–248.

§ I.24 On $\sum_{n\le x} \varphi(n)/n$

1) a) $\sum_{n\le x} \frac{\varphi(n)}{n} = \frac{6}{\pi^2} \cdot x + O((\log x)^{2/3}(\log\log x)^{4/3})$

A. Walfisz. *Weylsche Exponentialsummen in der neueren Zahlentheorie*. Berlin 1963.

b) $\sum_{n\le x} \left(\frac{\varphi(n)}{n}\right)^{\alpha} = Cx + O((\log x)^{2/3}(\log\log x)^{4/3})$

$0 < \alpha < 1$

I.I. Il'jasov. *An estimate of the remainder term of the sum* $\sum_{n\le x}(\varphi(n)/n)^{\alpha}$. Izv. Akad. Nauk Kazah. SSR Ser. Fiz. Mat. (1969), 77–79.

c) $\sum_{n\le x,(n,k)=1} \frac{\varphi(n)}{n} = \frac{k\varphi(k)}{\zeta(2)J_2(k)}x + O(\theta(k)\log x)$

where $\theta(k)$ is the number of squarefree divisors of k

D. Suryanarayana. *The greatest divisor of n which is prime to k*. Math. Student 37 (1969), 147–157.

2) Let $g(n)/\log\log\log n \to \infty$. Then we have:

a) $\sum_{m=n}^{n+g(n)} \frac{\varphi(m)}{m} = (1+o(1))\frac{6}{\pi^2}g(n)$

b) The number of integers $m \in (n, n+g(n))$ which satisfy

$\frac{\varphi(m+1)}{m+1} > \frac{\varphi(m)}{m}$ equals $(1+o(1))\frac{g(n)}{2}$

c) The number of integers $m \in (n, n+g(n))$ which satisfy $\frac{m}{\varphi(m)} \le c$ equals $(1+o(1)) \cdot g(n) \cdot f(c)$, where $f(x)$ is the distribution function of $\frac{n}{\varphi(n)}$

Remark. The existence of a distribution function for $n/\varphi(n)$ is due to I.J. Schoenberg.

I.J. Schoenberg. *Über die asymptotische Verteilung reeller Zahlen* mod 1. Math. Z. 28 (1928), 171–200.

P. Erdős. *Some remarks about additive and multiplicative functions*. Bull. A.M.S. 52 (1946), 527–537.

3) Let $v_N(x) = \text{card}\left\{n \le x : \frac{\varphi(n)}{n} \le x\right\}$. Then

$$\frac{1}{N}\cdot v_N(x) = v(x) + O\left(\frac{1}{\log\log N}\right)$$

uniformly in $x \in [0, 1]$ where is a certain function of x

A.S. Faĭnleĭb. *Distribution of values of Euler's function*. Mat. Zametki 1 (1967), 645–652.

§ I.25 On $\sum_{n\le x} J_k(n) - x^{k+1}/(k+1)\zeta(k+1)$

Let $E_k(x) = \sum_{n\le x} J_k(n) - \dfrac{x^{k+1}}{(k+1)\,\zeta(k+1)}$ for $k \ge 2$. Then:

1) $\sum_{n\le x} E_k(n) \sim \dfrac{x^{k+1}}{2(k+1)\,\zeta(k+1)}$

Remark. For $k = 1$ see

S.D. Chowla and S.S. Pillai. *On the error terms in some asymptotic formulae in the theory of numbers*. London

Ok as Set? Math. Soc. 5 (1930), 95–101 § I.23 $\sum_{n\le x}\varphi^*(n)$, $\sum_{n\le x}\varphi(n)\varphi(n+k)$ and related results on $\sum_{n\le x}\varphi(f(n))$, f a polynomial $\sum_{p\le x}\varphi(p-1)$ for p prime On $\sum_{n\le x}\dfrac{1}{\varphi(n)}$ and related questions An expansion of J_k.)

2) $\limsup_{n\to\infty} \dfrac{E_k(n)}{n^k} \ge \dfrac{1}{2\zeta(k+1)}$

$\liminf_{n\to\infty} \dfrac{E_k(n)}{n^k} \le -\dfrac{1}{2\zeta(k+1)}$

S.D. Adhikari and A. Sankaranarayanan. *On an error term related to the Jordan totient function $J_k(n)$*. J. Numb. Theory No. 2, 34 (1990), 178–188.

3) $E_k(n) > 0$

for all $n \ge n_0$

S.D. Adhikari and A. Sankaranarayanan. *Ibid.*

§ I.26 An expansion of J_k

Let $\gamma(n,m) = \begin{cases} 1 & \text{, for } (n,m)=1 \\ \prod\limits_{p|n,\, p|m} (1-p) & \text{, for } (n,m)>1 \end{cases}$

Then

$$J_k(n) = \frac{n^k}{\zeta(k+1)} \sum_{m=1}^{\infty} \frac{\gamma(m,n)}{m^{k+1}}$$

$k = 1, 2, 3, \ldots$

E. Krätzel. *Zahlentheorie*. Berlin, 1981 (pp. 147–148.)

§ I.27 On $\sum_{n \le x} 1/\varphi(n)$ and related questions

1) $$\sum_{3 \le n \le x} \frac{1}{\log \varphi(n)} = x \cdot \sum_{j=1}^{k} \frac{a_j}{\log^j x} + O\left(\frac{x}{\log^{k+1} x}\right)$$

where a_j are explicitely given constants. In particular, $a_1 = 1$ and $a_2 = 1 - \sum\limits_p \frac{1}{p} \log\left(1 - \frac{1}{p}\right)$, the sum being over all primes p.

T. Cai. *On a sum of Euler's totient function* (Chinese.) J. Shandong Univ., Nat. Sci. Ed. 24 (1989), 106–110 (by applying Perron's formula.)

2) a) $$\sum_{n \le x} \frac{1}{\varphi(n)} = A(\log x + B) + E_0(x)$$

where $\int_0^x E_0(t)\mathrm{d}t = -\log x + O(1)$

b) $$\sum_{n \le x} \frac{n}{\varphi(n)} = Ax - \frac{1}{2}\log x + E_1(x)$$

where $E_1(x) = O(\log^{2/3} x)$ and $\int_1^x E_1(t)\mathrm{d}t = \alpha x + O(x^{4/5})$, with $\alpha < 0$, a constant (A, B constants.)

R. Sitaramachandrarao. *On an error term of Landau*. II. Rocky Mt. J. Math. 15 (1985), 579–588.

c) $$\int_0^x E_1^2(t)\mathrm{d}t = Cx + O(x^{4/5}(\log x)^{3/5}(\log\log x)^{6/5})$$

where $C \approx 0.546$

W.G. Nowak. *On an error term involving the totient function*. Indian J. Pure Appl. Math. 20 (1989), 537–542.

3) a) $$\sum_{n\le x}\frac{1}{\varphi(n)}=\frac{315\zeta(3)}{2\pi^4}\left(\log x+\gamma-\sum_p\frac{\log p}{p^2-p+1}\right)+O(\log x/x)$$

E. Landau. *Göttinger Nachr.* 1900, 177–186, Jbuch 31, 179.

b) $$\sum_{n\le x}\frac{1}{\varphi(n)}=\frac{\zeta(2)\zeta(3)}{\zeta(6)}\log x+A+O(\log x/x)$$

where $A=\gamma\sum_{n=1}^{\infty}\frac{\mu^2(n)}{n\varphi(n)}-\sum_{n=1}^{\infty}\frac{\mu^2(n)\log n}{n\varphi(n)}$

H.L. Montgomery. *Primes in arithmetic progressions*. Mich. Math. J. 17 (1970), 33–39.

c) The same result with $O\left((\log x)^{2/3}/x\right)$

R. Sitaramachandrarao. *On an error term of Landau*. Indian J. Pure Appl. Math. 13 (1982), 882–885.

4) For any fixed integer $k\ge 1$

$$\sum_{1<n\le x}\frac{\log\varphi(n)}{\log\sigma(n)}=x\left(1+\sum_{j=1}^{k}\frac{a_j}{\log^j x}+O\left(\frac{1}{\log^{k+1}x}\right)\right)$$

where a_j are computable constants. In particular,

$$a_1=\sum_p\left(1-\frac{1}{p}\right)\sum_{r=1}^{\infty}\left(2\log\left(1-\frac{1}{p}\right)-\log\left(1-\frac{1}{p^{r+1}}\right)\right)\cdot p^{-r}$$

J.-M. de Koninck and A. Ivić. *Topics in arithmetical functions*. Notas de Matemática (72). North. Holland, Amsterdam, New York, Oxford 1980 (p. 106.)

§ I.28 $\sum_{p\le x}\varphi(p-1)$ for p prime

1) a) $$\sum_{p\le x}\varphi(p-1)=\sum_{n=1}^{\infty}\frac{\mu(n)}{n\varphi(n)}\mathrm{li}(x^2)+O(x^2/(\log x)^m)$$

for all given positive integer m

S.S. Pillai. *On the sum function connected whith primitive roots*. Proc. Indian Acad. Sci. Sect. A. 13 (1941), 526–529.

b) $$\sum_{p\le x}\frac{p-1}{\varphi(p-1)}=O(x/\log x)$$

K. Prachar. *Primzahlverteilung*. Berlin, Göttingen, Heidelberg, 1957 (see p. 41.)

§ I.29 On $\sum_{n\le x}\varphi(f(n))$, f a polynomial

1) $\sum_{n\le x}\varphi(f(n)) = a\cdot a_h\cdot(h+1)^{-1}x^{h+1} + O(x^h\log^h x)$

where $f(x)$ is a polynomial of degree $h>0$ with integer coefficients and $a_h>0$ its leading coefficient; $f(x)$ has no multiple roots and $f(n)>0$ for $n\ge 1$. Here

$$a = \sum_{n=1}^{\infty}\mu(n)\rho_f(n)/n^2$$

where $\rho_f(n)$ is the number of incongruent solutions of the congruence $f(x)\equiv 0(\mathrm{mod}\, n)$

2) $\sum_{n\le x}\frac{\varphi(f(n))}{f(n)} = ax + O(\log^h x)$

H.N. Shapiro. *Introduction to the theory of numbers*. John Wiley and Sons. INC. 1983.

Remark. For an extension of 1) and 2) for generalized totient functions see
J. Chidambaraswamy. *On the average of the generalized totient function over polynomial sequences*. Indian J. Pure Appl. Math. 19 (1988), 1149–1155.

§ I.30 $\sum_{n\le x}\varphi^*(n)$, $\sum_{n\le x}\varphi(n)\,\varphi(n+k)$ and related results

1) $\sum_{n\le x}\varphi^*(n) = \frac{1}{2}ax^2 + O(x\log^{5/3}x(\log\log x)^{4/3})$

(where $a>0$ constant.)

R. Sitaramachandrarao and D. Suryanarayana. *On $\sum_{n\le x}\sigma^*(n)$ and $\sum_{n\le x}\varphi^*(n)$*. Proc. Amer. Math. Soc. 41 (1973), 61–66.

2) a) $\sum_{n\le x}\varphi(n)\varphi^*(n) = \frac{2\alpha}{\pi^2}x^3 + O(x^2\log^3 x)$

where $\alpha = \prod_p(1-(p+1)^{-2})$

b) $\sum_{n\le x}\sum_{d\| n} d\varphi^*\left(\frac{n}{d}\right) = \frac{3\alpha}{\pi^2}x^2\log x + O(x^2)$

where $d\parallel n$ denotes that d is a unitary divisor of n

L. Tóth. *The unitary analogue of Pillai's arithmetical function*. Collect. Math. 40 (1989), 19–30.

3) a) $$\sum_{i=1}^{n-1} \varphi(i)\,\varphi(n-i) \sim \frac{1}{6}an^3 \prod_{p|n} \frac{p^3-2p+1}{p(p^2-2)}$$

where $a = \prod_p (1 - 2/p^2)$

A.E. Ingham. *Some asymptotic formulae in the theory of numbers*. J. London Math. Soc. 2 (1927), 202–208.

b) $$\sum_{n\le x} \varphi(n)\varphi(n+k) = \frac{1}{3}x^3 \prod_p (1-2/p^2) \cdot \prod_{p|k}\left(1+\frac{1}{p(p^2-2)}\right) + O(x^2\log^2 x)$$

c) $$\sum_{n\le x} \varphi(n)/\varphi(n+k) = x \prod_{p\nmid k}\left(1+\frac{1}{p^2(p-1)}\right) + O(\log^2 x)$$

d) $$\sum_{n\le x} \varphi(n+k)/\varphi(n) = x \prod_{p\nmid k}\left(1+\frac{1}{p^2(p-1)}\right) + O(\log^2 x)$$

L. Mirsky. *Summation formula involving arithmetic functions*. Duke Math. J. 16 (1949), 261–272.

§ I.31 Asymptotic formulae for generalized Euler functions

1) Let $F = \{f_1(x), \ldots, f_k(x)\}, k \ge 1$, be a set of polynomials with integral coefficients and let A be the set of all ordered k-tuples of integers $(a_1, \ldots, a_k)$, such that

$$0 \le a_1, \ldots, a_k \le n$$

Define $\varphi_F(n)$ as the number of elements in A such that g.c.d. $(f_1(a_1), \ldots, f_k(a_k), n) = 1$. See

H. Stevens. *Generalizations of the Euler φ-function*. Duke Math. J. 38 (1971), 181–186.

If each polynomial $f_i(x)$ has relatively prime coefficients, then

$$\sum_{n\le x} \varphi_F(n) = \frac{\lambda_F(k)x^{k+1}}{k+1} + O(R_k(x))$$

where

$$\lambda_F(s) = \prod_p \left(1 - \frac{N_1\cdots N_k}{p^{s+1}}\right)$$

where N_i denotes the number of incongruent solutions of $f_i(x) \equiv 0(\mathrm{mod}\,p)$($p$-prime); and $R_k(x) = x^k$ or $x^{1+\varepsilon}$ (for all $\varepsilon > 0$) according as $k \ge 2$ or $k = 1$.

J. Sándor and L.Tóth. *An asymptotic formula concerning a generalized Euler function*. Fib. Quart. 27 (1989), 176–180.

Remark. The Stevens totient function generalizes totients considered by Jordan, Schemmel, Nagell, etc.

2) Let $S \subset \{1, 2, 3, \ldots\}$ and $\rho_S(n)$ its characteristic function, i.e. $\rho_S(n) = 1$ if $n \in S; 0$ if $n \notin S$, and define $\zeta_S(z) = \sum_{n=1}^{\infty} \rho_S(n)/n^z$. Define also

(see E. Cohen. *Arithmetical functions associated with arbitrary sets of integers.* Acta Arith. 5 (1959), 407–415)

$\varphi_S(n)$ as the number of those positive integers $x (\bmod n)$ with $(x, n) \in S$ (e.g. when $S = \{1\}$, one gets the classical Euler function φ). Then

$$\sum_{n \le x} \varphi_S(n) = \frac{3}{\pi^2} \zeta_S(2) x^2 + O(x \log^2 x)$$

J. Sándor and L. Tóth. *On some arithmetical products.* Publ. Centre Rech. Math. Pures, Neuchâtel Série I, 20 (1990), 5–8.

§ I.32 On $\varphi(x, n) = \sum_{m \le x, (m,n)=1} 1$ and on Jacobstahl's arithmetic function

1) a) For $x \ge 1, n \ge 2$ and $([x], n) = 1$,

$$\frac{\Psi(n)}{n} - \frac{\theta(n)}{2} \le \varphi(x, n) - [x]\frac{\varphi(n)}{n} \le \frac{\theta(n)}{2} - \frac{\Psi(n)}{n} + 1$$

D. Suryanarayana. *On* $\Delta(x, n) = \varphi(x, n) - x\varphi(n)/n$. Proc. Amer. Math. Soc. 44 (1974), 17–21.

b) Let $\varphi(x, n) = \sum_{n \le x, (m,n)=1} 1$ and $\Delta(x, n) = \varphi(x, n) - x\frac{\varphi(n)}{n}$

Then

$$\left| \Delta(x, n) + \{x\}\frac{\varphi(n)}{n} - \frac{1}{2}\left[\frac{1}{(n, x)}\right] \right| \le \frac{\theta(n)}{2} - \frac{k\theta(k)\Psi(n)}{n\Psi(k)} + \frac{\theta(k)}{2}$$

where $k = (n, [x])$, $\theta(n)$ is the number of squarefree divisors of n and $\Psi(n)$ is Dedekind's function.

A. Sivaramasarma. Math. Student 46 (1978), 160–164 (1982.)

2) Let $\varphi_n(x, y)$ denote the number of integers $x < m \le y$ with $(m, n) = 1$ Then for every $\varepsilon > 0$ and $\eta > 0$ there exists an $A_0(\varepsilon, \eta)$ so that for every $A > A_0(\varepsilon, \eta)$ the number of integers x, $1 \le x \le n$, for which

$$(1 - \varepsilon)A < \varphi_n\left(x, x + A\frac{n}{\varphi(n)}\right) < (1 + \varepsilon)A$$

is not satisfied, is less than $\eta \cdot n$

P. Erdős. *On the integers relatively prime to n and on a number theoretic function considered by Jacobsthal.* Math. Scad. 10 (1962), 163–170.

3) Let $n \in \mathbb{I}$. E. Jacobsthal defined the function $g(n)$ as the least integer so that amongst any $g(n)$ consecutive integers

$$a, a+1, \ldots, a+g(n)-1$$

there is at least one which is relatively prime to n

E. Jacobsthal. *Über Sequenzen ganzer Zahlen von denen keine zu n teilerfremd ist I–III.* Norske Vid. Selsk. Forth. Trondheim 33 (1960), 117–139.

Then:

a) For almost all n,

$$g(n) = \frac{n}{\varphi(n)}\omega(n) + o(\log\log\log n)$$

b) For all n,

$$g(n) > \frac{n}{\varphi(n)}\omega(n)\left(1 - \frac{c \cdot \log\log\omega(n)}{\log\omega(n)}\right)$$

where $\omega(n)$ denotes the number of distinct prime factors of n

P. Erdős. *See 2).*

c) If $k^{(\log 3)/(\log 5)} < p_1 < \cdots < p_k$ (primes), then

$$g(p_1 \ldots p_k) \leq 2k$$

H.-J. Kanold. *Über einige Abschätzungen von $g(n)$.* J. Reine Angew. Math. 290 (1977), 142–153.

§ I.33 On the iteration of φ

We write $\varphi^{(n)}(x) = \varphi(\varphi^{(n-1)}(x))$, $\varphi^{(1)}(x) = \varphi(x)$. For $x > 2$ let $C(x) = n$ if $\varphi^{(n)}(x) = 2$. We define $C(1) = C(2) = 0$. Then:

1) a) $C(mn) = C(m) + C(n) + \varepsilon(m, n)$

where $\varepsilon(m, n) = 1$ if both m, n even; $= 0$ otherwise.

H.N. Shapiro. *An arithmetic function arising from the φ function.* Amer. Math. Monthly 50 (1943), 18–30.

b) $\dfrac{\log(x/2)}{\log 3} \leq C(x) \leq \dfrac{\log x}{\log 2}$

S.S. Pillai. *On a function connected with $\varphi(n)$.* Bull. Amer. Math. Soc. 35 (1929), 837–841.

2) $\limsup\limits_{x\to\infty}(C(x+1) - C(x)) = +\infty$

$\liminf\limits_{x\to\infty}(C(x+1) - C(x)) = -\infty$

I. Niven. *The iteration of certain arithmetic functions.* Canad. J. Math. 2 (1950), 406–408.

§ I.34 Iterates of φ and the order of $\varphi^{(k)}(n)/\varphi^{(k+1)}(n)$

We denote by $\varphi^{(n)}$ the n-th iterate of the Euler φ-function. Then:

1) $\varphi^{(n)}(ab) \geq \varphi^{(n)}(a)\varphi^{(n)}(b)$

for all $a, b = 1, 2, \ldots$

T. Popoviciu. *On indicators* (Romanian.) Gaz. Mat. (Bucureşti), 51 (1946), 306–313.

2) If $W(x) = n$ denotes the smallest n for which $\varphi^{(n)}(x) = 1$, then

$$\left[\frac{\log 3x}{\log 3}\right] \leq W(x) \leq \left[\frac{\log 2x}{\log 2}\right]$$

S.S. Pillai. *On a function connected with* $\varphi(n)$. Bull. Amer. Math. Soc. 35 (1929), 837–841.

See also

J.C. Parnami. *On iterates of Euler's φ-function.* Amer. Math. Monthly 74 (1967), 967–968.

3) Neglecting a sequence of density 0, we have for $k \geq 2$,

$$\lim_{k\to\infty} \frac{\varphi^{(k)}(n)\log\log\log n}{\varphi^{(k-1)}(n)} = e^{-\gamma}$$

P. Erdős. *Some remarks on number theory.* II (Hungarian.) Mat. Lapok 12 (1961), 161–169.

4) Let $V(x) = \text{card}\{m \leq x : m = \varphi^{(2)}(n) \text{ for some } n\}$. Then

$$V(x) \ll \frac{x}{\log^2 x} \cdot \exp\left(\frac{c \cdot \log\log x \cdot \log\log\log\log x}{\log\log\log x}\right)$$

($c > 0$ constant.)

P. Erdős and R.R. Hall. *Euler's function and its iterate.* Mathematika 24 (1977), 173–177.

5) Denote $N_\varphi(k, \alpha, x) = \text{card}\,\{n \leq x : \varphi^{(k)}(n) > \alpha n\}$. Then:

a) For every $\alpha < 1/2$, $\varepsilon > 0$, $t > 0$, and $x > x_0(\alpha, t, \varepsilon)$ we have

$$\frac{x}{\log x} \cdot (\log\log x)^t < N_\varphi(2, \alpha, x) < \frac{x}{\log x} \cdot (\log x)^\varepsilon$$

b) For every $\alpha > 0$, $\varepsilon > 0$, and $x > x_0(\alpha, \varepsilon)$ we have

$$N_\varphi(3, \alpha, x) < \frac{x}{\log^2 x} \cdot (\log x)^\varepsilon$$

P. Erdős. *Some remarks on the iterates of the φ and σ functions.* Colloq. Math. 17 (1967), 195–202.

c) For $\alpha < \alpha_1$ and $x > x_0(\alpha, t)$, we have

$$N_\varphi(3, \alpha, x) > \frac{x}{\log^2 x} \cdot (\log\log x)^t$$

where t is arbitarily large.

H. Maier. *On the third iterates of the* φ- *and* σ-*functions*. Colloq. Math. 34 (1984), 123–130.

6) a) Let $\varepsilon(x) \to 0+$ arbitrarily slowly as $x \to \infty$. If $k \le (\log\log x)^{\varepsilon(x)}$ then the normal order of $\varphi^{(k)}(n)/\varphi^{(k+1)}(n)$ for $n \le x$ is

$$k \cdot e^{\gamma} \cdot \log\log x$$

Corollary. The set $\{n : n/\varphi^{(k+1)}(n) \le u \cdot k! \cdot e^{k\gamma} \cdot (\log\log\log n)^k\}$ has asymptotic density for every integer $k \ge 0$ and every real u

Remark. This generalizes a result ($k = 1$) in

I.J. Schoenberg. *Über die asymptotische Verteilung reeller Zahlen* mod 1. Math. Z. 28 (1928), 171–200.

b) Let $\varepsilon(x)$ as above. Then if $k \le \dfrac{\varepsilon(x) \cdot \log\log\log x}{\log\log\log\log x}$, the normal order of $\varphi(n)/\varphi^{(k+1)}(n)$ for $n \le x$ is $k! \cdot e^{k\gamma} \cdot (\log\log\log x)^k$

c) There is an absolute constant $c > 0$ such that if $1 \le k \le c \cdot \log\log x$, then the number of $n \le x$ for which $\varphi^{(k)}(n)/\varphi^{(k+1)}(n) > k(\log\log\log x - \log k)$ fails is

$$O\left(\frac{x}{k}(\log\log\log x - \log k)^{-1}\right)$$

In particular,

d) $\max\limits_k \varphi^{(k)}(n)/\varphi^{(k+1)}(n) \gg \log\log n$

for a set of density 1

P. Erdős, A. Granville, C. Pomerance and C. Spiro. *On the normal behavior of the iterates of some arithmetic functions*. Analytic number theory. Proc. of a Conf. in Honor of P.T. Bateman, Birkhäuser Boston Inc., 1990; pp. 165–204.

§ I.35 Normal order of $\omega(\varphi(n))$

1) Let $\omega(m)$ denote the number of distinct prime factors of m. The normal order of $\omega(\varphi(n))$ is

$$\frac{1}{2}(\log\log n)^2$$

M. Ram Murty and V. Kumar Murty. *Prime divisors of Fourier coefficients of modular forms*. Duke Math. J. 51 (1984), 57–76.

2) $$\lim_{x\to\infty} \frac{1}{x} \operatorname{card}\{n \le x : \omega(\varphi(n)) - \frac{1}{2}(\log\log x)^2 \le \frac{u}{\sqrt{3}}(\log\log x)^{3/2}\} = \phi(u) = \frac{1}{\sqrt{2\pi}} \int_{\infty}^{u} e^{-t^2/2} dt$$

P. Erdős and C. Pomerance. *On the normal number of prime factors of* $\varphi(n)$. Rocky Mountain J. Math. 15 (1985), 343–352.

Remarks. (i) 2) is valid also for $\Omega(\varphi(n))$, where $\Omega(m)$ denotes the total number of prime factors of m

(ii) For more elementary proofs of normal order results see

M. Ram Murty and N. Saradha. *On the sieve of Eratosthenes*. Canad. J. Math. 39 (1987), 1107–1122.

(iii) In the paper by Erdős and Pomerance there is an error in the proof of a lemma. This is corrected in

P. Erdős, A. Granville, C. Pomerance and C. Spiro. *On the normal behavior of the iterates of some arithmetic functions*. Analytic number theory. Proc. of Conf. in Honor of P.T. Bateman, Birkhäuser Boston, Inc., 1990; pp. 165–204.

3) $$\lim_{x\to\infty} \frac{1}{x} \operatorname{card}\{n \le x : \omega(\varphi(\varphi(n))) - \frac{1}{6}(\log\log x)^3 \le \frac{u}{\sqrt{10}}(\log\log x)^{5/2}\} = \phi(u)$$

I. Kátai. *On the number of prime factors of* $\varphi(\varphi(n))$. Acta Math. Hung. 58 (1991), 211–225.

Chapter II

THE ARITHMETICAL FUNCTION *d(n)*, ITS GENERALIZATIONS AND ITS ANALOGUES

§ II. 1 The divisor functions at consecutive integers

1) $d(n) \leq 2\sqrt{n}$
$n = 1, 2, 3, \ldots$

W. Sierpiński. *Elementary theory of numbers*. Warsawa, 1964.

2) For all $h, m = 1, 2, 3, \ldots$, there exists $n > 1$ such that

$$\frac{d(n)}{d(n \pm i)} > m$$

for $i = 1, 2, \ldots, h$

A. Schinzel. *Sur une propriété du nombre de diviseurs*. Publ. Math. (Debrecen) 3 (1954), 261–262.

3) a) For each $c > 0$ there exists a natural number n such that

$$d(n+1) - d(n) > c$$

and

$$d(n-1) - d(n) > c$$

P. Turán. *Problem 71*. Mat. Lapok 5 (1954), 48.

b) For each $k = 1, 2, \ldots$ there exists $n = 1, 2, \ldots$ such that

$$d(n) > \prod_{i=1}^{k} d(n+i)\, d(n-i)$$

P. Erdős. *Proposed problem*. Mat. Lapok 5 (1955), 351.

For any positive intergers k, N, m, and $M > 1$ there exists an integer $n > N$ and prime $p > N$ such that:

c) $\log^{(m)} d(n) > M \cdot \prod_{i=1}^{k} d(n+i)\, d(n-i)$

d) $\log^{(m)} d(p+1) > M \cdot \prod_{i=1}^{k} d(p+1+i)d(p+1-i)$

where

$$\log^{(m)} x = \log\log^{(m-1)} x$$

and

$$\log^{(0)} x = x$$

P. C. Shao. *On the divisor problem of Erdős* (Chinese). Acta Math. Sinica 24 (1981), 797–800.

4) a) Let $N(x) = \text{card}\,\{n \le x : d(n) = d(n+1)\}$. Then

$$N(x) \gg x(\log x)^{-7}$$

Corollary. There are infinitely many positive integers n with $d(n) = d(n+1)$

D.R. Heath-Brown. *The divisor function at consecutive integers.* Mathematica 31 (1984), 141–149.

Remark. In 1981 C. Spiro proved that $d(n) = d(n+5040)$ infinitely often.

C. Spiro. *Thesis*, Urbana, 1981.

b) $N(x) \gg x/(\log\log x)^3$

A. Hildebrand. *The divisor function at consecutive integers.* Pacific J. Math. 129 (1987), 307–319.

c) $N(x) = O(x/(\sqrt{\log\log x}))$

P. Erdős, C. Pomerance and A. Sárközy. *On locally repeated values of certain arithmetic functions.* III. Proc. Amer. Math. Soc. 101 (1987), 1–7.

5) a) Let $T(x) = \text{card}\,\{n \le x : d(n) \mid d(n+1)\}$. Then

$$T(x) = x \sum_{j=1}^{N-1} R_j (\log x)^{1/2-j} + O_N(x \log^{1/2-N} x)$$

where N is an arbirary but fixed, $R_1 > 0, R_2, \ldots, R_{N-1}$ are certain computable constants.

b) Let $S(x) = \text{card}\,\{n \le x : d(n) \mid n\}$. Then

$$S(x) = (1+o(1))\, x(\log x)^{-1/2} (\log\log x)^{-1}$$

as

$$x \to \infty$$

C.A. Spiro. *How often does the number of divisors of an integer divide its succesor?* J. London Math. Soc. (2) 31 (1985), 30–40.

§ II. 2 On $d(n+i_1) > \cdots > d(n+i_r)$

1) a) If $r_n = c(\log n)^{1/2}/\log\log n$ and $\{i_1, \ldots, i_{r_n}\}$ is a permutation of $\{1, 2, \ldots, r_n\}$ then there exists $m < n$ with

$$d(m+i_1) > d(m+i_2) > \cdots > d(m+i_{r_n})$$

P. Erdős. *Remarks on two problems* (Hungarian) Mat. Lapok 11 (1960), 26–31.

b) Let $i_1, \ldots, i_r; j_1, \ldots, j_r$ be two permutations of $1, 2, \ldots, r$. Then for infinitely many n one has

$$d(n+i_1) > \cdots > d(n+i_r)$$

and

$$\omega(n+j_1) > \cdots > \omega(n+j_r)$$

P. Erdős, K. Győry and Z. Papp. *On some new properties of functions* $\sigma(n), \varphi(n), d(n)$, *and* $\nu(n)$. (Hungarian). Mat. Lapok 28 (1980), 125–131.

§ II. 3 Relations connecting d, ω, Λ, d_k

1) $\dfrac{d(n^2)}{d(n)} \geq \left(\dfrac{3}{2}\right)^{\omega(n)}$

2) $\Lambda(n)(d(n)-1) \leq \dfrac{d(n)\log n}{2}$

where $\Lambda(n)$ is von Mangoldt's function.

J. Sándor. *Some diophantine equations for particular arithmetic functions.* (Romanian). Seminarul de teoria structurilor, No. 53, Univ. Timişoara, 1989, pp. 1–10.

3) Let $n = \prod_{i=1}^{r} p_i^{\alpha_i}$ be the prime factorization of $n > 1$. Then

$$k^{\omega(n)} \leq \prod_{i=1}^{r}\left(1 + \frac{k-1}{\alpha_i}\right)^{\alpha_i} \leq d_k(n) \leq k^{\Omega(n)}$$

J. Sándor. *On the aritmetical function $d_k(n)$*. L'analyse Numér. Th. Approx. 18 (1989), 89–94.

Corollary. The normal order of magnitude of $\log d_k(n)$ is $\log k \cdot \log\log n$

Remark. For a short proof of the inequality

$$\log d_k(n) \leq \frac{(\log n)\cdot(\log k)}{\log\log n}\cdot$$
$$\cdot\left(1 + O\left(\frac{\log\log\log n + \log k}{\log\log n}\right)\right)$$

see

L.P. Usol'cev. *On the estimation of a multiplicative function* (Russian). Additive problems of number theory. Interuniv. Collect. sci. works. Kujbyshev 1985, 34–37.

§ II. 4 On $d(m\,n)$

1) $(d(m\,n))^2 \geq d(m^2)\,d(n^2)$
for all $m, n = 1, 2, 3, \ldots$

2) $\dfrac{d(m\,n)}{d(m)\,d(n)} \leq \dfrac{\sigma(m\,n)}{\sigma(m)\,\sigma(n)}$
for all $m, n = 1, 2, 3, \ldots$

3) $\dfrac{d(m^2\,n)\,d(k^2\,n)}{(d(m\,n\,k))^2} \geq \dfrac{d(m^2)\,d(k^2)}{(d(m\,k))^2}$
for all $m, n, k = 1, 2, 3, \ldots$

J. Sándor. *Some arithmetic inequalities.* Bull. Number Theory. Rel. Topics 11 (1987), 149–161.

4) $\dfrac{d(m\,n)}{d(m)} \geq \dfrac{\sigma(m\,n)}{n\,\sigma(m)}$
for all $m, n = 1, 2, 3, \ldots$

J. Sándor. *Corrections to: "Some arithmetic inequalities."* Bull. Number Theory Rel. Topics 12 (1988), 93–94.

§ II. 5 An inequality for $d_k(n)$

Let the non-negative arithmetical functions $f_i\ (i = 1, 2, \ldots, k)$ satisfy $f_i(m) \leq f_i(m\,n)$ for all positive integers m, n. Then

$$d_k(n) \prod_{i=1}^{k} (f_i * I)(n) \geq (d(n))^k\,(f_1 * \cdots * f_k)(n)$$

where $I(n) = 1$ for all n and "$*$" is the Dirichlet convolution.

J. Rutkowski. *On Čebyšev inequality for arithmetical functions.* Funct. Approximatio Comment. Math. (Poznań) 18 (1989), 99–104.

§ II. 6 Majorization for $\log d(n)/\log 2$

1) $d(n) \leq \left(\log \left(n \prod_{p|n} p \right)^{1/\omega(n)} \right)^{\omega(n)} \Big/ \prod_{p|n} \log p$

$(n > 1)$ where p runs over the prime divisors of n, and $\omega(n)$ denotes the number of distinct prime divisors of n

D. Somasundaram. *A divisor problem of Srinivasa Ramanujan in Notebook 3.* Math. Student 55 (1987), 175–176.

Remark. This inequality appears without proof in Ramanujan Notebook 3.

2) For $n \geq 3$ one has:

a) $$\frac{\log d(n)}{\log 2} \leq C_1 \frac{\log n}{\log \log n}$$
where $C_1 = 1.5\,379\ldots$ with equality for
$$n = 2^5 \cdot 3^3 \cdot 5^2 \cdot 7 \cdot 11 \cdot 13 \cdot 17 \cdot 19$$

J.L. Nicolas and G. Robin. *Majorations explicites pour le nombre de diviseurs de n*. Canad. Math. Bull. 26 (1983), 485–492.

b) $$\frac{\log d(n)}{\log 2} \leq \frac{\log n}{\log \log n} + C_2 \frac{\log n}{(\log \log n)^2}$$
where $C_2 = 1.9349$

c) $$\frac{\log d(n)}{\log 2} \leq \frac{\log n}{\log \log n} + \frac{\log n}{(\log \log n)^2} + C_3 \frac{\log n}{(\log \log n)^3}$$
where $C_3 = 4.7\,624$

d) $$\frac{\log n}{\log 2} \leq \frac{\log n}{\log \log n - C_4}$$
where $C_4 = 1.39177$, for $n \geq 56$

G. Robin. *Thèse d'état*. Université de Limoges, France, 1983.

e) Let $\varepsilon > 0$ fixed and $H_\varepsilon = \prod_{p \leq 2^{1/\varepsilon}} p^{\alpha(p,\varepsilon)}$, where
$\alpha(p, \varepsilon) = [1/(p^\varepsilon - 1)]$. Then for $n \geq 1$ one has
$$d(n)/n^\varepsilon \leq d(H_\varepsilon)/H_\varepsilon^\varepsilon \leq (2/(e\,\varepsilon\,\log 2))^{2^{1/\varepsilon}}$$

J.L. Nicolas. *Bornes effectives pour certaines functions arithmétiques*. Colloque de Théorie Analytique des nombres ≪Jean Coquet≫ (Marseille, 1985), 91–99. Publ. Math. Orsay, 80–02, Univ. Paris XI, Orsay, 1981.

3) There exists a positive constant $c > 0$ such that
$$d(p-1) > \exp\left(c \frac{\log p}{\log \log p}\right)$$
for infinitely many primes p

K. Prachar. *Primzahlverteilung*. Die Grundlehren der mathematischen Wissenschaften, Bd. 91, Berlin, Göttingen, Heidelberg, 1957 (p. 370.)

§ II. 7 $\max_{n\le x} d(n)$ and $\max_{n\le x} (d(n), d(n+1))$ and generalizations

1) $\max_{n\le x} d(n) = \exp\left(\log 2 \frac{\log x}{\log\log x} + O\left(\frac{\log x \log\log\log x}{(\log\log x)^2}\right)\right)$

S. Wigert. *Sur l'ordre de grandeur du nombre des diviseurs d'un entier.* Arkiv. för Math. 3 (1907), 1–9.

Corollary. $\limsup_{n\to\infty} \frac{\log d(n) \log\log n}{\log n} = \log 2$

Remark. For an intresting proof and a slightly stronger result that 1), see
S. Ramanujan. *Highly composite numbers.* Proc. London Math. Soc. 14 (1915), 347–409.

2) $\sum_{n\le x} \max(d(n), d(n+1)) = 2x \log x + O(x(\log x)^{-1-\varepsilon})$

$\varepsilon > 0$-constant

I. Kátai. *On the local behaviour of the function d(n)* (Hungarian.) Mat. Lapok 18 (1967), 297–302.

3) Define $f(n) = \frac{\log d(n) \cdot \log\log n}{\log 2 \log n}$, $(n \ge 2)$. Then the maximum of $f(n)$ is attained at

$$n = 2^5 \cdot 3^3 \cdot 5^2 \cdot 7 \cdot 11 \cdot 13 \cdot 17 \cdot 19$$

and

$$\max f(n) = 1.5379\ldots$$

J.-L. Nicolas and G. Robin. *Majorations explicites pour le nombre de diviseurs de N.* Canad. Math. Bull. 26 (1983), 485–492.

4) a) $\sum_{n\le x} \max\ (d(n), d(n+1), \ldots, d(n+k-1)) \sim kx \log x$

provided $k = o((\log x)^{\alpha})$, where $\alpha = 3 - 2\sqrt{2}$

P. Erdős and R.R. Hall. *Values of the divisor function on short intervals.* J. Number. Theory 12 (1980), 176–187.

b) Relation a) is true for $k \le (\log x)^{\log 4 - 1} \exp(-\xi(x) \cdot \sqrt{\log\log x})$, where

$$\xi(x) \to \infty$$

as $x \to \infty$

R.R. Hall. *The maximum value of the divisor function in short intervals.* Bull. London Math. Soc. 13 (1981), 121–124.

5) For $A \subset \mathbb{N}$ define $d_A(n) = \sum_{d|n, d\in A} 1$ and $D_A x = \max_{n\le x} d_A(n)$. Let $f_A(x) = \sum_{a\le x, a\in A} \frac{1}{a}$. Then:

a) $\limsup_{x\to\infty} D_A(x)\exp\left(-\left(\frac{e}{16}-\varepsilon\right)(\log f_A(x))^2\right)=+\infty$

if $f_A(x)\to\infty$

b) There exists a set $A\subset\mathbb{N}$ of density 1 for which

$$\limsup_{x\to\infty} D_A(x)\exp\left(-\left(\frac{1}{2}+\varepsilon\right)(\log f_A(x))^2\right)=0$$

c) For all $\lambda>0$ there exists an $x(\lambda)$ such that if $x>x(\lambda)$ and if $f_A(y)>22\log\log\log y$ with $y=\exp(\log x(\log\log x)^{-21})$, then

$$D_A(x)>\lambda\cdot f_A(x)$$

P. Erdős and A. Sárközy. *Some asymptotic formulas on generalized divisor functions.* IV. Studia Sci. Math. Hungar. 15 (1980), 467–479.

§ II. 8 Highly composite, superior highly composite, and largely composite numbers

A number n is called *highly composite* (after Ramanujan) if $d(m)<d(n)$ for all $m<n$.

1) a) Let $Q(x)$ denote the number of highly composite numbers not exceeding x. Then

$$\lim_{x\to\infty}\frac{Q(x)}{\log x}=+\infty$$

S. Ramanujan. *Highly composite numbers.* Proc. London Math. Soc. (2) 14 (1915), 347–409.

b) If n is highly composite, then the next highly composite number is

$$\leq 1+(\log n)^{-c}$$

for a certain constant c

P. Erdős. *On highly composite numbers.* J. London Math. Soc. 19 (1944), 130–133.

Corollary. $Q(x)>(\log x)^{1+c}$ for certain $c>0$

2) Let us define $c(x)$ by $Q(x)=(\log x)^{c(x)}$. Then

a) $\underline{\lim}\, c(x)\geq 1.13682\ldots$

J.-L. Nicolas. *Répartition des nombres hautement composés de Ramanujan.* Can. J. Math. 23 (1971), 116–130.

b) $\underline{\lim}\, c(x)\leq 1.44$

J.-L. Nicolas. *Nombres hautement composés.* Acta Arith. 49, N° 4, dedicated to P. Erdős on his 75th birthday.

c) $\overline{\lim}\, c(x) \leq 1.71$

Remark. Under certain very strong conjectures it can be proved that

$$\lim c(x) = (\log 30)/(\log 16) = 1.227\ldots$$

See

T.H. Tran. *Nombres hautement composés de Ramanujan généralisés.* C.R. Acad. Sci. Paris, Sér. A-B, 282, 1976, no. 17, pp. A939–A942.

3) A number N is called superior highly composite (after Ramanujan), see

S. Ramanujan: *Highly composite numbers*, 1915

if there exists $\varepsilon > 0$ such that for all n we have

$$\frac{d(n)}{n^\varepsilon} \leq \frac{d(N)}{N^\varepsilon}$$

a) Let $E = \left\{ \dfrac{\log(1+1/k)}{\log p} : k \geq 1, p \text{ prime} \right\}$. Then if $\varepsilon \notin E$, the maximum of $d(n)/n^\varepsilon$ is attained at only one integer N_ε, and we have

$$N_\varepsilon = \prod_{i=1}^{\infty} p_i^{\alpha_i}$$

with $\alpha_i = [1/(p_i^\varepsilon - 1)]$

S. Ramanujan. Ibid.

b) Let n be highly composite and N the superior highly composite number just preceding n. Let ε be any parameter such that $N = N_\varepsilon$ and $x = 2^{1/\varepsilon}$. We write

$$N = \prod_{p \leq x} p^{b_p}$$

with $b_p = [1/(p^\varepsilon - 1)]$ and define the benefit of n (relative to N and ε) by $\operatorname{ben} n = \log \dfrac{d(N)}{N^\varepsilon} - \log \dfrac{d(n)}{n^\varepsilon}$. Then there exists C such that

$$\operatorname{ben} n \leq C \cdot x^{-0.0307\ldots}$$

J.-L. Nicolas. See 2) a).

4) A natural number n is called *largely composite* if $d(n) \geq d(m)$ for all $n \geq m$. Let $Q_l(x)$ be the counting function of largely composite numbers. Then there exist $c, d > 0$ such that

$$\exp(\log^c x) \leq Q_l(x) \leq \exp(\log^d x)$$

for all large x

J.-L. Nicolas. *Répartition des nombres largement composés.* Acta Arith. 34 (1979), 379–390.

§ II. 9 Congruence property of $d(n)$

Let $S_k = \{n \in \mathbb{N} : k \mid d(n)\}$ and $S_k(x)$ the enumerative function of $S_k(x)$. Then:

1) For odd $k > 1$, $S_k(x) \sim \alpha_k x$, $(x \to \infty)$, where α_k is a positive constant.

L.G. Sathe. *On a congruence property of the divisor function.* Amer. J. Math. 67 (1945), 397–406.

2) a) For odd $k > 1$

$$S_k(x) = \alpha_k x + O(x^{1/2} \log^2 x)$$

E. Cohen. *Arithmetical notes. V. A divisibility property of the divisor function.* Amer. J. Math. 83 (1961), 693–697.

b) $S_p(x) = (1 - c_p)x + O(x^{1/(p-1)})$
where p is a fixed odd prime and $c_p = \zeta(p)/\zeta(p-1)$

E. Cohen. *Arithmetical notes. IV. A set of integers related to the divisor function.* J. Tennessee Acad. Sci. 37 (1962), 119–120.

§ II.10 $\Delta(x) = \sum_{n \le x} d(n) - x \log x - (2\gamma - 1)x$

$\sum_{n \le x} d(n) = x \log x + (2\gamma - 1)x + \Delta(x)$ where γ is Euler's constant, and:

1) $\Delta(x) = O(\sqrt{x})$

G.L. Dirichlet. *Sur l'usage des séries infinies dans la théorie des nombres.* Crelle's Journal 18 (1938), 259–274.

2) $\Delta(x) = O(x^{1/3} \log x)$

G. Voronoi. *Sur un problème du calcul des fonctions asymptotiques.* J. reine angew. Math. 126 (1903), 241–282.

3) $\Delta(x) \ne O(x^{1/4})$

G.H. Hardy. *The average order of the arithmetical functions $P(x)$ and $\Delta(x)$.* Proc. London Math. Soc. 15 (2) (1916), 192–213,

and

E. Landau. *Über die Gitterpunkte in einem Kreise.* Akademie der Wissenschaften. Göttingen Nachr. 5 (1915), 161–171.

4) $\Delta(x) = O(x^{27/82} (\log x)^{11/41})$

J.G. van der Corput. *Zum Teilerproblem.* Math. Ann. 98 (1928), 697–716.

5) $\Delta(x) = O(x^{15/46} (\log x)^{30/23})$

H. Richert. *Verschärfung der Abschätzung beim Dirichletschen Teilerproblem.* Math. Zeitschrift 58 (1953), 204–218.

6) a) $\Delta(x) = O(x^{7/22+\varepsilon})$

H. Iwaniec and C.J. Mozzochi. *On the divisor and circle problems.* J. Number Theory 28 (1988), 60–93.

b) $\Delta(x) = O(x^{139/429+\varepsilon})$

G. Kolesnik. Acta Arith. 45 (1985), 115–143.

7) $\Delta(x) = O(x^{7/22}(\log x)^{89/22})$

M.N. Huxley. *Exponential sums and lattice points.* J. London Math. Soc. (to appear).

8) a) $\int_1^T (\Delta(x))^2 x^{-3/2} \mathrm{d}x \sim c \cdot \log T$

as $T \to \infty (c > 0$, constant.)

R. Bellman. *The Dirichlet divisor problem.* Duke Math. J. 14 (1947), 411–417.

b) $\int_0^T (\Delta(x))^2 \mathrm{d}x = \alpha T^{3/2} + O(T \log^5 T)$

K.C. Tong. *On divisor problems.* II. Acta Math. Sinica 6 (1956), 139–152.

c) $\int_0^T (\Delta(x))^2 \mathrm{d}x = \alpha T^{3/2} + O(T(\log T)^4)$

where $\alpha = \frac{1}{6\pi^2} \sum_{n=1}^{\infty} (d(n))^2 n^{-3/2}$

E. Preismann. *Sur la moyenne quadratique du terme du reste du problème du cercle.* C.R. Acad. Sci. Paris Sér. I. Math. 306 (1988), 151–154.

d) The moments $\lim_{T\to\infty} T^{-1-k/4} \int_0^T |\Delta(x)|^k \mathrm{d}x$ exist for any real $k \in [0, 9]$

D.R. Heath-Brown. *The distribution and moments of the error term in the Dirichlet divisor problems.* Acta Arith. 60 (1992), 389–414

Remark. In the above paper Heath-Brown shows that $x^{-1/4} \cdot \Delta(x)$ has a distribution function. It follows that $x^{-1/4} \cdot \Delta(x)$ lies with a positive probability in a given interval of positive length. See also

D.R. Heath-Brown. *The Dirichlet divisor problem.* Proc. third conf. of Canad. Numb. Th. Assoc., Oxford: Clarendon Press, 31–35 (1993).

For estimates on $\int_0^T (\Delta(x))^3 \mathrm{d}x$ and $\int_2^T (\Delta(x))^4 \,\mathrm{d}x$ see also

K.-M. Tsang. *Higher power moments of $\Delta(x)$, $E(t)$, and $P(x)$.* Proc. London Math. Soc. III Ser. (to appear.)

9) $\Delta(x) = \Omega_-(x^{1/4} \exp(c/\log\log x)^{1/4}(\log\log\log x)^{-3/4})$

K. Corrádi and I. Kátai. *A comment on K.S. Gangadharan's paper entitled "Two classical lattice point problem"* (Hungarian.) Magyar Tud. Akad. Mat. Fiz. Oszt. Közl. 17 (1967), 89–97.

10) Suppose $H \cdot U \gg X^{1+\varepsilon}$ and $X^{\varepsilon} \ll U \ll \frac{1}{2}X^{1/2}$. Then

$$H \cdot U \cdot \log^3(X^{1/2}/U) \ll \int_X^{X+H} (\Delta(x+U) - \Delta(x))^2 dx \ll H \cdot U \cdot \log^3(X^{1/2}/U)$$

M. Jutila. *On the divisor problem for short intervals.* Ann. Univ. Turku Ser. A I, No. 186 (1984), 23–30.

11) $\Delta(x) = \Omega_+((x \log x)^{1/4} (\log\log x)^{\alpha} \exp(-A\sqrt{\log\log\log x}))$

where $\alpha = \dfrac{3 + 2\log 2}{4}$ $(A > 0$ absolute constant)

J.L. Hafner. *New omega theorems for two classical lattice point problems.* Invent. Math. 63 (1981), 181–186.

12) $\sum_{n \le x} \Delta(n) = \frac{1}{2} x \log x + \left(\gamma - \frac{1}{4}\right) x + O(x^{3/4})$

S.L. Segal. *A note on the average order of number-theoretic error terms.* Duke Math. J. 32 (1965), 279–284.

13) Let $D(x) = \sum_{n \le x} d(n)$ and $f(x)$ be positive and increasing with $f(x)/x$ decreasing for $x > 0$

a) If $f(x)/x \to 0$ and $f(x)/\log^2 x \to \infty (x \to \infty)$, then

$$D(x + f(x)) - D(x) \sim f(x)$$

for almost all $x > 0$

b) If $f(x)/x \to 0$ and $f(x)/\log^6 x \to \infty (x \to \infty)$, then

$$D(x + f(x)) - D(x) = f(x)\log x + 2\gamma f(x) + O(f(x))$$

for almost all $x > 0$

T. Chih. *A divisor problem.* Acad. Sinica Sci. Rec. 3 (1950), 177–182.

§ II.11 $\sum_{p \le x} d(p-1)$, p prime

1) a) $\sum_{p \le N} d(p-1) = \dfrac{315\,\zeta(3)}{2\pi^4} N + O\left(\dfrac{N}{(\log N)^{\alpha}}\right)$

where $\alpha \in (0, 1)$ is an arbitrary constant.

Yu. V. Linnik. *New versions and new uses of the dispersion method in binary additive problems* (Russian). Dokl. Akad. Nauk SSSR 137 (1961), 1299–1302.

Remark. The first result on $\sum_{p \le x} d(p-1)$ is due to Titchmarsh, who proved that

$$\sum_{p \le x} d(p-1) = O(x)$$

E.C. Titchmarsh. *A divisor problem.* Rendiconti Palermo 54 (1930), 414–429.

b) $\sum_{d\le x} d(p-1) = x\prod_p\left(1+\frac{1}{p(p-1)}\right) + O\left(\frac{x\log\log x}{\log x}\right)$

G. Rodriguez. *Sul problema dei divisori di Titchmarsh.* Boll. Un. Mat. Ital. (3) 20 (1965), 358–366,

and

H. Halberstam. *Footnote to the Titchmarsh-Linnik divisor problem.* Proc. Amer. Math. Soc. 18 (1967), 187–188.

c) $\sum_{p\le x} d(p-1) = c_1 x\log x + c_2 x + O_A(x/(\log x)^A)$

for any $A > 0$ (c_1, c_2 explicit constants.)

E. Bombieri, J.B. Friedlander and H. Iwaniec. *Primes in arithmetic progressions to large moduli.* Acta Math. 156 (1986), 203–251,

and

É. Fouvry. *Sur le problème des diviseurs de Titchmarsh.* J. Reine Angew. Math. 357 (1985), 51–76.

d) $A_k x(\log x)^{k-2} \le \sum_{p\le x} d_k(p-1) \le B_k x(\log x)^{k-2}$

for $k\ge 3$, where A_k and B_k are positive constants.

N.P. Ryzhova. *Order estimates in the generalized Titchmarsh problem* (Russian). Additive problems of number theory, Interuniv. Collect. sci. Works. Kujbyshev 1985, 51–55 (1985).

2) a)
$$\sum_{0<p-a\le n} d(p-a) = \frac{315\zeta(3)}{2\pi^4}\prod_{p\mid a}\frac{(p-1)^2}{p^2-p+1}n + O(n(\log n)^{-1+\varepsilon})$$

for $\varepsilon > 0$

B.M. Bredihin. *Binary additive problems of indeterminate type.* I (Russian) Izv. Akad. Nauk SSSR Ser. Mat. 27 (1963), 439–462.

b)
$$\sum_{a<p\le x} d_k(p-a) = H_k J_k(a)\, x\log^{k-2} x + O(x\log^{k-3} x(\log\log x)^2 + a\log^{k-2} x + \delta\log^{k-2} x)$$

If $0\le Q < Q+R\le x$ and $0<\delta\le\frac{1}{12}$ then the above relation holds for all but at most

$$C(R+x^{1-1/k})x\exp(-\delta\log x)\delta^{-2}x^{-1/k}\exp(-\delta\log x)(\log x)^{k^2-2k+2}$$

values of a with $Q < a \le Q+h$

R.C. Vaughan. *On the number of solutions of the equation $p = a + n_1 + \cdots + n_k$ with $a < p \le x$.* J. London Math. Soc. (2) 6 (1972), 43–55.

3) $$\sum_{p_1,p_2\le\sqrt{x}} d(p_1p_2-1) = \frac{630\,\zeta(3)}{\pi^4}\frac{x}{\log x} + O\left(\frac{x(\log\log x)^7}{\log^2 x}\right)$$

M.B. Barban. *Analogues of the divisor problem of Titchmarsh* (Russian). Vestnik Leningrad Univ. Mat. Meh. Astronom 18 (1963), 5–13.

4) $$\sum_{p^i \le x^{\alpha_i}} d(p_1 \cdots p_k - 1) = \frac{315\,\zeta(3)}{\pi^4} \frac{1}{\alpha_1 \cdots \alpha_{k-1}} \frac{x}{\log^{k-1} x} + O(x \log\log x / \log^k x)$$

where $\alpha_1 + \cdots + \alpha_k = 1, 0 < \alpha_1 < \cdots < \alpha_{k-1} < \alpha_k = \frac{1}{2}; \alpha_{k-1} > \frac{1}{4}$

A.K. Karšiev. *The generalized problem of divisors of Titchmarsh* (Russian). Izv. Akad. Nauk. Uz SSSR Ser. Fiz.-Mat. Nauk 11 (1967), 21–28.

5) $$\sum_{p \le x} \max(d(p-1), d(p+1)) = C(-1,1)x + O\left(\frac{x}{(\log x)^{\alpha}}\right)$$

($\alpha > 0$ constant), where $C(-1,1) = \lim_{x\to\infty} \frac{1}{x} \sum_{p \le x} (d(p-1) + d(p+1))$

I. Kátai. On the local behaviour of the function $d(n)$ (Hungarian). Mat. Lapok 18 (1967), 297–302.

Remark. The existence of $C(-1,1)$ was proved by Yu.V. Linnik.

Yu.V. Linnik. *The dispersion method in binary additive problems.* Leningrad 1961 (Russian).

6) $$\sum_{p_1 p_2 \le x} d(p_1 p_2 - 1) = \frac{315}{\pi^4} \zeta(3) x \log\log x + O(x)$$

A. Fuji. *On some analogues of Titchmarsh divisor problem.* Nagoya Math. J. 64 (1976), 149–158.

7) Let $S(x,y) = \{n \le x \colon P(n) \le y\}$, where $P(n)$ denotes the greatest prime factor of n. Denote $T(x,y) = \sum_{n \in S(x,y)} d(n-1)$. Then

$$T(x,y) = \Psi(x,y) \log x \left(1 + O\left(\frac{\log(u+1)}{\log y}\right)\right)$$

if $x^{\eta^{\log_3 x/\log_2 x}} \le y \le x$ ($\eta > 0$, $A > 0$ are fixed constants). Here $u = \frac{\log x}{\log y}$ for $x \ge y \ge 2$ and $\Psi(x,y) = \operatorname{card} S(x,y)$ and $\log_k$ denotes the k-fold iterated logarithm.

É. Fouvry and G. Tenenbaum. ≪*Diviseurs de Titchmarsh des entiers sans grand facteur premier*≫. Analytic Number Theory (Tokyo 1988), 86–103. Lecture Notes Math. 1434, Springer, Berlin 1990.

§ II.12 $\Delta_k(x) = \sum_{n \le x} d_k(n) - x \cdot P_{k-1}(\log x)$, $k \ge 2$

Let $\Delta_k(x) = \sum_{n \le x} d_k(n) - x P_{k-1}(\log x)$ for $k \ge 2$ (where $P_{k-1}(t)$ is a polynomial in t of degree $k-1$) and denote

$$\beta_k = \inf\left\{b : \int_1^x \Delta_k^2(t)dt = O(x^{1+2b})\right\}$$

Then:

1) $\beta_5 \leq \dfrac{119}{260} = 0.45769$

A. Ivić. *Topics in recent zeta-function theory.* Univ. Paris XI, Orsay, 1983.

2) $\beta_5 \leq \dfrac{9}{20} = 0.45$

W.P. Zhang. *On the divisor problem.* Kexue Tongbao (English Ed.) 33 (1988), 1484–1485.

Let $\alpha_k = \inf\{a : \Delta_k(x) = O(x^\alpha)\}$. Then:

3) $\alpha_{10} < 0.675$

$\alpha_{11} < 0.6957$

$\alpha_k \leq \dfrac{63k - 258}{64k}$

for $79 \leq k \leq 123$

$\alpha_k \leq \dfrac{59k - 348}{59k}$

for $k \geq 124$

4) $\beta_7 < 0.55469$

A. Ivić. and M. Ouellet. *Some new estimates in the Dirichlet divisor problem.* Acta Math. 52 (1989), 241–253.

5) a) $\alpha_k \leq 1 - c \cdot k^{-2/3}$

with $c > 0$ absolute constant.

H.-E. Richert. *Einführung in die Theorie der starken Rieszchen Summierbarkeit von Dirichletreihen.* Nachr. Akad. Wiss. Göttingen (Math.-Physik), 1960, 17–75.

b) Suppose that D is a constant such that

$$\zeta(\sigma + it) = O\left(t^{D(1-\sigma)^{3/2}} \log^{2/3} t\right)$$

whenever t is sufficiently large and $1/2 \leq \sigma \leq 1$ (Here ζ is the Riemann's zeta function.) Then

$$\alpha_k < 1 - \frac{1}{3} \cdot 2^{2/3}(Dk)^{-2/3}$$

and

$$\beta_k < 1 - \frac{2}{3} \cdot (Dk)^{-2/3}$$

A. Ivić. and M. Ouellet. Ibid.

c) $\alpha_k \leq 1 - (1 - a)/(1 + k\mu(a))$

for $0 < a < 1$; and if $\alpha_k \leq a < 1$, then

$$k\mu(a) \leq (1 - \alpha)/(1 - \alpha_k)$$

where $\mu(a) = \limsup_{t\to\infty} \dfrac{\log|\zeta(a+it)|}{\log t}$

V.I. Melnik. *A Tauberian theorem with remainder for the Laplace transform and its application in the theory of the Riemann zeta-function* (Russian). Ukr. Mat. Zb. 43 (1991), 1368–1378.

Remark. This result implies, as a Corollary, the estimate b) by A. Ivić. and M. Ouellet.

6) a) $\alpha_k \le \dfrac{k-1}{k+1}$

$k = 2, 3, 4, \ldots$

G. Voronoi. *Sur un problème du calcul des fonctions asymptotiques.* J. Reine Angew. Math. 126 (1903), 241–282,

and

E. Landau. *Über die Anzahl der Gitterpunkte in gewissen Bereichen.* Gött. Nachr. (1912), 687–771.

b) $\alpha_k \le \dfrac{k-1}{k+2}$

$k = 4, 5, \ldots$

G.H. Hardy and J.E. Littlewood. *The approximate functional equation in the theory of the zeta-function, with applications to the divisor problems of Dirichlet and Piltz.* Proc. London Math. Soc. (2) 21 (1922), 39–74.

c) $\alpha_k \ge \beta_k \ge \dfrac{k-1}{2k}$

$k = 2, 3, \ldots$

G.H. Hardy. *On Dirichlet's divisor problem.* Proc. London Math. Soc. (2) 15 (1915), 1–25.

Remark. The famous Lindelöf hypothesis on the ζ-function is equivalent with $\beta_k = \dfrac{k-1}{2k}$ for all $k = 2, 3, \ldots$. See

E.C. Titchmarsh. *The theory of the Riemann zeta function.* Oxford, 1951.

d) $\alpha_k \le \dfrac{3}{4} - \dfrac{1}{k}$

if $4 \le k \le 8$

$\alpha_k \le 1 - \dfrac{3}{k}$

if $k \ge 8$

D.R. Heath-Brown. *Mean values of the zeta-function and divisor problems.* Recent progress in analytic number theory. Vol. 1 (Durham, 1979), pp. 115–115 (London, 1981).

7) a) $\beta_2 = \dfrac{1}{4}$

$\beta_3 = \dfrac{1}{3}$

G.H. Hardy. *On the average order of the arithmetical functions P(n) and $\Delta(n)$*. Proc. London Math. Soc. (2) 15 (1915), 192–213;

H. Cramér. *Über das Teilproblem von Piltz.* Arkiv för Mat. Astr. och. Fysik. 16 (1922), No. 21;

E.C. Titchmarsh. *On divisor problems*. Quart. J. Math. 9 (1938), 216–220.

b) $\beta_4 = \frac{3}{8}$

D.R. Heath-Brown. Ibid.

8) a) $$\Delta_k(x) = \Omega^*\left((x\log x)^{(k-1)/(2k)}(\log\log x)^{\alpha_k}\cdot \cdot\exp(-A\sqrt{\log\log\log x})\right)$$

where $a_k = \frac{k-1}{2k}(k\log k - k + 1) + k - 1$

and $\Omega^* = \begin{cases}\Omega_+ & \text{if } k = 2, 3\\ \Omega_- & \text{if } k \geq 4\end{cases}$

J.L. Hafner. *On the average order of a class of arithmetical functions.* J. Number Theory 15 (1982), 36–72.

b) $$\int_1^T \Delta_k^2(x)\mathrm{d}x = \Omega(T^{(2k-1)/k})$$

for $k \geq 2$

A. Ivić. *The general divisor problem.* J. Number Theory 26 (1987), 73–91.

9) Let $1 \leq t_1 < \cdots < t_R \leq T$ and $|t_r - t_s| \geq V$ for $r \neq s \leq R$

a) If $\Delta_2(t_r) \gg V > T^{7/32+\varepsilon}$ for $r \leq R$, then
$$R \ll T^{\varepsilon}(TV^{-3} + T^{15/4}V^{-12})$$

b) If $\Delta_3(t_r) \gg V > T^{18/67+\varepsilon}$ for $r \leq R$, then
$$R \ll T^{\varepsilon}(T^2V^{-4} + T^{57/13}V^{-132/13})$$

A. Ivić. *Large values of the error term in the divisor problem.* Invent. Math. 71 (1983), 513–520.

10) a) Let $|z| \leq 1 (z \in \mathbb{C})$ and N be arbitrary, fixed positive integer. Then
$$\sum_{n\leq x} d_z(n) = c_1(z)x\log^{z-1}x + c_2(z)x\log^{z-2}x + \cdots \cdots + c_N(z)x\log^{z-N}x + O(x\log^{\mathrm{Re}z-N-1})$$

uniformly in z, where $c_j(z) = B_{j-1}(z)/\Gamma(z-j+1)(j = \overline{1, N})$ and each $B_j(z)$ is analytic for $|z| \leq 1$. Here $d_z(n)$ is defined for an arbitrary complex number z by the identity
$$\sum_{n=1}^{\infty} d_z(n)/n^s = (\zeta(s))^z$$

R.D. Dixon. *On a generalized divisor problem.* J. Indian Math. Soc. 28 (1964), 187–196.

Remark. The above result for $N = 1$ is due to A. Selberg.

A. Selberg. *Note on a paper by L.G. Sathe.* J. Indian Math. Soc. 18 (1954), 83–87.

b) $\sum_{n \le x} (d_z(n))^t \le x \cdot \exp\{(z^t - 1)\log\log x + z \log\log(3z) + O(z)\}$
uniformly for all real $x \ge 3$, $z > 1$ and $0 < t \le 1$

K.K. Norton. *Upper bounds fot sums of powers of divisor functions.* J. Number Theory 40 (1992), 60–85.

Remark. Similar, but more complicated estimates are given also for $t > 1$

§ II.13 $\sum_{n \le x} d_k^2(n)$

1) a) $$\sum_{n \le x} d^2(n) = \frac{1}{\pi^2} x(\log x)^3 + Bx(\log x)^2 + Cx + D + O(x^{1/2+\varepsilon})$$

where B, C, D are constants, e.g. $B = \dfrac{12\gamma - 3}{\pi^2} - \dfrac{36}{\pi^4}\zeta'(2)$

b) $$\sum_{n \le x} d^r(n) = x\Big(A_1(\log x)^{2^r-1} + A_2(\log x)^{2^r-2} + \cdots + A_{2^r}\Big) + O\big(x^{(2^r-1)/(2^r+2)+\varepsilon}\big)$$

for $r \ge 2$, integer and $A_1, \ldots, A_{2^r}$ constants.

B.M. Wilson. *Proofs of some formulae enunciated by Ramanujan.* Proc. London Math. Soc. 21 (1922), 235–255.

2) For all $k \in \mathbb{N}$ with $k = o(\sqrt{\log\log x})$,
$$\sum_{n \le x} d_k^2(n) = xQ_{k^2-1}(\log x) + O\left(\exp\left(\exp c_1 k^2\right)x^{1-c_2 \cdot 2^{k-4/3}} + \varepsilon\right)$$
where $Q_{k^2-1}(u)$ is a polynomial of degree $k^2 - 1$ and $c_1, c_2 > 0$ are constants.

V. Kalinka. *A variant of the divisor problem with a large number of components* (Russian). Litovsk. Mat. Sb. 14 (1974), 107–114, 237–238.

§ II.14 On $\sum_{n \le x} (g * d_k)(n)$

Let $g : \mathbb{N} \to \mathbb{C}$ and $f_k = g * d_k$ (where "$*$" denotes the Dirichlet convolution). If $\alpha = \sum_{n=1}^{\infty} \dfrac{g(n)}{n}$ is an absolute convergent series, then:

1) $\lim_{x\to\infty}\left(\frac{1}{x(\log x)^{k-1}}\sum_{n\le x} f_k(n)\right) = \frac{\alpha}{(k-1)!}$

E. Cohen. *Arithmetical notes.* I. Proc. Amer. Math. Soc. 12 (1961), 214–217.

Remark. The case $k = 1$ (when $d_k(n) = 1$ for all n), the above result appears in J.G. van der Corput. *Sur quelques fonctions arithmétiques élémentaires.* Nederl. Akad. Wetensch., Proc. 42 (1939), 859–866.

2) $\sum_{n\le x} f_k(n) = \frac{\alpha}{(k-1)!}x(\log x)^{k-1} + O(\beta x(\log x)^{k-2})$

where $\beta = |g(1)| + \sum_{n=2}^{\infty}\frac{|g(n)|}{n}\log n$

3) $\sum_{n\le x}\frac{f_k(n)}{n} = \frac{\alpha}{k!}(\log x)^k + O(\beta(\log x)^{k-1})$

S.A. Burr. *On uniform elementary estimates of arithmetic sums.* Proc. Amer. Math. Soc. 39 (1973), 497–502.

§ II.15 $\Delta_3(x)$

1) $\sum_{n\le x} d_3(n) = xP(\log x) + \Delta_3(x)$

If $\alpha_3 = \inf\{\alpha > 0 : \Delta_3(x) = O(x^\alpha)\}$, then:

a) $\alpha_3 \le \frac{37}{75}$ (Here P is a second degree polynomial.)

F.V. Atkinson. *A divisor problem.* Quart. J. Math. Oxford Ser. 12 (1941), 193–200.

b) $\alpha_3 \le \frac{14}{29}$

M.I. Yüh. *A divisor problem.* Sci. Record (N.S.) 2 (1958), 326–328.

c) $\alpha_3 \le \frac{8}{17}$

W.-L. Yin. *Piltz's divisor problem for $k = 3$.* Sci. Rec. (N.S.) 3 (1959), 169–173.

d) $\alpha_3 \le \frac{5}{11}$

J.-R. Chen. *On the divisor problem for $d_3(n)$.* Sci. Sinica 14 (1965), 19–29.

e) $\alpha_3 \le \frac{127}{282}$

W.L. Yin and Z.F. Li. *An improvement on the estimate for the error term in the divisor problem for $d_3(n)$* (Chinese). Acta Math. Sinica 24 (1981), 865–878.

f) $\alpha_3 \le \frac{43}{96}\left(<\frac{127}{282}\right)$

G. Kolesnik. *On the estimation of multiple exponential sums. Recent progress in analytic number theory.* Symp. Durham 1979, vol. 1, 231–246, London (1981).

2) $$\sum_{n\le x, n\equiv b (\mathrm{mod}\ a)} d_3(n) = M_3(x,b,a) + O(x^{86/107+\varepsilon}a^{-66/107})$$

if $a \le x^{21/41}$, where

$$M_3(x,b,a) = \frac{x}{\varphi(a/\delta)}\cdot \mathrm{Res}\left(\left(\sum_{m=1,(m,a)=\delta}^{\infty} d_3(m)m^{-s}\right)\frac{x^{s-1}}{s}; s=1\right)$$

where $(b,a)=\delta$

D.R. Heath-Brown. *The divisor function $d_3(n)$ in arithmetic progressions.* Acta Arith. 47 (1986), 29–56.

§ II.16 The divisor problem in arithmetic progressions

1) a) $$\sum_{n\le x, n\equiv b (\mathrm{mod}\ a)} d(n) = c_1(x\log x + 2\gamma - 1) - 2c_2 x + O(x^{1/2})$$

where c_1 and c_2 depend only on a and (a,b)

H.G. Kopetzky. *Über die Größenordnung der Teilerfunktion in Restklassen.* Montsh. Math. 82 (1976), 287–295.

b) For all $a \ge 1$, and all b natural numbers, $1 \le b \le a$ and $\forall\varepsilon > 0$,

$$\sum_{n\le x, n\equiv b (\mathrm{mod}\ a)} d(n) = \alpha x\log x + \beta x + O(x^{35/108+\varepsilon})$$

where α, β depend on a and b

W.G. Nowak. *On the divisor problem in an arithmetic progression.* Comment. Math. Univ. St. Paul 33 (1984), 209–217.

c) $$\sum_{n\le x, n\equiv b (\mathrm{mod}\ a)} d(n) = A_1(a)x\log x + A_2(a)x + O(a^{7/3}x^{1/3}\log x)$$

(with $(a,b)=1$ and $0<b<a$) where $A_1(a)=\varphi(a)/a^2$, and $A_2(a) = (2\gamma-1)\varphi(a)a^2 - (2/a)\sum_{d|a}\frac{\mu(d)\log d}{d}$

D.I. Tolev. *On the divisor problem in arithmetic progressions.* C.R. Acad. Bulgare Sci. 41 (1988), 33–36 (By elementary arguments.)

d) $$\sum_{n\le x, n\equiv b(\bmod a)} d_k(n) = \frac{x}{\varphi(a)} P_k(\log x) + O(x^{1-\gamma/k^\delta+\varepsilon} a^\delta)$$

where $\delta = 3/4 + \varepsilon$ (< 1), with some $\gamma > 0$. (Here $k \le 2$, $(a, b) = 1$ and P_k is a polynomial of degree $k - 1$)

Ž. Edgorov. *The divisor problem in special arithmetic progressions* (Russian). Izv. Akad. Nauk Uz SSR Ser. Fiz.-Mat. Nauk 1977, 9–13, 94. (Using Dirichlet L-functions.)

e) $$\sum_{n\le x, n\equiv b(\bmod a)} d_k(n) = \frac{x}{\varphi(a)} P_k(\log x) + F_k(0) + O\left(x^{(k-1)/(k+1)} (\log 2x)^{k-1} d_k(a)\right)$$

where $k \ge 2$, $x \geqslant 1$, $1 \leqslant b < a$, $(a, b) = 1$, $a \le x^{2/(k+1)}$,
and P_k is a polynomial of degree $k - 1$,
and $F_k(0) = O\left(a^{(k-1)/2+\varepsilon}\right)$, $\forall \varepsilon > 0$

R.A. Smith. See 2) a).

Remark. The estimate on $F_k(0)$ was obtained in

K. Matsumoto. *A remark on Smith's result on a divisor problem in arithmetic progressions*. Nagoya Math. J. 98 (1985), 37–42.

2) Denote by $d(n; b, a)$ the number of positive divisors of n in the residue class $b(\bmod a)$ and let $D(x; b, a) = \sum_{n\le x} d(n; b, a)$. Then:

a) $$D(x; b, a) = \frac{x \log x}{a} + \left(\gamma(b, a) - \frac{1}{a}(1 - \gamma)\right) x + O\left((ax)^{1/3} d(a) \log x\right)$$

if $(b, a) = 1$ and $x \ge a$. Here

$$\gamma(b, a) = \lim_{x\to\infty} \left(\sum_{n\le x, n\equiv b(\text{mode } a)} \frac{1}{n} - \frac{1}{k} \log x \right)$$

R.A. Smith. *The average number of divisors in an arithmetic progression*. Canad. Math. Bull. 24 (1981), 37–41.

b) $$\sum_{n\le x} d(n; b, a) = \frac{x \log x}{a} + \left(\gamma(b, a) - \frac{1-\gamma}{a}\right) x + O\left(\left(\frac{x}{a}\right)^\theta\right)$$

uniformly in b, a, and x, provided that $1 \le b \le a \le x$ and $(a, b) = 1$. Here $\theta < 1/3$ (constant.)

W.G. Nowak. *On a result of Smith and Subbarao concerning a divisor problem*. Canad. Math. Bull. 27 (1984), 501–504.

c) $$D(x; b, a) = \frac{x}{a} \log \frac{x}{a} + \frac{x}{a} c_0(1, k) + \frac{x}{b} + O\left(x^\varepsilon \cdot \left(\frac{x}{a}\right)^\alpha\right)$$

uniformly for b, a, and x satisfying $\sqrt{bx} \le a \le x^{1-\varepsilon}$ and for some $\alpha \le 1/3$ (Assume $1 \le b \le a$)

P.D. Varbanec and P. Zarzycki. *Divisors of integers in arithmetic progression.* Canad. math. Bull. 33 (1990), 129–134.

d) Let $f(n) = kn^2 + ln + s$ be an irreductible polynomial with integer coefficients and discriminant $\Delta = l^2 - 4ks; (\Delta, a) = 1$. Assume that for all $n, (f(n), a) = 1$ and $f(n) > 0$. Let $(a, b) = 1$. Then

$$\sum_{n \le x} d(f(n); b, a) = A_1 x \log x + O(x \log\log x)$$

where A_1 depends on k and on the coefficients of f

E.J. Scourfield. *The divisors of a quadratic polynomial.* Proc. Glasgow Math. Assoc. 5 (1961), 8–20.

3) Let $d'_k(n)$ denote the number of representations of n as $n = u_1 \cdots u_k$ satisfying $u_j \equiv l_j (\bmod m_j)(\overline{1,s})$, where $l_j, m_j, l_j < m_j, (\overline{1,s})$ are positive integers, and where $k = s + t$, $(s, t) \in \mathbb{N}$ with k given. Then for a given positive integer k we have

$$\sum_{n \le x} d'_k(n) - \sum_{i=0}^{k-1} c_i x (\log x)^i + O(1) =$$
$$= \Omega\left((x \log x)^{(k-1)/2k} (\log\log x)^{t-1} \cdot (\log\log\log x)^{-(k+2)(k-1)/4k}\right)$$

for $k \ge 2$ and $x \to \infty$

W.G. Nowak. *On the Piltz divisor problem with congruence conditions.* II. Abh. Math. Semin. Univ. Hamb. 60 (1990), 153–163.

§ II.17 On $\sum_{n \le x} 1/d_k(n)$

1)
$$\sum_{n \le x} \frac{1}{d_k(n)} = b_{k,1} x \log^{1/k-1} x + \cdots$$
$$\cdots + b_{k,N} x \log^{1/k-N} x + O(x \log^{1/k-N-1} x)$$

where $k \ge 2$ and N is arbitrary, fixed, natural number; the constants $b_{k,1}, \ldots, b_{k,N}$ depend only on k

A. Ivić. *On the asymptotic formulae for some functions connected with powers of the zeta-function.* Mat. Vesnik (Belgrade) 1 (14) (29) (1977), 79–90.

Remark. For $k = 2$ this was stated without proof by Ramanujan and proved in

B.M. Wilson. *Proofs of some formulae enunciated by Ramanujan.* Proc. London Math. Soc. (2) 21 (1922), 235–255.

2) $$\sum_{1<n \le x} \frac{1}{\log d_k(n)} = \frac{1}{\log k} \frac{x}{\log\log x} + O(x/(\log\log x)^2)$$

J.-M. de Koninck and A. Ivić. *Topics in arithmetical functions*. Notas de Matemática (72) North Holland, Amsterdam, New York, Oxford, 1980.

§ II.18 Average order of $d_k(n)$ over integers free of large prime factors

1) Let $P(n)$ be the greatest prime factor of n and $\Psi(x, y)$ the number of positive integers not exceeding x, all of whose prime factors do not exceed y. Then we have uniformly

$$\sum_{n\le x, P(n)\le y} d_k(n) = K^{u+O(u/\log u)} \cdot \Psi(x, y) \cdot (\log y)^{k-1}$$

where $x \ge 3$, $\exp((\log\log x)^{5/3+\varepsilon}) \le y \le x$ and $u = \log x/\log y$ and $k \ge 2$ is a fixed integer.

Remark. Recently, H. Smida proved that the above result is valid for all fixed positive real numbers k.

H. Smida. *Valeur moyenne des fonctions de Piltz sur les entiers sans grand facteur premier*. Acta Arith. 63 (1993), 21–50.

2) $$\sum_{n\le x} \frac{d_k(n)}{P(n)} = k^{(2\log x/\log\log x)^{1/2}(1+O(\log\log\log x)/(\log\log x))} \cdot \sum_{n\le x} \frac{1}{P(n)}$$

T.Z. Xuan. *The average order of $d_k(n)$ over integers free of large prime factors*. Acta Arith. LV (1990), 249–260.

Remark. Stronger results are also proved in the above paper. See also

T. Xuan. *The average order of the divisor function over integers free from large prime factors* (Chinese). Chin. Ann. Math. Ser. A12 (1991), Suppl., 28–33.

§ II.19 On a sum on d_k and Legendre's symbol

Let $\lambda \in (0, 1/4)$, $q > q_0(k)$ a prime number and $a, b \in \mathbb{N}$, where $ab\,(a-b) \not\equiv 0 \,(\mathrm{mod}\, q)$. Then

$$\sum_{n\le x} d_k(n)((n+a)(n+b)|q) = O(x^{1-0.01\lambda^2})$$

if $q^{0.75+\lambda} < x < q^2$, (where $(u|v)$ denotes the Legendre symbol.)

A.A. Karacuba. *The asymptotics of certain arithmetical sum (Russian)*. Mat. Zametki 24 (1978), 737–740, 892.

§ II.20 A sum on d_k, d and ω

$$\sum_{n\le x} d_k^t(n)\, d(n-a)\,\omega(n-b) = Dx(\log x)^{k^t} \log\log x(1+O(1/\log\log x))$$

where D is a constant and $a \neq b$

Yu.Yu. Goryunov. *Application of a theorem of Vinogradov-Bombieri type to additive problems in number theory.* Soviet Math. (Iz. VUZ) 32 (1988), 102–104.

§ II.21 On $\sum_{n\le x} d(n)\cdot d(n+N)$ and related problems

1) For $1 \le N \le \dfrac{x^{4/5}}{\log^2 x}$

$$\sum_{n\le x} d(n)\,d(n+N) = C_1(N)\,x\log^2 x + C_2(N)\,x\log x + $$
$$+ C_3(N)\,x + O(x^{2/3}\log^3 x\cdot N^{1/6}\cdot d(N)\cdot\log(N+1))$$

with

$$C_1(N) = \frac{\sigma_{-1}(N)}{\zeta(2)}$$

$$C_2(N) = C_1(N)\cdot\left\{4\gamma - 2 - 4\frac{\zeta'}{\zeta}(2) - 4\frac{\sigma'_{-1}}{\sigma_{-1}}(N)\right\}$$

$$C_3(N) = C_1(N)\cdot\left\{\left(2\gamma - 1 - 2\frac{\zeta'}{\zeta}(2) - 2\frac{\sigma'_{-1}}{\sigma_{-1}}(N)\right)^2 + 1 - \right.$$
$$\left. -4\frac{\zeta''}{\zeta}(2) + 4\left(\frac{\zeta'}{\zeta}(2)\right)^2 + 4\frac{\sigma''_{-1}}{\sigma_{-1}}(N) - 4\left(\frac{\sigma'_{-1}}{\sigma_{-1}}(N)\right)^2\right\}$$

N.V. Kuznetsov. *Convolution of the Fourier coefficients of the Einsenstein-Maass series.* J. Soviet Math. 29 (1985), 1131–1159.

2) For $0 < \alpha < 3/2$, and $1 \le N \le x^{2(2+\alpha)/(5+4\alpha)}/\log^2 x$

$$\sum_{n\le x} d(n)\,\sigma_\alpha(n+N) = C_1(N,\alpha)\,x^{1+\alpha}\log x + $$
$$+ C_2(N,\alpha)\,x^{1+\alpha} + C_3(N,\alpha)\,x\log x + C_4(N,\alpha)x + $$
$$+ O\left(x^{2(1+\alpha)^2/(3+2\alpha)}\log^3 x\cdot N^{1/(6+4\alpha)}\cdot d(N)\log(N+1)\right)$$

where the O-constant depends at most on ε for $\alpha \in \left(0, \dfrac{3}{2} - \varepsilon\right)$ and

$$C_1(N,\alpha) = \frac{\zeta(1+\alpha)}{(1+\alpha)\zeta(2+\alpha)}\sigma_{-1-\alpha}(N)$$

$$C_2(N, \alpha) = C_1(N, \alpha)\left\{2\gamma - (1+\alpha)^{-1} - 2\frac{\zeta'}{\zeta}(2+\alpha) - 2\frac{\sigma'_{-1-\alpha}}{\sigma_{-1-\alpha}}(N)\right\}$$

$$C_3(N, \alpha) = \frac{\zeta(1-\alpha)}{\zeta(2-\alpha)}\sigma_{\alpha-1}(N)$$

$$C_4(N, \alpha) = C_3(N, \alpha)\left\{2\gamma - 1 - 2\frac{\zeta'}{\zeta}(2-\alpha) - 2\frac{\sigma'_{\alpha-1}}{\sigma_{\alpha-1}}(N)\right\}$$

where $\sigma'_\nu(N)$ denotes the derivative of $\sigma_\nu(N)$ with respect to ν

U. Balakrishnan and J. Sengupta. *On the sum* $\sum d(n)\sigma(n+N)$. Manuscripta Math. 67 (1990), 367–378.

3) a) $\sum_{n\le x} d(n)\,d_3(n+N) \sim C_3(N)\,x\log^3 x$

b) $\sum_{n\le x} d(n)\,d_4(n+N) \sim C_4(N)\,x\log^4 x$

R. Bellman. *On some divisor sums associated with Diophantine equations*. Quart. J. Math. Oxford Ser. (2) 1 (1950), 136–146.

c) $\sum_{n\le x} d(n)\,(d_k(n+N))^\alpha = C(k, \alpha, N)\,x(\log x)^{k^\alpha}(1+o(1))$

if $2 \le k; k^\alpha > 5/2\,(k \in \mathrm{I}), \alpha > 0$ and $N \ne 0$

D. Wolke. *Über das summatorische Verhalten zahlentheoretischer Funktionen*. Math. Ann. 194 (1971), 147–166.

d) $\sum_{n=1}^{m-N} d(n)\,d(n+N) = m P_2(\log m) + O(m^{5/6}\log^{3+\varepsilon} M)$

$\varepsilon > 0$, where $P_2(x)$ is a polynomial of degree two.

D. Ismoilov. *Asymptotics of the representation of numbers as the difference of two products* (Russian). Dokl. Akad. Nauk Tadžik. SSR 22 (1979), 75–79.

e) $\sum_{n\le x} d_k(n)\,d(n+N) = x P_k(\log x) + O(x(\log\log x)^c/\log x)$

where P_k is a polynomial of degree k and the exponent c may depend upon k only.

Y. Motohashi. *An asymptotic series for an additive divisor problem*. Math. Z. 170 (1980), 43–63.

f) $$\sum_{n=1}^{m} d(n)\,d(n+N) = \frac{6}{\pi^2}\sigma_{-1}(N)\,m\log^2 m +$$

$$+ a_1 m\log m + a_2 m + O(m^{5/6+\varepsilon})$$

where $N = O(m^{5/6})$ and $\varepsilon > 0$ is arbitrary chosen (a_1, a_2 are constants depending on N)

G. Babaev, N. Gafurov and D. Ismoilov. *Some asymptotic formulas connected with divisors of polynomials* (Russian). Trudy Mat. Inst. Steklov 163 (1984), 10–18.

§ II.22 On $\sum_{n\le x} d_k(n)\cdot d(n+1)$ and related questions

1) $$\sum_{n\le x} d^2(n)\,d(n+1) = Cx(\log x)^4 + O(x(\log x)^3\log\log x)$$

where $C = \pi^{-2}\prod_p\left(1-\frac{1}{p}+\frac{1}{p}\left(1-\frac{1}{p}\right)^2\left(1+\frac{1}{p}\right)^{-1}\right)$

Y. Motohashi. *An asymptotic formula in the theory of numbers.* Acta Arith. 16 (1969/70), 255–264.

2) a) $\sum_{n\le x} d(n+1)\,d_k(n) \sim Ax(\log x)^k$ (A = constant.)

Ju.V. Linnik. *New versions and new uses of the dispersion method in binary additive problems* (Russian). Dokl. Akad. Nauk SSSR 137 (1961), 1299–1302.

b) $$\sum_{n\le x} d(n+1)\,d_z(n) = \prod_p\left(1-\frac{1}{p}+\frac{1}{p}\left(1-\frac{1}{p}\right)^{z-1}\right)\frac{1}{\Gamma(z)}\cdot$$
$$\cdot\, x(\log x)^z + O(x(\log x)^{\mathrm{Re}z-1}\log\log x)$$

uniformly for $|z|\le B\,(B>0)$, $x\geqslant 2$. (Here $z\in\mathbb{C}$ and $\Gamma(z)$ is the Euler gamma-function.)

A. Mercier. *Une formule asymptotique pour* $\sum_{n\le x} a_z(n)\,d(n+1)$. Math. Ann. 255 (1981), 369–378; Erratum: Math. Ann. 258 (1981/1982), 352.

c) $$\sum_{n\le x} d(n+1)\,d_k(n) = xP_k(\log x) + O(x\exp(-c_k\sqrt{\log x}))$$

(P_k is a polynomial of degree k.)

É. Fouvry and G. Tenenbaum. *Sur la corrélation des fonctions de Piltz.* Rev. Mat. Iberoamerican 1 (1985), 43–54.

Remark. A key tool in the proof is the following result: for almost all primes $q\le x$ we have

$$\omega(\sigma(q+1)) = \sum_{p|(q+1)}\omega(p+1) +$$
$$+\,O(\log\log x\cdot(\log\log\log\log x)^2)$$

(due to the authors.)

3) a) $$\sum_{n\le x} d(n+a)\,d_k(n) = x\sum_{j=0}^{k} c_{k,j}(a)(\log x)^{k-j} +$$
$$+\,O(x(\log x)^{-1-\varepsilon})$$

$k=2,3,\ldots$, where $c_{k,j}(a)$ are constants depending on k and a

Y. Motohashi. *On some additive divisor problems.* II. Proc. Japan Acad. 52 (176), 279–281.

b) $\sum_{n \le x} d(n+a)\, d_k(n) = x f_k(\log x) + O(x^{1-\delta_k})$

with $\delta_3 = 1/9 - \varepsilon$; $\delta_k = \dfrac{1}{6} - \dfrac{1}{2k}$ for $k \in \{4, 5\}$ and $\delta_k = \dfrac{1}{2k} - \varepsilon$ for $k \ge 6$

V.A. Bykovskiĭ and A.I. Vinogradov. *Inhomogenous convolutions* (Russian). Zap. Nauchn Sem. Leningrad Otdel. Mat. Inst. Steklov (LOMI) 160 (1987), Anal. Teor. Chisel i Teor. Funkt. 8, 16–30, 296.

4) $\sum_{n \le x} d(n)\, d_k(n+1) = c_k x \log^k x + E_k(x)$

where:

a) $E_3(x) = O(x(\log x \cdot \log\log x)^2)$

C. Hooley. *An asymptotic formula in the theory of numbers*. Proc. London Math. Soc. (3) 7 (1957), 396–413.

b) $E_k(x) = O(x(\log x)^{k-1}(\log\log x)^4)$

for $k \ge 4$

Ju.V. Linnik. *The dispersion method in binary additive problems* (Russian). Ch. III, Izdat. Leningrad Univ., 1961.

Remark. The details of proof are given in

B.M. Bredikhin. *Binary additive problems of indefinite type*. III. The additive problem of divisors. Izv. Akad. Nauk. SSSR Ser. Mat. 27 (1963), 777–794.

c) $E_k(x) = O\left(\dfrac{x}{\log x}(\log\log x)^c\right)$

for some $c = c(k)$

Y. Motohashi. *An asymptotic series for an additive divisor problem*. Math. Z. 170 (1980), 43–63.

5) $\sum_{n \le x} d(n)\, d(n+1) = x P(\log x) + R(x)$

where:

a) $R(x) = O(x \log x)$

A.E. Ingham. *Some asymptotic formulae in the theory of numbers*. J. London Math. Soc. 2 (1927), 202–208.

b) $R(x) = O\left(x^{11/12} \log^{17/16+\varepsilon}\right)$

T. Estermann. *Über die Darstellungen einer Zahl als Differenz von zwei Produkten*. J. Reine Angew. Math. 164 (1931), 173–182.

c) $R(x) = O(x^{5/6+\varepsilon})$

D.R. Heath-Brown. *The fourth power moment of the Riemann zeta-function*. Proc. London Math. Soc. (3) 38 (1979), 385–422.

d) $R(x) = O(x^{2/3+\varepsilon})$

J.-M. Deshouillers and A. Iwaniec. *An additive divisor problem*. J. London Math. Soc. (2) 26 (1982), 1–14.

(Here $P(t)$ is a quadratic polynomial with leading coefficient $6/\pi^2$)

Remark. The above result remains valid also for $\sum_{n\le x} d(n)\,d(n+a)$ when the coefficients of $P(t)$ depend on a

6) Let b be a $\mathbb{C}^\infty$ function with support in [1/2,1]. Then there exists a polynomial P of degree 3 and a real number $\delta > 0$ such that

$$\sum_{n\le x} b\left(\frac{n}{x}\right) d_2(n)\,d_3(n+1) = xP(\log x) + O(x^{1-\delta})$$

J.-M. Deshouillers. *Majorations en moyenne de sommes de Kloosterman*. Seminar on Number Theory, 1981/1982, Exp. No. 3, 5 pp. Univ. Bordeaux I, Talence, 1982.

§ II.23 Iteration *d*

Denote by $d^{(k)}(n) = d(d^{(k-1)}(n))$, where $d^{(0)}(n) = n (k \ge 1)$. Then:

1) $\sum_{n\le x} d(n) = (1+o(1))x\log x$

R. Bellman and H.N. Shapiro. *On the normal order of arithmetic functions*. Proc. Nat. Acad. Sci. 38 (1952), 884–886.

2) $\sum_{n\le x} d(n) \ge x\log x$
for $x \ge 1$ integer

G.A. Bottorf. *On divisor sums*. Indian J. Pure Appl. Math. 2 (1971), 501–503.

3) a) $\sum_{n\le x} d^{(2)}(n) = (1+o(1))\,c_2 x\log\log x$

b) $$\sum_{n\le x} d^{(2)}(n) = c_2 x\log\log x + \sum_{i=0}^{[\log x]} a_i x\log^{-i} x + O(x\exp(-c\sqrt{\log x}))$$

(c_2, a_i, c constants.)

E. Heppner. *Über die Iteration von Teilerfunktionen*. J. Reine Angew. Math. 265 (1974), 176–182.

c) $\sum_{n\le x} d^{(3)}(n) = (1+o(1))\,c_3\,x\log\log\log x$
where c_2, c_3 are constants.

I. Kátai. *On the iterations of the divisor function*. Publ. Math. Debrecen 16 (1969), 3–15.

d) $\sum_{n\le x} d^{(4)}(n) = (1+o(1))\,c_4 x\log\log\log\log x$

I. Kátai. *On the sum* $\sum d^{(4)}(n)$. Acta Sci. Math. Szeged 30 (1969), 313–324.

4) a) $\limsup_{n\to\infty} \frac{\log\log d^{(k)}(n)}{\log\log n} = \frac{1}{l_k}$

where l_k is the $(k+1)$-st term of the Fibonacci sequence $1, 1, 2, 3, 5, \ldots$

b) If $K(n)$ denotes the least positive integer k with $d^{(k)}(n) = 2$ then

$$0 < \limsup_{n\to\infty} \frac{K(n)}{\log\log\log n} < +\infty$$

P. Erdős and I. Kátai. *On the growth of* $d_k(n)$. Fib. Quart. 7 (1969), 267–274.

§ II.24 On $d(f(n))$ and $d(d(f(n)))$, f a polynomial

1) For any polynomial $f(x)$ with integral coefficients and any given integer $r > 0$, there exists a constant $c = c(f, r)$ such that

$$\sum_{n\le x} (d(|f(n)|))^r = O(x(\log x)^c)$$

J.G. van der Corput. *Une inégalité relative au nombre des diviseurs*. Indag. Math. 1 (1939), 177–183.

Remark. For a new proof see

B. Landreau. *A new proof of a theorem of van der Corput*. Bull. London Math. Soc. 21 (1989), 366–368.

2) For an irreducible polynomial $f(x)$, with integral coefficients:

a) $c_1 x \log x < \sum_{n\le x} d(|f(n)|) < c_2 x \log x$

for $x \ge 2$, (c_1, c_2 positive constants.)

P. Erdős. *On the sum* $\sum_{k=1}^{x} d(f(k))$. J. London Math. Soc. 27 (1952), 7–15.

b) $c_1 x(\log x)^L < \sum_{n\le x} (d(|f(n)|))^r < c_2 x(\log x)^L$

where $L = 2^r - 1$ and c_1, c_2 are positive constants.

F. Delmar. *Sur la somme de diviseurs* $\sum_{k\le x} (d(f(k)))^s$. C.R. Acad. Sci. Paris Ser. A-B 272 (1971), A849–A852.

c) If in addition $f(n) > 0$ for $n = 1, 2, 3, \ldots$ and $f(x)$ is not of the form cx ($c \in \mathbf{I}$), then

$$\sum_{n\le x} d(d(f(n))) = c_1 x \log\log x + O(x\sqrt{\log\log x})$$

if degree $f(x) \leqslant 3$

$$\sum_{p\le x} d(d(f(p))) = c_2 \operatorname{li} x \cdot \log\log x + O(\operatorname{li} x \sqrt{\log\log x \cdot \log\log\log x})$$

if degree $f(x) \le 2$ (here p denotes a prime, c_1, c_2 are positive constants.)

I. Kátai. *On the sum* $\sum dd(f(n))$. Acta Sci. Math. Szeged 29 (1968), 199–206.

3) Let $d_k(n)$ denote the number of ways of writing n as a product of k factors, $r \ge 1$ and $0 < \alpha, \beta < 1$. Then, as $x \to \infty$, we have

$$\sum_{\substack{x-y<n\le x \\ n\equiv a (\bmod s)}} (d_k(|f(n)|))^r \ll \frac{y}{s}\left(\frac{\varphi(s)}{s}\log x\right)^{k^r-1}$$

uniformly in a, s and y, with $a \le s$, $(s, f(a)) = 1$, $x^\alpha \le y \le x$ and $s \le y^{1-\beta}$ (the implied constant may depend on f, k, r, α and β).

M. Nair. *Multiplicative functions of polynomial values in short intervals*. Acta Arith. 62 (1992), 257–269.

§ II.25 On $\sum_{n\le x} d(n^2+a)$ and $\sum_{m,n\le x} d(m^2+n^2)$

1) $\sum_{n\le x} d(n^2+a) = A(a)x\log x + B(a)x + O(x^{8/9}(\log x)^3)$

where $-a$ is not a perfect square and $A(a)$ and $B(a)$ depend only on a

C. Hooley. *On the number of divisors of a quadratic polynomial*. Acta Math. 110 (1963), 97–114.

2) a) $\sum_{m,n\le x} d(m^2+n^2) = Cx^2\log x + O(x^2)$

where $C = \frac{1}{2}\pi\left(\sum_{n=1}^{\infty}\chi(n)\mu(n)/n^2\right)^{-1}$ with $\chi(n) = 0$ if $2|n$, $\chi(n) = 1$ if $n \equiv 1 (\bmod 4)$ and $\chi(n) = -1$ if $n \equiv 3 (\bmod 4)$

N. Gafurov. *The sum of the number of divisors of quadratic form* (Russian). Dokl. Akad. Nauk Tadzhik. SSR 28 (1985), 371–375.

b) $$\sum_{m,n\le x} d(m^2+n^2) = A_1x^2\log x + A_2x^2 + O(x^{5/3}\log^9 x)$$

where A_1, A_2 are constants.

N. Gafurov. *On the number of divisors of a quadratic form*. Proc. Steklov Inst. Math. 200 (1993), 137–148.

3) $\sum_{m,n\le x} d(a\,n^2 + b\,n\,m + c\,m^2) = 2Ax^2\log x + O(x^2)$

where A is a constant depending on $L(s, \chi_{4D})$, with s denoting the discriminant and χ_{4D} the Dirichlet real character modulo $4D$

N. Gafurov. *On the sum of divisors of irreductible quadratic forms* (Russian). Dokl. Akad. Nauk Tadzhik. SSR 33 (1990), 577–582.

§ II.26 $\sum_{|f(r,s)|\le N} d(|f(r,s)|)$, $f(x,y)$ a binary cubic form

Let $f(x,y)$ denote a fixed binary cubic form

$$f(x,y) = a_0x^3 + a_1x^2y + a_2xy^2 + a_3y^3$$

irreducible over the integers, having non-zero discriminant

$$D = a_1^2a_2^2 - 4a_0a_2^3 - 4a_1^3a_3 - 27a_0^2a_3^2 + 18a_0a_1a_2a_3$$

Then there exist constants c_1, c_2 (depending only on f) such that

$$\sum_{|f(r,s)|\le N} d(|f(r,s)|) = c_1N^{2/3}\log N + c_2N^{2/3} + O(N^{9/14+\varepsilon})$$

for any fixed $\varepsilon > 0$.

G. Greaves. *On the divisor-sum problem for binary cubic forms*. Acta Arith. 17 (1970), 1–28.

§ II.27 Weighted divisor problem

1) Let a, b be integers with $1 \le a \le b$ and consider $D(x;a,b) = \sum_{m^an^b\le x} 1$. Then

$$D(x;a,b) = \zeta(b/a)x^{1/a} + \zeta(a/b)x^{1/b} + \Delta_{a,b}(x)$$

where:

a) $\Delta_{a,b}(x) = \Omega_+\left(x^{1/(2(a+b))}(\log x)^{b/(2(a+b))}(\log\log x)\right)$

b) $\Delta_{a,b}(x) = \Omega_-\left(x^{1/(2(a+b))}\exp\left(B(\log\log x)^{b/(2(a+b))}\right)\cdot\right.$
$\left.\cdot(\log\log\log x)^{b/(2(a+b)-1)}\right)$

J.L. Hafner. *New omega results in a weighted divisor problem*. J. Number Theory 28 (1988), 240–257.

Remark. a) and b) improve results from

A. Schierwagem. *Über ein Teilerproblem*. Math. Nachr. 72 (1976), 151–168.

2) Let a, b, c integers with $1 \le a \le b \le c$ and let $d(a,b,c;k)$ be the number of representations of k in the form $n_1^an_2^bn_3^c$ with $n_1, n_2, n_3 \in \mathbb{N}$. Denote $D(a,b,c;x) = \sum_{1\le k\le x} d(a,b,c;k)$. Then

$$D(a,b,c;x) = \zeta\left(\frac{b}{a}\right)\zeta\left(\frac{c}{a}\right)x^{1/a} + \zeta\left(\frac{a}{b}\right)\zeta\left(\frac{c}{b}\right)x^{1/b} + \\ + \zeta\left(\frac{a}{c}\right)\zeta\left(\frac{b}{c}\right)x^{1/c} + \Delta(a,b,c;x)$$

where:

a) $\Delta(1,2,3;x) = O(x^{3/10}\log^{9/10} x)$

H.-E. Richert. *Über die Anzahl Abelscher Gruppen gegebener Ordnung (I).* Math. Z. 56 (1952), 21–32.

b) $\Delta(a,b,c;x) = \Omega\left(x^{1/(2(a+b))}\right)$
for $c > 2(a+b)$

c) $\Delta(a,b,c;x) = O\left(x^{\frac{1}{c}-\frac{3a}{c}\frac{3(a+b)-2c}{2c(a+3b)+9a(a+b)}}\right)$
for $7a \le 6b < 12a$, $2c < 3(a+b)$

d) $\Delta(a,b,c;x) = O\left(x^{\frac{1}{c}-\frac{3a}{c}\frac{5a+2b-2c}{3a(5a+2b)+2c(3a+2b)}}\right)$
for $b > 2a$, $2c < 5a+2b$

E. Krätzel. *Teilerprobleme in drei Dimension.* Math. Nachr. 42 (1969), 275–288.

§ II.28 On $\sum_{k<n^{1/t}} d(n-k^t)$

1) $\sum_{k<n^{1/t}} (d(n-k^t))^s < n^{1/t}(\log n)^{c_{t,s}}$
where n, t, s are positive integers.

S. McDonagh. *On the sum* $\sum_{t<N^{1/k}} d(N-t^k)$. Proc. Edinburgh Math. Soc. (2) 15 (1967), 215–219.

2) $\sum_{1\le k<\sqrt{n}} d(n-k^2) = O(\sqrt{n}\log^3 n)$
where n is not a perfect square.

H.W. Lu. *A divisor problem* (Chinese.) J. China Univ. Sci. Tech. 10 (1980), 131–132.

§ II.29 Divisor sums on squarefree or squarefull integers

1) Let $\theta(n)$ denote the number of squarefree divisors of n.

Remark. $\theta(n) = d^*(n)$—the number of unitary divisors of n

Then

$$\sum_{n\le x}\theta(n)=\frac{x}{\zeta(2)}\left(\log x+2\gamma-1-\frac{2\zeta'(2)}{\zeta(2)}\right)+S_2(x)$$

where:

a) $S_2(x)=O(x^{1/2}\log x)$

F. Mertens. *Über einige asymptotische Gesetze der Zahlentheorie*. Crelle's Journal 77 (1874), 289–338.

Remark. For a new proof of a) see

E. Cohen. *The number of unitary divisors of an integer*. Amer. Math. Monthly 67 (1960), 879–880.

b) $S_2(x)=O(x^{1/2})$

A.A. Gioia and A.M. Vaidya. *The number of squarefree divisors of an interger*. Duke Math. J. 33 (1966), 797–799.

2) Let $\theta_k(n)$ denote the number of k-free divisors of n, $(k\ge 2)$. Then

$$\sum_{n\le x}\theta_k(n)=\frac{x}{\zeta(k)}\left(\log x+2\gamma-1-\frac{k\zeta'(k)}{\zeta(k)}\right)+S_k(x)$$

where:

a) $S_k(x)=O(x^{1/3})$
for $k=3$ and $S_k=O(x^\alpha)$ for $k\ge 4$ where α is the exponent in the error term for the Dirichlet divisor problem.

D. Suryanarayana. (1968) MR 38 #4428.

b) $S_2(x)=O(x^{1/2}(x))$ and $S_3(x)=O(x^{1/3}\,\delta(x))$
where $\delta(x)=\exp(-A\log^{3/5}x(\log\log x)^{-1/5})$, $A>0$

D. Suryanarayana and V.S.R. Prasad. *The number of k-free divisors of an integer*. Acta Arith. 17 (1970/71), 345–354.

c) $S_3(x)=\Omega(x^{k-\varepsilon})$
with $k=1/4$ if $t\le 3/4$ and $k=1/3-w/9$ if $t>3/4$, where t is the supremum of the real parts β, where $\beta+i\gamma$ are non trivial zeros of Riemann's ζ-function, and w is the infimum of the real parts of those zeros ρ which satisfy $3/4\le\beta\le t$

B. Saffari. *An Ω-theorem of the "non-effective" type*. Proc. London Math. Soc. (3) 35 (1977), 181–192.

3) a) $$\sum_{n\le x,n\text{ square-full}}\theta(n)=x^{1/2}(a\log x+b)+O(x^{1/3}\log^2 x)$$

b) $$\sum_{n\le x,n\text{ square-full}}d(n)=x^{1/2}(A\log^2 x+B\log x+C)+$$
$$+O(x^{1/3}\log^5 x)$$

(where A, B, C, a, b are constants.)

V. Sitaramaiah and D. Suryanarayana. *On certain divisor sums over square-full integers.* Proc. Conf. Number Theory (Mysore, 1979), pp. 98–109, Madras, 1980.

c) Let u be a positive squarefree integer. Then

$$\sum_{\substack{n\le x,(n,u)=1\\ n \text{ square-free}}} \frac{d(n)}{n} = \frac{1}{2}A\log^2 x + (A+B)\log x +$$
$$+C + O\left(x^{-1/2+\varepsilon}\exp\frac{c\cdot\sqrt{\log 3u}}{\log\log 3u}\right)$$

where A, B, C depend only on u and c is an absolute constant.

B. Gordon, K. Rogers. *Sums of the divisor function.* Canad. J. Math. 16 (1964), 151–158.

§ II.30 Exponential divisors

1) Let $d^e(n)$ denote the number of exponential divisors of n for $n > 1, d^e(1) = 1$. (If $n = p_1^{a_1}\cdots p_r^{a_r}$, then d is an exponential divisor of n if $d \mid n$ and $d = p_1^{b_1}\cdots p_r^{b_r}$ where $b_j \mid a_j (1 \le j \le r)$.) Then:

a) $$\limsup_{n\to\infty} \frac{\log d^e(n)\log\log n}{\log n} = \frac{1}{2}\log 2$$

b) $$\sum_{n\le x} d^e(n) = Ax + O(x^{1/2}\log x)$$
where A is a constant.

M.V. Subbarao. *On some arithmetic convolutions. The theory of arithmetic functions.* (Proc. Conf. Western Michigan Univ., Kalamazao, Mich. 1971), pp. 247–271. Lecture Notes, Vol. 251, Berlin, 1972.

2) $$\sum_{1<n\le x} \frac{1}{\log d^e(n)} = x\int_{-\infty}^{0}\left(C(t) - \frac{6}{\pi^2}\right)dt + O(x^{1/2}\log^{1/2} x)$$

where $C(t) = \prod_p \left(1 + \sum_{k=2}^{\infty}(d^t(k) - d^t(k-1))p^{-k}\right)$, $d(k)$ denoting the number of divisors of k

J.-M. de Koninck and A. Ivić. *An asymptotic formula for reciprocals of logarithms of certain multiplicative functions.* Can. Math. Bull. 21 (1978), 409–413.

§ II.31 Bi-unitary divisors

1) a) card $\{n \le x : nd(n) \le x\} = (\alpha + o(1))x/\sqrt{\log x})$

b) card $\{n \le x : nd^*(n) \le x\} = (\beta + o(1))x/\sqrt{\log x})$

where $d^*(n)$ denotes the number of unitary divisors of n. (α, β are positive constants. For an expression of β as an infinite product, see

H.L. Albott and M.V. Subbarao. *On the distribution of the sequence* $(nd^*(n))$. Can. Math. Bull. 32 (1989), 105–108).

R. Balasubramanian and K. Ramachandra. *On the number of integers n such that* $nd(n)) \le x$. Acta Arith. 49 (1988), 313–322.

2) Let $d^{**}(n)$ denote the number of bi-unitary divisors of n (A divisor $d > 0$ of n is called bi-unitary if $d\delta = n$ and $(d, \delta)^{**} = 1$, where $(d, \delta)^{**}$ denotes the greatest unitary divisor of both d and δ.)

$$\sum_{n \le x} d^{**}(n) = ax(\log x + 2\gamma - 1 + b) + E(x)$$

where $E(x) = O(x^{1/2} \exp(-A \log^{3/5} x (\log\log x)^{-1/5}))$

where $A > 0,\ a = \prod_p \left(1 - \dfrac{p-1}{p^2(p+1)}\right),\ b = 2\sum_p \dfrac{(p^2 - p - 1)\log p}{P^4 + 2p^3 + 1}$

D. Suryanarayana and B. Sitavamachandrarao. *The number of bi-unitary divisors of an integer.* II. J. Indian Math. Soc. (N.S.) 39 (1975), 261–280.

§ II.32 Sums over $d(n) \cdot \omega(n)$, $d(n)/\omega(n)$, $\omega(d(n))$, $\Omega(d(n))$

1) a) $\sum_{n \le x} d(n)\,\omega(n) = 2x \log x(\log\log x) + Ax \log x + O(x)$

J.-M. de Koninck and A. Mercier. *Remarque sur un article de T.M. Apostol.* Math. Bull. 20 (1977), 77–78.

b) $$\sum_{1 < n \le x} \frac{d(n)}{\omega(n)} = x \log x \sum_{k=1}^{N} \frac{B_k}{(\log\log x)^k} + O\left(\frac{x \log x}{(\log\log x)^{N+1}}\right)$$

where N is arbitrary fixed natural number and B_i's are computable constants.

J.-M. de Koninck and A. Ivić. *Topics in arithmetical functions.* Notas de Matemática (72) North Holland, Amsterdam, New York, Oxford, 1980.

2) a) $\sum_{n \le x} \omega(d(n)) = Cx + O(\sqrt{x} \log^5 x)$

where $C > 0$ (constant.)

G.J. Rieger. *Über einige arithmetische Summen.* Manuscripta Math. 7 (1972), 23–34.

b)
$$\sum_{n\le x}\Omega(d(n)) = bx\log\log x + \sum_{k=0}^{[\sqrt{\log x}]}\frac{b_k x}{\log^k x} + O(x\exp(-c\log x))$$
$(b, c > 0$ constants.)

E. Heppner. *Über die Iteration von Teilerfunktionen.* J. Reine Angew. Math. 265 (1974), 176–182.

c) Let $k \ge 2$ be an integer and L be the set of squarefree numbers and $H_k = \{M \in \mathbb{N} : p|m \Rightarrow p^k|m\}$.
Denote $M_k = \{n \in \mathbb{N} : n = lm \text{ with } l \in L, m \in H_k\}$. Then
$$\sum_{n\le x, n\in M_k}\omega(d(n)) = c(k)x + O(x^{1/k}(\log x)/(\log\log x))$$
where $c(k) > 0$

S.B. Ablyalimov. *Two sums wich involve the divisor function* $\gamma(n)$ (Russian). Izv. Vyssh. Uchebn. Zaved. Mat. 1982, no. 10, 3–7.

§ II.33 $\sum_{n\le x} d(a(n))$, $a(n)$ the number of abelian groups with n elements

1) $\sum_{n\le x} d(a(n)) = \sum_{k=1}^{\infty} d(k)d_k x + O(x^{1/2}\log^4 x)$
where $d_k = \lim_{x\to\infty}\frac{1}{x}\sum_{n\le x, a(n)=k} 1$, and $a(n)$ denotes the number of nonisomorphic Abelian groups with n elements.

A. Ivić. *On the number of abelian groups of a given order and on certain related multiplicative functions.* J. Number Theory 16 (1983), 119–137.

Remark. That d_k exists was firstly shown in

D.G. Kendall and R.A. Rankin. *On the number of Abelian groups of a given order.* Quart. J. Math. Oxford Ser (2) 18 (1947), 197–208.

2) For all $\varepsilon \in (0, 1)$
$$\sum_{n\le x, d(n)>a(n)} 1 = x + O(x\log^{\varepsilon-1} x)$$

A. Ivić. Ibid.

§ II.34 $d(n)$ in short intervals

1) If $g(x) \le \log^2 x$, $g(x) \uparrow +\infty$ for $x \to \infty$, then for almost all $N \le x$ we have

$$\sum_{k \leq t} d(k+N) = t(\log N + 2\gamma) + o(t)$$

where $t = g(x)\log^3 x$

Y. Motohashi. *On the number of divisors in a short segment.* Acta Arith. 17 (1970), 249–253.

2) Let $\varepsilon > 0$ be an arbitrary constant. Then for $x^\varepsilon \leq y \leq x$

$$\sum_{x \leq n \leq x+y} d_r(n) < c(r,\varepsilon) y (\log x)^{r-1}$$

where $c(r,\varepsilon)$ is some positive constant depending on r and ε

J. Galambos, K.-H. Indlekofer and I. Kátai. *A renewal theorem for random walks in multidimensional time.* Trans. Amer. Math. Soc. 300 (1987), 759–769.

§ II.35 Number of distinct values of $d(n)$ for $1 \leq n \leq x$

Let $D(x)$ denote the number of distinct values assumed by $d(n)$ for $1 \leq n \leq x$ and $B(x)$ the number of integers $n \leq x$ wich have the form

$$n = p_1^{q_1 - 1} \cdots p_k^{q_k - 1}$$

for p_i, q_i primes and k arbitrary. Then:

1) $D(x) > B(x) + c_1 \log\log\log x$

P. Erdős and L. Mirsky. *The distribution of values of the divisor function d(n).* Proc. London Math. Soc. (3) 2 (1952), 257–271.

2) $D(x) > B(x) + \exp((\log x)^{1/3}/(\log\log\log x))$
for sufficiently large x, where $c = 2.999$

P. Shiu. *Note on a result of Erdős and Mirsky.* J. London Math. Soc. (2) 17 (1978), 228–230.

3) $D(x) > B(x) + \exp((\log x)^{1/2}/\log\log x)$

M. Nair and P. Shiu. *On some results of Erdős and Mirsky.* J. London Math. Soc. (2) 22 (1980), 197–203.

§ II.36 On the distribution function of $d(n)$

Let $r_n(\lambda) = \text{card}\{m : m \leq n, d(m) \leq 2^{e(n)}\}$
where $e(n) = \log\log n + \lambda(2\log\log n)^{1/2}$. Then:

1) $$\lim_{n\to\infty} \frac{r_n(\lambda)}{n} = \frac{1}{\sqrt{\pi}} \int_{-\infty}^{\lambda} e^{-u^2} du$$

M. Kac. *Note on the distribution of values of the aritmetical function d(m).* Bull. Amer. Math. Soc. 47 (1941), 815–817.

2) $r_n(\lambda) = n\frac{1}{\sqrt{2\pi}}\int_{-\infty}^{\lambda} e^{-u^2/2}du + O(n \log\log\log n/(\log\log n)^{1/2})$

W.J. Le Veque. *On the size of certain number-theoretic functions*. Trans. Amer. Math. Soc. 66 (1949), 440–463.

§ II.37 On $(n\,d(n), \sigma(n)) = 1$

Let $U(x) = \text{card}\{n \le x : (nd(n), \sigma(n)) = 1\}$. Then there exist positive constants c_1 and c_2 such that

$$c_1\left(\frac{x}{\log x}\right)^{1/2} < U(x) < c_2\left(\frac{x}{\log x}\right)^{1/2}$$

H.-J. Kanold. *Über das harmonische Mittel der Teiler einer natürlichen Zahl*. II. Math. Ann. 134 (1958), 225–231.

§ II.38 Average value for the number of divisors of sums $a + b$.

Let $\varepsilon > 0, n \in \mathbb{N}$ and $A, B \subseteq \{1, 2, \ldots, n\}$ with min $(|A|, |B|) > \varepsilon n$. Then there exist effectively computable constants c_0, c_1, c_2 such that if $n > c_0$ and $\exp(-c_1(\log n)^{1/2}) < \varepsilon < 1/8$, then

$$\frac{1}{|A|\cdot|B|}\cdot\sum_{a\in A, b\in B} d(a+b) > \frac{c_2 \log n}{(\log(1/\varepsilon))^5 \cdot \log\log(1/\varepsilon)}$$

A. Sárközy and C.L. Stewart. *On the average value for the number of divisors of sums $a + b$*. Illinois J. Math. 38 (1994), 1–18.

Chapter III

SUM-OF-DIVISORS FUNCTION, GENERALIZATIONS, ANALOGUES; PERFECT NUMBERS AND RELATED PROBLEMS

§ III. 1 Elementary inequalities on $\sigma(n)$ and $\sigma(n)/n$

1) a) If n is composite number, then

$$\sigma(n) > n + \sqrt{n}$$

W. Sierpiński. *Elementary theory of numbers*. Warszawa, 1964.

b) For any $n > 2$

$$\sigma(n) < n\sqrt{n}$$

C.C. Lindner. *Problem E 1888. Amer. Math. Monthly* 73 *(1966). Solution by A. Bager and S. Russ*. Amer. Math. Monthly 74 (1967), 1143.

c) For every natural number $n \neq 1, 2, 3, 4, 6, 8$ we have

$$\sigma(n) < \frac{6}{\pi^2} n\sqrt{n}$$

V. Annapurna. *Inequalities for $\sigma(n)$ and $\varphi(n)$*. Math. Mag. 45 (1972), 187–190.

d) If n is of the form $n = 2m + 1$ or $n = 2(2m + 1)$, $m \in \mathbb{N}$ then

$$\sigma(n) < n\sqrt{\gamma(n)}$$

where $\gamma(n)$ denotes the greatest squarefree divisor of n

E.S. Langford. *Private correspondence to D.S. Mitrinović*.

e) $$\prod_{p|n} \frac{p+1}{p} \leq \frac{\sigma(n)}{n} \leq \prod_{p|n} \frac{p}{p-1}, n > 1$$

where p are the prime divisors of n

O. Meissener. *Über einige zahlentheoretische Funktionen*. Arch. Math. Phys. (3), 12 (1907), 199–202.

f) $$\prod_{p|n} \frac{p^{2k}-1}{p^k-1} \left(\frac{1}{p}\right)^k \leq \frac{\sigma_k(n)}{n^k} < \prod_{p|n} \frac{p^k}{p^k-1}$$

where $k \geq 1$

J. Sándor. *On Jordan's arithmetical function*. Math. Student 52 (1984), 91–96 (1988.)

g) Denote $M(k,n) = \prod_{p|n} \frac{p^k - 1}{p - 1}$. Then

$$\frac{M(k,n)\sigma_k(n)}{n^k(\gamma(n))^{k-1}} < \left(\frac{3}{2}\right)^{\omega(n)}$$

(for n odd)

$$\frac{M(k,n)\sigma_k(n)}{n^k(\gamma(n))^{k-1}} < 2\left(\frac{3}{2}\right)^{\omega(n)-1}$$

(for n even.)

J. Sándor. *Ibid.*

Remark. The case $k = 1$ is due to

M. Satyanarayana. *Bounds of* $\sigma(N)$. Math. Student, 28 (1960), 79–81.

h) Let $1 < n = \prod_{i=1}^{r} p_i^{\alpha_i}$ be the prime factorisation of n, where $p_1 < p_2 < \cdots < p_r$. Then if n is odd, and:

if $3|n$, $5 \nmid n$ and $r > 2$, then

$$\frac{\sigma(n)}{n} < \frac{15}{8}\left(\frac{p_3 + 2r + \frac{13}{2}}{p_3 + 3}\right)^{\frac{1}{2}}$$

if $3|n$, $5 \nmid n$ and $r > 1$, then

$$\frac{\sigma(n)}{n} < \frac{3}{2}\left(\frac{p_2 + 2r + \frac{17}{2}}{p_2 + 3}\right)^{\frac{1}{2}}$$

if $3 \nmid n$, $5 \nmid n$ and $r > 1$, then

$$\frac{\sigma(n)}{n} < \frac{5}{4}\left(\frac{p_2 + 2r + \frac{19}{2}}{p_2 + 3}\right)^{\frac{1}{2}}$$

if $3 \nmid n$, $5 \nmid n$, then

$$\frac{\sigma(n)}{n} < \left(\frac{p_1 + 2r + \frac{21}{2}}{p_1 + 3}\right)^{\frac{1}{2}}$$

B. Satyanarayana and S. Vangipuram. *Bounds for* $\sigma(N)/N$. Math. Student 56 (1988), 242–248.

i) $\dfrac{\sigma(n)}{n} < \dfrac{7\omega(n)+10}{6}$

where $\omega(n) = r$ denotes the number of distinct prime factors of n

R.L. Duncan. *Some estimates for $\sigma(n)$*. Amer. Math. Monthly 74 (1967), 713–715.

§ III. 2 On $\sigma(n)/n \log\log n$

1) For $n \geq 7$

$$\sigma(n) < 2.59n \log\log n$$

A. Ivić. *Two inequalities for the sum of divisor function.* Univ. u Novom Sadu Zb. Rad. Prirod.-Mat. Fak. 7 (1977), 17–22.

2) a) $\limsup_{n\to\infty} \dfrac{\sigma(n)}{n \log\log n} = e^{\gamma}$

T.H. Gronwall. Trans. Amer. Math. Soc. 14 (1913), 113–122.

b) The inequality

$$\sigma(n) < e^{\gamma} n \log\log n$$

for all sufficiently large n (in fact, for $n \geq 5\,041$) is true if and only if the Riemann Hypothesis is true.

c) If (c_n) is the sequence of colossally abundant numbers, then the sequence

$$\left(\frac{\sigma(c_n)}{c_n \log\log c_n}\right)$$

contains an infinite number of local extrema.

G. Robin. *Sur l'ordre maximal de la fonction somme des diviseurs*. Seminar on number theory, Paris, 1981–82, 233–244, (1983.)

See also

P. Erdős and J.-L. Nicolas. *Répartition des nombres superabondantes*. Bull. Soc. Math. France 103 (1975), 65–90.

d) $\limsup_{n\to\infty} \left(\dfrac{\sigma(n)}{n} - e^{\gamma} \log\log n\right) \left(\sqrt{\log n}\right) \leq$

$\leq e^{\gamma}(4 - 2\sqrt{2} + \gamma - \log 4\pi) = 1.39\ldots$

S. Ramanujan. *Unpublished manuscript*. See

J.-L. Nicolas. *On highly composite numbers*. Ramanujan revisited (Urbana-Champaign III, 1987), 215–244, Academic Press, Boston, MA, 1988.

Remark. This result has been rediscovered in

G. Robin. *Grandes valeurs de la fonction somme des diviseurs et hypothèse de Riemann*. J. Math. pures et appl. 63 (1984), 187–213.

3) There exists a positive constant $c > 0$ such that

$$\sigma(n) < cn \log\log \varphi(n)$$

for all $n \geq 3$

J. Sándor. *Remarks on the functions* $\varphi(n)$ *and* $\sigma(n)$. Prepr., "Babeş-Bolyai" Univ., Fac. Math. Phys., Res. Semin. 1989, No. 7, 7–12 (1989.)

§ III. 3 On $\sigma_k(n)/n^k$

1) a) $\limsup\limits_{n\to\infty} \dfrac{\sigma_k(n)}{n^k} = \zeta(k)$

for $k > 1$

T.H. Gronwall. Trans. Amer. Math. Soc. 14 (1913), 113–122.

b) $\limsup\limits_{n\to\infty} \left(\log \dfrac{\sigma_k(n)}{n^k}\right) \dfrac{\log\log n}{(\log n)^{1-k}} = \dfrac{1}{1-k}$

for $0 < k < 1$

E. Krätzel. Zahlentheorie, Berlin 1981.

2) For $k > 1$ (fixed) let

$$f_k(x) = \frac{1}{x^k}\left(\sum_{n\leq x} \sigma_k(n) - \frac{x^{k+1}}{k+1}\zeta(k+1)\right)$$

($x \in \mathbb{R}$.) Then:

a) $\limsup\limits_{x\to\infty} f_k(x) = \dfrac{1}{2}\zeta(k)$

b) $\liminf\limits_{x\to\infty} f_k(x) = -\dfrac{1}{2}\zeta(k)$

R.A. MacLeod. *An extremal result for divisor functions*. J. Number Theory 23 (1986), 365–366.

c) Let $A = \{k > 1$: there is an $N_0(k)$ with $N^k f_k(N) > 0$ for some integer $N > N_0(k)\}$. Then $A = [a_0, +\infty)$, where $a_0 = 1.478751$ satisfies

$$\zeta(a_0) = 2\zeta(a_0 + 1)$$

R.A. MacLeod. *Ibid.*

3) a) $\lim\limits_{n\to\infty} \dfrac{\sigma_n(n+a)}{\sigma_n(n)} = e^a \ (n, a \in \mathbb{N})$

b) $\lim\limits_{n\to\infty} \dfrac{\sigma_n(n+a)J_n(n+a)}{\sigma_n(n)J_n(n)} = e^{2a}$

where J_n is Jordan's totient.

M. Sugunamma. *Certain results concerning $\sigma_k(n)$ and $J_k(n)$*. Annales Pol. Math 8 (1960), 173–176.

§ III. 4 $\sum_{n\leq x}\sigma(n)$, $\sum_{n\leq x, p|n}\sigma(n)$, $\sum_{n\leq x,(n,k)=1}\sigma(n)$

1) $$\sum_{n\leq x}\sigma(n)=\frac{\pi^2}{12}x^2+S(x)$$

where:

a) $S(x)=O(x\log x)$

G.L. Dirichlet. *Über die Bestimmung der mittleren Werthe in der Zahlentheorie*. Werke 2, (S. 59), G. Riemer, 1897, pp. 49–66.

b) $S(x)=O(x\log^{2/3}x)$

A. Walfisz. *Weylsche Exponentialsummen in der neuren Zahlentheorie*. Berlin, 1963.

c) $$S(x)=-x\sum_{d\leq\sqrt{x}}\frac{1}{d}p_1\left(\frac{x}{d}\right)-\frac{1}{2}x-\sum_{d\leq\sqrt{x}}dp_1\left(\frac{x}{d}\right)+O(\sqrt{x})$$

where $p_1(x)=\{x\}-\frac{1}{2}$, $p_2(x)=\{x\}^2-\{x\}+\frac{1}{6}$ ($\{x\}$ denotes the fractional part of x)

R.A. Macleod. *Fractional part sums and divisor functions*. J. Number Theory 14 (1982), 185–227.

d) $S(x)\neq o(x\log\log x)$

or equivalently

$$S(x)=\Omega(x\log\log x)$$

T.H. Gronwall. *Some asymptotic expressions in the theory of numbers*. Trans. Amer. Math. Soc. 14 (1913), 113–122.

e) $S(x)=\Omega_-(x\log\log x)$

Y.F.-S. Pétermann. *An Ω-theorem for an error term related to the sum-of-divisors function*. Monatsh. Math. 103 (1987), 145–157.

2) a) $$\sum_{n\leq x, p|n}\sigma(n)=\frac{\pi^2}{12}\left(\frac{1}{2}+\frac{1}{p^2}+\frac{1}{p^3}\right)x^2+O(x\log x)$$

where p is a fixed prime.

b) $$\sum_{n\leq x,(n,k)=1}\sigma(n)=\frac{\pi^2}{12}\prod_{p|k}\left(1+\frac{\sigma(p)}{p^2}+\frac{\sigma(p^2)}{p^4}+\cdots\right)^{-1}x^2+O(x\log x)$$

A. Mercier. *Sommes de fonctions additives restreintes à une classe de congruence*. Can. Math. Bull. 22 (1979), 59–73.

c) For any positive integer M

$$\sum_{2\le n\le x,\, p|n} \frac{1}{\log \sigma(n)} = x\sum_{i=1}^{M} a_i(\log x)^{-i} + O\left(\frac{x}{\log^{M+1} x}\right)$$

where $a_1 = \dfrac{1}{p}$(p a fixed prime) and $a_i(2 \le i \le M)$ computable constants, depending on p

J.M. Tourigny. *Master's thesis*. Univ. Laval, Québec, 1975.

d) $$\sum_{n\le x,(n,k)=1} \log \sigma(n) = \frac{\varphi(k)}{k} x \log x + Ax + O(\log^2 x)$$

See

A. Mercier (1979.)

§ III. 5 Sums over $\dfrac{\sigma(n)}{n}$

1) $$\sum_{n\le x} \frac{\sigma(n)}{n} = \frac{\pi^2}{6}x - \frac{1}{2}\log x + S_1(x)$$

where:

a) $S_1(x) = O(\log^{2/3} x)$

A. Walfisz. *Weylsche Exponentialsummen in der neuren Zahlentheorie*. Berlin, 1963.

b) $S_1(x) = \Omega_+(\log\log x)$

Y.F.-S. Pétermann. *About a theorem of Paolo Codecá's and Ω-estimates for arithmetical convolutions*. J. Number Theory 30 (1988), 71–85.

c) $S_1(x) = \Omega_-(\log\log x)$

Y.F.-S. Pétermann. *An Ω-theorem for an error term related to the sum-of-divisors function*. Monatsh. Math. 103 (1987), 145–157.

d) $$\int_0^T \left(S_1(x) + \frac{\gamma + \log 2\pi}{2}\right) \mathrm{d}x = O(T^{37/130+\varepsilon})$$

Y.F.-S. Pétermann. *Divisor problems and exponent pairs: on a conjecture by Chowla and Walum*. Prospects of mathematical science (Tokyo, 1986), 211–230, World Sci. Publishing, Singapore, 1988.

e) Let $F(x) = S_1(x) + \dfrac{\gamma}{2} + 1$. Then

$$\lim_{x\to\infty} \frac{1}{x}\int_1^x F^k(t)\mathrm{d}t$$

exists for all $k \in \mathbb{N}$ and

$$\lim_{x\to\infty} \frac{1}{x} \int_0^x |F(t)|^\lambda \mathrm{d}t$$

exists for all $\lambda > 0$

Y.F.-S. Pétermann. *Existence of all the asymptotic λ th means for certain arithmetical convolutions.* Tsukuba J. Math. 12 (1988), 241–248.

2) a) Let $\dfrac{g(n)}{\log\log n} \to \infty$. Then

$$\sum_{m=n}^{n+g(n)} \frac{\sigma(m)}{m} = [1+o(1)]\frac{\pi^2}{6} g(n)$$

b) Let $\dfrac{g'(n)}{\log\log\log n} \to \infty$. Then number of integers n in the interval $(n, n+g'(n))$ which satisfy $\dfrac{\sigma(n+1)}{n+1} > \dfrac{\sigma(n)}{n}$ equals

$$[1+o(1)]\frac{g'(n)}{2}$$

c) The number of integers $m \in (n, n+g_1(n))$ which satisfy $\dfrac{\sigma(m)}{m} < c$ equals

$$[1+o(1)]g_1(n)f(c)$$

where $\dfrac{g_1(n)}{\log\log\log n} \to \infty$ and $f(c)$ is the distribution function of $\dfrac{\sigma(n)}{n}$

P. Erdős. *Some remarks about additive and multiplicative functions.* Bull. A.M.S. 52 (1946), 527–537.

§ III. 6 Sums over $\sigma^k(n)$

1) a) $\displaystyle\sum_{n\le x} \sigma^2(n) = \frac{5}{6}\zeta(3)x^3 + O(x^2\log^2 x)$

S. Ramanujan. *Some formulae in the analytic theory of numbers.* Messenger Math. 45 (1916), 81–84.

b) $\displaystyle\sum_{n\le x} \sigma_k^2(n) = \frac{\zeta(2k+1)\zeta^2(k+1)}{(2k+1)\zeta(2k+2)} x^{2k+1} + O(C_k(x))$

where $C_k(x) = x^{2k}, x^2\log^2 x, x^{3/2}\log^2 x$ or $x^{k+1}\log x$ according as $k \ge 1$, $k=1, k=\frac{1}{2}$ or ($k<1$ and $k \ne \frac{1}{2}$)($k>0$).

L. Tóth. *Generalizations of an asymptotic formula of Ramanujan.* Studia Univ. "Babeş-Bolyai", 31 (1986), 9–15.

c) $\sum_{n\le x}\sigma^2(n) = \frac{5}{6}\zeta(3)x^3 + O(x^2\log^{5/3} x)$

d) $\sum_{n\le x} E(n) = \frac{1}{6}\zeta(0)x^3\log x + O(x^3)$

where $E(n) = \sum_{m\le n}\sigma^2(m) - \frac{5}{6}\zeta(3)n^3$

R.A. Smith. *An error term of Ramanujan*. J. Number Theory 2 (1970), 91–96.

2) $\sum_{n\le x, a|n}\sigma^k(n) = C_k(a)x^{k+1} + O\left(\left(\frac{\sigma(a)}{a}\right)^2 x^k(\log x)^{(3k-1)/3}\right)$

where $k, a \in \mathbb{N}$ and $C_k(a) \ll \frac{d(a)}{a}(\log a)^{k-1}$

V. Sitaramaiah and D. Suryanarayana. *An order result involving σ-function*. Indian J. Pure. Appl. Math. 12 (1981), 1192–1200.

§ III. 7 On sums over $\sigma_{-\alpha}(f(n))$, f a polynomial $(0 < \alpha < 1)$

1) a) $\sum_{k\le x}\sigma_{-1}(f(a^k)) = Ax + o(x)$

where $f(x)$ is any polynomial with integral coefficients, a any integer, and A is a constant.

P. Erdős. *On some problems of Bellman and a theorem of Romanoff*. J. Chinese Math. Soc. (N.S.) 1 (1951), 409–421.

b) $\sum_{n\le x}\sigma_{-\alpha}(n^2-1) = A(\alpha)x + \frac{1+\alpha}{\alpha^2}\frac{12}{\pi^2}x^{1-\alpha} + O\big(x^{m(\alpha)}(\log x)^{c(\alpha)}\big)$

for $0 < \alpha < 1$, where:

$$m(\alpha) = \max\left(1-2\alpha, \frac{8}{9}-\alpha, \frac{4}{5}(1-\alpha)\right)$$

$$\begin{aligned} c(\alpha) &= 2 \text{ if } 0 < \alpha < 1/9 \\ &= 3 \text{ if } 1/9 \le \alpha < 4/9 \\ &= 4 \text{ if } \alpha = 4/9 \\ &= 1 \text{ if } 4/9 < \alpha < 1 \end{aligned}$$

N.G. Gafurov. *On the sum of powers of the divisors of reducible quadratic polynomials* (Russian.) Dokl. Akad. Nauk Tadžik. SSR, 23 (1980), 355–358.

c) If $f(n)$ is a polynomial with integer coefficients, then

$$\sum_{n\le x, f(n)\ne 0}\sigma_{-\alpha}(f(n)) = c_f(\alpha)x + O(x^{1-\alpha}(\log x)^{c_0})$$

for a certain constant c_0, where $0 < \alpha \leq 1$, and $c_f(\alpha) = \sum_{k=1}^{\infty} p_f(k) k^{-1-\alpha}$,

with $p_f(k) = \sum_{f(n)\equiv 0(\mathrm{mod}\ k), 0<n\leq k} 1$

G. Babaev, N. Gafurov and D. Ismoilov. *Some asymptotic formulas connected with divisors of polynomials* (Russian.) Trudy Mat. Inst. Steklov 163 (1984), 10–18.

d) For any fixed $k \geq 1$ and $0 < \alpha < 1$

$$\sum_{n\leq x} \sigma_k^{-\alpha}(f(n)) = A_f(\alpha)x + O(x^{1-\alpha}\log^c x)$$

where $f(n) = n^2 + a$, $a \neq -n^2$, and c is a computable constant.

G. Babaev and N. Gafurov. *Improvement of a result on the sum of powers of divisors of polynomials* (Russian.) Dokl. Akad. Nauk Tadzhik SSR 31 (1988), 219–222.

2) $\sum_{m,n\leq x} \sigma_\alpha(n^2 + m^2) = B(\alpha)x^{2+2\alpha} + O(x^{2+\alpha}\log x)$

where $0 < \alpha < 1$ and $B(\alpha) = \int_0^1 \int_0^1 (t^2 + u^2)^\alpha \, dt\, du$, $A(\alpha) = \sum_{n=1}^{\infty} \frac{\rho(n)}{n^{2+\alpha}}$,

where $\rho(n)$ is the number of solutions of

$x^2 + y^2 \equiv 0(\mathrm{mod}\, n)$, $1 \leq x, y \leq n$

N. Gafurov. *Asymptotic formulas for the sum of powers of divisors of the quadratic form* (Russian.) Dokl. Akad. Nauk Tadzhik SSR 32 (1989), 427–431.

§ III. 8 On $\sum_{n\leq x} \sigma(f(n))$, f a polynomial

Let $f(x)$ be a polynomial with integer coefficients, of degree n, and such that $f(m) > 0$ for all positive integers m

$$\sum_{m\leq x} \sigma(f(m)) = \frac{\beta a_n}{n+1} x^{n+1} + O(x^n \log^n x)$$

where $\beta = \sum_{d=1}^{\infty} \frac{N(d)}{d^2}$ and a_n is the coefficient of x^n in $f(x)$($N(m)$ denotes the number of solutions, not counting multiplicities, of the congruence $f(x) \equiv 0(\mathrm{mod}\ m)$.)

H.N. Shapiro. *Introduction to the theory of numbers*. John Wiley and Sons, 1983 (See p. 181.)

§ III. 9 Sums on $\sigma_\alpha(n), \sigma_\beta(n+k)$

a) $\sum_{n\leq x} \sigma_\alpha(n)\sigma_\beta(n+k) \sim \frac{1}{\alpha+\beta+1} \frac{\zeta(\alpha+1)\zeta(\beta+1)}{\zeta(\alpha+\beta+2)} \sigma_{-\alpha-\beta-1}(k) x^{\alpha+\beta+1}$

where α and β are fixed positive integers.

A.E. Ingham. *Some asymptotic formulae in the theory of numbers*. J. London Math. Soc. 2 (1927), 202–208.

b) $$\sum_{n=1}^{m-1} \sigma_\alpha(n)\sigma_\beta(m-n) = A_1\sigma_{\alpha+\beta+1}(m) + A_2 m^\alpha \sigma_{-\alpha+\beta+1}(m) +$$
$$+ A_3 m^\beta \sigma_{\alpha-\beta+1}(m) + A_4 m^{\alpha+\beta}\sigma_{-\alpha-\beta+1}(m) + O(m^\omega \log^\lambda m)$$
where $A_i (i = \overline{1,4})$, ω, λ are constants depending on the positive constants α, β

H. Halberstam. *An asymptotic formula in the theory of numbers*. Trans. Amer. Math. Soc. 84 (1957), 338–351.

Remark. An asymptotic formula for $\sum \sigma_{\alpha_1}(n_1)\cdots\sigma_{\alpha_k}(n_k)$ is obtained in

L. Mirsky. *Note on theorem of Carlitz*. Duke Math. J. 15 (1948), 803–815.

§ III.10 Inequalities connecting σ_k, d, γ, Ω

1) Let $k \geq 1$. Then:

a) $$\frac{\sigma_k(n)}{d(n)} \geq \sqrt{n^k}$$

R. Sivaramakrishnan and C.S. Venkataraman. *Problem 5326*. Amer. Math. Monthly 72 (1965), 915.

b) $$\frac{\sigma_k(n)}{d(n)} \leq \frac{n^k+1}{2}$$

E.S. Langford. See D.S. Mitrinović and M.S. Popadić. *Inequalities in the number theory*. Niš, 1978 (p. 44)

c) $$\prod_{i=1}^{r}\left(\frac{p_i^k+1}{2}\right)^{a_i} \leq \frac{\sigma_k(n)}{d(n)} \leq \prod_{i=1}^{r}\left(\frac{p_i^{ka_i}+1}{2}\right)$$
where $n = \prod_{i=1}^{r} p_i^{a_i}$ (p_i–primes.)

J. Sándor. *An application of the Jensen–Hadamard inequality*. Nieuw Arch. Wiskunde (4) 8 (1990), 63–66.

d) If $r = \omega(n) \geq 2$, then
$$\frac{\sigma_k(n)}{d(n)} < \frac{n^k}{2}$$

J. Sándor. *Ibid.*

e) $$\frac{[\sigma_k(n)d(n)]^{1/2}}{\sigma_{k/2}(n)} \leq n^{-k/4}\left(\frac{n^{k/2}+1}{2}\right)$$
for $k \geq 0$

J. Sándor. *On the sum of powers of divisors of a natural number* (Hungarian.) Mat. Lapok (Cluj), 8/1989, 285–288.

f) $n^{(k-m)/2} \cdot [\gamma(n)]^{-(k+m-1)/2} < \dfrac{\sigma_{k+m}(n)}{\sigma_m(n)} < [n\gamma(n)]^k$

(k, m nonnegative integers, $\gamma(n)$-the greatest squarefree divisor of n (*core of* n))

J. Sándor. *Ibid.*

2) $\dfrac{\sigma(n)}{\Omega(n)} \le n$

if $n \ne$ prime number ($\Omega(n)$ denotes the total number of prime factors of n).

J. Sándor. *Remarks on two papers by K.T. Atanasov.* Bull. Number. Theory Rel. Topics 12 (1988), 56–59.

§ III.11 Sums over $\sigma(p-1)$, p a prime

1) a) $\displaystyle\sum_{p\le n} \sigma(p-1) = \frac{315\zeta(3)}{2\pi^4} n + O(n(\log n)^{-0.999})$

Yu.V. Linnik. *The dispersion method in binary additive problems.* Izdat. Leningrad Univ., Leningrad, 1961.

b) $\displaystyle\sum_{a<p\le x} \sigma_k(p-a) \sim \frac{cx^{k+1}}{(k+1)\log x}$

where $\displaystyle c = \sum_{n=1,(a,n)=1}^{\infty} \frac{1}{n^k \varphi(n)}$

W.A. Webb. *An asymptotic estimate for a class of divisor sums.* Duke Math. J. 38 (1971), 575–582.

c) $\displaystyle\sum_{p_1, p_2 \le \sqrt{x}} \sigma(p_1 p_2 - 1) = \frac{630\,\zeta(3)}{\pi^4} \frac{x}{\log x} + O\left(\frac{x(\log\log x)^7}{\log^2 x}\right)$

M.B. Barban. *Analogues of the divisor problem of Titchmarsh.* Vestnik Leningrad Univ. Ser. Mat. Meh. Astronom. 18 (1963), 5–13.

§ III.12 On $\sigma(mn)$

1) a) $n\,\sigma(m) \le \sigma(mn) \le \sigma(m)\,\sigma(n)$

(Simple consequence of the definition of σ)

b) $\dfrac{(d(mn))^2}{d(m^2)\,d(n^2)} \dfrac{4mn}{(mn+1)^2} \le \dfrac{(\sigma(mn))^2}{\sigma(m^2)\,\sigma(n^2)} \le \dfrac{(d(mn))^2}{d(m^2)\,d(n^2)}$

for all m, n positive integers.

J. Sándor. *Some arithmetic inequalities*. Bull. Number Theory Rel. Topics 11 (1987), 149–161.

c) Let $n \wedge m$ denote the property that there exists at least a prime t with $t|n, t \nmid m$. Then

$$\sigma(mn) \geq \sigma(m)\Psi(n)$$

for $n \wedge m$ where $\Psi(n)$ denotes Dedekind's arithmetical function (i.e. $\Psi(n) = n \prod_{p|n} \left(1 + \frac{1}{p}\right)$)

J. Sándor. *On Dedekind's arithmetical function*. Seminarul de teoria structurilor. Univ. Timişoara, 1988, No. 51, pp. 1–15.

d) If $m > 1, n > 1$ are natural numbers, then

$$\sigma(mn) > \sigma(m) + \sigma(n)$$

J.L. Hunsucker and J. Nebb. *Problem B 260 Fib. Quart. 11 (1973), 221 Solution by P.S. Bruckman*. Fib. Quart. 12 (1974), 223–224.

§ III.13 On $\sigma(n) \geq 4\varphi(n)$

1) a) If $\sigma(n) = 4\varphi(n)$ and n is even, then

$$\sigma(2n) \geq 4n$$

b) If $\sigma(n) \geq \varphi(n)$ and n is odd, then

$$\sigma(2n) > 4n$$

A. Mąkowski. *Remarks on some problems in the elementary theory of numbers*. Acta Math. Univ. Comenian 50/51 (1987), 277–281.

2) a) If n is odd, then

$$\sigma(n) \leq \varphi(n)P(n)$$

where $P(n)$ denotes the greatest prime factor of n

b) If n is even, then

$$\sigma(n) \leq 4\varphi(n)P(n)$$

K.T. Atanassov. *Remarks on φ, σ and other functions*. C.R. Acad. Bulgare Sci. 41 (1988), 41–44.

§ III.14 On $\sigma(n+i)/\sigma(n+i-1)$ and related theorems

1) Let $a_1, \ldots, a_h$ be any sequence of nonnegative integers or infinity. Then there exists an infinite sequence $n_1 < n_2 < \cdots$ of natural numbers such that

$$\lim_{k\to\infty} \frac{\sigma(n_k + i)}{\sigma(n_k + i - 1)} = a_i$$

$(1 \leq i \leq h)$

A. Schinzel. *On functions* $\varphi(n)$ *and* $\sigma(n)$. Bull. Acad. Polon. Sci. Cl. III. 3 (1955), 415–419.

2) a) Let $\{i_1, i_2, i_3, i_4\}$ and $\{j_1, j_2, j_3, j_4\}$ be two permutations of 1, 2, 3, 4. Then, for infinitely many natural numbers n we have simultaneously

$$\begin{cases} \varphi(n+i_1) > \varphi(n+i_2) > \varphi(n+i_3) > \varphi(n+i_4) \\ \sigma(n+j_1) > \sigma(n+j_2) > \sigma(n+j_3) > \sigma(n+j_4) \end{cases}$$

b) There is no natural number n with

$$\begin{cases} \varphi(n+1) \geq \varphi(n+2) \geq \varphi(n+3) \geq \varphi(n+4) \geq \varphi(n+5) \\ \sigma(n+1) \geq \sigma(n+2) \geq \sigma(n+3) \geq \sigma(n+4) \geq \sigma(n+5) \end{cases}$$

P. Erdős, K. Győry and Z. Papp. *On some new properties of functions* $\sigma(n)$, $\varphi(n)$, $d(n)$ *and* $\nu(n)$ (Hungarian.) Mat. Lapok 28 (1980), 125–131.

3) Let $\beta(x) = \text{card}\{n \leq x : \sigma(2n+1) \geq \sigma(2n)\}$
Then there exist $0 < \lambda < \mu < 1$ such that

$$\lambda x < \beta(x) < \mu x$$

for all sufficiently large x

M. Laub. *Problem 6555. Amer. Math. Monthly 94 (1987), 800. Solution by L.E. Mattics.* Amer. Math. Monthly 97 (1990), 351–353.

Remark. According to a note in the above solution, A. Hildebrand can show that the limit $\lim\limits_{x\to\infty} \dfrac{\beta(x)}{x}$ exists.

4) Given $a_1, \ldots, a_r$ unequal integers, any 0, and r real numbers $\alpha_1, \ldots, \alpha_r$ all ≥ 1, the set of $n \leq x$ for which

$$\alpha_i \leq \frac{\sigma(n+a_i)}{n+a_i} \leq \alpha_i + c$$

$(i = 1, 2, \ldots, r)$ is true has positive lower density.

M. Hausman and H.N. Shapiro. *On the denseness of arithmetic vectors.* Comm. Pure Appl. Math. 35 (1982), 185–196.

§ III.15 On $\sigma(\sigma(n))$; $\sigma^*(\sigma^*(n))$ and $\sigma^{(k)}(n)$, $\sigma(\varphi(n))$, $\varphi(\sigma(n))$

1) $\liminf\limits_{n\to\infty} \dfrac{\sigma(\sigma(n))}{n} = 1$

A. Schinzel. *Ungelöste Probleme, Nr.30.* Elem. Math. 14 (1959), 60–61.

Remarks: (i) For an elementary proof of this result, see

A. Mąkowski and A. Schinzel. *On the functions* $\varphi(n)$ *and* $\sigma(n)$. Colloq. Math. 13 (1964), 95–99.

(ii) The equality

$$\limsup_{n\to\infty} \frac{\sigma(\sigma(n))}{n} = +\infty$$

is trivial.

2) a) $\dfrac{\sigma^*(\sigma^*(n))}{\sigma^*(n)} \to 1$
on a set of density one.

b) $\dfrac{\sigma(\sigma(n))}{\sigma(n)} \to +\infty$
on a set of density one.

Remark. In the same way,

$$\frac{\varphi^*(\varphi^*(n))}{\varphi^*(n)} \to 1$$

except for a sequence of values of n of density zero.

P. Erdős and M.V. Subbarao. *On the iterates of some arithmetic functions. The theory of arithmetic functions.* (Proc. Conf., Western Michigan Univ., Kalamazoo, Mich. 1971), pp. 119–125. Lecture Notes in Math. vol.251, Berlin, 1972.

3) $\sigma^*(\sigma^*(n)) \neq 2n \pm 1$
for $n \notin \{1, 3\}$

J. Sándor. *On the composition of some arithmetic functions.* Studia Univ. Babeş-Bolyai 34 (1989), 7–14.

4) Denote by $N_\sigma(k, \alpha, x)$ the number of integers $n \leq x$ for which

$$\sigma^{(k)}(n) < \alpha n$$

where $\sigma^{(k)}(n)$ denotes the kth iterate of σ

a) For arbitrarily large t and for $x > x_0(t)$ we have

$$N_\sigma(2,2,x) > \frac{x}{\log x}(\log\log x)^t$$

b) For every $\alpha > 0$, $\varepsilon > 0$, and $x > x_0(\varepsilon, \alpha)$ we have

$$N_\sigma(2,\alpha,x) < \frac{x}{\log x}(\log x)^\varepsilon$$

$$N_\sigma(3,\alpha,x) < \frac{x}{\log^2 x}(\log x)^\varepsilon$$

P. Erdős. *Some remarks on the iterates of the φ and σ functions.* Colloq. Math. 17 (1967), 195–202.

c) For $\alpha > \alpha_0$ and $x > x_0(\varepsilon)$ we have

$$N_\sigma(3, \alpha, x) > \frac{x}{\log^2 x}$$

Corollary. $\liminf_{n\to\infty} \frac{\sigma^{(3)}(n)}{n} < \infty$

d) For $\alpha > \alpha_0$ and $x > x_0(\alpha, t)$ and t arbitrarily large

$$N_\sigma(3, \alpha, x) > \frac{x}{\log^2 x}(\log\log x)^t$$

H. Maier. *On the third iterates of the φ- and σ-functions*. Colloq. Math. 49 (1984), 123–130.

5) a) For every $\varepsilon > 0$

$$\varphi(\sigma(n)) > \varepsilon n$$

except for a set of density 0.

b) For every $c > 0$

$$\sigma(\varphi(n)) > cn$$

except for a set of density 0.

c) Except for a set of density zero

$$e^\gamma \varphi(\sigma(n)) \log\log\log n \sim \sigma(n)$$

and

$$e^{-\gamma}\sigma(\varphi(n))/\log\log\log n \sim \varphi(n)$$

L. Alaoglu and P. Erdős. *A conjecture in elementary number theory*. Bulletin A.M.S. 50 (1944), 881–882.

§ III.16 Divisibility properties of $\sigma_k(n)$

1) Let $N_k(x, p) = \text{card}\{n \le x : p \nmid \sigma_k(n)\}$, where p is an odd prime. Put $q = (p-1)/(k, p-1)$. Then there exist effective constants c_1, c_2 such that

a) $N_k(x, p) \sim c_1 x/(\log x)^{1/q}$
if q is even.

b) $N_k(x, p) \sim c_2 x$
if q is odd, as $x \to \infty$

C. Radoux. *Divisibilité de $\sigma_k(n)$ par un nombre premier*. Séminaire Delange-Pisot-Poitou (1977/78), Paris, Exp. No.3, 5pp.

2) Let $D_m(v_1, v_2; p, x) = \text{card}\{n \leq x : p^m \| \sigma_{v_1}(n), p^m \| \sigma_{v_2}(n)\}$, where $v_1, v_2 > 0$, $v_1 \neq v_2$, $m \geq 0$, and p-prime. Then:

a) $D_m(v_1, v_2; p, x) \sim A_1(m)x$
if p is odd and h_1 and h_2 are not both even

b) $D_m(v_1, v_2; p, x) \sim A_2(m)x(\log x)^{-\beta}(\log\log x)^{m'}$
if p is odd and both h_1 and h_2 are even

c) $D_m(v_1, v_2; 2, x) \sim A_3(m)x(\log x)^{-1}(\log\log x)^{m-1}$
Here $h_i = (p-1)/(v_i, p-1)$, $i \in \{1, 2\}$ and the positive constants $A_i(m)(i \in \{1, 2, 3\})$ depend only on v_1, v_2, p, m; and $m', \beta \geq 0$ are constants.

V.M. Želtonogov. *The divisibility of* $\sigma_v(n)$. Mat. Issled. 10 (1975), 2(36), 106–118, 283.

§ III.17 Divisibility and congruences properties of $\sigma_k(n)$

1) Let $\sigma^{(k)}(n) = \sigma(\sigma^{(k-1)}(n))$, $k \geq 2$, $\sigma^{(1)}(n) = \sigma(n)$. If p is a fixed odd prime, a prime number q is said to belong to the rth class if $\sigma^{(r)}(q) \equiv 0 (\text{mod } p)$ but $\sigma^{(k)}(q) \not\equiv 0 (\text{mod } p)$ whenever $k < r$. Let $g(p, r, x)$ denote the number of primes in the rth class smaller than x. Then

$$g(p, 3, x) \geq x/(\log x)^4$$

I. Kátai. *On a classification of primes*. Acta Sci. Math. (Szeged) 29 (1968), 207–212.

2) Let p be a prime and m a positive integer, and assume that both are fixed and independent of x. Let

$$D_m(v, p; x) = \text{card}\{n \leq x : p^m \| \sigma_v(n)\}$$

Define γ by $p^\gamma \| v$ and let $m' = (m/(\gamma + 1))$ and $h = (p-1)/(v, p-1)$

a) If p and h are both odd, then

$$D_m(v, p; x) \sim A_1^{(m)} x$$

as $x \to \infty$

b) If p is odd and h is even, then

$$D_m(v, p; x) \sim A_2^{(m)} x(\log\log x)^{m'}(\log x)^{-1/h}$$

c) $D_m(v, 2; x) \sim A_3^{(m)} x(\log\log x)^{m-1}(\log x)^{-1}$

where $A_1^{(m)}$, $A_2^{(m)}$, and $A_3^{(m)}$ are positive constants depending on v, p, and m

E.J. Scourfield. *On the divisibility of* $\sigma_v(n)$. Acta Arith. 10 (1964), 245–285.

Remark. The case $m = 0$ is due to

R.A. Rankin. *The divisibility of divisor functions*. Proc. Glasgow Math. Assoc. 5 (1961), 35–40.

3) Let $S(a) = \{n : \sigma(n) \equiv a (\bmod n)\}$, where $a \in \mathbb{I}$

a) The set $S(0)$ has density zero.

H.-J. Kanold. *Über die Verteilung der vollkommenen Zahlen und allgemeinerer Zahlenmengen*. Math. Ann. 132 (1956), 442–450.

b) $\operatorname{card}\{m \in S(a) : m \leq n\} = O\left(\frac{n}{\log n}\right)$

for all a

Corollary. The set $S(a)$ has density zero for all a

C. Pomerance. *On the congruences* $\sigma(n) \equiv a(\bmod n)$ *and* $n \equiv a(\bmod \varphi(n))$. Acta. Arith. 26 (1975), 265–272.

§ III.18 On $S(n) = \sigma(n) - n$

1) A number n is called *untouchable* if there is no solution of

$$s(n) = n$$

where $s(n) = \sigma(n) - n$

a) The lower density of untouchable numbers is positive.

P. Erdős. *Über die Zahlen der Form* $\sigma(n) - n$ *and* $n - \varphi(n)$. Elem. Math. 28 (1973), 83–86.

b) The lower density of untouchable numbers is >0.0324

H.J.J. te Riele. *A theoretical and computational study of generalized aliquot sequences (Dissertation)*. Mathematisch Centrum, Amsterdam, 1975.

2) Let $s(n) = \sigma(n) - n$, $s^{(k)}(n) = s(s^{(k-1)}(n))$, $k \geq 2$, $s^{(1)}(n) = s(n)$. Then:

a) For each k there is an n with

$$n < s(n) < s^{(2)}(n) < \cdots < s^{(k)}(n)$$

H.W. Lenstra, Jr. *Problem 6064. Amer. Math. Monthly 82 (1975), 1016; Solution by the proposer.* 84 (1977), 580.

b) For every fixed k and $\delta > 0$ and for all n except a sequence of density 0 one has

$$(1-\delta)n\left(\frac{s(n)}{n}\right)^i < s^{(i)}(n) < (1+\delta)n\left(\frac{s(n)}{n}\right)^i$$

for $1 \leq i \leq k$

P. Erdős. *On asymptotic properties of aliquot sequences*. Math. Comp. 30 (1976), 641–645.

c) The set of n with

$$\frac{s^{(i+1)}(n)}{s^{(i)}(n)} > \frac{s(n)}{n} - \varepsilon$$

for $i = 1, 2, \ldots, k$ (for each $\varepsilon > 0$ and k) has asymptotic density 1.

P. Erdős. *Ibid.*

d) For each $\varepsilon > 0$, the set of n with

$$\frac{s^{(2)}(n)}{s(n)} < \frac{s(n)}{n} + \varepsilon$$

has asymptotic density 1

e) Let $S^{(k)}(x)$ denote the number of odd numbers $m \leq x$ not in the range of the function $s^{(k)}$. There is a positive number δ such that

$$S^{(k)}(x) \ll x^{1-\delta}$$

uniformly for all natural numbers k and $x > 0$

P. Erdős, A. Granville, C. Pomerance, and C. Spiro. *On the normal behavior of the iterates of some arithmetic functions*. Proc. Conf. in Honor of P.T. Bateman, Birkhäuser Boston, Inc. 1990, pp. 165–204.

3) Let $s^*(n) = \sigma^*(n) - n$ and $s^{(k)*}(n) = s^*(s^{(k-1)*}(n))$, $k \geq 2$. For each k there is an n with

$$n < s^*(n) < s^{(2)*}(n) < \cdots < s^{(k)*}(n)$$

H.J.J. te Riele. *Unitary aliquot sequences*. MR 139/72, Mathematisch Centrum, Amsterdam, September 1972.

§ III.19 Number of distinct values of $\sigma(n)/n$, $n \leq x$

1) a) The number of integers m with $\sigma(m) \leq n$ equals

$$cn + o(n)$$

(c constant.)

P. Erdős. *Some remarks on Euler's φ function and some related problems*. Bull. Amer. Soc. 51 (1945), 540–544.

Remark. The same result is valid for the function φ. For improvements see Euler's function and related problems in Ch. I.

b) $c = \prod_p \left(1 - \frac{1}{p}\right) \sum_{i=0}^{\infty} \frac{p-1}{p^{i+1}-1}$

R.E. Dressler. *An elementary proof of a theorem of Erdős on the sum of divisors function.* J. Number Theory 4 (1972), 532–536.

2) a) The number of distinct numbers of the form $\frac{\sigma(n)}{n}$, $1 \le n \le x$ equals

$$cx + o(x)$$

where $6/\pi^2 < c < 1$

b) The number of solutions of the equation $\frac{\sigma(a)}{a} = \frac{\sigma(b)}{b}$, $a < b \le x$ equals

$$c'x + o(x)$$

$0 < c' < \infty$

P. Erdős. *Remarks on the number theory.* Acta Arith. 5 (1959), 171–177.

3) Let $\alpha \ge 1$, $\varepsilon > 0$ be real numbers. Then the inequality

$$\left|\frac{\sigma(n)}{n} - \alpha\right| \le n^{-2/5+\varepsilon}$$

has infinitely many solutions in natural numbers n

D. Wolke. *Eine Bemerkung über die Werte der Funktion o(n).* Monatsch. Math. 83 (1977), 163–166.

§ III.20 Frequency of integers $m \le N$ with $\log(\varphi(m)/m) \le x$, $\log(\sigma(m)/m) \le y$

Let $F_N(x, y)$ be the frequency of positive integers $m \le N$ for wich $\log\frac{\varphi(m)}{m} \le x$ and $\log\frac{\sigma(m)}{m} \le y$. Then

$$F_N(x, y) = F(x, y) + O\left(\log_2^2 N \log^{-1} N \log_3^{-1} N\right)$$

where $F(x, y)$ is the distribution function with the characteristic function

$$\prod_p (1 - p^{-1})\left(1 + \sum_{r=1}^{\infty} p^{-r}(1 - p^{-1})^{ls-it}(1 - p^{-irt})\right)$$

A.S. Badarëv. *A two-dimensional generalized Essen inequality and the distribution of the values of arithmetic functions* (Russian.) Taškent Gos. Univ. Naučn. Trudy Vyp. 418 Voprosy Mat. (1972), 99–110, 379.

§ III.21 On $\frac{\sigma(a^n - 1)}{a^n - 1}$ and related functions

1) a) $\frac{\sigma(F_n)}{F_n} \to \infty$

$(n \to \infty)$, where $F_n = 2^{2^n} + 1$ are the Fermat numbers.

P. Erdős. *Problem 4590*. Amer. Math. Monthly 61 (1954), 350.

b) $\dfrac{\sigma(M_P)}{M_P} \to 1$

$(p \to \infty)$, where $M_p = 2^p - 1$ (p-prime) are the Mersenne numbers.

c) $\dfrac{\sigma(2^{pq}+1)}{2^{pq}+1} \to \dfrac{\sigma(2^p+1)}{2^p+1}$

$(q \to \infty)$, where p, q are primes.

R. Bojanić. *Asymptotic evaluations of the sum of divisors of certain numbers* (Serbo-Croatian.) Bull. Soc. Math.-Phys, R.P. Macédoine 5 (1954), 5–15.

2) a) $\dfrac{\sigma(2^n-1)}{2^n-1} < c \log\log n$

$(c > 0$ constant.)

P. Erdős. *On the sum* $\sum_{d|2^n-1} 1/d$. Israel J. Math. 9 (1971), 43–48.

b) $\dfrac{\sigma(a^n-1)}{a^n-1} < c(a) \log\log n$

($c(a)$ positive constant depending on a)

P. Erdős. *Ibid.*

3) Let $A(n) = \sum_{d|2^n-1, d>n} \dfrac{1}{d}$. Then:

a) $A(n) = o(1)$

P. Erdős. *On some problems of Bellman and a theorem of Romanoff.* J. Chinese Math. Soc. (N.S.) 1 (1951), 409–421.

b) $A(n) \le \exp(-\log n \log\log\log n / 2 \log\log n)$

for $n \ge n_0$

C. Pomerance. *On primitive divisors of Mersenne numbers.* Acta Arith. 46 (1986), 355–367.

§ III.22 Normal order of $\omega(\sigma_k(n))$

1) The normal order of magnitude of $\omega(\sigma_k(n))$ is

$$\frac{1}{2} d(k)(\log\log n)^2$$

where $d(k)$ denotes the number of divisors of $k \in \mathbb{N}$

M. Ram Murty and V. Kumar Murty. *Prime divisors of Fourier coefficients of modular forms.* Duke Math. J. 51 (1984), 57–76.

2) The function $\omega(\sigma(p+1))$ (where p is a prime) has a normal limiting distribution with centering $\frac{1}{2}(\log\log x)^2$ and norming $\frac{1}{\sqrt{3}}(\log\log x)^{3/2}$

I. Katái. *Distribution of* $\omega(\sigma(p+1))$. Ann. Univ. Sci. Budap. Eötvös. Sect. Math. 34 (1991), 217–225.

§ III.23 Number of prime factors of $(\sigma(A_k), A_k)$

For every k, let $f(k)$ be the smallest index for which $\sigma(p_k \cdots p_{f(k)}) \geq 2p_k \cdots p_{f(k)}$ (where p_n is the nth prime) and denote $A_k = \prod_{k \leq s \leq f(k)} p_s$. Then

$$\lim_{k\to\infty} \omega((\sigma(A_k), A_k)) = +\infty$$

where (a, b) denotes the greatest common divisor of a and b, and $\omega(m)$ is the number of distinct prime factors of m

S.J. Benkoski and P. Erdős. *On weird and pseudoperfect numbers.* Math. Comp. 28 (1974), 617–623.

§ III.24 On $\sigma(p^a) = x^b$

If the prime p satisfies

$$\sigma(p^a) = X^b$$

with $X \in \mathbb{N}$, b an odd prime and $a \equiv -1 (\operatorname{mod} b)$, $b \nmid k$, where $b^k \| (a+1)$, then

$$p < a\, b^2 (2b)^{(a-1)b^a}$$

A. Takaku. *Prime numbers such that the sum of divisors of their powers are perfect power numbers.* Colloq. Math. 52 (1987), 319–323.

§ III.25 An inequality for $\sigma^*(n)$

$$\sigma^*(n) < \frac{28}{15} n \log\log n$$

for $n \geq 31$

A. Ivić. *Two inequalities for the sum of divisor function.* Univ. u Novom Sadu Zb. Rad. Prirod.-Mat. Fak. 7 (1977), 17–22.

§ III.26 Sums over $\sigma^*(n)$, $\frac{1}{\log \sigma^*(n)}$, $\sigma_k^{*^2}(n)$

1) $\sum_{n \le x} \sigma^*(n) = \frac{\pi^2 x^2}{12\zeta(3)} + S^*(x)$

where $S^*(x) = O(x \log^{2/3} x)$

R. Sitaramachandrarao and D. Suryanarayana. *On* $\sum_{n \le x} \sigma^*(n)$ *and* $\sum_{n \le x} \varphi^*(n)$. Proc. Amer. Math. Soc. 41 (1973), 61–66.

2) The mean $M(f)$ of an arithmetic function f is defined by $M(f) = \lim_{N \to \infty} \frac{1}{N} \sum_{n=1}^{N} f(n)$ provided the limit exists. For all complex numbers z, the mean $M\left(\left(\frac{\sigma^*(n)}{n}\right)^z\right)$ exists and

$$M\left(\left(\frac{\sigma^*(n)}{n}\right)^z\right) = \prod_p \left(1 - \frac{1}{p} + \sum_{k=1}^{\infty} \frac{p-1}{p^{k+1}}(1 + p^{-k})^z\right)$$

Corollary. Let $C(x)$ be the density of numbers n with $\frac{\sigma^*(n)}{n} \ge x$. Then $C(x)$ exists and is continuous for all real x

Ch.R. Wall. *Topics related to the sum of unitary divisors of an integer.* Ph.D. Thesis, 1970, Univ. of Tennessee.

3) a) $\sum_{2 \le n \le x} \frac{1}{\log \sigma^*(n)} = \frac{x}{\log x}\left(1 + O\left(\frac{\log\log\log x}{\log x}\right)\right)$

A. Ivić. The distribution of values of some multiplicative functions. Publ. de l'Inst. Math. (Belgrade) 22 (36) (1977), 87–94.

b) For every positive integer M

$$\sum_{2 \le n \le x} \frac{1}{\log \sigma^*(n)} = x \sum_{m=1}^{M} \frac{(-1)^{m-1} F^{(m-1)}(0)}{(\alpha \log x)^m} + O_M(x/\log^{M+1} x)$$

where for $-1/\gamma \le t \le 0$

$$F(t) = \frac{1}{\alpha t + 1} \prod_p \left(1 - \frac{1}{p}\right)\left(1 + \sum_{m=1}^{\infty} p^{-m} \left(\frac{f(p^m)}{p^{\alpha m}}\right)^t\right)$$

E. Brinitzer. *Eine asymptotische Formel für Summen über die reziproken Werte additiver Funktionen*. Acta Arith. 32 (1977), 387–391.

Remark. a) and b) are valid also for σ^* replaced by φ, σ, ψ, φ^*

4) For $k > 0$

$$\sum_{n \le x} \sigma_k^{*2}(n) = \frac{\zeta(k+1)\zeta(2k+1)\alpha_k}{2k+1} x^{2k+1} + O(B_k(x))$$

where

$$\alpha_k = \prod_p \left(1 + \frac{1}{p^{k+1}} - \frac{2}{p^{k+2}} - \frac{1}{p^{2k+2}} - \frac{2}{p^{3k+2}} + \frac{3}{p^{3k+3}}\right)$$

and

$$B_k(x) = x^{2k},\, x^2 \log^4 x,\, x^{k+1} \log^3 x,\, x^{3/2} \log^5 x,\ \text{or}\ x^{k+1} \log^4 x$$

according as $k > 1$, $k = 1$, $\frac{1}{2} < k < 1$, $k = \frac{1}{2}$, or $k < \frac{1}{2}$

L. Tóth. *An asymptotic formula concerning the unitary divisor sum function.* Studia Univ. Babeş-Bolyai, 34 (1989), 3–10.

§ III.27 Inequalities on σ_k^*, d^*, σ, ψ

1) a) $n^{k/2} \le \dfrac{\sigma_k^*(n)}{d^*(n)} \le \dfrac{n^k + 1}{2}$

b) $\dfrac{\sigma_{k+m}^*(n)}{\sigma_m^*(n)} \ge n^{k/2}$

for $k, m \ge 0$, real numbers.

J. Sándor and L. Tóth. *On certain number-theoretic inequalities.* Fib. Quart. 28 (1990), 255–258.

2) a) $\sigma(n) + \sigma^*(n) \le 2\Psi(n)$

b) Suppose that $n \parallel m$ and that $n \mid k$, $\omega(n) = \omega(k)$. Then

$$\frac{\sigma^*(k)}{k} \le \frac{\sigma^*(n)}{n} \le \frac{\sigma^*(m)}{m}$$

Ch.R. Wall. *Topics related to the sum of unitary divisors of an integer.* Ph.D. Thesis, 1970, Univ. Tennessee.

§ III.28 The sum of exponential divisors

1) Let $\sigma^e(n)$ denote the sum of exponential divisors of n. Then the sequence

$$(\sigma^e(n)/n)$$

is dense in $[1, +\infty)$

E.G. Straus and M.V. Subbarao. *On exponential divisors.* Duke Math. J. 41 (1974), 465–471.

2) $\limsup_{n\to\infty} \frac{\sigma^e(n)}{n \log\log n} = e^{\gamma} \frac{6}{\pi^2}$

where γ is Euler's constant.

J. Fabrykowski and M.V. Subbarao. *The maximal order and the average order of multiplicative function* $\sigma^e(n)$. Théorie des nombres (Quebec, PQ, 1987), 201–206, de Gruyter, Berlin-New York, 1989.

§ III.29 Average order of $\sigma^e(n)$

For any $\varepsilon > 0$

$$\sum_{n \le x} \sigma^e(n) = Bx^2 + O(x^{1+\varepsilon})$$

where

$$B = \frac{1}{2} \prod_p \left(1 + \frac{1}{p(p^2-1)} - \frac{1}{p^2-1} + \left(1 - \frac{1}{p}\right) \sum_{k=2}^{\infty} p^k/(p^{2k}-1)\right) = 0.568285\ldots$$

J. Fabrykowski and M.V. Subbarao. *The maximal order and the average order of multiplicative function* $\sigma^e(n)$. Théorie des nombres (Quebec, PQ, 1987), 201–206, de Gruyter, Berlin-New York, 1989.

§ III.30 Number of distinct prime divisors of an odd perfect number

1) If n is an odd perfect number, then:

a) $\omega(n) \ge 5$

J.J. Sylvester. *Sur l'impossibilité de l'existence d'un nombre parfait impair qui ne contient pas au moins 5 diviseurs premiers distincts*. Mathematical Papers, Cambridge Univ. Press, 1912, pp. 611–614.

b) $\omega(n) \ge 6$

I.S. Gradstein. *On odd perfect numbers*. Math. Sb. 32 (1925), 476–510; U. Kuhnel. *Verschärfung der notwendigen Bedingungen für die Existenz von ungeraden vollkommener Zahlen*. Math. Z. 52 (1949), 202–211;

G.C. Webber. *Nonexistence of odd perfect numbers of the form* $3^{2\beta} p^{\alpha} s_1^{2\beta_1} s_2^{2\beta_2} s_3^{2\beta_3}$. Duke Math. J. 18 (1951), 741–749.

Note. For the early history, see

P.J. McCarthy. *Odd perfect numbers*. Scripta Math. 23 (1957), No. 1–4, pp. 43–47.

c) $\omega(n) \ge 7$

N. Robbins. *The nonexistence of odd perfect numbers with less than seven distinct prime factors*. Notices Amer. Math. Soc. 19 (1972), A-52;

C. Pomerance. *Odd perfect numbers are divisible by at least seven distinct primes.* Acta Arith. 25 (1973/74), 265–300.

d) $\omega(n) \geq 8$

P. Hagis, Jr. *Every odd perfect number has at least 8 prime factors.* Notices Amer. Math. Soc. 22 (1975), A-60;

J.E.Z. Chein. *An odd perfect number has at least 8 prime factors.* Ph.D. Thesis, Pennsylvania State Univ., 1979.

2) If n is an odd perfect number, then:

a) The largest prime factor of n is greater than 100129

P. Hagis, Jr. and W.J. McDaniel. *On the largest prime divisor of an odd perfect number.* II. Math. Comp. 29 (1975), 922–924.

b) The largest prime factor of n is greater than 300000

J.T. Condict. *On an odd perfect number's largest prime divisor.* Senior Thesis, Middleburg College, May, 1978.

3) If n is an odd perfect number, then the second largest prime divisor must be at least:

a) 139

C. Pomerance. *The second largest prime factor of an odd perfect number.* Math. Comp. 29 (1975), 914–921.

b) 1000

P. Hagis, Jr. *On the second largest prime divisor of an odd perfect number*, in: Analytic number theory, Proc. Conf. Temple Univ., May 12–15, 1980.

4) If n is an odd perfect number, and $3 \nmid n$, then:

a) $\omega(n) \geq 9$

H.-J. Kanold. *Folgerungen aus dem Vorkommen einer Gauss'schen Primzahl in der Primfaktorzerlegung einer ungeraden Vollkommenen Zahl.* J. Reine Angew. Math. 186 (1944), 25–29.

b) $\omega(n) \geq 10$

M. Kishore. *Odd perfect numbers not divisible by* 3 *are divisible by at least ten distinct primes.* Math. Comp. 31 (1977), 274–279.

c) $\omega(n) \geq 11$

M. Kishore. *Odd perfect numbers not divisible by* 3. II. Math. Comp. 40 (1983), 405–411.

5) If n is an odd perfect number, then n is divisible by a prime power $> 10^{12}$.

J.B. Muskat. *On divisor of odd perfect numbers.* Math. Comp. 20 (1966), 141–144.

6) If n is an odd perfect number, then:

a) $n > 10^{50}$

P. Hagis, Jr. *A lower bound for the set of odd perfect numbers*. Math. Comp. 27 (1973), 951–953.

b) $n > 10^{160}$

R.P. Brent and G.L. Cohen. *A new lower bound for odd perfect numbers*. Math. Comp. 53 (1989), 431–437, S7–S24.

c) $n > 10^{300}$

R.P. Brent, G.L. Cohen and H.J.J. te Riele. *Improved techniques for lower bounds for odd perfect numbers*. Math. Comp. 57 (1991), 857–868.

d) If n is an odd perfect number with k distinct prime factors, then $n < 4^{4^k}$.

D.R. Heath-Brown. *Odd perfect numbers* (submitted), see
R.K. Guy. *Unsolved problems in Number Theory* (second edition, 1994), Springer Verlag, pp. 40–41.

§ III.31 Bounds for the prime divisors of an odd perfect number

1) a) If N is odd perfect number, then for the least prime divisor p one has

$$p \leq r$$

where $r = \omega(N)$

Cl. Servais. *Sur les nombres parfaits*. Mathesis 8 (1988), 92–93.

b) For an odd perfect number N the least prime divisor

$$p < \frac{2r}{3} + 2$$

where $r = \omega(N)$

O. Grün. *Über ungerade vollkommene Zahlen*. Math. Z. 55 (1952), 353–354.

c) For an odd perfect number $N = \prod_{i=1}^{r} p_i^{a_i}$ we have

$$p_i < (4r)^{2^{i(i+1)/2}}$$

for $1 \leq i \leq r$

C. Pomerance. *Multiply perfect numbers, Mersenne primes and effective computability*. Math. Ann. 266 (1977), 195–206.

d) With the same conditions,

$$p_i < 2^{2^{i-1}}(r - i + 1)$$

for $2 \leq i \leq 6$

M. Kishore. *On odd perfect, quasiperfect and odd almost perfect numbers*. Math. Comp. 36 (1981), 583–586.

2) a) If $\sigma(N)/N \geq \lambda$, then for the least prime factor of N we have

$$p < c(r \log r)^{1/\lambda}$$

where $r = \omega(N)$ is the number of distinct prime factors of N

H. Salié. *Über abundante Zahlen*. Math. Nachr. 9 (1953), 217–220 (1953).

Remark. For $\lambda = 2$ this sharpens Grün's result.

b) If $A(\lambda)$ denotes the density of numbers N with $\frac{\sigma(N)}{N} \geq \lambda$ then

$$A(\lambda)$$

exists and is continuous function of $\lambda \in \mathbb{R}$

H. Davenport. *Über numeri abundantes*. Preuss. Akad. Wiss. Sitzungsber 26/29 (1933), 830–837.

c) Define by $A(\lambda, j, k)$ the density of those positive integers n for which $j \mid n, \left(\frac{n}{j}, k\right) = 1$ and $\frac{\sigma(n)}{n} \geq \lambda$. Then $A(\lambda, j, k)$ exists and is a continuous function of λ. If k is squarefree and any prime divisor of j also divides k, then if $\sigma(j)/j \geq \lambda$, then

$$A(\lambda, j, k) \begin{cases} = \varphi(k)/jk & , \text{ for } \lambda \leq \sigma(j)/j \\ \leq \dfrac{\sigma(j)/j}{x - \sigma(j)/j} \dfrac{M}{j} & , \text{ for } \lambda > \sigma(j)/j \end{cases}$$

where $M = \frac{\varphi(k)}{k}\left(-1 + \frac{\pi^2}{6}\prod_{p|k}(1 - p^{-2})\right)$

F. Behrend. *Über numeri abundantes*. II. Preuss. Akad. Wiss. Sitzungesber 6 (1933), 830–837.

d) $0.2441 < A(2) < 0.2909$

(where A is defined in b))

Ch.R. Wall, Ph.L. Crew and D.B. Johnson. *Density bounds for the sum of divisors function*. Math. Comp. 26 (1972), 773–777.

Remark. F. Behrend obtained the bounds $0.241 < A(2) < 0.314$

3) a) Let N be an odd perfect number. If $N \equiv 1 (\text{mod } 12)$ and $5 | N$, then

$$\frac{1}{5} + \frac{1}{7} + \frac{\log 48/35}{11 \log 11/10} < \sum_{p|N} \frac{1}{p} = S < \frac{1}{5} + \frac{1}{2738} + \log 50/31$$

D. Suryanarayana. *On odd perfect numbers*. II. Proc. Amer. Math. Soc. 14 (1963), 896–904.

b) If $N \equiv 1 (\text{mod } 12)$, $5 \mid N$ and N is perfect, then

$$0.64738 < S < \frac{1}{5} + \log 50/31 \approx 0.67804$$

c) If $N \equiv 1 (\text{mod } 12)$, $5 \nmid N$ and N is perfect, then

$$0.66745 < S < 0.69315$$

d) If $N \equiv 9 \pmod{36}$, $5 \mid N$, N perfect, then

$$0.59606 < S < 0.67377$$

e) If $N \equiv 9 \pmod{36}$, $5 \nmid N$, N perfect, then

$$0.60383 < S < 0.65731$$

P. Hagis and D. Suryanarayana. *A theorem concerning odd perfect numbers*. Fib. Quart 8 (1970), 337–346, 374.

4) If p_n is smallest prime factor of odd perfect number N and $b < 3/5$, then N has at least

$$\text{li}\left(p_n^2\right) + O(n^2 \exp(-\log^b n))$$

prime factors and a prime factor at least as large as

$$p_n^2 + O(n^2 \exp(-\log^b n))$$

D. Suryanarayana. *On odd perfect numbers*. Math. Student 41 (1973), 153–154.

§ III.32 Density of perfect numbers

Let $V(x) = \text{card}\{n \le x : n \text{ perfect number}\}$. Then:

1) $V(x) < x^{1/2}$

B. Hornfeck. *Zur Dichte der Menge der vollkommenen Zahlen*. Arch. Math. 6 (1955), 442–443.

2) $\limsup\limits_{x\to\infty} \dfrac{V(x)}{\sqrt{x}} \le \dfrac{1}{2\sqrt{5}}$

B. Hornfeck. *Bemerkung zu meiner*. Note über vollkommene Zahlen. Arch. Math. 7 (1956), 273.

3) $V(x) < c\dfrac{x^{1/4}\log x}{\log\log x}$

H.-J. Kanold. *Über die Verteilung der vollkommenen Zahlen und allgemeinerer Zahlenmengen*. Math. Ann. 132 (1957), 442–450.

4) $V(x) < \exp\left(\dfrac{c\log x \log\log\log x}{\log\log x}\right)$

B. Hornfeck and E. Wirsing. *Über die Häufigkeit vollkommener Zahlen*. Math. Ann. 133 (1957), 431–438.

5) $V(x) < \exp\left(\dfrac{c\log x}{\log\log x}\right)$

E. Wirsing. *Bemerkung zu der Arbeit über vollkommene Zahlen*. Math. Ann. 137 (1959), 316–318.

Remark. 1) is valid also for $V_k(x) = \text{card}\{n \le x : \sigma(n) = kn\}$, where $k > 1$ is a fixed rational number.

§ III.33 Multiply perfect and multiperfect numbers

1) A number n is called *multiply perfect* if $n \mid \sigma(n)$ (i.e. $\sigma(n) = kn$ for some integer $k > 1$.) Let $P(x)$ denote the number of multiply perfect numbers not exceeding x

a) $P(x) = o(x)$
$(x \to \infty)$

H.-J. Kanold. *Über mehrfach vollkommene Zahlen.* J. Reine Angew. Math. 194 (1955), 218–220.

b) $P(x) < x^{3/4+\varepsilon}$
for every $\varepsilon > 0$ and all sufficiently large x

P. Erdős. *On perfect and multiply perfect numbers.* Ann. Mat. Pura Appl. (4) 42 (1956), 253–258.

c) $P(x) = o(x^{\varepsilon})$
for all $\varepsilon > 0$

B. Hornfeck and E. Wirsing. *Über die Häufigkeit vollkommener Zahlen.* Math. Ann. 133 (1957), 431–438.

Remarks. (i) More generally, the number of solutions of $\sigma(n) = a\,n$, where a is an arbitrary fixed rational number, for $n \le x$ is $o(x^{\varepsilon})$ for any $\varepsilon > 0$ uniformly for all a
(See B. Hornfeck and E. Wirsing.)

(ii) According to
P. Erdős and R.L. Graham. *Old and new problems and results in combinatorial number theory.* Monographies de l'Enseignement Mathématique, No. 28, Genéve (p. 103),
E. Wirsing can show that the number of solutions of $\sigma(n) = a\,n$, $(\forall a \in \mathbf{Q})$ for $n \le x$ is less than $cx^{c' \log\log\log x / \log\log x}$, (which is independent of a)

2) The number n is called *multiperfect* if $\sigma(n) = kn$ for $k > 2$. For an odd multiperfect number n:

a) The largest prime factor is ≥ 100129

b) The second largest prime factor is ≥ 1009

P. Hagis, Jr. and G.L. Cohen. Bull. Malaysian Math. Soc. (2) 8 (1985), 23–26.

c) The third largest prime divisor is ≥ 101

P. Hagis, Jr. *The third largest prime factor of an odd multiperfect number exceeds 100.* Bull. Malaysian Math. Soc. (2) 9 (1986), 43–49.

3) Let $n = \prod_{i=1}^{k} p_i^{a_i}$ be multiply perfect with

$$\max_{j=\overline{1,k}} a_j < 4$$

Then there are exactly six numbers with this property.

H.-J. Kanold. *Einige Bemerkungen über vollkommene und mehrfach vollkommene Zahlen.* Abh. Braunschw. Wiss. Ges. 42 (1990), 49–55.

§ III.34 k-perfect numbers

A positive integer n is called k-perfect if $\sigma(n) = k\,n$ for $k > 1$ a fixed rational number.

1) A positive integer n is called primitive if cannot written in the form $m = st$, where s is an even perfect number and $(s, t) = 1$

a) For any k there are only finitely many primitive k-perfect numbers with a fixed number of distinct prime factors.

H.-J. Kanold. *Über einen Satz von L.E. Dickson.* II. Math. Ann. 132 (1956), 246–255.

b) For every $k \geq 1$, rational, and every non-negative integer K, there is an effectively computable number $N(k, K)$ such that if $\omega(n) = K$ and n is primitive k-perfect, then $n \leq N(k, K)$

C. Pomerance. *Multiply perfect numbers, Mersenne primes, and effective computability.* Math. Ann. 226 (1977), 195–206.

2) If n is k-perfect, then

$$\omega(n) \geq k^2 - 1$$

P.J. McCarthy. *Note on perfect and multiply perfect numbers.* Portugal Math. 16 (1957), 19–21.

Remark. This result was improved by

W. McDaniel. *On odd multiply perfect numbers.* Boll. Un. Mat. Ital. 3 (1970), 185–190.

3) If n is k-perfect ($k \geq 1$), where $n = p_1^{a_1} \cdots p_r^{a_r} \geq 3$, then:

a) $$r\left(\sqrt[r]{3/2} - 1\right) < \sum_{i=1}^{r} \frac{1}{p_i} < r\left(1 - \sqrt[r]{6/k^2}\right)$$

for n even

b) $$r\left(\sqrt[3r]{k^2} - 1\right) < \sum_{i=1}^{r} \frac{1}{p_i} < r\left(1 - \sqrt[r]{8/(k\pi^2)}\right)$$

for n odd

M. Bencze. *On perfect numbers.* Studia Univ. Babeş-Bolyai 26 (1981), 14–18.

§ III.35 Primitive abundant numbers

Let $\alpha > 0$. A number n is called *primitive* α*-abundant* if $\frac{\sigma(n)}{n} \geq \alpha$, but $\frac{\sigma(d)}{d} < \alpha$ for all $d \mid m$, $d < m$

1) Let $A_\alpha(x) = \text{card}\ \{n \leq x : n \text{ primitive } \alpha\text{-abundant}\}$. Then:

a) $A_\alpha(x) = o\left(\frac{x}{\log x}\right)$

P. Erdős. *Remarks on number theory. I. On primitive α-abundant numbers.* Acta Arith. 5 (1958), 25–33.

b) $$x \exp\left(-(\sqrt{6}+\varepsilon)(\log x \log\log x)^{1/2}\right) < A_2(x) < \\ < x \exp\left(-(\sqrt{12}-\varepsilon)(\log x \log\log x)^{1/2}\right)$$

for $x > x_0(\varepsilon)$

A. Ivić. *The distribution of primitive abundant numbers.* Studia Sci. Math. Hungar 20 (1985), 183–187.

2) Let α be rational. A necessary and sufficient condition that there exist infinitely many primitive α-abunadnt numbers with k distinct prime factors is that α have a representation

$$\alpha = \frac{b\,\sigma\,(a)}{a\,\varphi\,(b)}$$

with $(a, b) = 1, b > 1$, where $\omega(a) + \omega(b) < k$

H.N. Shapiro. *Note on theorem of Dickson.* Bull. Amer. Math. Soc. 55 (1949), 450–452

(necessity) and

H.N. Shapiro. *On primitive abundant numbers.* Comm. Pure Appl. Math. 21 (1968), 111–118

(sufficiency.)

Remark. The above theorem generalizes the classical result: There exist at most finite number of odd perfect numbers with a given number of distinct prime factors.

L.E. Dickson. *Finiteness of odd perfect and primitive abundant numbers with n distinct prime factors.* Amer. J. Math. 35 (1913), 413–422

and rediscovered by

I.S. Gradstein. *On odd perfect numbers.* Mat. Sbornik 32 (1925), 476–510.

3) Let $x, y, z \in \mathbb{N}$ and let s^d be the largest prime power dividing n. If $x\sigma(n) = y\,n + z$, $z \geq 175$, n is not primitive (y/x)-abundant and $x\sigma(n/s) \neq y\,n/s$, then

$$n < 4(z + 1/2)^3/27y$$

G.L. Cohen. *On primitive abundant numbers.* J. Austral Math. Soc. Ser. A 34 (1983), 123–137.

§ III.36 Deficient numbers

1) A number n is called *deficient*, if $\sigma(n) < 2n$. Then:

a) If $\sum_{p|n} \frac{1}{p-1} \leq \log 2$, then n is deficient.

D. Rameswar Rao. *A sufficient condition for a number to be deficient.* Math. Student 36 (1968), 235–236.

b) If $n \geq n_0$, then there exists at least a deficient number between n and $n + (\log n)^2$

J. Sándor. *On a method of Galambos and Kátai concerning the frequency of deficient numbers.* Publ. Math. (Debrecen) 39 (1991), 155–7.

2) If (a, b) is deficient (where $(a, b) =$ g.c.d. (a, b)), then:

a) There exist infinitely many deficient integers n with $n \equiv a(\text{mod } b)$

b) There exist infinitely many abundant integer $n \equiv a(\text{mod } b)$

Ch.R. Wall. *Problem E3002.* Amer. Math. Monthly 90 (1983), 400. Solution by N.J. Fine. Amer. Math. Monthly 93 (1986), 814.

Note. The solution given by Fine shows that for b) the condition on the deficiency of (a, b) is not necessary.

§ III.37 Triperfect numbers

1) A positive integer n is called a *triperfect number* if $\sigma(n) = 3n$. If n is an odd triperfect number, then:

a) $\omega(n) \geq 9$

W. McDaniel. *On odd multiply perfect numbers.* Boll. Un. Mat. Ital. (1970), 185–190,

and

G.L. Cohen. *On odd perfect numbers II, multiperfect numbers and quasiperfect numbers.* J. Austral. Math. Soc. 29 (1980), 369–384.

b) $\omega(n) \geq 11$

E.A. Bugulov. *On the question of the existence of odd multiperfect numbers* (Russian.) Kabardino-Balkarsk Gos. Univ. Ucen. Zap. 30 (1966), 9–19,

and rediscovered by

M. Kishore. *Odd triperfect numbers are divisibile by eleven distinct prime factors.* Math. Comp. 44 (1985), 261–263.

c) $\omega(n) \geq 12$

H. Reidling. *Über ungerademehrfach vollkommene Zahlen.* Östereichische Akad. Wiss. Math.-Natur. 192 (1983), 237–266,

and

M. Kischore. *Odd triperfect numbers are divisibile by twelve distinct prime factors.* J. Austral. Math. Soc. (Series A), 42 (1987), 173–182.

2) If n is odd triperfect, then:

a) $n > 10^{50}$

W.E. Beck and R.M. Najar. *A lower bound for odd triperfect.* Math. Comp. 38 (1982), 249–251.

b) $n > 10^{60}$

L.B. Alexander. *Odd triperfect numbers are bounded below y* 10^{60} (M.A. Thesis, East Carolina Univ., 1984.)

c) $n > 10^{70}$

the largest prime factor of n is at least 100 129 and the second largest prime factor is at least 1 009

G.L. Cohen and P. Hagis, Jr. *Results concerning odd multiperfect numbers.*

§ III.38 Quasiperfect numbers

The number n is called *quasiperfect* if $\sigma(n) = 2n + 1$

1) If n is quasiperfect, then:

a) $\omega(n) \geq 5$ and $n > 10^{20}$

H.L. Abbott, C.E. Aull, E. Brown and D. Suryanarayana. *Quasiperfect numbers.* Acta Arith. 22 (1973), 439–447;

Correction: Acta. Arith. 29 (1976), 427–428.

b) $\omega(n) \geq 6$ and $n > 10^{30}$

M. Kishore. *Quasiperfect numbers are divisible by at least six distinct prime factors.* Notices Amer. Math. Soc. 22 (1975), p. A-441,

M. Kishore. *Odd integers n with five distinct prime factors for which* $2 - 10^{-12} < \sigma(n)/n < 2 + 10^{-12}$. Math. Comp. 32 (1978), 303–309.

c) $\omega(n) \geq 7$ and $n > 10^{35}$

G. Cohen and P. Hagis, Jr. *Some results concerning quasiperfect numbers.* J. Austral. Math. Soc. Ser.A 33 (1982), 275–286.

d) If a number of the form $\prod_{i=1}^{r} p_i^{6a_i+2}$ is quasiperfect, then

$$r \geq 230\,876$$

G.L. Cohen. *The nonexistence of quasiperfect numbers of certain forms.* Fib. Quart. 20 (1982), 81–84.

2) If n is quasiperfect, then if $n = \prod_{i=1}^{r} p_i^{a_i}$ (p_i-primes), then

$$p_i < 2^{2^{i-1}}(r - i + 1)$$

for $2 \leq i \leq 6$

M. Kishore. *On odd perfect, quasiperfect, and odd almost perfect numbers.* Math. Comp. 36 (1981), 583–586.

3) There are only finitely many quasiperfect numbers n for which

$$\omega(n) < S$$

where $\omega(n)$ is the number of distinct prime factors of n and S is an arbitrarily large fixed bound.

H.-J. Kanold. *Über "quasi-vollkommene Zahlen".* Abh. Braunschweig. Wiss. Ges. 40 (1988), 17–20.

§ III.39 Almost perfect numbers

A positive integer n is called *almost perfect*, if

$$\sigma(n) = 2n - 1$$

1) If n is an odd almost perfect number, then

$$\omega(n) \geq 6$$

M. Kishore. *Odd integers n with five distinct prime factors for which $2 - 10^{-12} < \sigma(n)/n < 2 + 10^{-12}$.* Math. Comp. 32 (1978), 303–309.

2) If n is an odd almost perfect number, then

$$p_i < 2^{2^{i-1}}(r - i + 1)$$

for $2 < i \leq 5$ and $p_6 < 23\,775\,427\,335(r - 5)$

M. Kishore. *On odd perfect, quasiperfect and odd almost perfect numbers.* Math. Comp. 36 (1981), 583–586.

§ III.40 Superperfect numbers

1) The number of even *superperfect numbers* i.e. $\sigma(\sigma(n)) = 2n$, $n \leq x$ is

$$o(\log x / \log\log x)$$

D. Bode. *Über eine Verallgemeinerung der vollkommenen Zahlen.* (Dissertation. Braunschweig 1971, 57 pp.)

2) The smallest odd superperfect number must be the greater than

$$7 \cdot 10^{24}$$

J.L. Hunsucker and C. Pomerance. *There are no odd superperfect number less than* $7 \cdot 10^{24}$. Indian J. Math. 17 (1975), 107–120.

3) $\sigma^{(k)}(n) \neq 2n$

for all $n \in \mathbb{N}$ and $k \geq 3$ (Here $\sigma^{(k)}$ denotes the kth iterate of the σ-function.) See

D. Bode

and for an elementary proof

G. Lord. *Even perfect and superperfect numbers.* Elem. Math. 30 (1975), 87–88.

§ III.41 Superabundant and highly abundant numbers

1) A number n is called *superabundant* if $\frac{\sigma(n)}{n} > \frac{\sigma(m)}{m}$ for all m with $1 \leq m < n$. Let $Q(x)$ be the counting function of superabundant numbers. Then:

a) If n and n' are two consecutive superabundant numbers then

$$\frac{n'}{n} < 1 + c(\log\log n)^2/\log n$$

Corollary. $Q(x) \geq c \log x \log\log x/(\log\log\log x)^2$

L. Alaoglu and P. Erdős. *On highly composite and similar numbers.* Trans. Amer. Math. Society 56 (1944), 448–469.

b) With the above notations,

$$\frac{n'}{n} \leq 1 + \frac{1}{\sqrt{\log n}}$$

for an infinity of n

J.-L. Nicolas. *Ordre maximal d'un élément du groupe S_n des permutations et ≪highly composite numbers≫.* Bull. Soc. Math. France 97 (1969), 129–191.

c) $\liminf(\log Q(x)/\log\log x) \geq 5/48$

P. Erdős and J.-L. Nicolas. *Répartition des nombres superabondantes.* Bull. Soc. Math. France 103 (1975), 65–90.

2) A number n is said to be *highly abundant* if

$$\sigma(n) > \sigma(m)$$

for all $m < n$

a) Define $f(x) = \max_{n \le x} \frac{\sigma(n)}{n}$. Then for $n =$ highly abundant we have

$$f(n) - \frac{\sigma(n)}{n} < c_1 \log\log n / \log n$$

b) Only a finite number of highly abundant numbers can be highly composite.

c) Let $H(x)$ be the counting function of highly abundant numbers. Then

$$H(x) > (1 - \varepsilon)\,(\log x)^2$$

for every $\varepsilon > 0$ and for large x

d) If n is highly abundant, then the largest prime factor of n is less that $c_2 \log n (\log\log n)^3$

See

L. Alaoglu and P. Erdős.

§ III.42 Amicable numbers

1) Let $B(x)$ denote the number of pairs of *amicable numbers* $a,\ b \in \mathbb{N}$ (i.e. $\sigma(a) = \sigma(b) = a + b$) with $a < b$ and $a < x$. Then:

a) $B(x) = o(x)$

P. Erdős. *On amicable numbers.* Publ. Math. (Debrecen) 4 (1955), 108–111.

b) $B(x) = O(x / \log\log\log x)$

P. Erdős and G.J. Rieger. *Ein Nachtrag über befreundete Zahlen.* J. Reine Angew. Math. 273 (1975), 220.

c) $B(x) \le x \exp(-(\log x)^{1/3})$

for x sufficiently large.

C. Pomerance. *On the distribution of amicable numbers.* II. J. Reine Angew. Math. 325 (1981), 183–188.

Corollary. The sum the reciprocals of the amicable numbers is finite.

d) $B(x) \ll x \exp(-c(\log x \log\log x)^{1/3})$

C. Pomerance. *Ibid.*

2) If a and b are relatively prime amicable numbers of opposite parity, such that $2 \mid b$ and $b > a$, then:

a) If $5 \mid m\,n$, then

$$\sum_{p \mid mn} \frac{1}{p} < 1.57549$$

b) If $5 \nmid mn$, then

$$\sum_{p|mn} \frac{1}{p} < 1.59862$$

These bounds also hold if $b < a$ and $2b > a$

c) If $(a, b) = 1$ are amicable pairs with opposite parity, then

$$\sum_{p|mn} \frac{1}{p} > 1.43151$$

if $5 \mid mn$ and $\sum_{p|mn} \frac{1}{p} > 1.45382$ if $5 \nmid mn$

P. Hagis, Jr. *Relatively prime amicable numbers of opposite parity.* Math. Mag. 43 (1970), 14–20.

3) The set of relatively prime amicable pairs *a, b* with $\omega(a) + \omega(b) \leq S$ (S-given) is finite.

W. Borho. *Befreundete Zahlen mit gegebener Primteileranzahl.* Math. Ann. 209 (1974), 183–193.

4) The positive integers *a* and *b* are called *quasi-amicable* if

$$\sigma(a) = \sigma(b) = a + b + 1$$

Let *a, b* be quasi-amicable.

a) If $b < a$ are of the same parity, then

$$b > 10^{10}$$

b) If $(a, b) = 1$ then

$$\omega(ab) \geq 4$$

If $(a, b) = 1$ and *a, b* are odd, then

$$\omega(ab) \geq 21$$

c) If $(a, b) = 1$ and *a, b* are of same parity, then

$$b > 10^{30}$$

$a > 10^{30}$

P. Hagis, Jr. and G. Lord. *Quasi-amicable numbers.* Math. Comp. 31 (1977), 608–611.

§ III.43 Weird numbers

A positive integer *n* is called *weird* if *n* is abundant (i.e. $\sigma(n) \geq 2n$) and not pseudoperfect (*n* is pseudoperfect, if *n* is the distinct sum of some of the proper divisors of *n*, see

W. Sierpiński. *Sur les nombres pseudoparfaits.* Mat. Vesnik 2 (17) (1965), 212–213.)

The density of weird numbers is positive.

S.J. Benkoski and P. Erdős. *On weird and pseudoperfect numbers.* Math. Comp. 28 (1974), 617–623.

§ III.44 Hyperperfect numbers

A positive integer m is called *n-hyperperfect* if there exists $n \in \mathbb{N}$ with

$$m = 1 + n(\sigma(m) - m - 1)$$

a) If $n = p_1^{\alpha_1} p_2^2$ is n-hyperperfect, then

$$p_1 > n + 1$$

(p_1, p_2-Primes) and

$$p_1 \leq (n+1)^2$$

b) If $n = p_1^{\alpha_1} p_2^{\alpha_2}$ is n-hyperperfect, then

$$\alpha_2 \leq \frac{1}{\log p_2} \log\left(n\left(p_1^{\alpha_1} + p_1^{\alpha_1 - 1} + \cdots + 1\right)(p_2 - 1)(n-1)\right)$$

D. Minoli. *Issues in nonlinear hyperperfect numbers.* Math. Comp. 34 (1980), 639–645.

c) There exist hyperperfect numbers with more than two different prime factors.

H.J.J. te Riele. *Hyperperfect numbers with three different prime factors.* Math. Comp. 36 (1981), 297–298.

§ III.45 Unitary perfect numbers, bi-unitary perfect numbers

1) A positive integer n is called *unitary perfect* if

$$\sigma^*(n) = 2n$$

where $\sigma^*(n)$ denotes the sum of the unitary divisors of n.
Suppose that $n = 2^a m$ is unitary perfect (it is immediate that n must be even), where m is odd. The old unitary perfect numbers are 6, 60, 90, 87 360, and 146 361 946 186 458 562 560 000. For any new unitary perfect number we have:

a) $a > 10$ and $\omega(m) > 6$

M.V. Subbarao, T.J. Cook, R.S. Newberry and J.M. Weber. *On unitary perfect numbers.* Delta 3, No. 1 (spring 1972), 22–26.

b) $\omega(m) > 8$

Ch.R. Wall. *New unitary perfect numbers have at least nine odd components.* Fib. Quart. 26 (1988), 312–317.

Note. The first four unitary numbers were discovered by M.V. Subbarao and L.J. Warren, see

M.V. Subbarao and L.J. Warren. *Unitary perfect numbers.* Canad. Math. Bull. 9 (1966), 147–153,

and the fifth by Ch.R. Wall, see

Ch.R. Wall. *The fifth unitary perfect number.* Canad. Math. Bull. 18 (1975), 115–122.

Remark. Any new unitary perfect number has a prime power unitary divisor larger than 2^{15}.

Ch.R. Wall. *On the largest odd component of unitary perfect number.* Fib. Quart. 25 (1987), 312–316.

2) A divisor d of an integer n is a *bi-unitary divisor* if the greatest common unitary divisor of d and n/d is 1. We say that n is *bi-unitary perfect* if

$$\sigma^{**}(n) = 2n$$

where $\sigma^{**}(n)$ denotes the sum of the bi-unitary divisors of n

a) $\sigma^{**}(n) \leq \sigma(n)$

b) The only bi-unitary perfect numbers are: 6, 60, 90

Ch.R. Wall. *Bi-unitary perfect number.* Proc. Amer. Math. Soc. 33 (1972), 39–42.

§ III.46 Primitive unitary abundant numbers

The positive integer n is said to be *primitive unitary* α*-abundant* if

$$\sigma^*(n) > \alpha n$$

but

$$\sigma^*(d) < \alpha n$$

for all $d \mid n, d < n$ ($\alpha \geq 2, \alpha \in \mathbb{R}$.) Let U_α be the set of these numbers.

a) If $n \in U_\alpha$, then

$$\sigma^*(n) - \alpha n < \alpha n / q^b$$

where $q^b \parallel n$ (q prime)

b) If $n \in U_\alpha$ then

$$\limsup_{n\to\infty} \frac{\sigma^*(n)}{n} = \alpha$$

V. Siva Rama Prasad and D.R. Reddy. *On primitive unitary abundant numbers.* Indian J. Pure Appl. Math. 21 (1990), 40–44.

§ III.47 Nonunitary perfect numbers

A natural number is called *nonunitary perfect*, if

$$\sigma^{\#}(n) = n$$

where $\sigma^{\#}(n)$ denotes the number of nonunitary divisors of n (a divisor $d \mid n$ is called nonunitary, if $\left(d, \frac{n}{d}\right) > 1$)

S. Ligh and Ch. Wall. *Functions of non-unitary divisors.* Fib. Quart. 25 (1987), 333–338.

Then:

a) If n is odd nonunitary perfect, then

$$n > 10^{15}$$

b) If n is odd nonunitary perfect, then

$$\omega(n) \geq 4$$

If $3 \nmid n$, then

$$\omega(n) \geq 7$$

P. Hagis, Jr. *Odd nonunitary perfect numbers.* Fib. Quart. 28 (1990), 11–15.

§ III.48 Exponentially perfect numbers

A number n is called *e-perfect* (*exponentially-perfect*) if

$$\sigma^{e}(n) = 2n$$

where σ^{e} denotes the sum of exponential divisors of n.

1) a) There are no odd e-perfect numbers. More generally,

$$\sigma^{e}(n) \neq kn$$

for any integer $k > 1$

b) For each r the number of e-perfect numbers with r prime factors is finite.

E.G. Straus and M.V. Subbarao. *On exponential divisors.* Duke Math. J. 41 (1974), 465–471.

2) The density of e-perfect numbers is 0.0087

P. Hagis, Jr. *Some results concerning exponential divisors.* Intern. J. Math. Math. Sci. 11 (1988), 343–349.

3) a) Any e-perfect number which is not divisible by 3 must be divisible by 2^{117}, greater than 10^{664} and divisible by at least 118 distinct prime divisors.

b) Let $S_p(x) = \text{card}\{k \leq x : \sigma^e(p^k) \not\equiv 0(\text{mod } 3)\}$, where p is a fixed prime. Then

$$\limsup_{x\to\infty} \frac{S_p(x)}{x} \leq 0.627$$

if $p \equiv 2(\text{mod } 3)$ and

$$\limsup_{x\to\infty} \frac{S_p(x)}{x} \leq 0.752$$

if $p \equiv 1(\text{mod } 3)$

J. Fabrykowski and M.V. Subbarao. *On e-perfect numbers not divisible by* 3. Nieuw Arch. Wisk. (4) 4 (1986), 165–173.

4) a) $\sigma^e(n) = 2n - \omega(n)$
if n is squarefree and $n = 4$

b) $\sigma^e(n) \neq 2n + \omega(n)$
for all n

J. Fabrykowski and M.V. Subbarao. *On some Diophantine equations involving exponentially multiplicative functions.* Congr. Numer. 56 (1987), 163–171.

5) n is called *e-multiperfect* if

$$\sigma^e(n) = kn$$

for some integer $k > 2$

If n is e-multiperfect, then

$$n > 2 \cdot 10^7$$

for $k = 3$

$$n > 10^{85}$$

for $k = 4$

$$n > 10^{320}$$

for $k = 5$

$$n > 10^{1210}$$

for $k = 6$

W. Aiello, G.E. Hardy and M.V. Subbarao. *On the existence of e-multiperfect numbers.* Fib. Quart. 25 (1987), 65–71.

§ III.49 Exponentially, powerful perfect numbers

Let $N_k(\sigma^e, x) = \text{card}\,\{n \leq x : \sigma^e(n) = kn, n \text{ powerful}\}$.

If $k > 1$, rational and $x \geq 3$, then there exists a constant c, not depending on k such that

$$N_k(\sigma^e, x) \leq \exp(c \log x / \log\log x)$$

L. Lucht. *On the sum of exponential divisors and its iterates.* Arch. Math. (Basel) 27 (1976), 383–386.

§ III.50 Practical numbers

A positive integer m is said to be a *practical number* iff every integer n, with $1 \leq n \leq \sigma(n)$, is a sum of distinct divisors of m. See

A.K. Srinivasan. *Practical numbers.* Current science, 1948, pp. 179–180,

and

B.M. Stewart. *Sums of distinct divisors.* Amer. J. Math. 76 (1954), 779–785.

1) Let $1 = d_1 < d_2 < \cdots < d_r = n$ be the divisors of n and for $k \leq r$ put $t_k = d_1 + \cdots + d_k$ with $t_0 = 0$. Then n is practical if $d_{k+1} \leq t_k + 1$ for all $k \in \{0, \ldots, r-1\}$

D.F. Robinson. *Egyptian fractions via Greek number theory.* New Zealand Math. Mag. 16 (1979), 47–52.

Corollary. If n is practical, then

$$\sigma(n) \geq 2n - 1$$

$(n \geq 1)$

2) Let $P(x) = \text{card}\{n \leq x : n, n \text{ practical number}\}$. Then:

a) $P(x) = O(x/(\log x)^\beta)$
for every fixed $\beta < \frac{1}{2}(1/\log 2 - 1)^2 = 0.0979\ldots$

M. Hausman and H.N. Shapiro. *On practical numbers.* Comm. Pure Appl. Math. 37 (1984), 705–713.

b) $$P(x) \geq Ax/\exp\left(\frac{1}{2\log 2}(\log\log x)^2 + 3\log\log x\right)$$

for x sufficiently large ($A = 2^{5/2}/5$).

M. Margenstern. *Les nombres pratiques: théorie, observations et conjectures.* J. Number Theory 37 (1991), 1–36.

3) For all $x \geq \frac{1}{3}$, the interval $(x, x + 2\sqrt{x})$ contains a practical number. See

M. Hausman and H.N. Shapiro.

4) a) If m is practical, distinct from a power of 2, then

$$\sigma(m) \geq 2m$$

See M. Margenstern.

Corollary. All even perfect numbers are practical.

b) $\liminf_{m\to\infty} \sigma(m)/m = 2$ and $\limsup_{m\to\infty} \sigma(m)/m = +\infty$
where m is practical, distinct from a power of 2
See
M. Margenstern.

§ III.51 Unitary harmonic numbers

1) A number n is called *harmonic* (called also *Ore number*) if $\sigma(n) \mid nd(n)$ (i.e., the harmonic mean of divisors of n is an integer).

The density of harmonic numbers is zero.
H.-J. Kanold. *Über das harmonische Mittel der Teiler einer natürlichen Zahl.* Math. Ann. 133 (1957), 371–374.

2) A number n is called *unitary harmonic*, if

$$\sigma^*(n) \mid nd^*(n)$$

Let $H^*(x)$ be the counting function of these numbers. Then for $\varepsilon > 0$ and large x

$$H^*(x) < 2.2x^{1/2} \cdot 2^{(1+\varepsilon)\log x/\log\log x}$$

P. Hagis Jr, and G. Lord. *Unitary harmonic numbers.* Proc. Amer. Math. Soc. 51 (1975), 1–7.

3) a) The set of n for which

$$d(n) \mid \sigma(n)$$

has density 1

b) Let $N(x) = \{n \le x : d(n) \nmid \sigma(n)\}$. Then

$$N(x) = x\exp(-(1+o(1))2\sqrt{\log 2}\sqrt{\log\log x})$$

c) The set of n which

$$d^2(n) \mid \sigma(n)$$

has density $\frac{1}{2}$

d) The number of rationals $r \le x$ of the form

$$r = \frac{\sigma(n)}{d(n)}$$

is $o(x)$

P.T. Bateman, P. Erdős, C. Pomerance and E.G. Straus. *The arithmetic mean of the divisors of an integer.* Analytic number theory (Philadelphia, Pa., 1980), pp. 197–220, Lecture Notes in Math., 899, Springer, Berlin-New-York, 1981.

§ III.52 Perfect Gaussian integers

For the Gaussian integer $\eta = \varepsilon \prod \pi_i^{a_i}$, where ε is a unit, Re $\pi_i > 0$, and Im $\pi_i \geq 0$, define the sum-of-divisors function by

$$\sigma(\eta) = \prod \left(\pi_i^{a_i+1} - 1\right)/(\pi_i - 1)$$

Then η is called *perfect* if

$$\sigma(\eta) = (1+i)\eta$$

and *norm-perfect* if

$$|\sigma(\eta)| = 2|\eta|$$

The number η is called *norm-abundant* if $|\sigma(\eta)| \geq 2|\eta|$ and *primitive norm-abundant* if no proper divisor of η satisfies the above inequality.

See

R. Spira. *The complex sum of divisors.* Amer. Math. Monthly 68 (1961), 120–124,

and

W.L. McDaniel. *Perfect Gaussian integers.* Acta Arith. 25 (1973), 137–144.

A number η is called even if $(1+i) \mid \eta$, and odd otherwise.

For odd primitive norm-abundant numbers and odd norm-perfect numbers, there are only finitely many with a fixed number of distinct prime factors.

M. Hausman. *On norm abundant Gaussian integers.* J. Indian Math. Soc. (N.S.) 49 (1985), 119–123 (1987.)

Chapter IV

P, p, B, β AND RELATED FUNCTIONS

§ IV. 1 Sums over $P(n)$, $p(n)$, $P(n)/p(n)$, $1/P^r(n)$

Let $P(k)$ be the largest prime divisor of k; $P(1) = 1$ and $p(k)$ the smallest prime divisor of k, $p(1) = 1$

1) $$\sum_{n\leq x} P(n) = \frac{\pi^2}{12}\frac{x^2}{\log x} + O\left(x^2 \log^{-3/2} x \log\log x\right)$$

Remark. Recently J. Lin has proved that

$$\sum_{2\leq n\leq x} P(n) = \int_{3/2}^{2} \zeta(\sigma)\frac{x^\sigma}{\sigma}\mathrm{d}\sigma + O(E(x)),$$

where $E(x) = x^2.\exp\{-c(\log x)^{3/5}.(\log\log x)^{-1/5}\}$ and ζ is Riemann's zeta function.

J. Lin. *Mean value estimates for arithmetic functions of prime factors* (Chinese). J. Shandong Univ., Nat. Sci. Ed. 28 (1993), 253–260.

2) $$\sum_{n\leq x} p(n) = \frac{1}{2}\frac{x^2}{\log x} + O\left(x^2 \log^{-2} x\right)$$

A.E. Brouwer. *Two number-theoretic sums.* Mathematisch Centrum, Amsterdam, 1974, ii + 3 pp.

Remark. In 1963 M. Kalecki proved that

$$\sum_{n\leq x} p(n) = \frac{1+o(1)}{2}\frac{x^2}{\log x}$$

M. Kalecki. *On certain sums extended over primes or prime factors* (Polish.) Prace Mat. 8 (1963/64), 121–129.

3) $$\sum_{n\leq x} \frac{p(n)}{P(n)} = \pi(x)(1+o(1))$$

as $x \to \infty$

4) $$\sum_{n\leq x} \frac{p(n)}{P(n)} = \frac{x}{\log x} + \frac{3x}{\log^2 x}(1+o(1))$$

P. Erdős and J.H. van Lint. *On the average ratio of the smallest and largest prime divisor of n.* Nederl. Akad. Wetensch. Indag. Math. 44 (1982), 127–132.

5) $$\sum_{n\le x}\frac{p(n)}{P(n)}=\frac{x}{\log x}+\frac{3x}{\log^2 x}+\frac{15x}{\log^3 x}+o\left(\frac{x}{\log^3 x}\right)$$

C.H. Jia. *A generalization of a theorem on prime numbers* (Chinese.) Adv. in Math. (Beijing) 16 (1987), 419–426.

6) $$\sum_{n\le x}\frac{1}{P(n)}=x\exp(-(2\log x\cdot\log\log x)^{1/2}+O(\log x\cdot\log\log x)^{1/2})$$

$(n>1)$

A. Ivić. *Sums of reciprocals of the largest prime factor of an integer.* Arch. Math. (Basel) 36 (1981), 57–61.

Remark. The same formula holds for $\sum_{n\le x}\frac{1}{B(n)}$ and $\sum_{n\le x}\frac{1}{\beta(n)}$

7) $$\sum_{2\le n\le x}\frac{1}{P^r(n)}=x\exp\big(-(2r)^{1/2}L_1(x)(1+g_{r-1}(x)+O(L_2(x)))\big)$$

where

$$L_1(x)=(\log x\cdot\log_2 x)^{1/2}$$
$$L_2(x)=\log_3^3 x/\log_2^3 x$$
$$f_r(x)=\log_3 x+\log(1+r)-\log 2$$
$$g_r(x)=(f_r(x)-2)\log_2^{-1}x\left(\frac{1}{2}+\log_2^{-1}x\right)-\frac{1}{8}f_r(x)^2\log_2^{-2}x$$

and $\log_n$ denotes the iterated logarithm.

A. Ivić and C. Pomerance. *Estimate for certain sums involving the largest prime factor of an integer.* Topics in classical number theory (Budapest, 1981), 769–789, North Holland, 1984.

8) $$\sum_{n\le x}\frac{1}{P(n)}=x\delta(x)\left(1+O((\log\log x/\log x)^{1/2}\right)$$

where $\delta(x)=\int_2^x t^{-2}\rho(\log x/\log t)\,dt$. Here $\rho(u)$ is the Dickman-de Bruijn function, defined by $\rho(u)=1, 0<u\le 1, u\rho'(u)=-\rho(u-1)$ for $u>1$

P. Erdős. A. Ivić and C. Pomerance. *On sums involving reciprocals of the largest prime factor of an integer.* Glas. Mat. Ser. III 21 (41) (1986), 283–300.

§ IV. 2 Sums over $\log P(n)$

1) a) $$\sum_{1\le n\le x}\log P(n)=ax\log x+O(x)$$

(a-constant)

N.G. de Bruijn. *On the number of positive integers $\leq x$ and free of prime factors greater than y.* Nederl. Akad. Wetensch. Proc. Ser. A 54 (1951), 50–60.

b) $$\sum_{n \leq x} \log P(n) = ax \log x - a(1-\gamma)x + O\big(x \exp(-(\log x)^{3/8-\varepsilon})\big)$$

where $a = 0.624\ldots$

G. Tenenbaum. *Introduction à la théorie analytique et probabiliste des nombres.* Publ. Inst. Elie Cartan Vol. 13, Nancy, 1990.

c) $$\sum_{2 \leq n \leq x} \frac{1}{n \log P(n)} = e^{\gamma} \log\log x + O(1)$$

J.-M. de Koninck and R. Sitaramachandrarao. *Sums involving the largest prime divisor of an integer.* Acta Arith. 48 (1987), no.1, 3–8.

Remark. The above formula with remainder term $O(\log\log\log x)$ was obtained firstly by

G.J. Rieger. *On two arithmetic sums.* Notices Amer. Math. Soc. 74T–A177.

2) If $u \geq 1$ and $\alpha \in \mathbb{R}$ are fixed, then

$$\sum_{1 \leq n \leq x, P_2(n) \leq (P(n))^{1/u}} (\log P(n))^{\alpha} = e^{\gamma} \cdot u^{\alpha} \cdot f(u,\alpha) \cdot x(\log x)^{\alpha} + O\big(x(\log x)^{\alpha-1}\big)$$

where $P_2(n)$ denotes the second largest prime factor of n and $f(u,\alpha)$ is a function of u and α

F.S. Wheeler. *Two differential-difference equations arising in number theory.* Trans. Amer. Math. Soc. 318 (1990), 491–523.

§ IV. 3 Sums over $P(n)^{-\omega(n)}$ and $P(n)^{-\Omega(n)}$

1) $$\sum_{n \leq x} P(n)^{-\omega(n)} = \exp\big((4+o(1))(\log x)^{1/2}/(\log\log x)\big)$$

2) $$\sum_{n \leq x} P(n)^{-\Omega(n)} = \log\log x + C + O(1/\log x)$$

(C-constant)

P. Erdős, A. Ivić and C. Pomerance. *On sums involving reciprocals of the largest prime factor of an integer.* Glas. Mat. Ser. III, 21 (41) (1986), 283–300.

§ IV. 4 Sums on $1/p(n)$, $\omega(n)/p(n)$, $d(n)/p(n)$

Let $p(n)$ denote the least prime divisor of n. Then

1) $\displaystyle\sum_{n\le x}\frac{1}{p(n)} = x\big(A + O\big(1/(\log x)^{1/14}\big)\big)$

2) $\displaystyle\sum_{n\le x}\frac{\omega(n)}{p(n)} = x\log\log x(A + O(1/\log\log x))$

Remark. The same formula holds by replacing $\omega(n)$ with $\Omega(n)$

3) $\displaystyle\sum_{n\le x}\frac{d(n)}{p(n)} = x\log x(B + O(1/(\log x)^{1/14}))$

where A and B are constants.

W.P. Zhang. *Average-value estimation of a class of number-theoretic functions.* Acta Math. Sinica 32 (1989), 260–267.

§ IV. 5 Density of reducible integers

1) The set of integers n such that $P(n) < 2\sqrt{n}$ has density

$$1 - \log 2$$

2) The density of the set of numbers n such that $P(n) > A\,n^{\alpha}$ (where $A \ge 1$, $\frac{1}{2} \le \alpha < 1$) is

$$\log(1/\alpha)$$

S.D. Chowla and J. Todd. *The density of reducible integers.* Canad. J. Math. 1 (1949), 297–299.

3) Let $L(n)$ be the number of $k \le n$ such that $P(k) < \sqrt{k}$ and $S(n) = \displaystyle\sum_{k=1}^{n}\frac{P(k)}{k}$. Then

$$\lim_{n\to\infty}\frac{\log n}{n}\cdot S(n) = \zeta(2)$$

and

$$\lim_{n\to\infty}\frac{1}{nL(n)} = \log 2$$

J.G. Kemeny. *Largest prime factor.* J. Pure. Appl. Algebra 89 (1993), 181–186.

§ IV. 6 On $p(n!+1)$, $P(n!+1)$, $P(F_n)$

1) a) Let $\varepsilon(n)$ be any positive function that decreases to 0 as $n \to \infty$. Then, for almost all integers n

$$p(n!+1) > n + \varepsilon(n)\sqrt{n}$$

b) For any positive integer n such that $n+1$ is composite we have

$$p(n!+1) > n + (1 - o(1)) \log n / \log\log n$$

c) $P(n!+1) > n + (1 - o(1)) \log n / \log\log n$

d) $\limsup_{n\to\infty} \frac{P(n!+1)}{n} > 2 + \delta$

for some $\delta > 0$

P. Erdős and C. L. Stewart. *On the greatest and least prime factors of n ! + 1*. J. London Math. Soc. (2) 13 (1976), 513–519.

2) If p_k denotes the k-th prime number, then for infinitely many n,

$$P(p_1 \cdots p_n + 1) > p_{n+k}$$

where $k > c \log n / \log\log n$ for some positive absolute constant c. See

P. Erdős and C.L. Stewart.

3) $P(F_n) > c \cdot n \cdot 2^n$

for all $n = 1, 2, \ldots$ where $F_n = 2^{2^n} + 1$ is Fermat's number, and c is a positive absolute constant.

C.L. Stewart. *The greatest prime factor of* $a^n - b^n$. Acta Arith. 26 (1975), 427–433;

See also

C.L. Stewart. *On divisors of terms of linear recurrence sequences*. J. Reine Angew. Math. 333 (1982), 12–31.

§ IV. 7 Greatest prime factor of an arithmetic progression

Let a, d, k be natural numbers, $(a, d) = 1$, $k \geq 3$ and let P denote the greatest prime factor of $a(a+d)\cdots(a+(k-1)d)$

1) If $a \geq d + k$, then

$$P > k$$

J.J. Sylvester. *On arithmetic series*. Messenger Math. 21 (1892), 1–19 and 87–120.

2) If $a > k$, then

$$P > k$$

M. Langevin. *Plus grand facteur premier d'entiers en progression arithmétique*. Séminaire Delange-Pisot-Poitou, 18e année, 1976/77, no. 3, 6 pp.

3) If $d \geq 2$ and $(a, k, d) \neq (2, 3, 7)$, then

$$P > k$$

T.N. Shorey and R. Tijdeman. *On the greatest prime factor of an arithmetical progression*. A tribute to Paul Erdős, pp. 385–389;
T.N. Shorey and R. Tijdeman. *On the number of prime factors of an arithmetic progression*. Sichuan Daxue Xuebao, 26 (1989), 72–74.

4) Let $x = a + (k-1)d$ and $\varepsilon > 0, x > k^{1+\varepsilon}$. Then there exists an effectively computable number $c > 0$ (depending only on ε) such that

$$P > c \cdot k \cdot \log\log x$$

T.N. Shorey and R. Tijdeman. *On the greatest prime factor of an arithmetical progression*. II. Acta Arith. 53 (1990), 499–504.

Remark. For part III see in
Approximations diophantiennes et nombres transcendants. C.-R. Colloque Lumigny (France 1990), 275–280 (1992).

5) Let $d = 1$ and $a = u + 1, k \geq 1$. If $k^{3/2} \leq u \leq k^{\log\log k}$, then

$$P > k^{1+2^{\lambda}(u,k)}$$

where $\lambda(u, k) = -((\log u)/(\log k) + 8)$

6) If $u \geq k^{3/2}$ and $u \notin \left[k^{\log k/(\log\log k)}, k^{(\log k)^{3+1/1000}}\right]$, then

$$P > (2 - \varepsilon)k \log k$$

for $k \geq k_0(\varepsilon)$
K. Ramachandra. *A note on numbers with a large prime factor*. II. J. Indian Math. Soc. (N.S.) 34 (1970), 39–48(1971) and Ibid. III. 19 (1971), 49–62.

Remark. The above result slightly improves a theorem by P. Erdős.
P. Erdős. *On consecutive integers*. Nieuw Arch. Wiskunde 3 (1955), 124–128.

7) For $k^{3/2} \leq u \leq k^{c_1(\log k)^{1/2}/\log\log k}$, we have

$$P \gg k^{1+c_2\Lambda(u,k)}$$

where $\Lambda(u, k) = ((\log k)/(\log u)^2)$ (c_1, c_2 positive constants.)
M. Jutila. *On numbers with a large prime factor*. II. J. Indian Math. Soc. (N.S.) 38 (1974), 125–130.

§ IV. 8 $P(n^2 + 1)$ and $P(n^4 + 1)$

1) a) $P(n^2 + 1) > n^{6/5}$

for infinitely many positive integers n

J.-M. Deshouillers and H. Iwaniec. *Kloosterman sums and Fourier coefficients of cusp forms.* Invent. Math. 70 (1982/83), 219–288.

b) For any $\varepsilon > 0$ there exist infinitely many integers n such that $n^2 + 1$ has a prime factor greater than $n^{\theta-\varepsilon}$, where $\theta = 1.202\ldots$ satisfies $2 - \theta - 2\log(2-\theta) = \frac{5}{4}$

J.-M. Deshouillers and H. Iwaniec. *On the greatest prime factor of* n^2+1. Ann Inst. Fourier (Grenoble) 32 (1982), 1–11.

Remark. The result b) can be generalized to

$$n^2 - D$$

where D is not a perfect square. The proof is based on C. Hooley's method.

C. Hooley. *On the greatest prime factor of a quadratic polynomial.* Acta Math. 117 (1967), 281–299.

2) $P(n^4 + 1) > 113$
for $n > 3$

M. Mureddu. A lower bound for $P(x^4 + 1)$. Ann. Fac. Sci. Toulouse Math. (5) 8 (1986/87), 109–119.

3) a) $P(x^3 + 1) \geq 31$
if $x > 69$

b) $P(x^3 + 2) \geq 11$
if $x > 2$

J. Buchmann, K. Győry, M. Mignotte and N. Tzanakis. *Lower bounds for* $P(x^3 + k)$, *an elementary approach.* Publ. Math. (Debrecen) 38 (1991), 145–163.

§ IV. 9 $P(a^n - b^n)$, $P(a^p - b^p)$

1) a) Let $a > b > 0$ be integers with $(a, b) = 1$. Let $f(n)$ be a strictly increasing and unbounded function of n depending only on a, b, k. Then

$$\frac{P(a^n - b^n)}{n} > f(n)$$

for all n with at most $k \log\log n$ prime factors, where

$$k < 1/\log 2$$

Corollary.

$$P(a^p - b^p) > \frac{1}{2} p(\log p)^{1/4}$$

and

$$P(a^{2p} - b^{2p}) > p(\log p)^{1/4}$$

for all primes $p > K$, where K is an effective constant, $K = K(a, b)$

C.L. Stewart. *The greatest prime factor of $a^n - b^n$*. Acta Arith. 26 (1974/75), 427–433.

Remark. The first result on $P(a^n - b^n)$ is due to Zsigmondy, who proved (on the above conditions) that

$$P(a^n - b^n) \geq n + 1$$

for $n > 2$

K. Zsygmondy. *Zur Theorie der Potenzreste*. Monatsh. Math. Phys. 3 (1892), 265–284.

b) $P(2^p - 1) \gg p \log p$
for all primes p, and

$P(2^p - 1) \gg p(\log p)^2 (\log\log p)^{-3}$
for almost all primes p

P. Erdős and T.N. Shorey. *On the greatest prime factor of $2^p - 1$ for a prime p and other expressions.* Acta Arith. 30 (1976), 257–265.

c) If $a > b \geq 1$ are integers, then

$$P(a^p - b^p) > c_1 p \log p$$

for all primes p and

$$P(a^p + b^p) > c_2 p \log p$$

(c_1, c_2 are positive numbers which are effectively computable in terms of $P(ab)$ only.)

T.N. Shorey and C.L. Stewart. *On divisors of Fermat, Fibonacci, Lucas and Lehmer sequences.* II J. London Math Soc. (2) 23 (1981), 17–23.

d) Given $\varepsilon > 0$, there exist positive constant n_0 and c depending only on ε such that for every $n \geq n_0$, the number of primes $n < p < 2n$ for which

$$\frac{P(2^p - 1)}{p} < c \cdot \frac{(\log p)^2}{(\log\log p)(\log\log\log p)}$$

is at most $\varepsilon n/\log n$. see

P. Erdős and T.N. Shorey.

§ IV.10 $P(u_n)$ for a recurrence sequence (u_n)

1) Let $r, s, u_0, u_1 \in \mathbb{I}$ with $r^2 + 4s \neq 0$; denote the two different roots of $x^2 - rx - s$ by α and β and define $a = (u_0\beta - u_1)/(\beta - \alpha)$, $b = (u_1 - u_0\alpha)/(\beta - \alpha)$. Assume that $\alpha\,\beta\,a\,b \neq 0$ and that α/β is not a root of unity.

Define the recurrence sequence $(u_n)_{n\geq 0}$ by $u_n = ru_{n-1} + su_{n-2}$ for $n \geq 2$. Then:

a) $P(u_n) > c(n/\log n)^{1/(d+1)}$
where $d = [Q(\alpha) : Q]$

b) If $\varepsilon : \mathbb{N} \to \mathbb{R}$ tends to 0 as $n \to \infty$, then for almost all n,

$$P(u_n) > \varepsilon(n) \cdot n \cdot \log n$$

C.L. Stewart. *On divisors of terms of linear recurrence sequences.* J. Reine Angew. Math. 333 (1982), 12–31.

2) The Lucas numbers are defined by

$$u_n = \frac{\alpha^n - \beta^n}{\alpha - \beta} \text{ (if } n \text{ is odd) and}$$

$$u_n = \frac{\alpha^n - \beta^n}{\alpha^2 - \beta^2} \text{ (if } n \text{ is even),}$$

where $(\alpha + \beta)^2$ and $\alpha\,\beta$ are relatively prime nonzero rational integers and α/β is not a root of unity.
If $|u_n| > \exp\exp(4C^3 \log C)$, where $C = e^{452} \cdot 4^{67}$, then

$$P(u_n) > \frac{1}{2}(\log\log|u_n|)^{1/3}$$

K. Győry. *On some arithmetical properties of Lucas and Lehmer numbers.* Acta Arith. 40 (1981/82), 369–373.

3) Let $u_m = ru_{m-1} + su_{m-2}$ with $u_0, u_1, r, s \in \mathbb{I}$. Let $\theta > 0$ be given. Then there exist positive constants c_1, c_2 such that for all n, k natural numbers with $n \geq c_1, \exp(e^e) < k \leq n^{1-\theta}$ we have

$$P(u_{n+1} \cdots u_{n+k}) \geq c_2 k \log k \cdot \log\log k (\log\log\log k)^{-1}$$

T.N. Shorey. *Applications of linear forms in logarithms to binary recursive sequences.* Seminar on Number Theory. Paris 1981–82, 287–301. Birkhäuser, 1983.

§ IV.11 Greatest prime factor of a product

Let (a_n) be a sequence of integers, $0 < a_1 < a_2 < \cdots$ and let $A \in \mathbb{N}^*$. Let P_x denote the greatest prime factor of $\prod_{n=1}^{x}(a_n^2 + A)$

1) If $\liminf_{n\to\infty} \frac{\log(a_1 \cdots a_n)}{a_n \log a_n} > \frac{1}{2}$ then

$$\lim_{n\to\infty} \frac{P_x}{x} = +\infty$$

2) If $\liminf\limits_{n\to\infty} \dfrac{\log(a_1 \cdots a_n)}{a_n \log n} > \dfrac{1}{2}$ then

$$\lim_{x\to\infty} \frac{P_x}{x} = +\infty$$

S. Knapowski. *On the greatest prime factors of certain products.* Ann. Polon. Math. 2 (1955), 56–63.

§ IV.12 $P(f(x))$, f a polynomial

1) Let $f(x)$ be any irreducible polynomial with degree > 1 and having integer coefficients.

a) $P(f(x)) \to \infty$
as $x \to \infty$

C.L. Siegel. *The integer solutions of equation* $y^2 = ax^n + bx^{n-1} + \cdots + k$. J. London Math. Soc. 1 (1926), 66–68.

b) $P(f(x)) > c \cdot \log\log x$
for all positive integers x, where $c = c(f) > 0$

S.V. Kotov. *The greatest prime factor of a polynomial* (Russian.) Mat. Zametki 13 (1973), 515–522.

Remark. The case of quadratic and cubic f was considered in

A. Schinzel. *On two theorems of Gelfond and some of their applications.* Acta Arith. 13 (1967), 177–236,

and

M. Keates. *On the greatest prime factor of a polynomial.* Proc. Edinb. Math. Soc. 16 (1969), 301–303.

2) Let $f(x)$ be as in 1) and denote $P_x = P\left(\prod\limits_{n\le x} f(n)\right)$. Then:

a) $P_x > c_1 x \log x$
where $c_1 > 0$ is a constant $(x \geq x_0(f))$

T. Nagell. *Généralization d'un théorème de Tchébycheff.* J. Math. Pure Appl. (8) 4 (1921), 343–356.

Note. The theorem of P.L. Tchébycheff (see E. Landau. *Handbuch.* I., Leipzig, 1909 (p. 559)) states that for $f(x) = x^2 + 1$, $P_x/x \to \infty$

b) $P_x > x(\log x)^{c_2 \log\log\log x}$
$x > x_0(f)$, where $c_2 > 0$

P. Erdős. *On the greatest prime factor of* $\prod_{k=1}^{x} f(k)$. J. London Math. Soc. 27 (1952), 379–384.

Remarks: For f = quadratic polynomial, one has:

(i) $P_x > x^{1+1/10}$

C. Hooley. *On the greatest prime factor of a quadratic polynomial.* Acta Math. 117 (1967), 2–16.

(ii) $P_x > x^{1.202}$

J.-M. Deshouillers and H. Iwaniec. *On the greatest prime factor of* n^2+1. Ann. Inst. Fourier (Grenoble) 32 (1982), 1–11.

(iii) $P_x > x \cdot \exp\exp\left(c_3(\log\log x)^{1/3}\right)$
$x > x_0(f)$, where $c_3 > 0$

P. Erdős and A. Schinzel. *On the greatest prime factors of* $\prod_{k=1}^{x} f(k)$. Acta Arith. 55 (1990), 191–200.

(iv) For any fixed $\alpha < 2 - \log 4 = 0.61370\ldots$

$$P_x > x\exp((\log x)^{\alpha})$$

$x \geq x_0(f, \alpha)$

G. Tenenbaum. *Sur une question d'Erdős et Schinzel.* II. Inventiones Math. 99 (1990), 215–224.

3) Let $f(x) \in \mathbb{I}[X]$ with at least two distinct roots and let $A > 0$. Then there exists an effective constant

$$\delta = \delta(A, f)$$

with the property:
If $P(f(n)) \leq \exp((\log\log n)^A)$ for $n > \exp(e^e)$, then

$$\omega(f(n)) \geq \delta \log\log n / \log\log\log n$$

where $\omega(n)$ denotes the number of distinct prime divisors of m.

T.N. Shorey and R. Tijdeman. *On the greatest prime factors of polynomials at integer points.* Compositio Math. 33 (1976), 187–195.

§ IV.13 Greatest prime factor of a quadratic polynomial

Let $D \in \mathbb{N}^*$, not a perfect square and let $P_x = P\left(\prod_{\sqrt{D}<n\leq x}(n-d)\right)$. Then

$$P_x > x^{11/10}$$

for all $x \geq x_0$

C. Hooley. *On the greatest prime factor of a quadratic polynomial.* Acta Math. 117 (1967), 281–299.

§ IV.14 $P(p+a)$, $p(p+a)$, p prime

1) Let P_x be the largest prime factor of $\prod_{0<p+a\le x}(p+a)$. Then

$$P_x > x^{\theta}$$

for all $x \ge x_0$, for every $\theta < \frac{5}{8}$

C. Hooley. *On the largest prime factor of $p + a$.* Mathematika 20 (1973), 135–143.

2) a) If $a \in \mathbb{I}^*$, then

$$P(p+a) > p^{21/32}$$

for infinitely many primes p

J.-M. Deshouillers and H. Iwaniec. *On the Brun-Titchmarsh theorem on averages.* Topics in classical number theory, vol. I, II (Budapest, 1981), 319–333, Colloq. Math. Soc. János Bolyai, 34, North Holland, 1984.

b) $P(p+a) < p^{0.35}$
($a \neq 0$) for infinitely many primes p

A. Balog. *$p + a$ without large prime factors.* Séminaire de Théorie des Nombres de Bordeaux 1983–84, Univ. Bordeaux, Talence, Exp. No. 31, 5 pp.

c) $P(p+a) < p^{0.303}$
($a \neq 0$) for a positive proportion of the primes.

J.B. Friedlander. *Shifted primes without large prime factors.* in: Number Theory and Applications (R.A. Mollin, ed.) Kluwer, pp. 393–401.

3) a) There are primes $q \le x$ such that

$$p(q+2) > x^{\theta}$$

for some $\theta > 2/7$, where $p(m)$ denotes the least prime factor of m

b) Let $\alpha > 3e^{0.3}/7$. Then there are primes $q \le x$ such that

$$P(q-1) \le x^{\alpha}$$

É. Fouvry and F. Grupp. *On the switching principle in sieve theory.* J. Reine Angew. Math. 370 (1986), 101–126.

§ IV.15 On $P(ax^m + by^n)$

1) a) The greatest prime factors of $ax^m + by^n$, $ab \neq 0$, $(x, y) = 1$, $m \ge 2$, $m \ge 3$, tends to infinity as $\max(|x|, |y|)$ tends to infinity.

K. Mahler. *On the greatest prime factor of $a\,x^m + b\,y^n$.* Nieuw Arch. Wisk. (3) 1 (1953), 113–122.

b) Let $n > 1$ be an integer and let $a, b \in \mathbb{I}^*$. Then for all $x, y, m \in \mathbb{I}^*$, $|x| > 1$, $(ax, by) = 1$, $m > e^e$, we have:

$$P(ax^m + by^n) \gg_{a,b,n} ((\log m)(\log\log m))^{1/2}$$

T.N. Shorey. *On the greatest prime factor of $a\,x^m + b\,y^n$*. Acta Arith. 36 (1980), 21–25.

2) Let $a, b \in \mathbb{I}, a \neq 0, b^2 - 4ac \neq 0$. Then the greatest prime factor of $ax^2 + bx + c$ tends to infinity with $x \in \mathbb{I}$

G. Pólya. *Zur arithmetischen Untersuchungen der Polynome*. Math. Z. 1 (1918), 143–148.

§ IV.16 Intervals containing numbers without large prime factors

1) The interval $(x, x + x^{1/2}]$ contains a number n such that:

a) $P(n) \geq n^{15/26}$
for all $x \geq x_0$

K. Ramachandra. *A note on numbers with a large prime factor*. J. London Math. Soc. 1 (1969), 303–306.

b) $P(n) \geq n^{5/8}$
for $x \geq x_0$

K. Ramachandra. *A note on numbers with a large prime factor*. II. J. Indian Math. Soc. 34 (1970), 39–48.

c) $P(n) \geq n^{0.66}$
for $x \geq x_0$

S.W. Graham. *The greatest prime factor of the integers in an interval*. J. London Math. Soc. (2) 24 (1981), 427–440.

d) $P(n) \geq n^{7/10}$
for $x \geq x_0$

R.C. Baker. *The greatest prime factor of the integers in an interval*. Acta Arith. 47 (1986), 193–231.

e) $P(n) \geq n^{0.71}$
for $x \geq x_0$

C. Jia. *The greatest prime factor of the integers in short intervals*, II. (Chinese) Acta Math. Sinica 32 (1989), 188–199.

f) $P(n) \geq n^{0.723}$
for $x \geq x_0$

H. Liu. *The greatest prime factor of the integers in an interval*. Acta Arith. 65 (1993), 301–328.

2) The interval $(x, x + x^{1/2+\varepsilon}]$ contains a number n with:

a) $P(n) \geq n^{\beta}$

for every $\beta < \dfrac{2}{3}$

M. Jutila. *On numbers with a large prime factor.* J. Indian Math. Soc. (N.S.) 37 (1973), 43–53.

Remark. For intervals of type $(x, x + x^{\alpha}]$ with a fixed number $\alpha < 1/2$, close to 1/2, see

K. Ramachandra, I, II, and M. Jutila. II. J. Indian Math. Soc. (N.S.) 38 (1974), 125–130.

b) $P(n) \geq n^{0.772}$

A. Balog. *Numbers with a large prime factor.* II. Coll. Math. Soc. János Bolyai 34. Topics in classical number theory (Budapest, 1981), North-Holland, 1984.

3) a) The interval $(x, x + x^{1/2} \cdot \log^{B} x]$, $x > x_0(B)$, ($B > 0$ absolute constant) always contains a number n with

$$P(n) > n^{2/3}$$

A. Balog, G. Harman and J. Pintz. *Numbers with a large prime factor.* III, Quart. J. Math. Oxford (2), 34 (1983), 133–140.

b) For any $\varepsilon > 0$ and $x \geq x_0(\varepsilon)$ the interval $(x, x + x^{1/4+\varepsilon}]$ contains an integer free of prime factors exceeding $\sqrt{x}$.

J.B. Friedlander and J.C. Lagarias. *On the distribution in short intervals of integers having no large prime factor.* J. Number Theory 25 (1987), 249–273.

c) For any fixed $\varepsilon > 0$ and $x \geq x_0(\varepsilon)$ the interval $(x, x + x^{1/2+\varepsilon}]$ contains an integer free of prime factors exceeding x^{ε}.

A. Balog. *On the distribution of integers having no large prime factor.* In: Journées Arithmétiques, Besançon 1985, Astérisque 147/148, pp. 27–31.

d) The bound x^{ε} in b) can be replaced by $\exp((\log x)^{2/3+\varepsilon})$

G. Harman. Short intervals containing numbers without large prime factors. Math. Proc. Cambridge Philos. Soc. 109 (1991), 1–5.

4) The interval $(x, x + x^{1/2})$ contains at least one number n with $p(n) > x^{1/(10-\varepsilon)}$, where $p(n)$ denotes the least prime factor of n

W.E. Mientka. *An application of the Selberg sieve method.* J. Indian Math. Soc. (N.S.) 25 (1961), 129–138.

§ IV.17 On $P(n)/P(n+1)$

1) Let $\varepsilon(x) > 0$ and set $y = x^{\varepsilon(x)}$. Let $P(n, x, y)$ denote the largest prime divisor of n that one $\leq y(x)$. Assume that $\varepsilon(x) \to 0$ as $x \to \infty$ and $y(x) \to \infty$. Then the density of those $n \leq x$ with $P(n, x, y) < P(n+1, x, y)$ equals 1/2

J. Galambos. *On a problem of Erdős on large prime divisors of n and n + 1*. J. London Math. Soc. (2) 13 (1976), 360–362.

2) a) For each $\varepsilon > 0$, there is a $\delta > 0$ such that for sufficiently large x, the number of $n \leq x$ with $x^{-\delta} < P(n)/P(n+1) < x^{\delta}$ is less than $\varepsilon \cdot x$

b) The lower density of integers n for which $P(n) > P(n+1)$ is at least 0.0 099. The same is true for integers n for which $P(n) < P(n+1)$

c) There are infinitely many n with $P(n) < P(n+1) < P(n+2)$

P. Erdős and C. Pomerance. *On the large prime factors of n and n + 1*. Aequationes Math. 17 (1978), 311–321.

§ IV.18 Consecutive prime divisors

If $p_1 < p_2 < \cdots < p_r$ are the consecutive prime factors of n, then for every $\varepsilon > 0, \eta > 0$ there is an $a = a(\varepsilon, \eta)$ so that the density of integers n for which for every $a < k \leq r$

$$k \cdot (1 - \varepsilon) < \log\log p_k < (1 + \varepsilon) \cdot k$$

is greater than $1 - \eta$

P. Erdős. *On the distribution function of additive functions*. Annals of Math. 47 (1946), 1–20

and

P. Erdős. *Some unconventional problems in number theory*. Soc. Math. France Astérisque 61 (1979), 73–82.

§ IV.19 Greatest prime factor of consecutive integers

Let $P(n, k) = \max\{P(n+i) : i = 0, 1, \ldots, k-1\}$. Then:

1) $P(n, k) \gg \dfrac{k \log k \cdot \log\log k}{\log\log\log k}$

for $n \geq k^{3/2}$

T.N. Shorey. *On gaps between numbers with a large prime factor*. II. Acta Arith. 25 (1973/74), 365–373.

2) For any $\varepsilon > 0$, all sufficiently large k and $n \geq n_0(\varepsilon, k)$, one has

$$P(n, k) \geq (1 - \varepsilon)k \log\log n$$

M. Langevin. *Plus grande facteur premier d'entiers voisins*. C.R. Acad. Sci. Paris Sér. A–B281, A491–A493.

Note. See also

J. Turk. *Prime divisors of polynomials at consecutive integers*. J. Reine Angew. Math. 319 (1980), 142–152.

§ IV.20 Frequency of numbers containing prime factors of a certain relative magnitude

1) Let $P(x,\theta) = \text{card}\{n \le x : P(n) > x^\theta\}$, where $0 < \theta < 1$. Then:

$$\lim_{x\to\infty} \frac{P(x,\theta)}{x} = A(\theta)$$

where $A(\theta)$ is a strictly decreasing continuous function in the interval $0 < \theta < 1$, with $A(1-0) = 0$ and $A(+0) = 1$

K. Dickman. *On the frequency of numbers containing prime factors of a certain relative magnitude.* Ark. Math. Astr. Fys. 22 (1930), no. A10.

Remark. The theorem has been rediscovered e.g. in

S. Chowla and T. Vijayaraghavan. On the largest prime divisor of numbers. J. Indian Math. Soc. (N.S.) 11 (1947), 31–37.

§ IV.21 Integers without large prime factors. The function $\Psi(x, y)$ and Dickman's function

1) a) For any fixed $u > 0$

$$\lim_{y\to\infty} \Psi(y^u, y)\cdot y^{-u} = \rho(u)$$

where $\rho(u)$ is defined as the (unique) continuous solution to the differential difference equation

$$\begin{cases} u\rho'(u) = -\rho(u-1), & (u > 1) \\ \rho(u) = 1, & (0 \le u \le 1) \end{cases}$$

(Dickman's function.)

K. Dickman. *On the frequency of numbers containing prime factors of a certain relative magnitude.* Ark. Math. Astr. Fys. 22 (1930), 1–14.

b) Let $c > 0$. Then

$$\psi(x, x^c) = x\cdot f(c) + O(x/\log x)$$

where $f(c) > 0$ and the O-term is uniform for $c \ge \delta > 0$. Moreover, $f(c)$ is a continuous function of c for all $c \in \mathbb{R}$ and increasing for $0 < c < 1$

V. Ramaswami. *On the number of positive integers less than x and free of prime factors greater than x^c.* Bull. A.M.S. 55 (1949), 1122–1127.

Remark. By substituting $c = 1/u$, $y = x^{1/c}$, we get

$$\psi(y^u, y) = y^u\cdot f(1/u) + O(y^u/\log y)$$

$u > 0$

2) a) For $x \geq 1$, $y \geq 2$ we have

$$\psi(x, y) \ll x \cdot e^{-u/2} \cdot \log y$$

where $u = \log x / \log y$

R.A. Rankin. *The difference between consecutive prime numbers*. J. London Math. Soc. 13 (1938), 242–247.

b) With the same conditions

$$\psi(x, y) \ll x \cdot e^{-u/2}$$

G. Tenenbaum. *Introduction à la théorie analytique et probabiliste des nombres*. Publ. Inst. Elie Cartan Vol. 13, Nancy (1990), (Th. III. 5.1.)

c) $\psi(x, y) = x \cdot \rho(u)\,(1 + O(\log(u+1)/\log y))$
uniformly in the range $y \geq 2$, $1 \leq u \leq (\log y)^{3/5-\varepsilon}$

N.G. de Bruijn. *On the number of positive integers $\leq x$ and free of prime factors $> y$*. Nederl. Akad. Wetensch. Proc. Ser. A54, 50–60.

d) For any fixed $\varepsilon > 0$ the relation in c) holds uniformly in the range $y \geq 2$, $1 \leq u \leq \exp((\log y)^{3/5-\varepsilon})$

A. Hildebrand. *On the number of positive integers $\leq x$ and free prime factors $> y$*. J. Number Theory 22 (1986), 289–307.

Remark. It can be shown that

$$\psi(x, y) = x\rho(u)\exp(O(\log(u+1)/\log y))$$

holds uniformly in the range $y \geq 2$, $1 \leq u \leq y^{1/2-\varepsilon}$, for any fixed $\varepsilon > 0$, if and only if the Riemann Hypothesis is true.

A. Hildebrand. *Integers free of large prime factors and the Riemann Hypothesis*. Mathematika 31 (1984), 258–271.

3) a) For $2 < y \leq x$ we have

$$\psi(x, y) \geq \binom{\pi(y) + [u]}{[u]}$$

where $u = \log x / \log y$

b) For $2 < y \leq x$

$$\log \psi(x, y) \sim \log \binom{\pi(y) + [u]}{[u]}$$

uniformly in y, as $x \to \infty$

N.G. de Bruijn. *On the number of positive integers $\leq x$ and free of prime factors $> y$. II*. Proc. Kon. Ned. Akad. v Wetensch. 69 (1966), 335–348.

Remark. For a short proof of a) and b) see

P. Erdős and J.H. van Lint. *On the number of positive integers $\leq x$ and free of prime factors $> y$*. Simon Stevin 40 (1956), 73–76.

c) For any fixed $\varepsilon > 0$ we have

$$\log(\psi(x,y)/x) = (1 + O(\exp(-(\log u)^{3/5-\varepsilon})))\log\rho(u)$$

uniformly in the range $y \geq 2$, $1 \leq u \leq y^{1-\varepsilon}$

Moreover, the lower bound in the above formula is valid uniformly for all $x \geq y \geq 2$

For the upper bound, see

N.G. de Bruijn (1966.)

For the lower bound, see

A. Hildebrand (1986) from 1) d).

Corollary. For any fixed $\varepsilon > 0$ we have

$$\psi(x,y) = x \cdot u^{-(1+o(1))u}$$

as $y, u \to \infty$, uniformly in the range $u \leq y^{1-\varepsilon}$

d) Uniformly for $x \geq y \geq 2$ we have

$$\log\psi(x,y) = Z\left(1 + O\left(\frac{1}{\log y}\right) + O\left(\frac{1}{\log\log x}\right)\right)$$

where

$$Z = Z(x,y) = \frac{\log x}{\log y}\log\left(1 + \frac{y}{\log x}\right) + \frac{y}{\log y}\log\left(1 + \frac{\log x}{y}\right)$$

See N.G. de Bruijn (1966) and G. Tenenbaum (1990) from 2) b).

4) a) Uniformly in the range $2 \leq y \leq (\log x)^{1/2}$ we have

$$\psi(x,y) = \frac{1}{\pi(y)!}\prod_{p\leq y}\frac{\log x}{\log y}\left(1 + O\left(\frac{y^2}{(\log x)(\log y)}\right)\right)$$

V. Ennola. *On numbers with small prime divisors.* Ann. Acad. Sci. Fenn. Ser. A I 440, 16 pp.

b) Uniformly in the range $x \geq y \geq 2$ we have (as u and y tend to infinity)

$$\psi(x,y) = \frac{x^{\alpha}\zeta(\alpha,y)}{\alpha\sqrt{2\pi\phi_2(\alpha,y)}}\left(1 + O\left(\frac{1}{u}\right) + O\left(\frac{\log y}{y}\right)\right)$$

where $\zeta(s,y) = \prod_{p\leq y}(1-p^{-s})^{-1}$; $\phi_k(s,y) = \frac{d^k}{ds^k}\phi(x,y)$ $(k \geq 1)$, and α is the unique positive solution to the equation

$$-\phi_1(\alpha,y) = \sum_{p\leq y}\frac{\log p}{p^{\alpha}-1} = \log x$$

A. Hildebrand and G. Tenenbaum. *On integers free of large prime factors.* Trans. Amer. Math. Soc. 296 (1986), 265–290.

Remark. It can be shown that

$$\alpha = \alpha(x,y) = \frac{\log(1 + y/\log x)}{\log y}\left(1 + O\left(\frac{\log\log(1+y)}{\log y}\right)\right)$$

uniformly in $x \geq y \geq 2$. (See A. Hildebrand and G. Tenenbaum.)

Corollary. Uniformly for $x \geq y \geq 2$ and $1 \leq c \leq y$ we have

$$\psi(cx, y) = \psi(x, y)c^{\alpha(x,y)}\left(1 + O\left(\frac{1}{u}\right) + O\left(\frac{\log y}{y}\right)\right)$$

c) For any fixed $\varepsilon > 0$ and uniformly for $y \geq y_0(\varepsilon)$ and $1 \leq u \leq \exp((\log y)^{3/5-\varepsilon})$ we have

$$\psi(x, y) = \Lambda(x, y)(1 + O(\exp((\log y)^{3/5-\varepsilon})))$$

where

$$\Lambda(x, y) = \begin{cases} x\int_{-\infty}^{+\infty} \rho(u - v)\, d([y^v] \cdot y^{-v}) \ , & (x \notin \mathbb{N}) \\ \Lambda(x + 0, y) & , (x \in \mathbb{N}) \end{cases}$$

where $\rho(u)$ is defined in 1) a) for $u \geq 0$ and $\rho(u) = 0$ for $u < 0$

E. Saias. *Sur le nombre des entiers sans grand facteur premier.* J. Number Theory 32 (1989), 78–99.

5) a) For any fixed $\varepsilon > 0$ and uniformly in the range $y \geq 2$, $1 \leq u \leq \exp((\log y)^{3/5-\varepsilon})$ and for $xy^{-5/12} \leq z \leq x$ we have

$$\psi(x + z, y) - \psi(x, y) = z\rho(u)\left(1 + O\left(\frac{\log(u+1)}{\log y}\right)\right)$$

(See A. Hildebrand (1986), 2) d).)

b) For any fixed $\varepsilon > 0$ and uniformly for $x \geq y \geq 2$ and $1 \leq z \leq x$ we have

$$\psi(x + z, y) - \psi(x, y) = \frac{z\alpha}{x}\psi(x, y)\left(1 + O\left(\frac{z}{x} + \frac{1}{u} + \frac{\log y}{y}\right)\right) + O(\psi(x, y) \cdot R_\varepsilon)$$

where $\alpha = \alpha(x, y)$ is defined in 4) b), and

$$R_\varepsilon = R_\varepsilon(x, y) = \exp(-(\log y)^{3/2-\varepsilon}) + (\log y)\exp(-c \cdot u(\log(u + 2))^{-2})$$

with a suitable positive constant c. (See A. Hildebrand and G. Tenenbaum (1986), 4) b).)

6) a) There exists a positive constant c such that for any fixed $\alpha \in (0, 1)$ and $\beta > 1 - \alpha - c\alpha(1 - \alpha)$ and for all sufficiently large x,

$$\psi(x + x^\beta, x^\alpha) - \psi(x, x^\alpha) \gg_{\alpha,\beta} x^\beta$$

J.B. Friedlander and J.C. Lagarias. *On the distribution in short intervals of integers having no large prime factor.* J. Number Theory 25 (1987), 249–273.

b) For any fixed $\varepsilon > 0, 0 < \beta \leq \alpha \leq 1$ and for all sufficiently large X the estimate

$$\psi(x + x^\beta, x^\alpha) - \psi(x, x^\alpha) \gg \beta\, \rho(1/\alpha)x^\beta$$

holds for all $x \in [1, X]$ with the exception of a set of measure $\ll X \cdot \exp(-(\log X)^{1/3-\varepsilon})$

c) For any fixed $\varepsilon > 0$, for all sufficiently large X and for y and z satisfying

$$\begin{aligned} \exp((\log X)^{5/6+\varepsilon}) &\le y \le X \\ y\exp((\log X)^{1/6+\varepsilon}) &\le z \le X \end{aligned} \qquad (*)$$

the estimate

$$\psi(x+z, y) - \psi(x, y) \gg \rho\left(\frac{\log X}{\log y}\right)\cdot z$$

holds for all $x \in [1, X]$ with the exception of a set of measure $\ll X\exp\left(-\frac{1}{2}(\log X)^{1/6}\right)$. (See J.B. Friedlander and J.C. Lagarias.)

Remark. Under R.H. the conclusion of the above result holds for

$$L(X) \le y \le x$$

and

$$\sqrt{L(X)} \le z \le X$$

where $L(x) = \exp(\sqrt{\log x \cdot \log\log x})$

J.L. Hafner. *On smooth numbers in short intervals under the Riemann Hypothesis.* preprint, 1991.

d) For any fixed $\varepsilon > 0$, for all sufficiently large X and y and z satisfying (*) from c), the estimate

$$\psi(x+z, y) - \psi(x, y) = z\,\rho(u)\cdot\left(1 + O\left(\frac{\log(u+1)}{\log y}\right)\right)$$

holds for all $x \in [1, X]$ with the exception of a set of measure $\ll X\exp(-(\log X)^{1/6-\varepsilon})$

A. Hildebrand and G. Tenenbaum. *Integers without large prime factors.* J. Théor. Nombres Bordeaux 5 (1993), 411–484.

Remark. For asymptotic results on the Dickman function (called also as the Dickman-de Bruijn function), see

T. Xuan. *On the asymptotic behavior of the Dickman-de Bruijn function.* Math. Ann. 297 (1993), 519–533.

e) Let $x \ge y \ge \exp((\log x)^{5/6} + \varepsilon)$ and $x \ge z \ge x^{1/2}\cdot y^2\cdot\exp((\log x)^{1/6})$. Then for $\varepsilon > 0$ and $x \to \infty$ we have

$$\psi(x+z, y) - \psi(x, y) \sim \frac{z}{x}\cdot\psi(x, y)$$

J.B. Friedlander and A. Granville. *Smoothing "smooth" numbers.* Philos. Trans. R. Soc. London Ser. A345, No. 1676 (1993), 339–347.

§ IV.22 Function $\psi(x, y; a, q)$. Integers without large prime factors in arithmetic progressions

1) For positive integers a and q define $\psi(x, y; a, q)$ as the number of positive integers $\leq x$, free of prime factors $> y$, and satisfying $n \equiv a(\bmod q)$

a) For fixed $u = (\log x)/(\log y)$ and positive integers q and $a, (a, q) = 1$, one has

$$\psi(x, y; a, q) = \frac{1}{q}\rho(u)x \cdot (1 + O((\log x)^{-1/2}))$$

A.A. Buchstab. *On those numbers in an arithmetic progression all prime factors of which are small in magnitude* (Russian.) Dokl. Akad. Nauk SSSR, 67 (1949), 5–8.

b) The asymptotic relation from a) remains valid in the range

$$u \ll (\log\log x)^{1-\varepsilon}$$

V. Ramaswami. *Number of integers in an assigned $a \cdot p$, $\leq x$ and prime to primes greater than x^c*. Proc. Amer. Math. Soc. 2 (1951), 318–319.

c) For any fixed positive integer k and positive real numbers ε and A, and uniformly in the range $y \geq y_0(\varepsilon)$,

$k + 1 + \varepsilon \leq u \leq (\log y)^{3/5-\varepsilon}$, $q \leq (\log x)^A$, $(a, q) = 1$ the estimate

$$\psi(x, y; a, q) = x\left(\frac{\rho(u)}{q} + \frac{1}{\varphi(q)}\sum_{i=1}^{k} a_i(q)\frac{\rho^{(i)}u}{(\log y)^i} + O_{k,\varepsilon,A}\left(\frac{u\rho^{(k)}(u)}{(\log y)^{k+1}}\right)\right)$$

holds, where $a_i(q)$ is the ith Taylor coefficient at the origin of the function

$$s\zeta(s+1)\cdot(s+1)^{-1}\prod_{p|q}(1-p^{-s-1})$$

B.V. Levin and A.S. Faĭnleĭb. *Application of some integral equations to problems in number theory*. Russian Math. Surveys 22 (1967), 119–204.

Remark. Let $(a, q) = 1$, for q fixed. Then $\Psi(x, y; a, q) > 0$ for all but $o(q)$ residue classes $a(\bmod q)$, provided $q = o(x)$ and $q \leq y^{2-\varepsilon}$

J.B. Friedlander. *Integers without large prime factor*. III. Arch. Math. 43 (1984), 32–36.

d) Let A be a fixed positive number. Then, with a suitable constant $c = c(A) > 0$, the estimate

$$\psi(x, y; a, q) = \frac{1}{\varphi(q)}\Lambda_q(x, y)\,(1 + O(\exp(-c\sqrt{\log y})))$$

holds uniformly in the range $x \geq 3, 1 \leq u \leq \exp(c\sqrt{\log y})$, $q \leq (\log x)^A$, $(a, q) = 1$. Here

$$\Lambda_q(x,y) = \begin{cases} x\displaystyle\int_{-\infty}^{+\infty} \rho(u-v)\,dR_q(v) & ,(x \notin \mathbb{N}) \\ \Lambda_q(x+0,y) & ,(x \notin \mathbb{N}) \end{cases}$$

where $R_q(x) = \text{card}\{n \le x : (n,q) = 1\} - x \cdot \dfrac{\varphi(q)}{q}$

É. Fouvry and G. Tenenbaum. *Entiers sans grand facteur premier en progressions arithmétiques*. Proc. London Math. Soc., III Ser. 63 (1991), 449–494.

e) Let $\psi_q(x,y) = \text{card}\{n \le x\colon P(n) \le y, (n,q) = 1\}$. Then for any fixed $\varepsilon > 0$ and uniformly for $x \ge y \ge 2, 1 \le q \le y^{1-\varepsilon}$ and $(a,q) = 1$, we have

$$\psi(x,y;a,q) = \frac{1}{\varphi(q)}\psi_q(x,y)\left(1 + O\left(\frac{\log q}{u^c \log y} + \frac{1}{\log y}\right)\right)$$

where c is positive constant.

A. Granville. *Integers, without large prime factors, in arithmetic progressions*. II. Philos. Trans. R. Soc. London, Ser. A345, No. 1676 (1993), 349–362.

f) For any fixed $\varepsilon > 0$ and uniformly in the range $y \ge 2, q \le y^{4/3-\varepsilon}, (a,q) = 1, x \ge \max(y^{3/2+\varepsilon}, y \cdot q^{3/4+\varepsilon})$ we have

$$\psi(x,y;a,q) \asymp \frac{1}{\varphi(q)}\psi_q(x,y)$$

A. Granville. *Integers without large prime factors in arithmetical progressions*. I. Acta Math. 170 (1993), 255–273.

Remark. The proof uses the estimate

$$\psi(x,y;a,q) = \frac{x}{q}\exp\{-u(\log u + \log\log u) + O(1)\}$$

with $u = \log x/\log y$ uniformly in the range $x \ge 2$,
$\exp\{(\log\log x)^2\} \le y \le x^{2/3-\varepsilon}$
$1 \le q \le \min\{y^{4/3-\varepsilon}, (x/y)^{4/3-\varepsilon}\}, (a,q) = 1$, for any given $\varepsilon > 0$, due to A. Balog and C. Pomerance.

A. Balog and C. Pomerance. *The distribution of smooth numbers in arithmetic progressions*. Proc. Amer. Math. Soc. 115 (1992), 33–43.

2) a) Let A be a given positive number. Then there is a constant $B = B(A)$ such that uniformly for $x \ge y \ge 2$ and $Q = \sqrt{x}(\log x)^{-B}$ we have

$$\sum_{q \le Q} \max_{z \le x} \max_{(a,q)=1} \left|\psi(z,y;a,q) - \frac{\psi_q(z,y)}{\varphi(q)}\right| \ll \frac{x}{(\log y)^A}$$

where $\psi(q)$ is defined in 1) e).

É. Fouvry and G. Tenenbaum. *Entiers sans grand facteur premier en progressions arithmétiques*. Proc. London Math. Soc., III Ser. 63 (1991), 449–494.

b) Let A be a fixed positive integer.

Then there exist positive constants $B = B(A)$ and $C = C(A)$ such that uniformly for $y \geq 100$ and

$1 \leq Q \leq \min\left(\exp\left(\dfrac{C \log y \cdot \log\log y}{\log\log\log y}\right), \sqrt{x}(\log x)^{-B}\right)$ we have

$$\sum_{q \leq Q} \max_{z \leq x} \max_{(a,q)=1} \left| \Psi(z, y; a, q) - \frac{\Psi_q(z, y)}{\varphi(q)} \right| \ll_A \frac{\Psi(x, y)}{(\log y)^A}$$

(See A. Granville. 1) f).)

§ IV.23 On $(n, \beta(n)) = 1$

Let $\beta(n) = \sum_{p|n} p$ and denote $T(x) = \text{card}\{n \leq x : (n, \beta(n)) = 1\}$. Then:

1) $T(x) = \dfrac{6}{\pi^2} \cdot x + O\left(x/(\log_3 x)^{1/4}(\log_4 x)^{3/4}\right)$

where $\log_n x$ denotes the iterated logarithm.

R.R. Hall. *On the probability that n and $f(n)$ are relatively prime*. Acta Arith. 17 (1970), 169–183.

2) $T(x) - \dfrac{6}{\pi^2} \cdot x \sim cx/(\log x)^{1/2}$

($c > 0$, constant)

R.R. Hall. *On the probability that n and $f(n)$ are relatively prime*. III. Acta Arith. 20 (1972), 267–289.

§ IV.24 Sums over $\beta_k(n)$, $B_k(n)$, $B(n) - \beta(n)$, $\dfrac{B(n)}{\beta(n)}$, $\dfrac{B(n) - \beta(n)}{P(n)}$

1) If $n = p_1^{a_1} \cdots p_r^{a_r}$, define $B_k(n) = \sum_{i=1}^{r} a_i p_i^k$ and $\beta_k(n) = \sum_{i=1}^{r} p_i^k$. (Clearly, $B_1(n) = B(n)$ and $\beta_1(n) = \beta(n)$)

a) $$\sum_{m=1}^{n} \beta_k(m) \sim \sum_{m=1}^{n} B_k(m) \sim \begin{cases} \dfrac{\zeta(k+1)}{k+1} \dfrac{n^{k+1}}{\log n}, & \text{if } k > 0 \\[2ex] n \displaystyle\sum_{p=2}^{\infty} p^k/(p-1), & \text{if } k < 0 \end{cases}$$

S.M. Kerawala. *A note on the orders of two arithmetic functions $F(n, k)$ and $F^*(n, k)$*. J. Natur. Sci. and Math. 10 (1970), 105–107.

2) a) $$\sum_{2\le n\le x} \frac{B^r(n)}{\beta^r(n)} = x + O\left(x \exp\left(-c\left(\log x \cdot \log\log x\right)^{1/2}\right)\right)$$

where $c > 0$ ($r > 0$, fixed number.)

Remark. The same formula holds for $\sum_{2\le n\le x} \frac{\beta^r(n)}{B^r(n)}$

b) $$\sum{}' \frac{1}{B(n) - \beta(n)} = Ax + O(x^{1/2}\log x)$$

where the sum is over $n \le x$ such that $B(n) \ne \beta(n)$

J.-M. de Koninck, P. Erdős and A. Ivić. *Reciprocal of certain large additive functions.* Canad. Math. Bull. 24 (1981), 225–231.

c) $$\sum_{n\le x} (B(n) - \beta(n)) = x \log\log x + O(x)$$

K. Alladi and P. Erdős. *On an additive arithmetic function.* Pacific J. Math. 71 (1977), 275–294.

d) $$\sum_{2\le n\le x} \frac{1}{\beta(n)} = (D + O((\log\log\log x)^2/\log\log x)) \cdot \sum_{2\le n\le x} \frac{1}{P(n)}$$

where $1/2 < D < 1$ is an absolute constant.

T. Xuan. *On sums involving reciprocals of certain large additive functions.* Publ. Inst. Math. (N.S.) 45 (59), (1989), 41–55.

e) $$\sum_{n\le x} \beta(n) = \frac{\pi^2}{12}\frac{x^2}{\log x} + O\left(\frac{x^2}{\log^2 x}\right)$$

Remarks. (i) The same result is valid replacing $\beta(n)$ by $B(n)$ (See K. Alladi and P. Erdős.)

(ii) In 1963 M. Kalecki proved that

$$\sum_{2\le n\le N} \beta(n) = \frac{\pi^2 + o(1)}{12}\frac{N^2}{\log n}$$

M. Kalecki. *On certain sums extended over primes or prime factors* (Polish.) Prace Mat. 8 (1963/64), 121–129.

f) $$\sum_{2\le n\le x} \frac{B(n)}{\beta(n)} = x + \frac{1}{2} D \cdot \log x \cdot \left(1 + O\left(\frac{(\log\log\log x)^2}{\log\log x}\right)\right) \cdot \sum_{2\le n\le x} \frac{1}{P(n)}$$

where $1/2 < D < 1$ is an absolute constant.

g) $$\sum_{2\le n\le x} \frac{B(n) - \beta(n)}{P(n)} = \frac{1}{2}\log x \cdot \left(1 + O\left(\frac{\log\log\log x}{\log\log x}\right)\right) \cdot \sum_{2\le n\le x} \frac{1}{P(n)}$$

T. Xuan. *On sums involving reciprocals of certain large additive functions.* II. Publ. Inst. Math. (N.S.) 46 (60) (1989), 35–32.

§ IV.25 Sums over $\frac{\beta(n)}{P(n)}$, $\frac{P(n)}{\beta(n)}$, $B(n) - P_1(n) - \cdots - P_{n-1}(n)$

1) a) $$\sum_{2 \le n \le x} \frac{\beta(n)}{P(n)} = x + O(x \log\log x / \log x)$$

Remark. The same formula is valid for $\sum_{2 \le n \le x} \frac{B(n)}{P(n)}$

K. Alladi and P. Erdős. *On the asymptotic behaviour of large prime factors of integers*. Pacific J. Math. 82 (1979), 295–315.

b) $$\sum_{2 \le n \le x} \frac{P(n)}{\beta(n)} = x + O(x \log\log x / \log x)$$

Remark. The same holds for $\sum_{2 \le n \le x} \frac{P(n)}{B(n)}$

P. Erdős and A. Ivić. *Estimates for sums involving the largest prime factor of an integer and certain related additive functions*. Studia Sci. Math. 15 (1980), 183–199.

2) Let $P_k(n)$ be the kth largest prime factor of n. Then:

a) $$\sum_{n \le x} (B(n) - P_1(n) - \cdots - P_{m-1}(n)) \sim \sum_{n \le x} P_m(n) \sim \frac{k_m \cdot x^{1+1/m}}{(\log x)^m}$$

($m \ge 1$, integer), where $k_m > 0$ is a constant depending only on m, and is a rational multiple of $\zeta(1 + 1/m)$

Remark. The same is valid when $B(n)$ is replaced by $\beta(n)$

K. Alladi and P. Erdős. *On an additive arithmetic function*. Pacific J. Math. 71 (1977), 275–294.

Note. For other asymptotic relations on $\sum_{n \le x} P_m(n)$ and $\sum_{n \le x} \frac{P_m(n)}{P(n)}$ see

J.-M. de Koninck and A. Ivić. *Sommes de réciproques de grandes fonctions additives*. Publ. Inst. Math. (Belgrade) (N.S.) 35 (1984), 41–48.

and

J.-M. de Koninck and A. Ivić. *The distribution of the average prime divisor of an integer*. Arch. Math. 43 (1984), 37–43.

b) $$\sum_{n \le x} P_m(n) = k_m \frac{x^{1+1/m}}{(\log x)^m} O(x^{1+1/m} / \log^N x)$$

for any fixed N

A. Ivić. *On the k-th prime factor of an integer*. Zb. Rad. Pric.-Mat. Fak. Univ. u Novom Sadu, Ser. Mat. 20 (1990), 67–73.

Remark. For similar results involving $P(n)$ and slowly oscillating functions, see
J.-M. de Koninck and A. Ivić. *On the average prime factor of an integer and some related problems.* Ric. Mat. 39 (1990), 131–140.

c) $$\sum_{\substack{n\le x\\ \Omega(n)\ge k}} \frac{1}{P_k(n)} = \lambda_k \cdot \frac{x(\log\log x)^{k-2}}{\log x} \cdot \left(1 + O\left(\frac{1}{\log\log x}\right)\right)$$

for $k \ge 3$; and

$$\sum_{\substack{n\le x\\ \Omega(n)\ge 2}} \frac{1}{P_2(n)} = x\sum_{j=1}^{A} \frac{\lambda_2^{(j)}}{\log^j x} + O\left(\frac{x}{\log^{A+1} x}\right)$$

where A ≥ 1 is a fixed integer, $\lambda_2^{(j)} (j = 1, \ldots, A)$ are constants and $\lambda_k = \lambda_2^{(1)}/(k-2)!$

J.-M. de Koninck. *Sur les plus grands facteurs premiers d'un entier.* Monatsh. Math. 116 (1993), 13–37.

§ IV.26 Distribution of $\frac{B(n)}{\beta(n)}$

1) Let $0 < \varepsilon < 1$ be fixed. Then for any fixed $C > 0$ the number of integers $2 \le n \le x$ such that

$$(1 - (\log\log x)^{\varepsilon-1}) \cdot \frac{\Omega(n)}{\omega(n)} \le \frac{B(n)}{\beta(n)} \le 1 + \exp\left(-\frac{1}{2}(\log x \cdot \log\log x)^{1/2}\right) \cdot \frac{\Omega(n)}{\omega(n)}$$

holds is

$$x + O(x(\log\log x)^{-c})$$

A. Ivić. *The distribution of quotients of small and large additive functions.* Boll. Un. Mat. Ital. B(7) 2 (1988), 79–97.

2) For $\varepsilon > 0$ and $x \ge 4$,

$x^{1-\varepsilon} \ll_\varepsilon A(x) - Q(x) \ll x(\log x)^2 \cdot \exp(-(1/5)(2\log x \cdot \log\log x)^{1/2})$

where $Q(x)$ is the number of squarefree n with $1 < n \le x$ and $A(x)$ is the number of n with $1 < n \le x$ and $B(n)/\beta(n) = \Omega(n)/\omega(n)$

R. Warlimont. *On a problem of A. Ivić about the quotients of certain additive arithmetic functions.* Arch. Math. 54 (1990), 376–379.

§ IV.27 On $(-1)^{B(n)}$

Let $\alpha(n) = (-1)^{B(n)}$. Then:

a) $\sum_{n\leq x} \alpha(n) = O(x\exp(-c(\log\log\log x)^{1/2}))$

$c > 0$

b) $\sum_{n=1}^{\infty} \frac{\alpha(n)}{n} = 0$

K. Alladi and P. Erdős. *On an additive arithmetic function*. Pacific J. Math. 71 (1977), 275–294.

§ IV.28 Sums over $B^1(n)$, $P(n)/B^1(n)$, $B^1(n)/B(n)$, $1/B^1(n)$, etc.

1) Denote $B^1(n) = \sum_{p^\alpha \| n} p^\alpha$. Then:

a) $\sum_{n\leq x} B^1(n) = \frac{\pi^2}{12}\frac{x^2}{\log x} + O(x^2/\log^2 x)$

T.Z. Xuan. *On some sums of large additive number theoretic functions* (Chinese.) Beijing Shifan Daxue Xuebao 1984, No. 2, 11–18.

b) $\sum_{2\leq n\leq x} \frac{P(n)}{B^1(n)} = x + O(x\log\log x/\log x)$

Remark. The same result is true when $P(n)/B^1(n)$ is replaced by

$$\beta(n)/B^1(n)$$

and

$$B(n)/B^1(n)$$

c) $\sum_{2\leq n\leq x} \frac{B^1(n)}{P(n)} = e^\gamma \cdot x\log\log x + O(x)$

Remark. The same holds for

$$\sum_{2\leq n\leq x} \frac{B^1(n)}{\beta(n)}$$

d) $\sum_{2\leq n\leq x} \frac{B^1(n)}{B(n)} = C\cdot x + O(x\log^{-1/3} x)$

where

$$C = \int_1^\infty \frac{1}{u} \sum_{0\leq s\leq u-1} \frac{\rho(u-[u]+s)}{[u]-s} du > 1$$

where $\rho(u)$ is defined by $\rho(u) = 1$ for $0 \leq u \leq 1$, $u\rho'(u) = -\rho(u-1)$ for $u \geq 1$

P. Erdős and A. Ivić. *Estimates for sums involving the largest prime factor of an integer and certain related additive functions*. Studia Sci. Math. Hungar. 15 (1980), 183–199.

e) $$\sum_{2\le n\le x} \frac{B^1(n)}{\beta(n)} \ge (1+o(1))\cdot x\log\log x$$

J.-M. de Koninck and A. Ivić. *Topics in arithmetical functions*. Notas de Matemática (72). North-Holland, 1980, (p. 173.)

2) a) $$\sum_{2\le n\le x} \frac{1}{B^1(n)} = x\exp\left(-(2\log x\log\log x)^{1/2} - \left(\frac{1}{2}\frac{\log x}{\log\log x}\right)^{1/2}\cdot\log\log\log x + O\left(\left(\frac{\log x}{\log\log x}\right)^{1/2}\right)\right)$$

b) $$\sum_{2\le n\le x}\left(\frac{1}{\beta(n)} - \frac{1}{B^1(n)}\right) = x\exp\left(-(3\log x\cdot\log\log x)^{1/2} - \left(\frac{3}{4}\frac{\log x}{\log\log x}\right)^{1/2}\cdot\log\log\log x + O\left(\left(\frac{\log x}{\log\log x}\right)^{1/2}\right)\right)$$

T.Z. Xuan. *On the sums of reciprocals of a class of additive number-theoretic functions* (Chinese.) J. Math. (Wuhan) 5 (1985), 33–40.

c) Let $0 < r < 1$. Then

$$\sum_{2\le n\le x} (B^1(n))^r/(P(n))^r = x + O(x/\log x)$$

Remark. $P(n)$ can be replaced with $\beta(n)$, $B(n)$ (and the same formula holds true.)

d) $$\sum_{2\le n\le x} \frac{(P(n))^r}{(B^1(n))^r} = x + O(x/\log x)$$

for $r > 0$

Remark. $P(n)$ can be replaced with $\beta(n)$ and $B(n)$

T.Z. Xuan. *On sums of powers of quotients of certain number–theoretic functions*. Beijing Shifan Daxue Xuebao 1986, 1–10.

§ IV.29 Numbers n with the property $B(n) = B(n+1)$

a) For every $\varepsilon > 0$, the number of $n \le x$ for which

$$B(n) = B(n+1)$$

is

$$O(x/(\log x)^{1-\varepsilon})$$

b) For every $\varepsilon > 0$, there is a $\delta > 0$ such that for sufficiently large x there are at least $(1-\varepsilon)x$ choices for $n \leq x$ such that

$$P(n) < B(n) < (1 + x^{-\delta})P(n)$$

P. Erdős and C. Pomerance. *On the largest prime factors of n and n + 1*. Aequationes Math. 17 (1978), 311–321.

§ IV.30 On greatest prime divisors of sums of integers

1) For any finite set A of positive integers there exist $a, b \in A$ such that

$$P(a+b) > c \cdot \log |A|$$

for a positive constant c.

P. Erdős and P. Turán. *On a problem in the elementary theory of numbers*. Amer. Math. Monthly 41 (1934), 608–611.

2) Let A, B be non-empty subsets of $\{1, 2, \ldots, N\}$. If $N > N_0$ and $|A| \cdot |B| > 100N(\log N)^2$, then there exist $a \in A$ and $b \in B$ such that

$$P(a+b) > \frac{(|A| \cdot |B|)^{1/2}}{16 \log N}$$

Corollary. If $|A| \gg N$ and $|B| \gg N$, then there exist $a \in A$ and $b \in B$ with

$$P(a+b) \gg \frac{N}{\log N}$$

(Here $|A|$ denotes the cardinality of the set A)

A. Balog and A. Sárközy. *On sums of sequences of integers*. II. Acta Math. Hung. 44 (1984), 73–86. (For I. see Acta Arith. 44 (1984), 73–86.)

3) Let A, B as above and $\varepsilon > 0$. If $(A \cdot B)^{1/2} > N^{5/6+\varepsilon}$ for N sufficiently large, then there exist $a \in A$ and $b \in B$ such that

$$P(a+b) > \frac{c_1(|A| \cdot |B|)^{1/2}}{\log R \cdot \log\log R}$$

and there exist $a_1 \in A$, $b_1 \in B$, $a_1 \neq b_1$, with

$$P(a_1 - b_1) > \frac{c_2(|A| \cdot |B|)^{1/2}}{\log R \cdot \log\log R}$$

where c_1, c_2 are positive constants, and $R = \dfrac{3N}{(|A| \cdot |B|)^{1/2}}$

Corollary. If $|A| \gg N$ and $|B| \gg N$, then there exist $a \in A$, $b \in B$ such that

$$P(a+b) \gg N$$

$a_1 \in A, b_1 \in B, a_1 \neq b_1$ with

$$P(a_1 - b_1) \gg N$$

A. Sárközy and C.L. Stewart. *On divisors of sums of integers*. II. J. Reine Angew. Math. 365 (1986), 171–191.

4) Let $A \subset \{1, \ldots, N\}$, $A \neq \emptyset$ and $\varepsilon > 0, k > 1$, integer.
If $|A| > (1+\varepsilon)N/k$ and p is a prime number with $N < p < (1+\varepsilon/2)N$ then there exist $a_1, \ldots, a_k \in A$ such that

$$P(a_1 + \cdots + a_k) = p$$

for N sufficiently large in terms of ε and k.
Further, if $|A| > 8\sqrt{N}\log N$ then there exist $a_1, \ldots, a_k \in A$ such that

$$P(a_1 + \cdots + a_k) > |A|/N^{1/k+\varepsilon}$$

for sufficiently large in terms of ε and k.

A. Sárközy and C.L. Stewart. *On divisors of sums of integers*. I. Acta Math. Hung. 48 (1986), 147–154.

5) Let A, B be sets of integers, $|A| = |B| = k$.
Let $N = \max\{|a+b| : a \in A, b \in B\}$. Let $S = \prod_{a \in A, b \in B} (a+b)$.
Assume $S \neq 0$ and let P be the largest prime divisor of S. Then

$$P \geq c \cdot \frac{k \log k}{\log N} \cdot \log \frac{\log N}{\log k}$$

with an absolute constant $c > 0$. Moreover, for $k > c(\varepsilon)\sqrt{N}$ we have

$$P > \left(\frac{2}{e} - \varepsilon\right) k$$

I.Z. Ruzsa. *Large prime factors of sums*. Studia Sci. Math. Hungar. 27 (1992), 463–470.

§ IV.31 On $\sum_{n \leq x} f(P(n))$, f a certain arithmetic function

Let $L(x)$ be a slowly oscillating function (i.e. positive, measurable for $x \geq x_0$ and $\lim_{x\to\infty} \frac{L(cx)}{L(x)} = 1$ for any $c > 0$) and $R(x) = x^\rho \cdot L(x)$, where $\rho \in \mathbb{R}$. Set $f(n) = \sum_{p|n} R(p)$ and denote $S_1 = \sum_{2 \leq n \leq x} f(P(n))$, $S_2 = \sum_{2 \leq n \leq x} f(n)$. If $\rho > 0$, then

$$S_1 \sim S_2 \sim \frac{x^{\rho+1} \cdot \zeta(\rho+1)}{\rho+1} \frac{L(x)}{\log x}$$

$(x \to \infty)$

J.-M. de Koninck, I. Kátai and A. Mercier. *Les fonctions arithmétiques et le plus grand facteur premier*. Acta Arith. 52 (1989), 25–48.

Remark. For connected results see

J.-M. de Koninck, I. Kátai and A. Mercier. *Additive functions and the largest prime factor of integers.* J. Number Theory 33 (1989), 293–310.

§ IV.32 On $\Phi(x, y)$ and Buchstab's function

Let $\phi(x, y)$ be the number of positive integers $\leq x$ with no prime factor $< y$.

1) a) For any fixed $u > 1$

$$\phi(x, y) \sim \frac{x}{\log y}\omega(u)$$

($x = y^u, x \to \infty$), where $\omega(u)$ is defined to be the continuous solution of the differential-difference equation.

$$(u\omega(u))' = \omega(u-1)$$

$u \geq 2$

$$\omega(u) = 1/u$$

$1 \leq u \leq 2$ (ω is called *Buchstab's function.*)

A.A. Buchstab. *Asymptotic estimates of a general number-theoretic function* (Russian.) Mat. Sb. (N.S.) 2 (44) (1937), 1239–1246.

b) $$\phi(x, y) = x \cdot \prod_{p<y}(1 - 1/p) \cdot (e^{\gamma} \cdot \omega(u) + O(1/\log y)) + O(y/\log y)$$

uniformly in the range $x \geq y \geq 2$, where $u = (\log x)/(\log y)$

N.G. de Bruijn. *On the number of uncancelled elements in the sieve of Eratosthenes.* Nederl. Akad. Wetensch. Proc. 53 (1950), 803–812.

c) $$\phi(x, y) = x \cdot \prod_{p<y}(1 - 1/p) \cdot (1 + O(1/\log x))$$

$x > 0, y \geq 2$

R. Warlimont. *Eine Bemerkung zu einem Ergebnis von N.G. de Bruijn.* Monatsh. Math. 74 (1970), 273–276.

2) Let $\phi_q(x, y)$ be the number of positive integers $\leq x$ all of whose prime factors which are $< y$ also divide q (q is a given positive integer).
If $f(x)$ is a positive-valued function tending to 0 as $x \to \infty$ then

$$\phi_q(x, y) = x \prod_{p<y, p \nmid q} (1 - 1/p)\left(e^{\gamma} \cdot \omega(u) + O\left(f(y) + \frac{\log\log y}{\log y}\right)\right)$$

uniformly in the range $u, y \geq 2, x = y^u$, with q any positive integer which has less than $\exp(f(y)\log y)$ distinct prime factors smaller than y.

J.B. Friedlander and A. Granville. *Limitations to the equi-distribution of primes.* I. Ann. Math. 129 (1989), 363–382.

3) Let $(l, k) = 1, l, k \in \mathbb{N}$ and $\phi_l(k, x, y)$ be the number of positive integers in the arithmetical progressions $kn + l$, not exceeding x with no prime divisors $< y$. Then

$$\phi_l(k, x, y) = u \cdot \omega(u) \cdot \frac{x}{\varphi(k) \log x} + O(x/\log^{3/2} x)$$

where $x = y^u$ (See 1) a)). (See A.A. Buchstab.)

4) Let $W(u) = \omega(u) - e^{-\gamma}(u \geq 1)$. Then:

a) $W(u) \to 0$
as $u \to \infty$. (See N.G. de Bruijn.)

b) $|W(u)| \leq \rho(u-1)/u$
for all $u \geq 1$, where ρ is Dickman's function (defined by $\rho'(u) = -\rho(u-1)(u \geq 1), \rho(u) = 1(0 \leq u \leq 1)$.)

W.B. Jurkat and H.-E. Richert. *An improvement of Selberg's sieve method.* I. Acta Arith. 11 (1965), 217–240.

c) Let $W^*(u) = \max_{v \geq u} |W(v)|$ and let u^* be the smallest value $\geq u$ at which the maximum is attained. Then

$$u \leq u^* \leq u + 2$$

for all $u \geq 1$ and in every interval of unit length there are either one or two zeros of $W(u)$, and either one or two extrema. Moreover each extremum is either a maximum with $W(u) > 0$ or a minimum with $W(u) < 0$

A.Y. Cheer and D.A. Goldston. *A differential delay equation arising from the sieve of Eratosthenes.* Math. Comp. 55 (1990), 129–161.

d) $W^*(u) = \rho(u)^{1+o(1)}$

A. Hildebrand and H. Maier. *Irregularities in the distributions of primes in short intervals.* J. Reine Angew. Math. 397 (1989), 162–193.

e) There exists a decreasing function $W_0(u)$ satisfying

$$W_0(u) = \exp(-u \cdot \xi_u + O(u))$$

$(u \geq u_0)$, where $\xi_u =$ the positive solution of the equation

$$e^x = u \cdot x + 1$$

(it is easy to see that

$$\xi_u = \log u + \log\log u + O((\log\log u)/(\log u))$$

for $u \geq 3$) and a function $\theta(u)$ satisfying

$$\theta(u + t) - \theta(u) = t \cdot \pi \cdot (1 + O(1/\log u))$$

$(u \geq u_0, 0 \leq t \leq 1$ such that, as $u \to \infty$

$$W(u) = W_0(u) \cdot (\cos\theta(u) + O(1/\log u))$$

J. Friedlander, A. Granville, A. Hildebrand and H. Maier. *Oscillations theorems for primes in arithmetic progressions and for sifting functions*. J. Amer. Math. Soc. 4 (1991), 25–86.

§ IV.33 On the partition of primes into two subsets with nearly the same number of products

Let $\psi(x, P)$ be the number of integers $\leq x$ whose prime factors all belong to the set P of primes. There exists a partition of the primes into two sets $P \cup Q$ such that

$$|\psi(x, P) - \psi(x, Q)| < x/\log^A x$$

for $x > x_A$, for any given $A > 0$.

B. Birch and E. Scourfield. *Dividing the primes into two subsets with nearly the same number of products*. Proc. London Math. Soc. III Ser. 67, No. 1 (1993), 53–74.

Remark. The above result answers a question posed by P. Erdős.

Chapter V

ω (*n*), Ω (*n*) AND RELATED FUNCTIONS

§ V. 1 Average order of ω, Ω, $\Omega - \omega$, Ω_k

1) a) $\sum_{n \le x} \omega(n) = x \log\log x + Ax + O\left(\frac{x}{\log x}\right)$

$(A = \gamma + \sum_p (\log(1 - 1/p) + 1/p))$

b) $\sum_{n \le x} \Omega(n) = x \log\log x + Bx + O\left(\frac{x}{\log x}\right)$

$(B = A + \sum_p (1/(p(p-1)))$

c) $\sum_{n \le x} (\Omega(n) - \omega(n)) = x \sum_p \frac{1}{p(p-1)} + O\left(\frac{x}{\log x}\right)$

G.H. Hardy and S. Ramanujan. *The normal number of prime factors of a number n*. Quart. J. Math. 48 (1917), 76–92.

2) If $n = \prod_{i=1}^{r} p_i^{\alpha_i}$, denote $\Omega_k(n) = \alpha_1^k + \alpha_2^k + \cdots + \alpha_r^k (k \ge 0$, integer) (Clearly, $\Omega_0 = \omega$, $\Omega_1 = \Omega$). Then

$$\sum_{n \le x} \Omega_k(n) = x \log\log x + B_k x + O\left(\frac{x}{\log x}\right)$$

R.L. Duncan. *A class of additive arithmetical functions*. Amer. Math. Monthly 69 (1962), 34–36.

§ V. 2 Sums over $\omega^2(n)$, $\omega^k(n)$

1) a) $\sum_{n \le x} \omega^2(n) = x(\log \log x)^2 + O(x \log \log x)$

G.H. Hardy and S. Ramanujan. *The normal number of prime factors of a number n*. Quart. J. Math. 48 (1917), 76–92.

b) $\sum_{n\le x}\omega^2(n) = x(\log\log x)^2 + ax\log\log x + bx + O\left(\frac{x\log\log x}{\log x}\right)$

H.N. Shapiro. *Introduction to the theory of numbers*. John Wiley and Sons, 1983 (p. 347).

2) The Riemann hypothesis is true iff

$$\sum_{n\le x}\omega^k n = x\cdot\sum_{0\le j\le(\log x)/2} R_{jk}(\log\log x)\log^{-j}x + O(x^{1/2+\varepsilon})$$

holds for every $\varepsilon > 0$, where R_{jk} are suitable polynomials satisfying the condition $\deg R_{0k}\le k$, $\deg R_{jk}\le k-1 (j\ge 1)$

D. Wolke. *Über die zahlentheoretische Funktion* $\omega(n)$. Acta Arith. 55 (1990), 323–331.

§ V. 3 Sums over $(\omega(n) - \log\log x)^2$

1) a) $\sum_{n\le x}(\omega(n)-\log\log x)^2 = O(x\log\log x)$

P. Turán. *On a theorem of Hardy and Ramanujan*. J. London Math. Soc. 9 (1934), 274–276.

b) $\sum_{n\le x}(\omega(n)-\log\log x)^m = m!x\cdot A_m(x) + O(x(\log\log x)^{m/2}/\log x)$

where $A_m(x)$ is the coefficient of ζ^m in the Maclaurin expansion of $F(e^\zeta)\exp((e^\zeta - 1 - \zeta)\log\log x)$, where $F(z) = (1/\Gamma(z))\prod_p(1-1/p)^z(1+z/(p-1))$ ($x>0$ and $z\in\mathbb{C}$)

H. Delange. *Sur des formules dues à Atle Selberg*. Bull. Sci. Math. (2) 83 (1959), 101–111.

c) Assume $\log\log x\le h\le x$. Then for all integers $m\le x$ with the exception of at most $O(h\log\log x/x)$ values, the inequality

$$\sum_{m<n\le m+h}(\omega(n)-\log\log x)^2 = O(h\cdot\log\log x)$$

holds.

V.A. Plaksin. *The order of* $\omega(n)$ *on short intervals* (Russian.) Analytic number theory (Russian), 69–73, 92, Petrozavodsk, Gos. Univ., Petrozavodsk, 1986.

d) $\sum_{n\le x}(\omega(n)-\log\log x)^2 = ax\log\log x + bx + O\left(\frac{x\log\log x}{\log x}\right)$

(where a and b are constants).

H.N. Shapiro. *Introduction to the theory of numbers*. John Wiley and Sons, 1983 (p. 347).

2) Let $g:\mathbb{N}\to\mathbb{R}$, $g(0)=0$, $g(1)\ne 1$ and define $f(1)=0$, $f(n)=\sum_{i=1}^{r} g(a_i)$, where $n=p_1^{a_1}\cdots p_r^{a_r}$. Then:

a) If $g(n) = O(b^{n/2})$ for $0 < b < 2$, then

$$\sum_{n \le x} (f(n) - g(1) \log\log x)^2 \le cx \log\log x$$

b) If $g(n) = O(b^{n/2}), 0 < b < 2$ and $g(n) \ge t > 0$, then there exist c_1, c_2 positive constants with

$$c_1 x/\log\log x < \sum_{n \le x} \frac{1}{f(n)} \le c_2 x/\log\log x$$

c) If $g(n) = O(n^k)$, then

$$\sum_{n \le x} \frac{f(n)}{\omega(n)} = xg(1) + O(x/\log\log x)$$

R.L. Duncan. *Some applications of the Turán-Kubilius inequality*. Proc. A.M.S. 30 (1971), 69–72.

Remark. For a slightly stronger result see

C.H.-Zhong. *A sum related to a class o arithmetical functions*. Utilitas Math. 44 (1993), 231–242.

§ V. 4 $\sum_{2 \le n \le x} \frac{1}{\omega(n)}, \sum_{2 \le n \le x} \frac{\Omega(n)}{\omega(n)}$, etc.

1) a) $$\sum_{2 \le n \le x} \frac{1}{\omega(n)} = O(x/\log\log x)$$

R.L. Duncan. *On the factorization of integers*. Proc. Amer. Math. Soc. 25 (1970), 191–192.

b) $$\sum_{2 \le n \le x} \frac{1}{\omega(n)} = x \cdot \sum_{i=1}^{k} \frac{a_i}{(\log\log x)^i} + O\left(\frac{x}{(\log\log x)^{k+1}}\right)$$

where $a_i (1 \le i \le k)$ are constants, e.g. $a_1 = 1, a_2 = 1 - \rho$, where $\rho = \gamma + \sum_p (\log(1 - 1/p) + 1/p)$

c) $$\sum_{2 \le n \le x} \frac{1}{\Omega(n)} = x \sum_{i=1}^{k} \frac{b_i}{(\log\log x)^i} + O\left(\frac{1}{(\log\log x)^{k+1}}\right)$$

where $b_i (1 \le i \le k)$ are constants, e.g. $b_1 = 1$,
$b_2 = 1 - \rho - \sum_p (1/(p(p-1)))$

Corollary.

$$\sum_{2 \le n \le x} \left(\frac{1}{\omega(n)} - \frac{1}{\Omega(n)}\right) = \sum_p \frac{1}{p(p-1)} \frac{x}{(\log\log x)^2} + O\left(\frac{x}{(\log\log x)^3}\right)$$

J.-M. de Koninck. *On a class of arithmetical functions*. Duke Math. J. 39 (1972), 807–818.

d) For every fixed integer $N \geq 1$, there exist computable constants $a_1, a_2, \ldots, a_N$, and slowly oscillating functions $L_1(x), \ldots, L_N(x)$ (asymptotic to $1/\log\log x$) which admit an extension in terms of negative powers of $\log\log x$, such that

$$\sum_{2 \leq n \leq x} \frac{1}{\omega(n)} = a_1 x L_1(x) + \cdots + a_N x L_N(x) \log^{1-N} x + O(x \log^{-N} x)$$

J.-M. de Koninck and A. Ivić. *Topics in arithmetical functions*. Notas de Matemática (72). North-Holland, 1980.

Remark. The same result holds when ω is replaced by Ω

2) a) $$\sum_{2 \leq n \leq x} \frac{1}{\omega^2(n)} = x \sum_{i=2}^{k} \frac{c_i}{(\log\log x)^i} + O\left(\frac{x}{(\log\log x)^{k+1}}\right)$$
where $k \geq 2; c_i (2 \leq i \leq k)$ are (computable) constants, e.g. $c_2 = 1$, $c_3 = 3 - 2\rho$.

b) $$\sum_{2 \leq n \leq x} \frac{1}{\Omega^2(n)} = x \sum_{i=2}^{k} \frac{d_i}{(\log\log x)^i} + O\left(\frac{x}{(\log\log x)^{k+1}}\right)$$
where $k \geq 2, d_i (2 \leq i \leq k)$ are constants, e.g. $d_2 = 1$, $d_3 = 3 - 2\rho(1/(p(p-1)))$

(See J.-M. de Koninck.)

3) a) $$\sum_{2 \leq n \leq x} \frac{\Omega(n)}{\omega(n)} = x + O\left(\frac{x}{\log\log x}\right)$$

(See R.L. Duncan.)

b) $$\sum_{2 \leq n \leq x} \frac{\Omega(n)}{\omega(n)} = x + x \sum_{i=1}^{k} \frac{A_i}{(\log\log x)^i} + O\left(\frac{x}{(\log\log x)^{k+1}}\right)$$
where A_i are computable constants,
e.g. $A_1 = \sum_p (1/(p(p-1)))$

J.-M. de Koninck. *Sums of quotients of additive functions*. Proc. Amer. Math. Soc. 44 (1974), 35–38.

c) $$\sum_{2 \leq n \leq x} \frac{\Omega(n)}{\omega(n)} = x\left(a_1 + b_1 L_1(x) + \frac{a_2 + b_2 L_2(x)}{\log x} + \cdots \right.$$
$$\left. \cdots + \frac{a_N + b_N L_N(x)}{\log^{N-1} x}\right) + O\left(\frac{x}{\log^N x}\right)$$
where a_i, b_i are constants and $L_1, \ldots, L_N$ as in 1) d).

(See J.-M. de Koninck and A. Ivić.)

4) a) $$\sum_{\substack{2\le n\le x\\ n\ne\text{squarefree}}} \frac{1}{\Omega(n)-\omega(n)} = a_0x + a_1x^{1/2}L_1(x)\log^{-1}x + \cdots$$
$$\cdots + a_Nx^{1/2}L_N(x)\log^{-N}x + O(x^{1/2}\log^{-N-1}x)$$
(See J.-M. de Koninck and A. Ivić.)

b) If $h \ge x^{105/407}\log^3 x$, then
$$\sum_{\substack{x<n\le x+h\\ n\ne\text{ squarefree}}} \frac{1}{\Omega(n)-\omega(n)} = \left(\int_0^1 \left(F(z)-\frac{6}{\pi^2}\right)\cdot z^{-1}dz + o(1)\right)\cdot h$$
(as $x\to\infty$), where $F(z) = \dfrac{6}{\pi^2}\prod_p \dfrac{1-z/(p-1)}{1-z/p}$

(See J.-M. de Koninck and A. Ivić.)

5) a) $$\sum_{\substack{2\le n\le x\\ n\equiv 0(\mathrm{mod}\,k)}} \frac{1}{\omega(n)} = \frac{x}{k}\sum_{i=1}^{M}\frac{a_i}{(\log\log x)^i} + O\left(\frac{x}{(\log\log x)^{M+1}}\right)$$
where k and M are fixed positive integers.

Remark. The similar result holds replacing ω by Ω

b) $$\sum_{\substack{2\le n\le x\\ n\equiv 0(\mathrm{mod}\,k)}} \frac{\Omega(n)}{\omega(n)} = \frac{x}{k} + x\sum_{i=1}^{M}\frac{b_i}{(\log\log x)^i} + O\left(\frac{x}{(\log\log x)^{M+1}}\right)$$

(See J.-M. de Koninck and A. Ivić.)

6) $$\sum_{2\le n\le x} \frac{1}{\omega(n)\cdot\Omega(n)} = x\sum_{i=2}^{M}\frac{c_i}{(\log\log x)^i} + O\left(\frac{x}{(\log\log x)^{M+1}}\right)$$
($M\ge 2$ is a fixed integer, c_i computable constants).
(See J.-M. de Koninck and A. Ivić.)

§ V. 5 $\sum_{p\le n}\omega^k(p-1)$ (p prime)

1) Let $\omega_n(m)$ denote the number of all prime divisors of m less than n. Then
$$\sum_{p\le x}(\omega_n(p-1)-\log\log n)^2 = O(\pi(x)\log\log n)$$
where $n\le\log x$.

Y. Motohashi. *On a property of p* − 1. Proc. Japan Acad. 45 (1969), 639–640.

2) $\sum_{p\le n} \omega^k(p-1) = \frac{n}{\log n}(\log\log n)^k + O\left(\frac{n}{\log n}(\log\log n)^{k-1}\right)$

N.P. Ryzhova. *Asymptotic formulae in a binary problem of shifted prime numbers* (Russian.) Additive problems of number theory, Interuniv. Collect. sci. Works, Kujbyshev 1985, 25–31 (1985).

§ V. 6 $\sum_{p\le n} \omega(f(p)$, f polynomial (p prime)

1) Let $f(x)$ be an irreducible polynomial with integer coefficients which is not of the form ax ($a \in \mathbb{I}$). Then

$$\sum_{p\le x} \omega(f(p)) > c\cdot x \log\log x/\log x$$

K. Prachar. *On the sum* $\sum_{p\le x} \omega(f(p))$. J. London Math. Soc. 28 (1953), 236–239.

2) If $f(x)$ is an irreducible polynomial with integer coefficients,

a) $\sum_{p\le n} \omega(f(p)) \sim \frac{n\log\log n}{\log n}$
$(n\to\infty)$

b) Let

$f_n(u) = \text{card}\{p \le n : \omega(f(p)) < \log\log n + u\cdot(\log\log n)^{1/2}\}$

Then

$$\frac{f_n(u)}{\pi(n)} \to \phi(u)$$

where $\phi(u) = \frac{1}{\sqrt{2\pi}}\int_{-\infty}^{u} e^{-t^2/2}dt$ $(n\to\infty)$

H. Halberstam. *On the distribution of additive arithmetic functions*. III. J. London Math. Soc. 31 (1956), 14–27.

3) Suppose $f_1(x), \dots, f_n(x) \in \mathbb{I}[x]$ are primitive, non-constant, with $f_i(0) \ne 0$. Suppose further that $f_i(x)$ does not divide any power of the l.c.m. $(f_1(x), \dots, f_{i-1}(x))$ for $i = 2, \dots, n$.
Let $a_2, \dots, a_n > 0$ and let $K > 0$. Then there exists a prime p such that

$$\Omega(f_1(p)) > K$$

and

$$\Omega(f_i(p)) - a_i\Omega(f_{i-1}(p)) > K$$

for $i = 2, \dots, n$

A. Turull and A. Zame. *Number of prime divisors and subgroup chains*. Arch. Math. (to appear.)

§ V. 7 $\sum_{n\le x} z^{\omega(n)}$ and related sums

1) $\sum_{n\le x} z^{\omega(n)} = z\cdot F(z)x(\log x)^{z-1} + O(x(\log x)^{\mathrm{Re}(z-2)})$
where, for $R>0$, the O-constant is uniform for $|z|\le R$. Here

$$F(z) = \frac{1}{\Gamma(z+1)}\prod_p\left(1+\frac{z}{p-1}\right)\left(1-\frac{1}{p}\right)^z$$

A. Selberg. *Note on a paper by L.G. Sathe.* J. Indian Math. Soc. 18 (1954), 83–87.

2) a) $\sum_{n\le x} 2^{\Omega(n)} = (r+1)c_1 x\log(2^{-r/2}x) - (r+1)c_2x + O(x^c)$

where $r = [\log x/\log 2]$, $c_1 = \frac{1}{4}\prod_{p\ge 3}\left(1+\frac{1}{p^2-2p}\right)$,

$$c_2 = c_1\left(2\sum_{p\ge 3}\log p/(p-1)(p-2) - 2\log 2 - 2\gamma + 1\right),\ c<0.84$$

E. Grosswald. *The average order of an arithmetic function.* Duke Math. J. 23 (1956), 41–44.

b) The above relation is true with $c<\log 2/\log 3$

P.T. Bateman. *Proof of a conjecture of Grosswald.* Duke Math. J. 25 (1957), 67–72.

3) For fixed $\alpha\ge 1$ and arbitrary $0<\varepsilon<1$,

$$\sum_{\substack{n\le x\\ p|n\Rightarrow\alpha<p\le y}} \alpha^{\Omega(n)} = x(\log x)^{\alpha-1}\rho_\alpha(u)(1+o(1))$$

uniformly for $1\le u\le(1-\varepsilon)\log\log x/\log\log\log x$ as $x\to\infty$, where $\rho_\alpha(u)$ satisfies a differential equation.

D. Hensley. *The sum of $\alpha^{\Omega(n)}$ over integers $n\le x$ with all prime factors between α and y.* J. Number Theory 18 (1984), 206–212.

4) Let $0<\delta<1$, $\varepsilon>0$ fixed. Then, uniformly for $\delta\le a\le 2-\delta$,

$$\sum_{2\le n\le x} a^{\Omega(n)/\omega(n)} = ax + O(x(\log\log x)^{\varepsilon-1})$$

A. Ivić. *The distribution of quotients of small and large additive functions.* Boll. Un. Mat. Ital. B(7) 2 (1988), 79–97.

5) Denote $\Omega(n,y) = \sum_{p^\alpha|n,\,p>y}\alpha\ (n\ge 1,\ y\ge 2)$. Then

$$\sum_{n\le x}\alpha^{\Omega(n,y)} \ll_\alpha x\cdot u^{\alpha-1}$$

for $x\ge y\ge 2$, $1\le\alpha<p_1(y)$, where $p_1(y)$ is the least prime number $>y$.

K.K. Norton. *On the number of restricted prime factors of an integer.* I, Ill. J. Math. 20 (1976), 681–705.

§ V. 8 Sums over $\lambda(n) = (-1)^{\Omega(n)}$

1) a) $\sum_{n\le x}(-1)^{\omega(n)} = o(x)$
as $x \to \infty$

b) $\sum_{n=1}^{\infty}(-1)^{\omega(n)}/n = 0$

J. van de Lune and R.E. Dressler. *Some theorem concerning the number theoretic function* $\omega(n)$. J. Reine Angew. Math. 277 (1975), 117–119.

c) $\sum_{n\le x}(-1)^{\omega(n)} = O(x\exp(-c\sqrt{\log x}))$

W. Schwarz. *A remark on multiplicative functions*. Bull. London Math. Soc. 4(1972), 136–140.

d) $\sum_{n\le x}\lambda(n) \equiv \sum_{n\le x}(-1)^{\Omega(n)} = O(x\exp(-c\sqrt{\log x}))$

R. Ayoub. *An introduction to the analytic theory of numbers*. Amer. Math. Soc. Publ. 1963 (p. 123.)

2) If $\lambda(n) = (-1)^{\Omega(n)}$ and $L(x) = \sum_{n\le x}\lambda(n)$, then

$$\limsup_{x\to\infty}\frac{L(x)}{\sqrt{x}} > 0.023$$

R.J. Anderson and H.M. Stark. *Oscillation theorems*. Analytic number theory (Philadelphia, Pa., 1980), pp. 79–106. Lectures Notes in Math. 899, Springer 1981.

Remark. Pólya conjectured that

$$L(x) \le 0$$

for all $x \ge 2$. This was disproved by Haselgrove, who showed also that the conjecture

$$\sum_{n\le x}\lambda(n)/n > 0$$

for $x \ge 1$, stated by Turán, is false.

C.B. Haselgrove. *A disproof of a conjecture of Pólya*. Mathematika 5 (1958), 141–145.

P. Turán. *On some approximative Dirichlet-polynomials in the theory of the zeta-function of Riemann*. Danske Vid. Selsk. Mat.-Fyz. Medd 24 (1948), no. 17, 36 pp.

§ V. 9 Sums over $n^{-1/\omega(n)}$, $n^{-1/\Omega(n)}$

a) $\sum_{2\le n\le x} n^{-1/\omega(n)} = x\exp(-(c_1 + o(1))(\log x \cdot \log\log x)^{1/2})$

b) $\sum_{2\le n\le x} n^{-1/\Omega(n)} = x\exp\left(-(c_2+o(1))(\log x)^{1/2}\right)$

where c_1, c_2 are positive constants.

T.Z. Xuan. *On a problem of Erdős and Ivić*. Publ. Inst. Math. (Beograd) (N.S.) 43 (57) (1988), 9–15.

Note. a) and b) were conjectured by Erdős and Ivić.

P. Erdős and A. Ivić. Same journal (N.S.) 32 (46) (1982), 49–56.

c)
$$\sum_{2\le n\le x} n^{-1/\omega(n)} = x\exp\left(-\sqrt{2}(\log\log x)^{1/2} + O\left(\left(\frac{\log x}{\log\log x}\right)^{1/2}\log\log\log x\right)\right)$$

d) $\sum_{2\le n\le x} n^{-1/\Omega(n)} = cx(\log x)^{5/4}\cdot\exp(-2(\log 2\log x)^{1/2})\cdot(1+O((\log x)^{-1/4}))$

T.Z. Xuan. *On a problem of Erdős and Ivić* (Chinese.) J. Math. (Wuhan) 9 (1989), 375–380.

§ V.10 Sums on $d(n)\,\omega(n-1)$, $d_k(n)\,\omega(n)$

1)
$$\sum_{n\le x} d(n)\omega(n-1) = x\log x\cdot\log\log x + a_1 x\log x + a_2 x\log\log x + a_3 x + O\left(\frac{x}{\log x}(\log\log x)^6\right)$$

B.V. Levin and N.M. Timofeev. *An additive problem* (Russian.) Mat. Zametki 46 (1989), 25–33.

2)
$$\sum_{n\le x} d_k(n)\,\omega(n) = x\sum_{j=1}^{k}\left((a_{k,j}\cdot\log\log x + b_{k,j})\log^{k-j} x\right) + x\sum_{j=k+1}^{N} c_{k,j}\log^{k-j} x + O(x\log^{k-N-1} x)$$

A. Ivić. *Sums of product of certain arithmetical functions*. Publ. Inst. Math. (Beograd) (N.S.) 41 (55) (1987), 31–41.

§ V.11 Sums on $\frac{\omega(n)}{P(n)}$, $\frac{\omega(n)}{\beta(n)}$

1) $\sum_{n\le x}\frac{\omega(n)}{P(n)} = \left(\frac{2\log x}{\log\log x}\right)^{1/2}\left(1+O\left(\frac{\log\log\log x}{\log\log x}\right)\right)\sum_{n\le x}\frac{1}{P(n)}$

where $P(n)$ denotes the greatest prime divisor of n.

A. Ivić. *On some estimates involving the number of prime divisors of an integer*. Acta Arith. 49 (1987), 21–33.

2) $$\sum_{2\le n\le x} \frac{\omega(n)}{\beta(n)} = D\left(\frac{2\log x}{\log\log x}\right)^{1/2} \cdot \left(1 + O\left(\frac{(\log\log\log x)^2}{\log\log x}\right)\right) \sum_{2\le n\le x} \frac{1}{P(n)}$$

where $\beta(n) = \sum_{p|n} p$ and $D \in (1/2, 1)$ is an absolute constant.

T. Xuan. *On sums involving reciprocals of certain large additive functions*. Publ. Inst. Math., Nouv. Sér. 45 (49) (1989), 41–55.

Remark. The result remains valid if $\omega(n)$ is replaced by $\Omega(n)$

§ V.12 $\omega(a(n))$, $\omega(d(n))$, etc.

1) Let $a(n)$ denote a number of nonisomorphic Abelian groups of order n. Then:

a) $$\sum_{n\le x} \omega(a(n)) = \sum_{k=1}^{\infty} \omega(k) d_k x + O(x^{1/2}\log^3 x/(\log\log x)^2)$$

b) $$\sum_{n\le x} \Omega(a(n)) = \sum_{k=1}^{n} \Omega(k) d_k x + O(x^{1/2}\log^3 x/(\log\log x))$$

where $d_k = \lim_{x\to\infty} \frac{1}{x} \sum_{n\le x, a(n)=k} 1 \ (k \ge 1)$

c) $$\sum_{n\le x, \omega(n)>a(n)} 1 = x + O(x(\log\log x)^{-k})$$

A. Ivić. *On the number of Abelian groups of a given order and on certain related multiplicative functions*. J. Number Theory 16 (1983), 119–137.

Remark. The existence of d_k was first proved in
D.G. Kendall and R.A. Rankin. *On the number of Abelian groups of a given order*. Quart. J. Math. Oxford Ser. (2) 18 (1947), 197–208.

2) a) $$\sum_{n\le x} \omega(d(n)) = cx + O(x^{1/2}\log^5 x)$$

b) $$\sum_{n\le x} \Omega(d(n)) = bx\log\log x + \sum_{k=0}^{[\sqrt{\log x}]} b_k \cdot x\log^{-k} x + O(x\exp(-A\log x))$$

where $d(n)$ is the number of divisors n. (c, b, b_k, A are positive constants).

E. Heppner. *Über die Iteration von Teilerfunktionen*. J. Reine Angew. Math. 265 (1974), 176–182.

§ V.13 $\dfrac{\Omega(n)-\omega(n)}{P(n)}$, $\dfrac{\Omega(n)-\omega(n)}{\beta(n)}$, etc.

1) a) $$\sum_{2\le n\le x}\frac{\Omega(n)-\omega(n)}{P(n)} = x\cdot\exp(-\sqrt{2}\cdot L_1(x)(1+g_0(x)+O(L_2(x)))$$

where $P(n)$ denotes the greatest prime divisor of n and

$L_1(x) = (\log x \cdot \log\log x)^{1/2}$

$L_2(x) = (\log\log\log x)^3/(\log\log x)^3$

$$g_0(x) = (f_o(x)-2)\frac{1}{\log\log x}\left(\frac{1}{2}+\frac{1}{\log\log x}\right)-\frac{1}{8}f_0^2(x)\frac{1}{(\log\log x)^2}$$

where $f_0(x) = \log\log\log x - \log 2$

A. Ivić and C. Pomerance. *Estimates for certain sums involving the largest prime factor of an integer.* Topics in classical number theory, vol. I, II (Budapest, 1981), 769–789. Colloq. Math. Soc. János Bolyai, 34, North Holland, 1984.

b) $$\sum_{n\le x}\frac{\Omega(n)-\omega(n)}{P(n)} = \left(c+O\left(\left(\frac{\log\log x}{\log x}\right)^{1/2}\right)\right)\sum_{n\le x}\frac{1}{P(n)}$$

A. Ivić. *On some estimates involving the number of prime divisors of an integer.* Acta Arith. 49 (1987), 21–23.

2) a) $$\sum_{n\le x, P(n)\le y}(\Omega(n)-\omega(n)) = \Psi(x,y)(c+O(\log\log x/\log y))$$

where $y \le x$ and $\log y/\log\log x \to \infty$ as $x \to \infty$. Here $\Psi(x, y)$ denotes the number of positive integers $\le x$ without prime factors $> y$, and $c = \sum_p (1/(p^2-p))$.

(See A. Ivić.)

3) $$\sum_{2\le n\le x}\frac{\Omega(n)-\omega(n)}{\beta(n)} = \left(D\sum_p\frac{1}{p^2-p}+O\left(\frac{(\log\log\log x)^2}{\log\log x}\right)\right)\sum_{2\le n\le x}\frac{1}{P(n)}$$

where $\beta(n) = \sum_{p|n} p$ and $D \in (1/2, 1)$ an absolute constant.

T. Xuan. *On sums involving reciprocals of certain large additive functions.* Publ. Inst. Math. N.S. 45 (59) (1989), 41–55.

§ V.14 On the number of integers $n \le x$ with $\Omega(n) - \omega(n) = k$

1) Let $V_k(x) = \text{card}\{n \le x : \Omega(n)-\omega(n) = k\}$, $k = 0, 1, 2, \ldots$. Then:

a) $\lim_{x\to\infty} \frac{V_k(x)}{x} = d_k$
$(0 < d_k < \infty)$, for all k

A. Rényi. *On the density of certain sequences of integers.* Acad. Serbe Sci. Publ. Inst. Math. 8 (1955), 157–162.

b) $V_k(x) = d_k \cdot x + o(\sqrt{x}(\log\log x)^k)$

H. Delange. *Sur un théorème de Rényi.* Acta Arith. 11 (1965), 241–252.

Remark. The result b) for $k = 0$ is due to Landau, and for $k = 1$ to Cohen.

E. Landau. *Handbuch*, vol. II, Leipzig: Teubner 1909;

R. Cohen. *Arithmetical notes. VIII. An asymptotic formula of Rényi.* Proc. Amer. Math. Soc. 13 (1962), 536–539.

c) $V_k(x) - d_k \cdot x \sim -4\zeta\left(\frac{1}{2}\right)\frac{1}{\Gamma(k)}\sqrt{x}(\log x)^{-2}(\log\log x)^{k-1}$
for $k \geq 1$

H. Delange. *Sur un théorème de Rényi.* III. Acta Arith. 23 (1973), 153–182.

d) $$V_k(x) - d_k \cdot x = x^{1/2}\sum_{j=1}^{N} P_j(\log\log x)\log^{-j-1} x + + O\left(\sqrt{x}\frac{(\log\log x)^{k-1}}{\log^{N+2} x}\right)$$
for every fixed integer $k \geq 1$, where each $P_j(t)$ is a polynomial in t of degree $\leq k - 1$, and N is an arbitrary positive fixed integer.

(See H. Delange. III.)

e) Let $N(x) = (\log x)^{1/3} \cdot (\log\log x)^{-1/3}$, and let b_j denote positive absolute constants. Then uniformly for integers k such that $0 \leq k \leq b_1 N(x)$ one has
$$V_k(x) = d_k \cdot x + x^{1/2}\sum_{1\leq j\leq N(x)} P_{j,k}(\log\log x)(\log x)^{-j-1} + + O(x^{1/2}\exp(-b_2 N(x)))$$
where the polynomials $P_{j,k}$ (of degree $\leq k - 1$) satisfy
$$|P_{j,k}(\log\log x)| \leq b_3 \cdot b_4^j \cdot (j+1)!k\log x$$
$(x \geq 10)$

D. Wolke. *On a problem of Rényi.* Monatsh. Math. 111 (1991), 323–330.

2) Let $V(x, \lambda) = \text{card}\{n \leq x : \Omega(n) > \lambda\omega(n)\}$
$(\lambda > 1)$

a) If $\lambda = a/q$, $(a, q) = 1$, then

$$V(x,\lambda) \sim \frac{\log 2}{q(2^{1/q}-1)} H(\alpha) x (\log x)^{\alpha-1}$$

$(x \to \infty)$

b) If $\lambda \notin Q$, then

$$V(x,\lambda) \sim H(\alpha) x (\log x)^{\alpha-1}$$

where $\alpha = 2^{1-\lambda}$,

$$H(\alpha) = (\alpha \cdot 2^{-\alpha} / \Gamma(\alpha) \log 4) \prod_{p>2} (1-1/p)^{\alpha} (1+\alpha/(p-2))$$

G. Tenenbaum. *Sur la distribution conjointe des fonctions ≪nombre de facteurs premiers≫*. Aequationes Math. 35 (1988), 55–68.

3) If $h \geq x^{105/407} \log^3 x$, then as $x \to \infty$

$$\sum_{\substack{\Omega(n)-\omega(n)=k \\ x<n\leq x+h}} 1 = (d_k + o(1))h$$

uniformly in k, where $k \geq 0$ is a fixed integer and

$$\sum_{k=0}^{\infty} d_k z^k = \frac{6}{\pi^2} \prod_p \left(\frac{1 - z/(p+1)}{1 - z/p} \right)$$

J.-M. de Koninck and A. Ivić. *Topics in arithmetical functions*. Notas de Matemática (72). North-Holland, 1980.

§ V.15 Estimates of type $\omega(n) \leq c \cdot \log n / \log\log n$

1) a) $\omega(n) \leq \dfrac{\log n}{\log\log n - 1.1714}$

for $n \geq 26$, with equality when n is the product of the first 189 primes.

G. Robin. *Estimation de la fonction de Tchebychef θ sur le k-ième nombre premier et grandes valeurs de la fonction $\omega(n)$ nombre de diviseurs premiers de n.* Acta Arith. 42 (1983), 367–389.

b) $\omega(n) \leq \dfrac{\log n}{\log\log n} \cdot c_1$

(where $c_1 = 1.38402\ldots$) for $n \geq 3$

c) $\omega(n) \leq \dfrac{\log n}{\log\log n} \cdot \left(1 + \dfrac{c_2}{\log\log n}\right)$

(where $c_2 = 1.45743\ldots$) for $n \geq 3$

d) $\omega(n) \leq \dfrac{\log n}{\log\log n} \cdot \left(1 + \dfrac{1}{\log\log n} + \dfrac{c_3}{(\log\log n)^2}\right)$

$$(\text{where } c_3 = 2.89726\ldots), \quad \text{for } n \geq 3$$

(See G. Robin above and G. Robin: *Sur la différence* Li $(\theta(x)) - \pi(x)$. Annales Fac. Sci. Toulouse, 6 (1984), 257–268.)

2) Let $W(n) = \omega(n)\log\log n/\log n$. Then

$$\sup\{W(n) : n \in \mathbb{N}\} = W\left(\prod_{2\leq p\leq 23} p\right)$$

H.-J. Kanold. *Über einige Ergebnisse aus der kombinatorische Zahlentheorie*. Abh. Braunschweig, Wiss. Ges. 36 (1984), 203–229.

Remark. It is well known that $\limsup\limits_{n\to\infty} W(n) = 1$. See

G.H. Hardy and E.M. Wright. *An introduction of the theory of numbers*. Oxford, 1960.

3) Let $\Omega(n; y, z) = \sum\limits_{p^\alpha|n, y<p\leq z} \alpha (n \geq 1, 0 < y \leq z)$. Then:

a) $$\max_{3/2\leq z\leq n} \left|\Omega(n; z, n) - \log\left(\frac{\log n}{\log z}\right)\right| <$$
$$< 2\sqrt{(\log\log n)(\log\log\log\log n)}$$
for almost all n

b) Let $0 < \varepsilon < 1, \lambda > 0$. Then

$$\max_{2\leq z\leq x} (\Omega(n; 1, z) - (1+\varepsilon)\log\log z) \leq \lambda$$

for all positive integers $n \leq x$, excepting at most
$\ll \frac{1}{\varepsilon^2}(1+\varepsilon)^{-\lambda}x$ values.

c) Let $0 < \varepsilon < 1, \lambda > 0$. The number of integers $n \leq x$ not satisfying

$$\max_{\substack{2\leq 2y\leq z\leq x\\ \log z\geq u^{10}\log x}} \left(\Omega\left(n; \frac{z}{y}, z\right) - (1+\varepsilon)\log\frac{1}{u}\right) \leq \lambda$$

is

$$\ll \varepsilon^{-4}(1+\varepsilon)^{-\lambda} \cdot x$$

M. Mendes France and G. Tenenbaum. *Systèmes de points, diviseurs et structure fractale*. Les prépublications de l'Institut Élie Cartan 91/No. 15, pp. 1–22.

§ V.16 On $\omega(n) - \omega(n+1)$ or $\omega(m) - \omega(n)$

1) For any m,

$$\Omega(n) \equiv \Omega(n+1) (\text{mod } m)$$

for infinitely many n

V.V. Glazkov. *Distribution of the values of characters* (Russian). Studies in Number Theory. No. I, pp. 12–20, Izdat. Saratov Univ. Saratov, 1966.

2) a) There is an x_1 such that for $x > x_1$ the equation

$$n + \omega(n) = m + \omega(m)$$

$n \le x, m \le x, m \ne n$, has more than
$x \cdot \exp(-4000 \log\log x \cdot \log\log\log x)$ solutions.

P. Erdős, A. Sárközy and C. Pomerance. *On locally repeated values of certain arithmetic functions*. I.J. Number Theory 21 (1985), 319–332.

Remark. The analogous result holds with $\Omega(n)$ replacing $\omega(n)$. (Also with $d(n)$ — the number of divisors of n)

b) $\mathrm{card}\,\{(m, n) : m < n \le x, m + \omega(m) = n + \omega(n)\} = O(x)$

P. Erdős, C. Pomerance and A. Sárközy. *On locally repeated values of certain arithmetic functions*. III. Proc. A.M.S. 101 (1987), 1–7.

3) a) There are absolute constants $c_1, c_2 > 0$ such that for $x \ge 3$

$$\mathrm{card}\,\{n \le x : |\omega(n) - \omega(n+1)| \le c_1\} \ge c_2 x/\sqrt{\log\log x}$$

P. Erdős, C. Pomerance and A. Sárközy. *On locally repeated values of certain arithmetic functions*. II. Acta Math. Hungar. 49 (1987), 251–259.

Remark. $\omega(n)$ can be replaced with $\Omega(n)$. The result is valid for $c_1 = 3$

b) $\mathrm{card}\,\{n \le x : \omega(n) = \omega(n+1)\} = O(x)/\sqrt{\log\log x}$
The same is valid with $\Omega(n)$ replacing $\omega(n)$

(See P. Erdős, C. Pomerance and A. Sárközy. III.)

c) Let $g(n) = \mathrm{card}\{m \le n : m + \omega(m) > n\}$. The average order of $g(n)$ is

$$\log\log n + O(1)$$

and the normal order is

$$\log\log n$$

(See P. Erdős, C. Pomerance and A. Sárközy. III.)

§ V.17 The values of ω on consecutive integers

1) If $\omega(n) < \omega(n+1) < \cdots < \omega(n + f(n))$, then

$$\limsup_{n\to\infty} f(n)\frac{\log\log n}{\log n} \le \frac{1}{2}$$

P. Erdős. *Remarks on two problems* (Hungarian.) Mat. Lapok 11 (1960), 26–32.

2) $\Omega(n) + \Omega(n+1) \leq \frac{\log n}{\log 2} \cdot (1 + o(1))$
as $n \to \infty$

P. Erdős and J.-L. Nicolas. *Sur la fonction: nombre de facteurs premiers de n.* L'Enseign. Math. 27 (1981), 3–27.

3) a) With the possible exception of a finite number of cases amongst $p_1 \cdots p_{k-1} p_{k+1}$ consecutive integers there is always one number n with $\omega(n) \geq k (p_i$ denotes the i-th prime.)

A. Schinzel. *Problem 31.* Elem. Math. 14 (1959), 82–83.

b) $\liminf\limits_{n\to\infty} \sum_{i=0}^{k-1} \omega(n+i) \geq k + \pi(k) - 1$

P. Erdős. *Some problems on the prime factors of consecutive integers.* Ill. J. Math. 11 (1967), 428–430.

§ V.18 Local growth of ω at consecutive integers

For every $\varepsilon > 0$:

a) $O_k(n) = \max\{\omega(n+j) : j = 1, 2, \ldots, k\} \leq$
$\leq (1+\varepsilon)\, \rho(\log k / \log\log n) \cdot \log\log n$

where ρ denotes the inverse function to f such that

$$f(x) = x \log(x/e) + 1$$

$(x \geq 1)$

I. Kátai. *Local growth of the number of the divisors of consecutive integers.* Publ. Math. Debrecen 18 (1971), 171–175.

b) For all n excepting a set of density zero,

$$(1-\varepsilon)\rho\left(\frac{\log k}{\log\log n}\right) \cdot \log\log n \leq O_k(n) \leq (1+\varepsilon)\rho\left(\frac{\log k}{\log\log n}\right) \cdot \log\log n$$

for all $k = 1, 2, \ldots$

P. Erdős and I. Kátai. *On the growth of some additive functions of small intervals.* Acta Math. Hungar. 33 (1979), 345–359.

§ V.19 Normal order of ω(φ(n))

1) a) The normal order of magnitude of $\omega(n)$ and $\Omega(n)$ is

$$\log\log n$$

G.H. Hardy and S. Ramanujan. *The normal number of prime factors of a number n.* Quart. J. Math. 48 (1917), 76–92.

b) The normal order of magnitude of $\Omega(n)/\omega(n)$ is

$$1$$

(See G.H. Hardy and S. Ramanujan.)

2) a) The normal order of magnitude of $\omega(p \pm 1)$ (p-prime) is

$$\log\log p$$

P. Erdős. *On the normal number of prime factors of $p-1$ and some related problems concerning Euler's φ function.* Quart. J. Math. Oxford 6 (1935), 205–213.

Note. For the normal order of magnitude of $\omega(\varphi(n))$ and $\omega(\sigma(n))$ see Chapters I and III.

b) For all $\varepsilon > 0$,

$$\text{card}\,\{p \le x : (1-\varepsilon)\log\log x < \omega(p-1) < (1+\varepsilon)\log\log x\} = $$
$$= x/(\log x) + o(x/\log x)$$

(See P. Erdős.)

3) Denote by $N(n)$ the number of prime twins with first elements $p \le n$ such that

$$|\omega(p+1) - \log\log n| \ge (\log\log n)^{1/2+\varepsilon}$$

Then

$$N(n) = o(n/\log^2 n)$$

for all $\varepsilon > 0$

M.B. Barban. *On the number of divisors of "translations" of the prime number-twins* (Russian.) Acta Math. Hungar. 15 (1964), 285–288.

§ V.20 Function $\omega(n; u, v)$

Let $\omega(n; u, v) = \text{card}\,\{p : p|n, u < p < v\}$

a) Let $u = u(x)$, $v = v(x)$ and assume that $\log\log v - \log\log u \to \infty$. Then for all but $o(x)$ integers $n < x$,

$$\omega(n; u, v) = (1 + o(1))(\log\log v - \log\log u)$$

P. Turán. *On a theorem of Hardy and Ramanujan.* J. London Math. Soc. 9 (1934), 274–276.

b) Assume $(\log\log v - \log\log u)/\log\log\log n \to \infty$. Then we have for almost all n uniformly in u and v

$$\omega(n; u, v) = (1 + o(1))\,(\log\log v - \log\log u)$$

P. Erdős. *On the distribution of prime divisors.* Aequationes Math. 2 (1969), 177–183.

c) There are two continuous functions $f_1(c)$ and $f_2(c)$, $f_1(0) = \infty$, $f_1(\infty) = 1$, $f_2(c)$ is strictly decreasing for $0 < c < \infty$; $f_2(c) = 0$ for $0 \le c \le 1$,

$f_2(\infty) = 1$, $f_2(c)$ is strictly increasing in $1 < c < \infty$, satisfying for almost all n and for every $c > 0$

$$\max \omega(n; u, v) = (1 + o(1)) f_1(c)(\log\log v - \log\log u)$$

and

$$\min \omega(n; n, v) = (1 + o(1)) f_2(c)(\log\log v - \log\log u)$$

where the max and min is taken over the values $1 \le u < v \le n$ satisfying

$$\log\log v - \log\log u > c \log\log\log n$$

(See P. Erdős.)

§ V.21 On the number of values $n \le x$ with $\omega(n) > f(x)$

a) Let $0 < c < 1$. Then

$$\text{card}\left\{n \le x : \omega(n) > \frac{c \log x}{\log\log x}\right\} = x^{1-c+o(1)}$$

b) For $\alpha > 1$,

$$\text{card}\{n \le x : \omega(n) > [\alpha \log\log x]\} = \frac{1}{\sqrt{2\pi}} \cdot \frac{F(\alpha)}{\alpha - 1} \cdot$$

$$\cdot \alpha^{1/2+\{\alpha \log\log x\}} \cdot \frac{x(1 + O(1/\log\log x))}{(\log x)^{1-\alpha+\alpha \log x} \cdot \sqrt{\log\log x}}$$

where $F(\alpha) = \dfrac{1}{\Gamma(\alpha+1)} \prod_p \left(1 + \dfrac{\alpha}{p-1}\right) \cdot \left(1 - \dfrac{1}{p}\right)^{\alpha}$

c) For $0 < \alpha < 1$ the same formula is valid (replacing $F(\alpha)/(\alpha - 1)$ by $F(\alpha)/(1 - \alpha)$) for estimating

$$\text{card}\{n \le x : \omega(n) \le \alpha \log\log x\}$$

P. Erdős and J.-L. Nicolas. *Sur la fonction: nombre de facteurs premiers de n*. L'Enseign. Math. 27 (1981), 3–27.

§ V.22 On $\omega(2^p - 1)$, $\Omega(a^n - 1)/n$

1) If $p > 27$ is a prime for which $P(2^p - 1) \le p^2$ (where $P(n)$ denotes the greatest prime factor of n), then there is an effectively computable constant c such that

$$\omega(2^p - 1) \ge c \frac{\log p}{\log\log p}$$

P. Erdős and T.N. Shorey. *On the greatest prime factor of $2^p - 1$ for a prime p and other expressions*. Acta Arith. 30 (1976), 257–265.

2) $$\lim_{n\to\infty} \frac{\Omega(a^n - 1)}{n} = 0$$
for every $a > 1$, integer.

A. Turull and A. Zame. *Number of prime divisors and subgroup chains*. Arch. Math. (to appear).

§ V.23 ω-highly composite, ω-largely composite and ω-interesting numbers

1) A number n is called *ω-highly composite* if $m < n \Rightarrow \omega(m) \le \omega(n)$.
Let $Q_h(x) = \{\text{card } n \le x : n \text{ is an } \omega\text{-highly composite number}\}$. Then
$$Q_h(x) \sim \frac{\log x}{\log\log x}$$

P. Erdős and J.-L. Nicolas. *Sur la fonction: nombre de facteurs premiers de n*. L'Enseign. Math. 27 (1981), 3–27.

2) A number $n \ge 2$ is called *ω-largely composite* if
$$1 \le m \le n \Rightarrow \omega(m) \le \omega(n)$$
Let $Q_l(x) = \{\text{card } n \le x : n \text{ is an } \omega\text{-largely composite number}\}$. Then
$$\exp\left(c_1\sqrt{\log x}\right) \le Q_l(x) \le \exp\left(c_2\sqrt{\log x}\right)$$
where $0 < c_1 < c_2$ are constants.

(See P. Erdős and J.-L. Nicolas.)

3) A number is called *ω-interesting* if $m > n \Rightarrow \omega(m)/m < \omega(n)/n$.
There exist infinitely many *strangulation points* (n_k) for the function $n - \omega(n)$, i.e. such that
$$m < n_k \Rightarrow m - \omega(m) < n_k - \omega(n_k)$$
and
$$m > n_k \Rightarrow m - \omega(m) > n_k - \omega(n_k)$$

(See P. Erdős and J.-L. Nicolas.)

§ V.24 On $\omega(n)/n$

The set $\{n : \omega(n)|n\}$ has density 0

C.N. Cooper and R.E. Kennedy. *Chebyshev's inequality and natural density*. Amer. Math. Monthly 96 (1989), 118–124.

Remark. The determination of the density of the set $\{n : \Omega(n) | n\}$ is an open problem.

§ V.25 On $(n, \omega(n)) = 1$ and $(n, \Omega(n)) = 1$

1) Let $T(x) = \text{card}\{n \leq x : (n, \omega(n)) = 1\}$. Then

$$T(x) \sim \frac{6}{\pi^2} \cdot x$$

$(x \to \infty)$

V.E. Vol'kovič. *Numbers that are relatively prime to their number of prime divisors* (Russian.) Izv. Akad. Nauk UzSSR Ser. Fiz.-Mat. Nauk 1976, no. 4, 3–7, 86.

2) Let $Q(x) = \text{card}\{n \leq x : (n, \Omega(n)) = 1\}$. Then

$$Q(x) = \frac{6}{\pi^2} \cdot x + O(x(\log\log\log x)^{-1/3}(\log\log\log\log x)^{-1})$$

K. Alladi. *On the probability that n and $\Omega(n)$ are relatively prime*. Fib. Quart. 19 (1981), 228–233.

§ V.26 On $\omega((n, \varphi(n))) = k$

1) Let $N_k(m) = \sum_{d|m, \Omega(d)=k} 1$. Then

a) $N_{k-1}(m) \cdot N_{k+1}(m) \leq N_k^2(m)$
for $0 < k < \Omega(m)$

b) $N_{k+1}(m) \leq (\Omega(m) - k) \cdot N_k(m)$
for $0 \leq k < \Omega(m)/2$

I. Anderson. *On the divisors of a number*. J. London Math. Soc. 43 (1968), 410–418.

2) Let $A_k(x) = \sum_{n \leq x, \omega((n, \varphi(n)))=k} 1; k = 0, 1, 2, \ldots$. Then

$$A_k(x) = (1 + o(1)) \cdot x \cdot e^{-\gamma} \cdot (\log\log\log\log x)^k / (k! \log\log\log x)$$

M.R. Murty and V.K. Murty. *Some results in number theory*. I. Acta Arith. 35 (1979), 367–371.

Remark. The case $k = 0$ is due to Erdős.

P. Erdős. *Some asymptotic formulas in number theory*. J. Indian Math. Soc. (N.S.) 12 (1948), 75–78.

§ V.27 Gaussian law of errors for ω

1) Let $\lambda \in \mathbb{R}$ and $e(n) = \log\log n + \lambda(\log\log n)^{1/2}$. Then

a) $\mathrm{card}\{m \le n : \omega(m) < e(n)\} = n \cdot \phi(\lambda) + o(n)$

P. Erdős and M. Kac. *On the Gaussian law of errors in the theory of additive functions.* Proc. Nat. Acad. Sci. 25 (1939), 206–207,

and

P. Erdős and M. Kac. *On the Gaussian law of errors in the theory of additive number theoretic functions.* Amer. J. Math. 62 (1940), 738–742.

b)
$$\mathrm{card}\{m \le n : \omega(m) < e(n)\} = n \cdot \phi(\lambda) + O(n \log\log\log n/(\log\log n)^{1/4})$$

where $\phi(\lambda) = \frac{1}{\sqrt{2\pi}} \int_{-\infty}^{\lambda} e^{-t^2/2} dt$

W.J. Le Veque. *On the size of certain number-theoretic functions.* Trans. Amer. Math. Soc. 66 (1949), 440–463.

c) $\mathrm{card}\{m \le n : \Omega(m) < e(n)\} = n \cdot \phi(\lambda) + O(n(\log\log n)^{-1/2})$

A. Rényi and P. Turán. *On a theorem of Erdős-Kac.* Acta Arith. 4 (1958), 71–84.

Note. This result was a conjecture of Le Veque. For a generalization see

J. Galambos. *On the distribution of prime-independent additive number-theoretical functions.* Arch. Math. (Basel) 19 (1968), 296–299.

2) Let $\Delta(m) = (\omega(m) - \log\log x)/(\log\log x)^{1/2}$. Then:

a) For fixed distinct integers $a_1, \ldots, a_s \ge 0$ and for a fixed integer $a > 0$, we have

$$\frac{1}{x} \cdot \mathrm{card}\{m \le x : \Delta(m + a_1) < \lambda_1, \ldots, \Delta(m + a_s) < \lambda_s\} = \phi(\lambda_1) \cdots \phi(\lambda_s) + R_1$$

and

b)
$$\frac{1}{x} \cdot \mathrm{card}\{m \le x : \omega(m) < \lambda(m + a) + \lambda\sqrt{2\log\log x}\} = \phi(\lambda) + R_2$$

where $R_1 = O((\log\log x)^{-1/2}(\log\log\log x)^2)$,
$R_2 = O((\log\log x)^{-1/3}(\log\log\log x)^{5/3})$ uniformly for all real $\lambda_1, \ldots, \lambda_s$, and λ, as $x \to \infty$

I. P. Kubilius. *On asymptotic distribution laws of certain number-theoretic functions* (Russian.) Vilniaus Valst. Univ. Mosklo Darbai Mat. Fiz. Chem. Mosklu Ser. 4 (1955), 45–59.

Remark. For $s = 1, a_1 = 0, R_1 = o(1)$ we reobtain the famous theorem of Erdős and Kac.

P. Erdős and M. Kac. *The Gaussian law of errors in the theory of additive number theoretic functions.* American J. Math. 62 (1940), 738–742.

c) $$\frac{1}{N}\sum_{n\le N}|\Delta(n)|^k \sim 2^{k/2}\cdot\pi^{-1/2}\cdot\Gamma\left(\frac{k+1}{2}\right)$$

as $N \to \infty$

A. Ghosh. *An extension of the method of moments for additive functions.* H. Delange Colloquium (Orsay, 1982), 65–73, Publ. Math. Orsay, 83–74, Orsay, 1983.

3) Let A and B two finite sets of integers, and let $A \oplus B$ denote their direct sum. Then we have, uniformly for $x \ge 3, t > 0$,

$$(|A|\cdot|B|)^{-1}\sum_{\substack{n\in A\oplus B\\ \Omega(n)\le t}} = \phi\left(\frac{t-\log\log x}{\log\log x}\right) + O\left(\frac{x}{\sqrt{|A|\cdot|B|\log\log x}}\right)$$

where $|A|$ denotes the cardinality of the set A, and

$$\phi(u) = \frac{1}{\sqrt{2\pi}}\int_{-\infty}^{u} e^{-t^2/2}dt$$

G. Tenenbaum. *Facteurs premiers de sommes d'entiers.* Proc. Amer. Math. Soc. 106 (1989), 287–296.

Remarks: (i) This theorem generalizes the Erdős-Kac theorem which corresponds to the case $A = B = \{n : n \le x\}$

(ii) The result has been conjectured by Erdős, Maier and Sárközy. They obtained an O term with $(\log\log x)^{-1/4}$ in place of $(\log\log x)^{-1/2}$, which is optimal.

P. Erdős, H. Maier and A. Sárközy. *On the distribution of the number of prime factors of sums $a+b$.* Trans. Amer. Math. Soc. 302 (1987), 269–280.

§ V.28 On the statistical property of prime factors of natural numbers in arithmetic progressions

1) Let

$N(n, x; k, l) =$ card $\{m \le n : \Omega(m) < \log\log n + x(\log\log n)^{1/2}$ and $m \equiv l(\text{mod} k)\}$

Then

$$\frac{1}{n}\cdot N(n, x; k, l) = \frac{1}{k}\cdot\phi(x) + O((\log\log n)^{-1/2})$$

G.J. Rieger. *Zur Statistik der Primfaktoren der natürlichen Zahlen in arithmetischen Progressionen.* J. Reine Angew. Math. 206 (1961), 26–30.

2) Let $V_{k,l}$ denote the number of prime divisors of m, with

$$V_{k,l} \equiv l(\text{mod } k)$$

counting multiplicity. Let $(k, l) = 1$. Then

$$\operatorname{card}\{m \le n : V_{k,l}(m) < \log\log n/\varphi(k) + x(\log\log n/\varphi(k))^{1/2}\} = \\ = \phi(x) + O((\log\log n)^{-1/2})$$

F. Gyapjas and I. Kátai. Zur Statistik der Primfaktoren der natürlichen Zahlen. Ann. Univ. Sci. Budapest 7 (1964), 59–66.

§ V.29 Distribution of values of ω in short intervals

Let $\{b(n)\}$ denote a sequence of integers with $1 \le b(n) \le n$ and

$$b(n) \ge n^{a(n)(\log\log n)^{-1/2}}$$

where $1 \le a(n) \le (\log\log n)^{1/2}$ and $a(n) \to \infty$ as $n \to \infty$. Then

$$\frac{1}{b(n)} \cdot \operatorname{card}\{m \in (n, n + b(n)] : \omega(m) - \log\log m < \\ < x(\log\log m)^{1/2}\} \to \frac{1}{\sqrt{2\pi}} \int_{-\infty}^{x} e^{-t^2/2} dt$$

as $n \to \infty$

G.J. Babu. *Distribution of the values of ω in short intervals.* Acta Math. Acad. Sci. Hungar. 40 (1982), 135–137.

See also

G.J. Babu. *On the mean values and distributions of arithmetic functions.* Acta Arith. 40 (1981/82), 63–77.

§ V.30 Distribution of ω in the sieve of Eratosthenes

Define $S(x, y) = \{n \le x :$ least prime divisor of n is $\ge y\}$, $\phi_\lambda(x, y) = \operatorname{card}\{n \in S(x, y) : \omega(n) - \log\alpha < \lambda\sqrt{\log\alpha}\}$, where $\alpha = \log x/\log y$. Then for $2 \le y \le x$,

$$\left|\phi_\lambda(x, y)/|S(x, y)| - \frac{1}{\sqrt{2\pi}} \int_{-\infty}^{\lambda} e^{-u^2}/2du\right| \ll \frac{1}{\sqrt{\log\alpha}}$$

K. Alladi. *The distribution of ω(n) in the sieve of Eratosthenes.* Quart. J. Math. Oxford Ser. (2) 33 (1982), 129–148.

§ V.31 Number of $n \le x$ with $\Omega(n) = i$

1) Denote $\pi_i(x) = \sum_{n \le x, \Omega(n)=i} 1 =$ number of integers $n \le x$ satisfying $\Omega(n) = i$.

Then:

a) For all $\delta > 0$ there exists a constant $c_1 > 0$ such that for all sufficiently large x

$$\pi_i(x) < \frac{c_1}{\delta}\frac{x}{\log x}\frac{(\log\log x)^{i-1}}{(i-1)!}$$

for $1 \le i \le (10/9 - \delta)\log\log x$

G.H. Hardy and S. Ramanujan. *The normal number of prime factors of a number n.* Quart. J. Math. 48 (1920), 76–92.

b) For all $\delta > 0$ there exists a constant $c_2(\delta) > 0$ such that

$$\pi_i(x) < c_2(\delta)\frac{x}{\log x}\frac{(\log\log x)^{i-1}}{(i-1)!}$$

for $x \ge 3$, $1 \le i \le (2 - \delta)\log\log x$

L.G. Sathe. *On a problem of Hardy on the distribution of integers having a given number of prime factors.* J. Indian Math. Soc. 17 (1953), 63–141 and 18 (1954), 27–81.

c) For all $\delta > 0$ we have

$$\pi_i(x) \sim c_3(x\log x)2^{-i}$$

for $(2+\delta)\log\log x \le i \le c_4\log\log x$

A. Selberg. *Note on a paper by L.G. Sathe.* J. Indian Math. Soc. 18 (1954), 83–87.

d) $\pi_i(x) = F\left(\frac{i-1}{\log\log x}\right)\frac{x}{\log x}\cdot\frac{(\log\log x)^{i-1}}{(i-1)!}\cdot\left(1 + O_\delta\left(\frac{1}{\log\log x}\right)\right)$ uniformly for $x \ge 3,\ 0 < \delta < 1, 1 \le i \le (2-\delta)\log\log x$, where $F(z) = (1/\Gamma(z+1))\prod_p(1 - z/p)^{-1}(1 - 1/p)^z(|z| < 2)$

(See A. Selberg.)

e) For all $\delta > 0$ we have

$$\pi_i(x) < \begin{cases} c_5(\delta)\dfrac{x}{\log x}\cdot\dfrac{(\log\log x)^{i-1}}{(i-1!)}, & \text{for } 1 \le i \le (2-\delta)\log\log x \\ c_6 i^4\dfrac{x\log x}{2^i}, & \text{for } 1 \le i \end{cases}$$

and for all $x \ge 3$.

P. Erdős and A. Sárközy. *On the number of prime factors of integers.* Acta Sci. Math. 42 (1980), 237–246.

Corollary. If $\delta > 0$ and $1 < y < 2 - \delta$, then we have

$$\sum_{i\ge j}\pi_i(x) < c_7(\delta)\frac{1}{y-1}\frac{x}{\log x}\frac{(\log\log x)^{j-1}}{(j-1)!}$$

for $y \le j/\log\log x \le 2 - \delta$, $x > x_0(y, \delta)$

Corollary. If $y > 1$ and $\varepsilon > 0$ then for $y\log\log x \le j$, $x > x_0(\varepsilon)$ we have

$$\sum_{i \geq j} \pi_i(x) < \begin{cases} \dfrac{x}{(\log x)^{\varphi(y)-\varepsilon}} & , \text{ if } \quad 1 < y < 2 \\ \dfrac{x}{(\log x)^{(1-\varepsilon)y\log 2 - 1}} & , \text{ if } \quad 2 \leq y \end{cases}$$

where $\varphi(y) = 1 + y \log y - y$

(See P. Erdős and A. Sárközy.)

2) If $\pi_i(x, y) = \text{card}\{n \leq x : \Omega(n, y) = i\}$, where $\Omega(n, y) = \sum_{p^\alpha \| n, p > y} \alpha$, then

$$\pi_i(x, y) \ll_\varepsilon \frac{x}{u} \frac{(\log 2u)^k}{k!}$$

for all $\varepsilon > 0$ and $k \in \mathbb{N}$ with $0 \leq k \leq (2 - \varepsilon) \log u$

G. Halász. *Remarks to my paper: "On the distribution of additive and the mean value of multiplicative arithmetic functions."* Acta Math. Hungar. 23 (1972), 425–432

(The above result is a particular case.)

3) Let $\pi_i(x, \alpha) = \sum_{n \leq x, \Omega(n) = i} e(n\alpha)$, where $e(t) = \exp(2\pi i t)$

a) Let $x \geq 2, 0 < \delta < 1$ and $Q = x/(\log x)^{10}$. For all $\alpha \in \mathbb{R}$ and all integers a, q, i such that $(a, q) = 1, 1 \leq q \leq Q, |\alpha - a/q| \leq 1/qQ$, $2 \leq i \leq (2 - \delta) \log \log x$, we have

$$\frac{\pi_i(x, \alpha)}{\pi_i(x)} \ll_\delta \frac{\log \log \log x}{\log \log x} + q^{-\delta/2}$$

Y. Dupain, R.R. Hall and G. Tenenbaum. Sur l'équirépartition modulo 1 de certain fonctions de diviseurs. J. London Math. Soc. (2) 26 (1982), 394–411.

b) If $E(x, \alpha) = \frac{1}{x} \sum_{n \leq x} e(\alpha n)$, we have uniformly for $x \geq 1, i \geq 1, \alpha \in \mathbb{R}$,

$$\pi_i(x, \alpha) = \pi_i(x) \cdot (E(x, \alpha) + O(\delta_i(x)))$$

where

$$\delta_i(x) = \frac{1}{\sqrt{\log \log x}} + \frac{|i - \log \log x|}{i + \log \log x}$$

G. Tenenbaum. *Facteurs premiers de sommes d'entiers.* Proc. Amer. Math. Soc. 106 (1989), 287–296.

Remark. 3) a) and b) remain valid if Ω is replaced by ω.

4) a) For all $\varepsilon > 0, \eta > 0$, if $x \geq$ e and
$(2 + \varepsilon) \log \log x \leq k \leq (\log x)/\log 2 - \eta$, then
$\pi_i(x) = C(x/2^i) \cdot \log(x/2^i) \cdot (1 + O(\log \log(3x/2^i))^{-1/4})$
where $C = 0.378694$

J.-L. Nicolas. *Autour de formules dues à A. Selberg.* H. Delange Colloquium (Orsay, 1982), 122–134, Publ. Math. Orsay, 83–4, Orsay, 1983.

b) Let $P_n(x) = \sum_{j=0}^{n} x^j/j!$, $y = x \cdot 2^{-i}$.

$$r = 0$$

if $i = 1$;

$$r = 2 \cdot (P_{i-2}(2 \log\log y)/P_{i-1}(2 \log\log y))$$

if $i \geq 2$, $y \geq 3$;

$$f(z) = \frac{1}{2^{z-1} \cdot \Gamma(z+1)} \prod_{p \geq 3} \left(1 - \frac{1}{p}\right)^z \left(1 - \frac{z}{p}\right)^{-1}$$

$$Q(\lambda) = \lambda \log \lambda - \lambda + 1$$

$$R_\lambda(y) = \inf((\log\log y)^{-1}, (\log\log y)^{-1/2} \cdot (\log y)^{-2Q(\lambda)})$$

if $\lambda \geq 1$, $y \geq 3$.

Let $B < 3/2$. Then uniformly for $y \geq 3$,

$$\pi_i(x) = f(r) \cdot \frac{y}{\log y} \cdot P_{i-1}(2 \log\log y) \cdot (1 + O_B(R_\lambda(y)))$$

where $\lambda = \sup(1, \inf(B, (i-1)/2 \log\log y))$

M. Balazard, H. Delange and J.-L. Nicolas. *Sur le nombre de facteurs premiers des entiers.* C.R. Acad. Sci. Paris Sér. I, Math. 306 (1988), 511–514.

c) For sufficiently large x, the function $\pi_i(x)$ is unimodal in i, i.e. there exists an i_0 such that $\pi_i(x)$ is nondecreasing for $i < i_0$ and nonincreasing for $i \geq i_0$.

M. Balazard. *Comportement statistique du nombre de facteurs premiers des entiers.* Séminaire de Théorie des Nombres, Paris, 1987–88, 1–21, Progr. Math., 81, Birkhäuser Boston, MA, 1990.

§ V.32 Number of $n \leq x$ with $\omega(n) = i$

Let $\rho_i(x) = \sum_{n \leq x, \omega(n) = i} 1$

1) a) $\rho_i(x) \sim \frac{x}{\log x} \frac{(\log\log x)^{i-1}}{(i-1)!}$

$(x \to \infty)$

E. Landau. *Handbuch der Lehre von der Verteilung der Primzahlen.* Vol. I, Leipzig, 1909 (p. 211.)

b) $\rho_i(x) = F(y) \frac{x}{\log x} \frac{(\log\log x)^{i-1}}{(i-1)!} \left(1 + O\left(\frac{1}{\log\log x}\right)\right)$

uniformly for $x \geq 3$ and $1 \leq i \leq C \log\log x$, for any given fixed $C > 0$, where $F(z) = \frac{1}{\Gamma(z+1)} \prod_p \left(1 + \frac{z}{p-1}\right)\left(1 - \frac{1}{p}\right)^z$ and $y = \frac{i}{\log\log x}$

L.G. Sathe. *On a problem of Hardy and Ramanujan on the distribution of integers having a given number of prime factors*. J. Indian Math. Soc. 17 (1953), 63–141 and 18 (1954), 27–81.

and

A. Selberg. *Note on a paper by L.G. Sathe*. J. Indian Math. Soc. 18 (1954), 83–87.

c) $\rho_i(x) \leq c_1 \frac{x}{\log x} \frac{(\log\log x + c_2)^{i-1}}{(i-1)!}$
for all $x \geq 3$

G.H. Hardy and S. Ramanujan. *The normal number of prime factors of a number n*. Quart. J. Math. 48 (1920), 76–92.

d) If $L = \log\log x - \log i - \log\log i$ and $(\log\log x)^2 \leq i \leq \log x/(3 \log\log x)$, then

$$\rho_i(x) = \frac{x}{i!} \exp(i(\log L + (\log L)/L + O(1/L)))$$

C. Pomerance. *On the distribution of round numbers*. Number theory (Oatacamund, 1984), Lecture Notes 1122, Springer, 1985, 173–200.

Remark. This result is in connection with a theorem of D. Hensley:

$$\rho_i(x) = F(y) \frac{x}{\log x} \frac{(\log\log x)^{i-1}}{(i-1)!} \exp\left(-\frac{1}{2} i \left(\frac{\log\log\log x}{\log\log x}\right)^2\right) \cdot \left(1 + O\left(\frac{1}{\sqrt{\log\log\log x}}\right)\right)$$

uniformly in the range $x \geq x_0$,

$$i \ll \frac{(\log\log x)^2}{(\log\log\log x)^{3/2} (\log\log\log\log x)}$$

D. Hensley. *The distribution of round numbers*. Proc. London Math. Soc. (3) 54 (1987), 412–444.

e) $\rho_i(x) = \frac{\lambda^{-1} x^\alpha F(\lambda, \alpha)}{(\log x) i w(i) w(\lambda)} (1 + O(1/L))$
uniformly for $x \geq x_0, 1 \leq i \ll \log x/(\log\log x)^2$, where $F(z, s) = \prod_p (1 + z/p^s - 1)$, $w(t) = \Gamma(t)(t/e)^t (t > 0)$ and λ, α are the (unique) solutions of

$$\sum_p \frac{1}{p^\alpha - 1 + \lambda} = \frac{i}{\lambda}$$

$$\sum_p \frac{\lambda \log p}{(1 - p^{-\alpha})(p^\alpha - 1 + \lambda)} = \log x$$

the summation being over all primes

A. Hildebrand and G. Tenenbaum. *On the number of prime factors of an integer.* Duke Math. J. 56 (1988), 471–501.

f) $\dfrac{\rho_{i+1}(x)}{\rho_i(x)} = \dfrac{L}{k}\left(1 + O\left(\dfrac{\log L}{L}\right)\right)$
uniformly for $x \geq 3$, $1 \leq i \ll \log x/(\log\log x)^2$.
Here $L = \log\log x - \log i - \log\log(i+2)$ (as in e))
(See A. Hildebrand and G. Tenenbaum.)

g) $$\rho_i(x) = \sum_{j=0}^{N} x P_j(\log\log x)\log^{-j-1} x + O(x\log^{-N-2} x(\log\log x)^{k-1})$$
for any fixed integer $k \geq 1$, where each P_j is a polynomial of degree not exceeding $k-1$, and N is arbitrary non-negative, fixed integer.

H. Delange. *Sur des formules de Atle Selberg.* Acta Arith. 19 (1977), 105–146.

Remark. A similar result holds, when ω is replaced by Ω.

2) Let $A_i(x,h) = \text{card}\{n : \omega(n) = i, x \leq n \leq x+h\}$, where $1 \leq h \leq x$. Then
$$A_i(x,h) = (1+o(1))(h(\log\log x)^{i-1}/(i-1)!\log x)$$
uniformly in $1 \leq i \leq \log\log x + c_x(\log\log x)^{1/2}$, where c_x is a sequence of real numbers tending to infinity "quite slowly."

Corollary. $$\frac{1}{h}\text{card}\left\{n \in [x, x+h] : \frac{\omega(n) - \log\log x}{\sqrt{\log\log x}} < y\right\} = (1+o(1))\,\phi(y)$$
as $h \to \infty$, where ϕ is the Gaussian distribution function.

I. Kátai. *A remark on a paper by K. Ramachandra.* Number theory (Octacamund, 1984), 147–152. Lecture Notes, 1122, Springer, 1985.

3) Let $f_k(n)$ be the characteristic function of the set of positive integers with exactly k different prime factors. Then
$$\sum_{d \leq x^{1/2-\varepsilon}} \max_{(a,d)=1} \max_{y \leq x} \left| \sum_{\substack{n \leq y \\ n \equiv a(d)}} f_k(n) - \frac{1}{\varphi(d)} \sum_{\substack{n \leq y \\ (n,d)=1}} f_k(n) \right| \ll$$
$$\ll \rho_z(n) \cdot (\log x)^{-A}$$
uniformly in the interval $1 \leq k \leq \eta \log x(\log_2 x)^{-2}$,
$\eta = \eta(A, \varepsilon) > 0$

D. Wolke and T. Zhan. *On the distribution of integers with a fixed number of prime factors.* Math. Z. 213 (1993), 133–144.

§ V.33 The functions $\omega(n; E)$ and $S(x, y; E, \omega)$

1) Let E be a set of primes, $E(x) = \sum_{p \le x, p \in E} 1/p$ and $\omega(n; E)$ — the number of distinct primes in E that divide n.
If $0 < \alpha \le \beta < 1$ and $E(x) > 0$, then
$$\operatorname{card}\{n \le x : |\omega(n, E) - E(x)| > \alpha E(x)\}$$
$$\le c(\beta)\frac{1}{\alpha} \cdot x \cdot E(x)^{-1/2} \cdot \exp(Q(\alpha) \cdot E(x))$$
where $Q(\alpha) = \alpha - (1 + \alpha)\log(1 + \alpha)$

K.K. Norton. *The number of restricted prime factors of an integer.* I. Illinois J. Math. 20 (1976), 681–705.

Similar results appear in

K.K. Norton. *Ibid.* II. Acta Math. 143 (1979), 9–38.

2) Let $S(x, y; E, \omega) = \operatorname{card}\{n \le x : \omega(n, E) > y\}$. Suppose that there exists a real number $\gamma(E) > 0$ such that
$$\sum_{p \le x, p \in E} 1 = \gamma(E)(x/\log x)(1 + O_E(1/\log x))$$
for $x \ge 2$. Let $\varepsilon > 0$ and let $x \ge c_1(E, \varepsilon)$.

a) If
$$c_2(E) \le y \le (\log x)(\log\log x)^{-1} +$$
$$+ (1 + \log\gamma(E) - \varepsilon) \cdot (\log x) \cdot (\log\log x)^{-2}$$
then
$$S(x, y; E, \omega) \ge x \exp(-y(\log y + \log\log y - \log\gamma(E) - 1) +$$
$$+ O_E(y(\log\log y)/\log y))$$

b) If $x \ge 3$ and $y \ge \gamma(E)\log\log x$, then
$$S(x, y; E, \omega) \le x \exp(-y(\log y - \log\log\log x - \log\gamma(E) - 1) -$$
$$- \gamma(E)\log\log x + O_E(y/\log\log x))$$

K.K. Norton. *On the number of restricted prime factors of an integer.* III. Enseign. Math. (2) 28 (1982), 31–52.

3) Let $\Omega_E(n) = \operatorname{card}\{(p, k) \in E \times \mathbb{N}^* : p^k | n\}$ (see the notations in 1).) Then, if $\sum_{p \in E}(1/p) = +\infty$, $p_1 < p_2 < \cdots$ is the ordered sequence of elements of E, then:

a) $\Omega_E(n) \le \log n/\log p_1$

b) $\sum_{n \le x} \Omega_E(n) \sim x \cdot E(x)$
as $x \to \infty$

c) $\Omega_E(n) \sim E(n)$

almost everywhere.

M. Balazard. *Distribution des valeurs de la fonction* $\Omega_E(N)$. Colloque de Théorie Analytique des Nombres ≪Jean Coquet≫. (Marseille, 1985), 49–54, Publ. Math. Orsay, 88–02, Univ. Paris XI, Orsay, 1988.

4) Let $N(m, x) = \sum_{n \le x, \Omega_E(n) = m} 1$. Then:

a) $N(m, x) = x(E^m(x)/m!)e^{-E(x)}\cdot$

$\cdot (1 + O(|m - E(x)|/E(x)) + O\left(1/\sqrt{E(x)}\right)$

uniformly in m and x for $\delta \le m/E(x) \le 2 - \delta$, $E(x) \ge 2$, with any fixed $\delta > 0$

G. Halász. *On the distribution of additive and the mean values of multiplicative arithmetic functions*. Studia Sci. Math. Hung. 6 (1971), 211–233.

Remark. This improves a result by Kubilius:

I. P. Kubilius. *Probabilistic methods in the theory of numbers*. Providence. R.I. 1964.

b) $N(m + 1, x) \sim N(m, x)$

for $m - E(x) = o(E(x))$

(See G. Halász.)

Note. See also:

G. Halász. *Remarks to my paper "On the distribution..."* Acta Math. Hungar. 23 (1972), 425–432,

and:

A. Sárközy. *Remarks on paper of G. Halász "On the distribution..."* Period. Math. Hungar. 6 (1971), 211–233.

§ V.34 Sumsets with many prime factors

Suppose that A and B are subsets of $\{n \le N/2\}$. Let $m = m(N)$ be the maximal value of $\omega(n)$ for $n \le N$. Then for each $\varepsilon > 0$ there is a $c(\varepsilon)$ such that if $|A| \cdot |B| > \varepsilon N^2$, then we have

$$\max_{a \in A, b \in B} \omega(a + b) > m - c(\varepsilon)\sqrt{m}$$

P. Erdős, C. Pomerance, A. Sárközy, C.L. Stewart. *On elements of sumsets with many prime factors*. J. Number Theory 44 (1993), 93–104.

§ V.35 On the integers n for which $\Omega(n) = k$

Let $S(x,k) = \{n \leq x : \Omega(n) = k\}$ and for any prime p, let $V_p(n)$ denote the exponent of p in the prime factorization of n. For $k \geq (2+\varepsilon)\log\log x$, the numbers

$$\frac{V_2(n) - k + 2\log\log(x \cdot 2^{-k})}{\sqrt{2\log\log(x \cdot 2^{-k})}}, n \in S(x,k)$$

have asymptotic normal distribution.

H. Delange. *On the integers n for which $\Omega(n) = k$*. I. Progr. Math. 85 (1990), 119–132.

Remark. For a slightly stronger, and more general results, see

H. Delange. Ibid. II. Monatsh. Math. 116 (1993), 175–196.

Chapter VI

FUNCTION μ; k-FREE AND k-FULL NUMBERS

§ VI. 1 Average order of $\mu(n)$

Let $M(x) = \sum_{n \leq x} \mu(n)$. Then:

1) $M(x) = o(x)$

H. von Mangoldt. *Beweis der Gleichung* $\sum_{k=1}^{\infty} \mu(nK)/k = 0$. Sitz. Königlichen Preuss. Akad. Wiss. Berlin, 1897, 835–852.

2) $M(x) = O(x \cdot \exp(-(\log x)^{1/2}))$

E. Landau. *Beiträge zur analytischen Zahlentheorie.* Rend. Circ. Mat. Palermo 26 (1908), 169–302.

3) $M(x) = O(x \exp(-A(\log x \cdot \log\log x)^{1/2}))$

E. Landau. *Vorslesungen über Zahlentheorie*, II. Band. Aus der analytischen und geometrischen Zahlentheorie. Leipzig 1927.

4) $M(x) = O(x \exp(-A \log^{3/5} x \cdot (\log\log x)^{-1/5}))$

A. Walfisz. *Weylsche Exponentialsummen in der neuren Zahlentheorie.* Berlin, 1963.

§ VI. 2 Estimates for $M(x)$. Mertens' conjecture

1) a) $|M(x)| < x/80$
for $x \geq 1119$ and

$$|M(x)| \leq \frac{x+1}{80} + \frac{11}{2}$$

for $x > 0$

R.A. Mac Leod. *A new estimate for the sum* $M(x) = \sum_{n \leq x} \mu(n)$. Acta Arith. 13 (1967/68), 49–59. Errata: 16 (1969/70), 99–100.

b) $|M(x)| < a \cdot x/(\log x)^{\alpha}$
for $x > 1$, where
$(a, \alpha) = (1.2, 2/3); (12, 1); (26, 10/9)$

L. Schoenfeld. *An improved estimate for the summatory function of the Möbius function.* Acta Arith. 15 (1968/69), 221–233.

c) $|M(x)| < x/143.7$
for $x \geq 3297$

F. Dress. *Majorations de la fonction sommatoire de la fonction de Möbius.* Bull. Soc. Math. France, mémoire N° 49–50, 1977, pp. 47–52.

d) $|M(x)| < x/1036$
if $x \geq 120727$

N. Costa Pereira. *Elementary estimates for the Chebyshev function $\Psi(x)$ and for the Möbius function $M(x)$.* Acta Arith. 52 (1989), 307–337.

2) a) $\liminf\limits_{x\to\infty} \dfrac{M(x)}{x} < -1$
and

$$\sqrt{x} \liminf \frac{M(x)}{\sqrt{x}} \leq M(x) + 2M^*\left(\frac{1}{x}\right) \leq \sqrt{x} \limsup \frac{M(x)}{\sqrt{x}}$$

where $M^*(x) = 1 + \sum_{n=1}^{\infty} (-1)^n (2\pi x)^{2n} \dfrac{1}{2n(2n)!\zeta(2n+1)}$

W.B. Jurkat. *On the Mertens conjecture and related general Ω-theorems.* Proc. Symp. Pure Math. 24 (1972), St. Louis Univ. MO., 147–158, A.M.S. Providence R.I., 1973.

Remark. The first relation disproves an old conjecture of R.D. von Sterneck from 1901 (that $|2M(x)| \leq \sqrt{x}$ for all large x.)

b$_1$) $\limsup \dfrac{M(x)}{\sqrt{x}} > 0.557$

R.J. Anderson and H.M. Stark. *Oscillation theorems.* Analytic Number Theory (Philad. Pa. 1980), pp. 79–106, Lecture Notes, 899, Springer, 1981.

b$_2$) $\limsup\limits_{x\to\infty} \dfrac{M(x)}{\sqrt{x}} > 0.86$

H.J.J. te Riele. *Computations concerning the Mertens conjecture.* J. Reine Angew. Math. 312 (1979), 356–360.

c) $\limsup\limits_{x\to\infty} \dfrac{M(x)}{\sqrt{x}} > 1.06$

d) $\liminf\limits_{x\to\infty} \dfrac{M(x)}{\sqrt{x}} < -1.009$

A.M. Odlyzko and H.J.J. te Riele. *Disproof of the Mertens conjecture.* J. Reine Angew. Math. 357 (1985), 138–160.

Remark. The famous conjecture by Mertens stated

$$|M(x)| < \sqrt{x}$$

for $x > 1$

e) $|M(x)| > \sqrt{x}$
holds for at least one $x \le \exp 10^{65}$

J. Pintz. *An effective disproof of the Mertens conjecture.* Astérisque No. 147–148(1987), 325–333, 346.

§ VI. 3 μ in short intervals

$$M(x+h) - M(x) = o(h)$$

as $x \to \infty$, uniformly for h, $x^{\theta} \le h \le x$, provided that $\theta > 7/12$

Y. Motohashi. *On the sum of the Möbius function in a short segment.* Proc. Japan Acad. 52 (1976), 477–479.

§ VI. 4 Sums involving $\mu(n)$ with $p(n) > y$ or $P(n) < y$, $n \le x$. Squarefree numbers with restricted prime factors

1) Let $M(x,y) = \sum\limits_{n\le x,\, p(n)>y} \mu(n)$ and $M^*(x,y) = \sum\limits_{n\le x,\, P(n)<y} \mu(n)$ where $p(n)$ and $P(n)$ denote the least and the largest prime factor of n, respectively. Then:

a) $M(x,y) = x\,\rho'(\alpha)/\log y + y/\log y + O(x\alpha^2/\log^2 y)$
uniformly for $2 \le y = x^{1/\alpha} < x$. Here $\rho(\alpha) = \lim\limits_{x\to\infty} \dfrac{\psi(x, x^{1/\alpha})}{x}$ and $\psi(x,y) = \sum\limits_{n\le x,\, P(n)<y} 1$

K. Alladi. *Asymptotic estimates of sums involving the Möbius function.* J. Number Theory 14 (1982), 86–98.

2) a) Let $R(y)$ be any function such that
$|\pi(y) - \operatorname{li} y| \le y\, R(y)/\log y$ and denote
$f(\beta) = \int_1^{\beta} M(t)/t^2 \mathrm{d}t$. Then:
$M^*(x,y) = xf(x/y)/\log y - (x/\log^2 y)(1 + O(R(x/y))) + O(x/\log^3 x)$

for $\sqrt{x} \le y \le x$ (where $y = x^{1/\alpha}$);

$M^*(x, y) = xw'(\alpha)/\log^2 y + O(xR(x/y^2)/\log^2 y) + O(x\alpha^2/\log^3 y)$

for $2 \le y \le \sqrt{x}$.

Here $\lim\limits_{x\to\infty} \dfrac{\Phi(x, x^{1/\alpha})\log x}{x} = \alpha w(\alpha)$, $\Phi(x, y) = \sum\limits_{n\le x, p(n)>y} 1$

K. Alladi. *Asymptotic estimates of sums involving the Möbius function.* II. Trans. Amer. Math. Soc. 272 (1982), 87–105.

3) Let $\psi_2(x, y) = \sum\limits_{n\le x, P(n)\le y} |\mu(n)|$. Then:

a) $\psi_2(x, y) = \left(\dfrac{6}{\pi^2} + O((\log\log x)^{-\varepsilon})\right) \cdot \psi(x, y)$

for $\varepsilon > 0$ and $\exp((\log\log x)^{2+\varepsilon}) \le y \le x$

A. Ivić. *On squarefree numbers with restricted prime factors.* Studia Sci. Math. Hungar. 20 (1985), 189–192.

b) The inequality from a) holds if and only if

$$\lim_{x\to\infty} \log y/\log\log x = +\infty$$

A. Ivić and G. Tenenbaum. *Local densities over integers free of large prime factors.* Quart. J. Math. (Oxford) (2) 37 (1986), 401–417,

and

M. Naimi. *Les entiers sans facteurs carrés $\le x$ dont leurs facteurs premiers $\le y$*, in: Groupe de travail en théorie analytique des nombres 1986–87, Publ. Math. Orsay 88–01, pp. 69–76.

c) For $y \ge \log^{1+\varepsilon} x$,

$$\frac{\psi(x, y)}{\zeta(2\beta, y)}\left(1 + O\left(\frac{\log y}{\log x}\right)\right) \le \psi_2(x, y) \le \frac{\psi(x, y)}{\zeta(2\alpha, y)}\left(1 + O\left(\frac{\log y}{\log x}\right)\right)$$

where $\zeta(s, y) = \prod\limits_{p\le y}(1 - p^{-s})^{-1}$, and α, β are the solutions of $\log x = \sum\limits_{p\le y} \log p/(p^\alpha - 1)$, $\log x = \sum\limits_{p\le y} \log p/(p^\beta + 1)$, respectively.

(See M. Naimi.)

§ VI. 5 Oscillatory properties of $M(x)$ and related results

1) If $\displaystyle\int_1^T \frac{|M(x)|}{x}dx < a \cdot T^{1/2}$, for $T \ge 1$, a being independent of T, then

$$\int_1^T \frac{|M(x)|}{x}dx > T^{1/2} \cdot \exp\left(-\frac{\log T}{(\log\log T)^{1/2}}\right)$$

for $T > \max(c, e^a)$, c-constant.

S. Knapowski. *On the mean values of certain functions in prime number theory.* Acta Math. Hungar. 10 (1959), 375–390.

2) Assuming the Riemann Hypothesis,

$$\int_{A(T)}^{T} \frac{|M(x)|}{x} dx > \sqrt{T} \exp\left(-12 \frac{\log T}{\log_2 T} \log_3 T\right)$$

$T > c_1$, where $A(T) = T \exp\left(-c \frac{\log T}{\log_2 T} \log_3 T\right)$ with $c = 100$ and c_1 explicit constant (Here $\log_k$ denotes the k times iterated logarithm.)

S. Knapowski. *On oscillations of certain means formed from the Möbius series.* I. Acta Arith. 8 (1963), 311–320.

3) For $T > c_2$ (positive constant), we have

$$\int_{1}^{T} \frac{|M(x)|}{x} dx > c_3 T^{\alpha/2}$$

with $\alpha = (2 - \sqrt{3})^2 = 0.07\ldots$, and $\max\limits_{T^{\alpha} \leq x \leq T} |M(x)| \geq T^{\alpha/2}$ (unconditionally.)

I. Kátai. *Comparative theory of prime numbers* (Russian). Acta Math. Acad. Hungar. 18 (1967), 133–149.

Remark. The relations in c) hold also with $\alpha = 0.36$, see

I. Kátai. *On oscillations of number-theoretic functions.* Acta Arith. 14 (1967), 107–122.

4) Supposing the existence of a zeta-zero $\rho_0 = \beta_0 + i\gamma_0$, for $T > T_0(\rho_0, \varepsilon)$ (ineffective constant) one has

$$\int_{T^{1-\varepsilon}}^{T} \frac{|M(x)|}{x} dx > T^{\beta_0 - \varepsilon}$$

I. Kátai. *Ω-theorems for the distribution of prime numbers* (Russian). Ann. Univ. Sci. Budapest Eötvös Sect. Mat, 9 (1966), 87–93.

5) If $\rho_0 = \beta_0 + i\gamma_0$ is a zeta-zero, then for $T > e^{|\gamma_0|+4}$ we have

$$\frac{1}{T} \int_{T/(100 \log T)}^{T} |M(x)| dx > \frac{1}{6} \frac{T^{\beta_0}}{|\rho_0|^3}$$

J. Pintz. *Oscillatory properties of* $M(x) = \sum_{n \leq x} \mu(n)$. I. Acta Arith. 42 (1982), 49–55.

6) If $T > c_4 > 0$, then $M(x)$ changes sign in the interval

$$\left[T \exp\left(-3 \log_2^{3/2} T\right), T\right]$$

J. Pintz. *Oscillatory properties of* $M(x) = \sum_{n \leq x} \mu(n)$. II. Studia Sci. Math. Hungar. 15 (1980), 491–496.

7) If $\rho_1 = \beta_1 + i\gamma_1$ is a zeta-zero with $\beta_1 \geq \frac{1}{2}$, and

$T > \max(\exp(2|\gamma_1|), c_5)$ ($c_5 > 0$-constant), then there exist $x', x'' \in \left[T\exp\left(-5\log_2^{3/2} T\right), T\right]$ such that

$$M(x') > \frac{(x')^{\beta_1}}{48|\rho_1|^3}$$

and

$$M(x'') < \frac{-(x'')^{\beta_1}}{48|\rho_1|^3}$$

Corollary. For $T > c_6 > 0$, there exist $x', x'' \in \left[T\exp\left(-5\log_2^{3/2} T\right), T\right]$ such that

$$M(x') > \frac{\sqrt{x'}}{136\,000}$$

$$M(x'') < -\frac{\sqrt{x''}}{136\,000}$$

J. Pintz. *Oscillatory properties of* $M(x) = \sum_{n \le x} \mu(n)$. III. Acta. Arith. 43 (1984), 105–113.

§ VI. 6 The function $M(n, T) = \sum_{d|n, d \le T} \mu(n)$

Let $M(n, T) = \sum_{d|n, d \le T} \mu(d)$. Then:

1) a) $|M(n, T)| \le \binom{\omega(n)}{[\omega(n)/2]} = o(2^{\omega(n)})$

where $\omega(n)$ denotes the number of distinct prime factors of n.

P. Erdös. *On a problem in elementary number theory.* Math. Student 17 (1949), 32–33.

Note. For a generalization, see

N.G. de Bruijn, C.T. van Ebbenhorst and D. Kruyswijk. *On the set of divisors of a number.* Nieuw Arch. Wiskunde (2) 23 (1951), 191–193.

b) For almost all n, we have

$$M(n) = \max_T |M(n, T)| < A^{\omega(n)}$$

for any fixed $A > 3/e$.

R.R. Hall. *A problem of Erdös and Kátai.* Mathematika 21 (1974), 110–113.

c) For almost all n, we have

$$M(n) \ge (\log\log n)^{\gamma}$$

for $\gamma < 0.28754\ldots$, and

$$M(n) \leq f(n) \log\log n$$

for any function $f(n)$ tending to ∞.

H. Maier. *On the Möbius function.* Trans. Amer. Math. Soc. 301 (1987), 649–664.

2) a) For every $\varepsilon > 0$, there exists a T_0 such that for fixed $T > T_0$, the density of integers n such that $M(n,T) \neq 0$ is $\ll (\log T)^{-\delta}$ where $\delta = 1 - (e/2)\log 2$.

b) Let q be fixed, $q \geq 2$, $u = q/(q-1)$ and $\beta = 0$ or 1 according as $q > 2$ or not. Then for almost all n, we have

$$\sum_{m \leq n} \frac{1}{m} |M(n,m)|^q \leq f(n)(F(u))^{(q-1)\,\omega(n)} (\log\log n)^{\beta}$$

provided that $f(n) \to \infty$ as $n \to \infty$. Here

$$F(u) = \frac{2^u}{\sqrt{\pi}} \cdot \frac{\Gamma((u+1)/2)}{\Gamma((u+2)/2)}$$

P. Erdős and R.R. Hall. *On the Möbius function.* J. Reine Angew. Math. 315 (1980), 121–126.

§ VI. 7 Möbius function of order k

1) For a positive integer k, define $\mu_k(1) = 1$; $\mu_k(n) = 0$, if $p^{k+1} | n$ for some prime p; $\mu_k(n) = (-1)^r$ if $n = p_1^k \cdots p_r^k \cdot \prod_{i>r} p_i^{a_i}$ with $0 \leq a_i < k$ for each $i > r$, and $\mu_k(n) = 0$ otherwise. Then

$$\sum_{n \leq x} \mu_k(n) = A_k \cdot x + O(x^{1/k} \cdot \log x)$$

where $A_k = \dfrac{1}{\zeta(k)} \displaystyle\sum_{n=1}^{\infty} \frac{\mu(n)}{n^k} \cdot \prod_{p|n} \frac{(1-p^{-1})}{(1-p^{-k})}$ and $k \geq 2$.

Remark. $\mu_1(n) = \mu(n)$

T.M. Apostol. *Möbius function of order k.* Pacific J. Math. 32 (1970), 21–27.

2) Define $\mu_k^1(n) = (-1)^{\Omega(n)}$, if n is k-free, and $\mu_k^1(n) = 0$ otherwise. Let $\mu_\infty^1(n) = (-1)^{\Omega(n)}$ for all n. (Thus $\mu_\infty^1 \equiv \lambda$-the Liouville function.) For $2 \leq k \leq \infty$ let $M_k^1(x) = \sum_{n \leq x} \mu_k^1(n)$. Then

$$M_k^1(x) = B_k\sqrt{x} + \Omega_{\pm}(\sqrt{x})$$

where $B_2 = 0$, $B_\infty = 1/\zeta(1/2)$, $B_k = \zeta(k/2)/\zeta(1/2)\zeta(k)$ if k is odd, $B_k = 1/\zeta(1/2)\zeta(k/2)$, if k is even.

M. Tanaka. *On the Möbius and allied functions.* Tokyo J. Math. 3 (1980), 215–218.

§ VI. 8 Sums on $\mu(n)/n$, $\mu(n)/n^2$, $\mu^2(n)/n$

1) a) $$\sum_{n\le x} \frac{\mu(n)}{n} = o(1)$$

H. von Mangoldt. *Beweis der Gleichung* $\sum_{k=1}^{\infty} \mu(nK)/k = 0$. Sitz. Königlichen Preuss. Akad. Wiss. Berlin 1987, 835–853.

Remark. For a new proof, see

E. Landau. *Neuer Beweis der Gleichung* $\sum_{k=1}^{\infty} \mu(k)/k = 0$. Inauguraldissertation, Berlin, 1899.

b) $$\sum_{n\le x} \frac{\mu(n)}{n} = O\left(\frac{1}{\log x}\right)$$

C.J. de la Vallée Poussin. *Sur la fonction $\zeta(s)$ de Riemann et le nombre des nombres premiers infèrieurs a une limite donnée.* Mémoires couronnées et autres mémoires. Acad. Royal Sci. Lettres Beaux-Arts Belgique 59, 1899–1900.

c) $$\sum_{n\le x} \frac{\mu(n)}{n} = O(\exp(-(\log x)^{1/\gamma}))$$

for $\gamma > 2$

E. Landau. *Beiträge zur analytischen Zahlentheorie.* Rend. Circ. Mat. Palermo 26 (1908), 169–302.

d) $$\sum_{n\le x} \frac{\mu(n)}{n} = o((\log x)^{-k})$$

for every $k > 0$

S.L. Segal. *A general Tauberian theorem of Landau-Ingham type.* Math. Z. 111 (1969), 159–167.

e) $$\sum_{n\le x} \frac{\mu(n)}{n} = O(\exp(-B(\log x)^{3/5} \cdot (\log\log x)^{-1/5}))$$

R.Q. Jia. *Estimation of partial sums of series* $\sum \mu(n)/n$. Kexue Tongbao 30 (1985), 575–578.

2) a) $$\sum_{n\le x} \frac{\mu(n)}{n^2} = \frac{1}{\zeta(2)} + O(x^{-1}\log^{-k} x)$$

for any $k > 0$

A.A. Gioia and A.M. Vaidya. *The number of square free divisors of an integer.* Duke Math. J. 33 (1966), 797–799.

b) $$\sum_{n\le x,(n,k)=1} \frac{\mu(n)}{n^2} = \frac{k^2}{\zeta(2)J_2(k)} + O(1/x)$$

D. Suryanarayana. *The greatest divisor of n which is prime to k.* Math. Student 37 (1969), 147–152.

3) $$\sum_{n\le x} \frac{\mu^2(n)}{n} = a\log x + b + O(\sqrt{x})$$

D. Suryanarayana. *Asymptotic formula for* $\sum_{n\le x}\mu^2(n)/n$. Indian J. Math. 9 (1967), 543–545.

§ VI. 9 Sums on $\mu(n)\log n/n$, $\mu(n)\log n/n^2$

1) $$\sum_{n\le x} \frac{\mu(n)\log n}{n} = -1 + R(x)$$

where:

a) $R(x) = O(\exp(-c(\log\log x)^{1/2}))$
for a specific constant c

E. Landau. *Über die asymptotischen Werthe einiger Zahlentheoretischer Funktionen.* Math. Ann. 54 (1901), 570–591.

b) $R(x) = O(\exp(-(\log x)^{1/\gamma}))$
for $\gamma > 2$

E. Landau. *Beiträge zur analytischen Zahlentheorie.* Rend. Circ. Math. Palermo 26 (1908), 169–302.

2) a) $$\sum_{n\le x} \frac{\mu(n)\log n}{n^2} = \frac{\zeta'(2)}{\zeta^2(2)} + O(x^{-1}\cdot\log^{-k} x)$$

for any $k > 0$

A.A. Gioia and A.M. Vaidya. *The number of square free divisors of an integer.* Duke Math. J. 33 (1966), 797–799.

b) $$\sum_{n\le x,(n,k)=1} \frac{\mu(n)\log n}{n^2} = \frac{k^2}{\zeta(2)J_2(k)}\left(\beta_{(k)} + \frac{\zeta'(2)}{\zeta(2)}\right) + O\left(\frac{\log x}{x}\right)$$

where $\beta(k) = \sum_{p|k} \log p/(p^2-1)$

D. Suryanarayana. *The greatest divisor of n which is prime to k.* Math. Stud. 37 (1969), 147–152.

3) $$\sum_{d|n} \frac{\mu^2(d)\log d}{d} = O((\log\log 3n)^2)$$

S. Uchiyama. *On the distribution of almost primes in an aritmetic progression.* J. Fac. Sci. Hokkaido Univ. 18 (1964), 1–22.

§ VI.10 Selberg's formula

1) a) $$\sum_{n\le x}\frac{\mu(n)}{n}\cdot\log^2(x/n)=2\log x+O(1)$$

A. Selberg. *An elementary proof of the prime number theorem.* Ann. Math. 50 (1949), 305–313.

b) $$\sum_{n\le x}\frac{\mu(n)}{n}\cdot\log^k\frac{x}{n}=k\log^{k-1}x+\sum_{i=1}^{k-2}c_i^{(k)}\log^i x+O(1)$$

where $k\ge 2$ is an integer, and $c_i^{(k)}$ are certain constants depending on k

H.N. Shapiro. *On a theorem of Selberg and generalization.* Ann. Math. 51 (1950), 485–497.

Remark. a) is equivalent with

$$\sum_{p\le x}\log^2 p+\sum_{pq\le x}\log p\cdot\log q=2x\log x+O(1)$$

which was used by Selberg to give an elementary proof of the prime number theorem. (An other method was obtained by P. Erdős, in the same year, to obtain the first elementary proof of the prime number theorem.)

c) $$\sum_{\substack{n\le x\\(n,k)=1}}\frac{\mu(n)\log^r(x/n)}{n}=$$

$$=\frac{rk}{\varphi(k)}\cdot\log^{r-1}x+\sum_{m-1}^{r-2}b_{m,r}\log^m x+O(1)$$

where $r\ge 2$ is an integer and $b_{m,r}$ are constants depending on r and k

G.J. Rieger. *Bemerkungen zu einem zahlentheoretischen Satz von Shapiro.* Arch. Math. 8 (1957), 251–254.

2) a) $$\sum_{n\le x}\sum_{d|n}\mu(d)\log^k(n/d)=(k\log^{k-1}x+a_{1,k}\log^{k-2}x+\cdots$$

$$\cdots+a_{k-1,k})x+O(x\exp(-c_k(\log x)^{3/5}(\log\log x)^{-1/5}))$$

where the $a_{i,k}$ and c_k are constants, $c_k>0$

A. Ivić. *On the asymptotic formulas for a generalization of von Mangoldt's function.* Rend. Mat. (6) 10 (1977), 51–59.

b) $$\sum_{n\le x,n\equiv l(\text{mod }k)}\sum_{d|n}\mu(d)\log^2(n/d)\sim\frac{2x\log x}{\varphi(k)}$$

$k\ge 1,\ (k,l)=1$

J.B. Friedlander. *Selberg's formula and Siegel's zero. Recent progress in analytic number theory.* Durham (1979), pp. 15–23, London, 1981.

§ VI.11 A sum on $\mu(n)\left[\frac{x}{n}\right]$

$$\sum_{\substack{1 \le n \le x \\ (n,m)=1}} \mu(n)\left[\frac{x}{n}\right] = \frac{1}{k!}\prod_{i=1}^{k}\alpha_i + O\left(\prod_{i=1}^{k-1}\alpha_i\right)$$

where $\alpha_i = i + \log_{pi} x (p_1 < \cdots < p_k$ are the prime factors of $m)$

H. Gupta. *A sum involving the Möbius function.* Proc. Amer. Math. Soc. 19 (1968), 445–447.

§ VI.12 A sum on $\mu(n)f(n)/n$, f-multiplicative, $0 \le f(p) \le 1$

Let $F(x) = \sum_{n \le x} \mu(n)f(n)/n$

where f is multiplicative and $0 \le f(p) \le 1$ for all primes p

1) If $\sum_{p \le x} \frac{f(p)\log p}{p} = o(\log x)$, then

$$F(x) = \prod_{p \le x}\left(1 - \frac{f(p)}{p}\right) + o(1)$$

uniformly for any set of functions f for which the first estimate holds uniformly.

2) $F(x) \ll \exp\left(-\sum_{p \le x} \frac{f(p)}{p}\right)$

with the implied constant being independent of f

L. Lucht. *Summen mit der Möbius-Funktion.* Math. Z. 159 (1978), 123–131.

§ VI.13 Gandhi's formula

Let Q be the product of all primes less than the odd prime p. Then

$$1 < 2^p\left(-\frac{1}{2} + \sum_{d|Q} \frac{\mu(d)}{2^d - 1}\right) < 2$$

J.M. Gandhi. *Abstracts of brief scientific communications.* Section 3: Theory of numbers (Russian), p. 5. Internat. Congr. Mathematicians, Moscow, 1966.

See also

J.M. Gandhi. *Formulae for the n-th prime.* Proc. Washington State Univ. Conf. Number Theory 1971, pp. 96–106.

Remark. This shows that if the first n primes $p_1, \ldots, p_n$ are known, then the next prime p_{n+1} is given "explicitly" by the above formula. Generalizations of Gandhi's formula and similar relations where obtained also by Golomb. Golomb proves e.g. that

$$p_{n+1} = \lim_{s\to\infty} (P_n(s) \cdot \zeta(s) - 1)^{-1/s}$$

where $P_n(s) = \prod_{p_i \mid Q} (1 - p_i^{-s})$ and $\zeta(s)$ is the Riemann zeta function.

S.W. Golomb. *Formulas for the next prime.* Pacific J. Math. 63 (1976), 401–404.

§ VI.14 An extremal property of μ

1) Let N be a positive squarefree number and denote $A(x) = -\mu(N) \sum_{d \mid N} \mu(d)\{xd\}$ where $\{\alpha\}$ denotes the fractional part of α. Then:

a) $|A(x)| \le d(N)/2$

b) $\int_0^1 |A(x)|^2 \mathrm{d}x \ge \frac{1}{12} \frac{\varphi(N)}{N} \cdot d(N)$

where $d(N)$ is the number of divisors of N.

Corollary. For any squarefree N there exists an x with

$$|A(x)| \gg \left(\frac{d(N)}{\log\log d(N)} \right)^{1/2}$$

A. Perelli and U. Zannier. *An extremal property of Möbius function.* Arch. Math. 53 (1989), 20–29.

2) Let $I_k = \liminf_{n \in \mathbb{N}\, n \to \infty} \sum_{d=1}^{\infty} \frac{\mu(d)}{d^k} \cdot \left(\frac{1}{2} - \left\{ \frac{n}{d} \right\} \right)$. Then

$$\frac{1}{2\zeta(k)} - \frac{1}{(k-1)N^{k-1}} < I_k < \frac{1}{2\zeta(k)}$$

where $N = N(k) \to \infty$ as $k \to \infty$

Y.-F.S. Pétermann. *Oscillations d'un terme d'erreur lié à la fonction totient de Jordan.* Sémin. Théor. Nombres Bordx., Sér. II 3 (1991), 311–335.

§ VI.15 On a sum connected with the Möbius function

Let $S(x) = \sum_{m,n \le x} \frac{\mu(m)\mu(n)}{[m,n]}$. Then

$$\lim_{x \to \infty} S(x) = L$$

where $L > 0$

F. Dress, H. Iwaniec and G. Tenenbaum. *Sur une somme liée à la fonction de Möbius.* J. Reine Angew. Math. 340 (1983), 53–58.

§ VI.16 Sums over $\frac{\mu^2(n)}{\omega(n)}, \frac{\mu^2(n)}{\omega^2(n)}, \frac{\mu^2(n)}{\varphi(n)}, \frac{\mu(n)}{nd(n)}$

1) a) $$\sum_{2 \le n \le x} \frac{\mu^2(n)}{\omega(n)} = x \sum_{i=1}^{k} \frac{a_i}{(\log\log x)^i} + O\left(\frac{x}{(\log\log x)^{k+1}}\right)$$

where a_i $(1 \le i \le k)$ are computable constants, e.g. $a_1 = 6/\pi^2$, $a_2 = 6/\pi^2 \left(1 - \rho + \sum_p \frac{1}{p(p+1)}\right)$, where $\rho = \gamma + \sum_p \left(\log\left(1 - \frac{1}{p}\right) + \frac{1}{p}\right)$ and γ is Euler's constant.

b) $$\sum_{2 \le n \le x} \frac{\mu^2(n)}{\omega^2(n)} = \sum_{i=2}^{k} \frac{b_i}{(\log\log x)^i} + O\left(\frac{x}{(\log\log x)^{k+1}}\right)$$

where $k \ge 2$, $b_2 = 6/\pi^2$,

$$b_3 = 6/\pi^2 \left(3 - 2\rho + 2\sum_p 1/(p(p+1))\right)$$

and the remaining b_i's are computable constants.

J.M. De Koninck. *On a class of arithmetical functions.* Duke Math. J. 39 (1972), 807–818.

2) a) $$\sum_{n \le x} \frac{\mu^2(n)}{\varphi(n)} = \log x + c + o(1)$$

where $c = \gamma + \sum_p \frac{\log p}{p(p-1)} = 1.33258\ldots$

D.R. Ward. *Some series involving Euler's function.* J. London Math. Soc. 2 (1927), 210–214.

Remarks. (i) For the approximation of c see

J.B. Rosser and L. Schoenfeld. *Approximate formulas for some functions of prime numbers.* Illinois J. Math. 6 (1962), 64–94.

(ii) For the sum $\sum_{n\le x, P(n)\le y} \mu^2(n)/\varphi(n)$ see

J.H. van Lint and H.-E. Richert. *Über die Summe* $\sum_{n\le x, P(n)\le y} \mu^2(n)/\varphi(n)$. Nederl. Akad. Wetensch. Proc. Ser A 67, 582–587.

b) $$\sum_{n\le x,(n,k)=1} \frac{\mu^2(n)}{\varphi(n)} = \frac{\varphi(k)}{k}\cdot \log x + O(\log\log 3k)$$

S. Uchiyama. *On the distribution of almost primes in an arithmetic progression.* J. Fac. Sci. Hokkaido Univ. 18 (1964), 1–22.

3) a) If $x \ge 1$, then

$$0 < \sum_{n\le x} \frac{\mu(n)}{n\,d(n)} \le 1$$

S. Selberg. *Über die Summe* $\sum_{n\le x} \frac{\mu(n)}{n\,d(n)}$. C.R. Douzième Congrès Math. Scand. Lund, 10–15 août 1953, pp. 268–272.

b) If $x \ge 1$ and $a \mid b$, then

$$0 < \sum_{n\le x,(n,a)=1} \frac{\mu(n)}{n\,d(n)} \le \sum_{n\le x,(n,b)=1} \frac{\mu(n)}{n\,d(n)} \le 1$$

(See S. Selberg.)

c) If $x \ge 2$, then there exist positive constants c_1, c_2 such that

$$c_1(\log x)^{-0.7} < \sum_{n\le x} \frac{\mu(n)}{n\,d(n)} < c_2\cdot(\log x)^{-0.5}$$

(See S. Selberg.)

§ VI.17 The distribution of integers having a given number of prime factors

1) $$\sum_{n\le x,\omega(n)=k} |\mu(n)| \sim \frac{x(\log\log x)^{k-1}}{(k-1)!\log x}$$
$(k \ge 1)$

E. Landau. See G.H. Hardy and E.M. Wright. *An introduction to the theory of numbers.* Fourth ed. 1960, Oxford (Theorem 437.)

2) $$\sum_{n\le x,\omega(n)=k} |\mu(n)| = (1+o(1))f(m)\frac{x}{\log x}\frac{(\log\log x)^{k-1}}{(k-1)!}$$
where $k = [m\log\log x]$, $m < e$ and

$$f(m) = \frac{1}{\Gamma(m+1)} \left(\prod_p \left(1 - \frac{1}{p}\right) e^{1/p} \right)^m \prod_p \left(1 + \frac{m}{p} \cdot e^{-k/p}\right)$$

L.G. Sathe. *On a problem of Hardy on the distribution of integers having a given number of prime factors. I.* J. Indian Math. Soc. (N.S.) 17 (1953), 63–82.

§ VI.18 Number of squarefree integers $\leq x$

Let $Q(x)$ denote the number of squarefree integers $\leq x$

1) a) $Q(x) = \frac{6}{\pi^2}x + O(x^{1/2})$

L. Gegenbauer. *Asymptotische Gesetze der Zahlentheorie.* Denkschriften Akad. Wien 49 (1) (1885), 37–80.

b) $Q(x) = \frac{6}{\pi^2}x + o(x^{1/2})$

E. Landau. *Über den Zusammenhang eininger neuren Sätze der analytischen Zahlentheorie.* Sitz. mat.-naturwiss Kl. Kaiserlichen Akad. Wiss. Wien 115 (1906), 589–632.

c) Assuming the Riemann hypothesis,

$$Q(x) = \frac{6}{\pi^2}x + O(x^{2/5+\varepsilon})$$

A.M. Vaidya. *On the order of the error function of the square free numbers.* Proc. Nat. Inst. Sci. India Part A, 32 (1966), 196–201.

Note. For more general and better results see the section with k-free integers ($k \geq 2$)

d) $Q(x) \geq \frac{53x}{88}$
with equality only for $x = 176$

K. Rogers. *The Schnirelmann density of the squarefree integers.* Proc. Amer. Math. Soc. 15 (1964), 515–516.

2) Let $R(x) = Q(x) - \frac{6}{\pi^2} \cdot x$. Then

a) $|R(x)| < 0.5\sqrt{x}$
for $x \geq 8$

L. Moser and R.A. Mac Leod. *The error term for the squarefree integers.* Canad Math. Bull. 9 (1966), 303–306.

b) $|R(x)| < 0.1333\sqrt{x}$
for $x \geq 1664$

H. Cohen and F. Dress. *Estimations numériques du reste de la fonction sommatoire relative aux entiers sans facteur carré.* Colloque de Théorie Analy. Nombres ≪Jean Coquet≫ (Marseille, 1985), 73–76, Publ. Math. Orsay, 88–02, Univ. Paris XI, Orsay 1988.

c) $|R(x+y) - R(x)| < 1.6749\sqrt{y} + 0.6212x/y$
for all x and all $y \geq 1$

Corollary. For all $x \geq 1$ and all $y > 1.911x^{2/3}$,

$$|R(x+y) - R(x)| < 2\sqrt{y}$$

(See H. Cohen and F. Dress.)

§ VI.19 On squarefree integers

Let $Q(x)$ denote the number of squarefree numbers $\leq x$.
Let $R(x) = Q(x) - 6x/\pi^2$. Then

$$R(x) = \Omega_{\pm}(x^{1/4})$$

R. Balasubramanian and K. Ramachandra. *On square-free integers.* Proc. Ramanujan Centennial Intern. Conf. (Annamalainagar, 1987), 27–30, RMS Publ., 1, — Ramanujan Math. Soc., Annamalainagar, 1988.)

Remark. The same result appeared in an earlier paper of the authors in
Studia Sci. Math. Hungar. 14 (1979), 193–202,
but the proof given here is effective and simpler. An ineffective version of this result was proved by
M. Vaidya, J. Indian Math. Soc. (N.S.) 32 (1968), 105–111.

§ VI.20 Intervals containing a squarefree integer

1) a) There exists a positive constant c such that for all $x \geq 1$, the interval

$$x < n \leq x + cx^{1/3}$$

always contains a square–free integer n

K.F. Roth. *On the gaps between consecutive squarefree numbers.* J. London Math. Soc. 26 (1951), 263–268.

b) Intervals of length

$$O(x^{3/13}(\log x)^{4/13})$$

must contain a squarefree integer.

(See K.F. Roth.)

c) Intervals of length

$$O(x^{2/9} \cdot \log x)$$

must contain a squarefree integer.

H.E. Richert. *On the difference between consecutive squarefree numbers.* J. London Math. Soc. 29 (1954), 16–20.

d) The result from b) is true for

$$O(x^{0.22198215+\varepsilon})$$

R.A. Rankin. *Van der Corput's method and the theory of exponent pairs.* Quart. J. Math. 6 (1955), 147–153.

e) For $x \geq x_0$, there is a squarefree number in the interval

$$(x, x + x^{\theta}]$$

where $\theta = 17/77$

O. Trifonov. *On the squarefree problem.* II. Math. Balcanica (N.S.) 3 (1989), 284–295.

f) The result from e) is true for

$$\theta = \frac{3}{14}$$

O. Trifonov and M. Filaseta. *On gaps between squarefree numbers.* Number Theory at Allerton Park, Proc. Conf. in Honor of P.T. Batemann, Birkhäuser, 1990.

g) The same is true with

$$\theta = \frac{1057}{4785} = 0.2208986\ldots$$

S.W. Graham and G. Kolesnik. *On the difference between consecutive squarefree integers.* Acta Arith. 49 (1988), 435–447.

Remark. The above proof is based on exponential sum estimates, while (the weaker) result f) is based on a more elementary method (the "second difference" technique.)

h) The same, with

$$\theta = 47/217 = 0.21658\ldots$$

M. Filaseta. *Short interval results for squarefree numbers.* J. Number Theory 35 (1990), 128–149.

Remark. It can be proved that the conjecture that for every $\varepsilon > 0$ and $x \geq x_0(\varepsilon)$ there is a squarefree number in $(x, x + x^{\varepsilon}]$ is equivalent with

$$(*) \sum_{S_{n+1} \leq x} (S_{n+1} - S_n)^{\varepsilon} \sim B(\varepsilon) \cdot x \qquad \text{as } x \to \infty$$

where (S_n) is the sequence of squarefree numbers and $B(\varepsilon)$ is a suitable constant. This results and the fact that (*) holds for all $0 \leq \varepsilon < 29/9$ is due to M. Filaseta. The first result that (*) holds for $0 \leq \varepsilon < 2$ is attributed to P. Erdős.

M. Filaseta. *On the distribution of gaps between squarefree numbers.* Mathematika 40 (1993), 88–101.

i) For $x > x_0$ there is at least a squarefree number in the interval $(x, x + cx^{1/5} \log x)$.

M. Filaseta and O. Trifonov. *On gaps between squarefree numbers.* II. J. London Math. Soc. II. Ser. 45 (1992), 215–221.

2) a) If $f(x)$ is any function for which $f(x) \to \infty (x \to \infty)$, for almost all positive integers n, the interval

$$(n, n + f(n))$$

contains a squarefree integer.

b) Let $Q(u, v)$ denote the number of squarefree integers q such that $u < q \leq v$ and $f(x) \nearrow \infty (x \to \infty)$, then for any function $k(n)$ such that $k(n) \to \infty, n \to \infty$, for almost all n we have

$$\frac{6}{\pi^2} \cdot f(n) - k(n) (f(n))^{1/2} \leq Q(n, n + f(n)) \leq \frac{6}{\pi^2} f(n) + k(n) (f(n))^{1/2}$$

Corollary. $Q(n, n + f(n))$ has normal order $\dfrac{6f(n)}{\pi^2}$

R. Bellman and H.N. Shapiro. *The distribution of squarefree integers in small intervals.* Duke Math. J. 21 (1954), 629–638.

3) Almost all intervals of the form $[n, n + (\log n)^c)$ contain squarefree numbers with two odd prime factors, where:

a) $c = 5 \cdot 10^6$

D. Wolke. *Fast-Primzahlen in kurzen Intervallen.* Math. Ann. 244 (1979), 233–242.

b) $c = 7 + \eta$
for any $\eta > 0$

G. Harman. *Almost–primes in short intervals.* Math. Ann. 258 (1981), 107–112.

Remark. The above results improve earlier theorems obtained in

D.R. Heath-Brown. *Almost primes in arithmetic progressions in short intervals.* Math. Proc. Cambridge Philos. Soc. 83 (1978), 357–375.

and

Y. Motohashi. *A note on almost primes in short intervals.* Proc. Japan Acad. Ser A Math. Sci. 55 (1979), 225–226.

§ VI.21 Distribution of squarefree numbers

1) Let $N(n, k)$ be the number of squarefree numbers in the interval $[n, n + k]$ and $D(n, k) = N(n, k) - b \cdot k/\pi^2$. Let $d_k(X)$ be the density of the integers n with $|D(n, k)| > X$. Then:

a) $d_k(k^{1/4} \cdot m) \leq c_0 \cdot m^{-2}$
for $m \geq 1$

b) $d_k(k^{1/2} \cdot m) \leq k^{-1/2} \cdot \exp(c_1 \cdot \log 3m - ((\log 3m)/(\log 2)) \log\log 3m)$
for $m \geq 1$, where c_0, c_1 are suitable absolute constants.

Corollary. If $\frac{1}{4} \leq \alpha \leq \frac{1}{2}$, then

$$\limsup_{k\to\infty} \frac{\log d_k(k^\alpha)}{\log k} \leq \frac{1}{2}(1-4\alpha)$$

R.R. Hall. *The distribution of squarefree numbers.* J. Reine Angew. Math. 394 (1989), 107–117.

2) Let $N_k(n)$ be the number of m with $1 \leq m \leq n$, $N(m, k-1) = j$, and let $p_k(j) = \lim_{n\to\infty} N_k(n)/n$ (which exists). Put $M_k = \sup_j p_k(j)$. Then

$$\left(\frac{1+o(1)}{3\sqrt{3}\sigma}\right) k^{-1/4} \leq M_k \leq ck^{-1/4}(\log k)^{1/2}$$

as $k \to \infty$ with σ, c absolute constants.

G. Grimmett. *Statistics of sieves and square-free numbers.* J. London Math. Soc. II. Ser. 43 (1991), 1–11.

Remark. The fact that M_k has the order $k^{-1/4}$ as $k \to \infty$ has been conjectured by Hall.

§ VI.22 On the frequency of pairs of squarefree numbers

Let $E(n) = \begin{cases} 1, & \text{if } n \text{ is squarefree} \\ 0, & \text{otherwise} \end{cases}$

1) $\sum_{n\leq x} E(n) \cdot E(n+1) = x \cdot \prod_p (1 - 2/p^2) + R(x)$
where:

a) $R(x) = O(x^{2/3+\varepsilon})$

L. Carlitz. *On a problem in additive arithmetic.* II. Quart. J. Math. Oxford 3 (1932), 273–290.

b) $R(x) = O(x^{2/3} \cdot \log^{4/3} x)$

L. Mirsky. *On the frequency of pairs of squarefree numbers with a given difference.* Bull. Amer. Math. Soc. 55 (1949), 936–939.

c) $R(x) = O(x^{2/3} \cdot \log^{2/3} x)$

R.R. Hall. *Squarefree numbers in short intervals.* Mathematika 29 (1982), 7–17.

d) $R(x) = O(x^{7/11} \cdot \log^7 x)$

D.R. Heath-Brown. *The square sieve and consecutive squarefree numbers.* Math. Ann. 266 (1984), 251–259.

2) $\sum_{n\le x} E(n)\cdot E(n+1)\cdot E(n+2) = x\prod_{p}(1-3/p^2) + O(x^{2/3+\varepsilon})$

L. Mirsky. *Arithmetical pattern problems relating to divisibility by r-th powers.* Proc. London Math. Soc.(2) 50 (1949), 497–508.

3) Let $S = \sum_{\substack{ka-lb=1,\, lb\le x\\ (k,a)=(l,b)=1}} E(k)\cdot E(l)$, where $k, l = 1, 2, \ldots$

Then uniformly for $1 \le b \le x^{1/2}$

$$S = \frac{6x}{\pi^2\cdot ab}\prod_{p|ab}\left(1-\frac{1}{p}\right)\prod_{p\nmid ab}\left(1-\frac{2}{p^2}\right)+$$

$$+O\left(x^{3/4}(ab)^{-1/2}2^{\omega(ab)} + (x/b)^{1/2} + (x/a)^{1/2}\cdot\frac{1}{b}\right)$$

P. Erdős and A. Ivić. *The distribution of values of a certain class of arithmetic functions at consecutive integers.* Coll. Math. Soc. János Bolyai 51, Number Theory, Budapest, 1987, pp. 45–91.

§ VI.23 Smallest squarefree integer in an arithmetic progression

1) Denote by $q(k,l)$ the smallest squarefree integer in the arithmetic progression $km + l(m = 0, 1, 2, \ldots)$. Then:

a) $q(k,l) \le c_1 \cdot k^{3/2} \cdot 2^{\omega(k)}$
where $c_1 > 0$ is a constant.

K. Prachar. *Über die kleinste quadratfrei Zahl einer arithmetischen Reihe.* Monat. Math. 62 (1958), 173–176.

b) $q(k,l) \le c_2 \dfrac{k^{3/2}}{\log k}$

P. Erdős. *Über die kleinste quadratfrei Zahl einer arithmetischen Reihe.* Monat. Math. 64 (1960), 314–316.

c) Given $\varepsilon > 0$ there exists some $c = c(\varepsilon) > 0$ such that

$$q(k,l) \le ck$$

for at least $(1-\varepsilon)\,\varphi(k)$ numbers $l, l \le k, (l,k) = 1$.

K. Prachar. *Sätze über quadratfreie Zahlen.* Monatsh. Math. 66 (1962), 306–312.

d) Given $c \ge 1$ there exists some $\varepsilon = \varepsilon(c) > 0$ such that for infinitely many k's the inequality

$$q(k,l) \ge ck$$

holds for at least $\varepsilon\,\varphi(k)$ numbers.

R. Warlimont. *On squarefree numbers in arithmetic progressions.* Monath. Math. 73 (1969), 433–448.

e) If (k, l) is squarefree, then

$$q(k, l) \ll (d(k) \cdot \log k)^6 \cdot k^{13/9}$$

f) If $p(k, l)$ denotes the smallest squarefree integer in $l (\mathrm{mod}\ k)$ with at most 4 prime factors, and $(k, l) = 1$, then

$$p(k, l) \ll (d(k) \cdot \log k)^6 \cdot k^{13/9}$$

D.R. Heath-Brown. *The least squarefree number in an arithmetic progression.* J. Reine Angew. Math. 332 (1982), 204–220.

2) a) Let $1 \leq l \leq k$, $(l, k) = 1$ and $0 < \alpha < 4/3$.

Define $S(k, \alpha) = \sum_{l=1,\,(l,k)=1}^{k} (q(k, l))^\alpha$. Then

$$\sum_{k \leq x} S(k, \alpha) \sim c_1(\alpha) \cdot x^{2+\alpha}$$

$(x \to \infty)$

R. Warlimont. *Über die kleinsten quadratfrei Zahlen in arithmetischen Progressionen.* J. Reine Angew. Math. 250 (1971), 99–106.

b) $\sum_{l=1}^{p-1} (q(p, l))^\alpha \sim c_2(\alpha) \cdot p^{1+\alpha}$

for all $0 \leq \alpha < 1$, p-prime.

R. Warlimont. *Progression mit primen Differenzen.* J. Reine Angew. Math. 253 (1972), 19–23.

3) The smallest squarefree number in the sequence $km^2 + l$

$(0 < l < k, (k, l) = 1, k$ a perfect square)

is at most

$$k^{3+\varepsilon}$$

for $k \geq k_0(\varepsilon)$

W. Fluch. *Notiz zu den quadratfreien Zahlen in arithmetischen Progressionen.* Monatsh. Math. 82 (1976), 269–274.

4) The least squarefree integer $\equiv l (\mathrm{mod}\ k), 0 < l < k, (l, k) = 1$ having at most r prime factors is

$$< \begin{cases} c \cdot k^{5/3+\varepsilon}, & \text{if } r = 3 \\ c \cdot k^{3/2+\varepsilon}, & \text{if } r = 4 \end{cases}$$

W. Fluch. Bemerkung *über quadratfreie Zahlen in arithmetischen Progressionen.* Monatsh. Math. 72 (1968), 427–430.

§ VI.24 The greatest squarefree divisor of n

1) a) $\sum_{n\le x}\gamma(n)=\frac{1}{2}\alpha x^2+O(x^{3/2})$
($\alpha>0$, constant)

E. Cohen. *Arithmetical functions associated with the unitary divisors of an integer.* Math. Z. 74 (1960), 66–80.

b) Assuming the Riemann Hypothesis, one has

$$\sum_{n\le x}\gamma(n)=\frac{1}{2}\alpha x^2+O(x^{7/5+\varepsilon})$$

where $\alpha=\prod_p\left(1-\frac{1}{p(p+1)}\right)$

D. Suryanarayana. *On the core of an integer.* Indian J. Math. 14 (1972), 65–74.

2) $\sum_{n\le x}\frac{\gamma(n)}{n^2}=A\log x+B+O(\log 2x/\sqrt{x})$
($x>1$) where $A=F(0)$, $B=\gamma F(0)+F'(0)$ with

$$F(s)=\prod_p\left(1-\frac{p-1}{p^{s+1}(p^{s+2}-1)}\right)$$

$\left(s=\sigma+it,\sigma>-\frac{1}{2}\right)$

S. Uchiyama. *On the sum* $\sum_{n\le x}\tilde{n}/n^2$. Proc. Japan Acad. 47 (1971), 39–41.

Remarks. (i) The above result has been rediscovered in

H.Q. Liu. *Two asymptotic formulas for* $\gamma(n)$ *and* $\varphi(n)$ (Chinese.) J. China Univ. Sci. Tech. 17 (1987), 98–104.

(ii) For sums of type $\sum_{n\le x}\gamma(n)/n^k$, see

E. Brinitzer. *Eine Bemerkung zu dem größten quadratfreien Teiler einer natürlichen Zahl.* Monatsh. Math. 84 (1977), 13–19.

3) $\sum_{n\le x}\log\gamma(n)\sim x\log x$

($x\to\infty$)

L. Panaitopol. *Über einige arithmetische Funktionen.* Bull. Math. Soc. Sci. Math. R.S. Roumanie (N.S.) 26 (74) (1982), 269–274.

4) $$\sum_{n\le x,\,Q(n)\in S}(\gamma(n))^\alpha=\frac{6x^{\alpha+1}}{\pi^2(\alpha+1)}\cdot\sum_{n=1,n\in S}^{\infty}n^{-2\alpha-2}+O\big(x^{\alpha+1/2}\cdot R_\alpha(x,S)\big)$$

uniformly in S where $R_\alpha(x, S) = \sum_{n \le \sqrt{x}, n \in S} n^{-2\alpha-1}$ if the sum is non-empty, otherwise $R_\alpha(x, S) = 1$, and α is arbitrary ≥ 0 real number and S is any non-empty set of positive integers.

E. Cohen. *Some asymptotic formulas in the theory of numbers.* Trans. Amer. Math. Soc. 112 (1964), 214–227.

§ VI.25 Estimates involving the greatest squarefree divisor of n

1) a) Let $\gamma(n)$ denote the greatest squarefree divisor of n ("core of n.") Then

$$\sum_{n \le x} \frac{1}{\gamma(n)} \sim \exp\left(\sqrt{\frac{8 \log x}{\log\log x}}\right)$$

as $x \to \infty$

N.G. de Bruijn. *On the number of integers $\le x$ whose prime factors divide n.* Illinois J. Math. 6 (1962), 137–141.

b) $$\sum_{n \le x} \frac{1}{\gamma(n)} \sim (4\pi)^{-1/2} \cdot 2^{-1/4} \cdot \left(\frac{\log\log x}{\log x}\right)^{1/4} \cdot Q(x)$$

where $Q(x) = \min_{0 < s < \infty} x^s \cdot \sum_{n=1}^{\infty} \frac{1}{\gamma(n) \cdot n^s}$

W. Schwarz. *Einige Anwendungen Tauberscher Sätze in der Zahlentheorie.* B.J. Reine Angew. Math. 219 (1965), 157–179.

c) There is a squence of polynomials $(P_j)_{j=1,2,\ldots}$ with $\deg P_j \le j$, such that for all integer $N \ge 1$ we have uniformly for $x \ge 16$,

$$\sum_{n \le x} \frac{1}{\gamma(n)} = \exp\left(\sqrt{\frac{8 \log x}{\log\log x}}\left(1 + \sum_{j=1}^{N} \frac{P_j(\log\log\log x)}{(\log\log x)^j} + \right.\right.$$
$$\left.\left. + O_N\left(\left(\frac{\log\log\log x}{\log\log x}\right)^{N+1}\right)\right)\right)$$

H. Squalli. *Sur la répartition du noyau d'un entier.* Thèse 3ème cycle. Univ. Nancy I (1985).

2) a) $$\sum_{2 \le n \le x} \frac{1}{n \log \gamma(n)} = \log\log x + c + O\left(\frac{\log\log x}{\log x}\right)$$

where c is a constant.

G.J. Rieger. *Zahlentheorie.* Vanderhoeck and Ruprecht, Göttingen, 1976 (p. 85.)

b) $$\sum_{2 \le n \le x} \frac{1}{n \log \gamma(n)} = \log\log x + \sum_{m=0}^{k-1} \frac{b_m}{(\log x)^m} + O(\log^{-k} x)$$

for each positive integer k, where b_i $(0 \le i \le k-1)$ are constants.

c) $$\sum_{2\le n\le x} \frac{q_s(n)}{n\log \gamma_r(n)} = \frac{1}{\zeta(s)}\log\log x + \sum_{m=0}^{k-1}\frac{c_m}{(\log x)^m} + O(\log^{-k} x)$$

where $\gamma_r(n)$ denotes the largest r-free divisor of n and q_s is the characteristic function of the set of s-free numbers.

J.-M. De Koninck and R. Sitaramachandrarao. *Sums involving the largest prime divisor of an integer.* Acta Arith. 48 (1987), 1–8.

§ VI.26 Estimates for $N(x, y) = \text{card}\,\{n \le x : \gamma(n) \le y\}$

Let $N(x, y) = \text{card}\,\{n \le x : \gamma(n) \le y\}$, where $\gamma(n)$ denotes the greatest squarefree divisor of n. Then:

1) For $x \ge y \ge 2$ let $\nu = \log(x/y)$. Then

$$N(x, y) = yF(\nu)\left(1 + O\left(\sqrt{\frac{\log(\nu+2)}{\log x}}\right)\right)$$

for all couples (x, y) with $\exp(5(\log x \cdot \log\log\log x)^{3/4}) \le y \le x, x \ge 16$.

Here $F(t) = e^t - \dfrac{6}{\pi^2}\displaystyle\sum_{m\le e^t}\left(\frac{e^t}{m} - 1\right)\prod_{p\mid m}\frac{1}{p+1}$

2) We have uniformly for all (x, y) satisfying the conditions from a) and $x/y \to \infty$,

$$N(2x, y) \sim N(x, y)$$

H. Squalli. *Sur la répartition du noyau d'un entier.* Thèse 3ème cycle, Univ. Nancy I (1985.)

§ VI.27 Number of non-squarefree odd, positive integers $\le x$

Let $A(x)$ denote the number of non-squarefree odd, positive integers $\le x$. Then

$$A(x) < \frac{x}{2}\cdot\left(1 - \frac{8}{\pi^2}\right) + \frac{\sqrt{x}}{2} + \frac{1}{8}$$

C.B. Lacampagne, C.A. Nicol and J.L. Selfridge. *Sets with non-squarefree sums.* Number theory, Proc. 1-st Conf. Can. Number Theory Assoc., Banff (Alberta, Can.) 1988, 299–311 (1990.)

§ VI.28 Number of squarefree numbers $\leq X$ which are quadratic residues (mod p)

Let p be a prime, $0 < a \leq 1/128$ and $X > p^{1/4+b}$ (with $b = b(a) > 0$.) Then the number of squarefree numbers $\leq X$ which are quadratic residues(mod p) equals

$$\frac{3}{\pi^2} \cdot X + O(X/p^a)$$

O.V. Popov. *On quadratic and nonresidues in the sequence of squarefree numbers* (Russian.) Vestn. Mosk. Univ., Ser. I 1989, No. 5, 81–83.

§ VI.29 Squarefree integers in nonlinear sequences

1) Let the sequence (u_n) be defined by the recurrence relation $u_n = ru_{n-1} + su_{n-2}$ (u_0, u_1, r, s integers). Then for each $\varepsilon > 0$ and all $n \in \mathbb{N}$, except perhaps a set of density zero,

$$\gamma(u_n) > n^{(\log n)^{1+\log 2-\varepsilon}}$$

C.L. Stewart. *On divisors of Fermat, Fibonacci, Lucas and Lehmer numbers.* III. J. London Math. Soc. (2) 28 (1983), 211–217.

2) The number of squarefree integers in sequences $[f(n+x)]$, $(n = 1, 2, \dots, N)$ is

$$\left(\frac{6}{\pi^2} + o(1)\right) N$$

for almost all $x \geq 0$, where f is a polynomial function of degree $k \geq 2$ or an exponential function.

F. Roesler. *Squarefree integers in nonlinear sequeces.* Pacific J. Math. 123 (1986), 223–225.

3) Let $S_c(x)$ denote the number of squarefree numbers of the form n^c which are $\leq x$

a) For $1 < c < 3/2$

$$S_c(x) = \frac{6x}{\pi^2} + O_c(x^{(2c+1)/4})$$

b) For $1 < c < 3/2$, let $M_c(x)$ be the number of representations of m as $m = q + [n^c]$ with squarefree q and natural number n. Then

$$M_c(x) = \frac{6}{\pi^2} \cdot m^{1/c} + O_c(m^{(2c+1)/4c})$$

G.J. Rieger. *Remark on a paper of Stux concerning squarefree numbers in non-linear sequences.* Pacific J. Math. 78 (1978), 241–242.

§ VI.30 Sumsets containing squarefree and k-free integers

1) a) For $n > n_0$, there exists a subset $A \subset \{1, 2, \ldots, n\}$ such that

$$|A| > \frac{1}{248} \log n$$

and $a + a'$ is squarefree for all $a \in A, a' \in A$.

b) If $n > n_1, A \subset \{1, 2, \ldots, n\}$ and $a + a'$ is squarefree for all $a \in A$, $a' \in A$ then we have

$$|A| < 3n^{3/4} \cdot \log n$$

P. Erdős and A. Sárközy. *On divisibility properties of integers of the form $a + a'$.* Acta Math. Hungar. 50 (1987), 117–122.

Remark. The autors note that they can obtain analogous results for k-th power free numbers.

2) Let $B \subset \{1, 2, \ldots, n\}$ such that $a + a' \notin Q_k$ for all $a, a' \in B, a \neq a'$ (where Q_k denotes the set of k-free numbers.)

a) Let $k \geq 3$ and $n = 2^k \cdot m$. Then either

$$|B| \leq m$$

or

$$n = 2^k$$

and

$$B = \{a, n - a\}$$

for some $a \in \{1, \ldots, n\} \backslash \{n/2\}$

b) Let $F_k(n)$ denote the cardinality of the largest such subset B such that B satisfies both

$$B \not\subseteq \{a \equiv 0 (\text{mod } 2^k)\} \text{ and } B \not\subseteq \{a \equiv 2^{k-1} (\text{mod } 2^k)\}$$

Then, for any $k \geq 2$,

$$\limsup_{n \to \infty} F_k(n)/n \leq 1 - 2^k/(2^k - 1)\zeta(k)$$

M.B. Nathanson. *Sumsets containing k-free integers.* Number theory Proc. 15-th Journ. Arith., Ulm FRG 1987, Lect. Notes Math. 1380, 179–184 (1989.)

§ VI.31 On the Möbius function

Define:

$$f(x) = \sum_{k \le x, \mu(q_k)=+1, \mu(q_{k+1})=-1} 1$$

$$g(x) = \sum_{k \le x, \mu(q_k)=-1, \mu(q_{k+1})=+1} 1$$

$$h(x) = \sum_{k \le x, \mu(q_k)=+1, \mu(q_{k+1})=+1} 1$$

$$k(x) = \sum_{k \le x, \mu(q_k)=-1, \mu(q_{k+1})=-1} 1$$

where q_k denotes the k-th squarefree number. Then:

a) $f(x) \ge x(\log x)^{-7-\varepsilon}$
and
$g(x) \ge x(\log x)^{-7-\varepsilon}$
for all sufficiently large x (for any $\varepsilon > 0$)

b) $h(x) > \frac{x}{60}(1 + o(1))$
and
$k(x) > \frac{x}{60}(1 + o(1))$

G. Harman, J. Pintz and D. Wolke. *A note on the Möbius and Liouville functions.* Studia Sci. Math. Hungar. 20 (1985), 295–299.

§ VI.32 Number of k-free integers $\le x$

Let $Q_k(x)$ denote the number of k-free integers $\le x$ ($k \ge 2$, integer.)

1) a) $Q_k(x) = \frac{x}{\zeta(k)} + O(x^{1/k})$

L. Gegenbauer. *Asymptotische Gesetze der Zahlentheorie.* Denkschriften Akad. Wien 49 (1) (1885), 37–80.

b) $Q_k(x) = \frac{x}{\zeta(k)} + O(x^{1/k} \cdot \exp(-A \cdot k^{-3/2} \cdot \log^{1/2} x))$

A. Axer. *Über einige Grenzwertsätze.* Sitz. kaiserlichen Akad. Wiss. Wien, math.-natur. Kl. 120, Abt. 2a (1911), 1253–1298.

c) $Q_k(x) = \frac{x}{\zeta(k)} + O(x^{1/k} \cdot \exp(-Ak^{-3/2}(\log\log\log x)^{1/2}))$

C.J.A. Evelyn and E.H. Linfoot. *On a problem in the additive theory of numbers* (Fourth paper.) Ann. Math. 32 (1931), 261–270.

d) $Q_k(x) = \dfrac{x}{\zeta(k)} + O(x^{1/k} \cdot \exp(-A \cdot k^{-8/5} \cdot \log^{3/5} x(\log\log x)^{-1/5}))$

A. Walfisz. *Weylsche Exponentialsummen in der neuren Zahlentheorie*, Berlin, 1963.

2) Let $R_k(x) = Q_k(x) - \dfrac{x}{\zeta(k)}$

a) Assuming the Riemann Hypothesis, one has

$$R_k(x) = O_{k,\varepsilon}(x^{1/(k+1)+\varepsilon})$$

H.L. Montgomery and R.C. Vaughan. *The distribution of squarefree numbers.* Recent progress in analytic number theory. Vol. 1, Academic Press, New York, 1981, pp. 247–256.

b) Assuming the Riemann Hypothesis,

$$R_2(x) = O_\varepsilon(x^{9/28+\varepsilon})$$

(See H.L. Montgomery and R.C. Vaughan.)

c) Assuming the Riemann Hypothesis,

$$R_2(x) = O(x^{8/25})$$

S.W. Graham. *The distribution of squarefree numbers.* J. London Math. Soc.(2) 25 (1981), 54–64.

d) Assuming the Riemann Hypothesis

$$R_2(x) = O(x^{7/22+\varepsilon})$$

R.C. Baker and J. Pintz. *The distribution of squarefree numbers.* Acta Arith. 46 (1985), 71–79.

Remark. This result has been rediscovered in

C.H. Jia. *The distribution of squarefree numbers* (Chinese.) Beijing Daxue Xuebao 1987, no. 3, 21–27.

e) Assuming the Riemann Hypothesis,

$$R_k(x) \ll x^{a(k)+\varepsilon}$$

with the implied constant depending on k and $\varepsilon > 0$, where:

$$a(k) = \frac{7}{8k+6}$$

if $2 \le k \le 5; a(6) = \dfrac{67}{514}$;

$$a(k) = \frac{11(k-4)}{12k^2 - 37k - 41}$$

if $7 \le k \le 12$;

$$a(k) = \frac{23(k-1)}{24k^2 + 13k - 37}$$

if $13 \le k \le 20$;

$$(a(k))^{-1} \sim k + \log k/(2\log 2)$$

for $k \to \infty$

S.W. Graham and J. Pintz. *The distribution of r-free numbers.* Acta Math. Hungar. 53 (1989), 213–236.

f) On R.H., one has

$$R_k(x) \ll_k x^{b(k)}$$

where $b(k) = \dfrac{1}{k + ck^{1/3}}$ for some constant c

(See S.W. Graham and J. Pintz.)

g) $R_k(x) \ll x^{c(k)}$

where $c(k) = \max\left(\dfrac{2}{2k+3}, \dfrac{9}{10k+8} + \varepsilon\right)$ (under R.H.)

H.Z. Li. *The distribution of k th-power-free numbers* (Chinese.) Chinese Quart. J. Math. 3 (1988), 10–16.

3) Let $D_k(Y) = \dfrac{1}{Y}\displaystyle\int_1^Y \dfrac{R_k(x)}{x}\mathrm{d}x$. Then:

a) $Y \cdot D_k(Y) > \dfrac{c}{k} \cdot Y^{0.36/(2k)}$

I. Kátai. *On oscillations of number-theoretic functions.* Acta Arith. 13 (1967), 107–122.

b) $\dfrac{1}{Y}\displaystyle\int_1^Y R_k(t)\mathrm{d}t = O(Y^{3/(4k+1)+\varepsilon})$

$\forall \quad \varepsilon > 0$

D. Suryanarayana and R. Sitaramachandraro. *On the order of the error function of the k-free integers.* Proc. Amer. Math. Soc. 28 (1971), 53–58.

c) $\dfrac{1}{Y}\displaystyle\int_1^Y R_k(t)\mathrm{d}t = O(Y^{1/2k+\varepsilon})$

for all $\varepsilon > 0$, assuming the Riemann Hypothesis.

V.S. Joshi. *On the order of some error functions related to k-free integers.* Proc. Amer. Math. Soc. 35 (1972), 325–332.

d) Suppose that $\zeta(s)$ has a zero $\rho_0 = \beta_0 + i\gamma_0 (\beta_0 \geq 1/2)$ of multiplicity ν, and that

$$a_k(\rho_0) := 2\pi i \cdot \operatorname*{Res}_{s=\rho_0/k}\left(\frac{\zeta(s)}{(s-\rho_0/k)^\nu}\right) \neq 0$$

Then

$$D_k(Y) > \frac{c_1(k) \cdot |a_k(\rho_0)|}{|\rho_0|^{k+3}} Y^{\beta_0/k}$$

for $Y > c_2(k) \cdot |\rho_0|^{k+3} / |a_k(\rho_0)|$

Corollary. $D_k(Y) > c_3(k) \cdot Y^{1/(2k)}$
for $Y > c_4(k)$

J. Pintz. *On the distribution of squarefree numbers.* J. London Math. Soc.(2) 28 (1983), 401–405.

4) a) If $h \ll x^{3/13}$, then there exists a squarefree number in the interval $(x, x+h]$ for $x \geq x_0$

b) If $h \ll x^{5/(10k+1)}$, then there exists a k-free number in the interval $(x, x+h]$ for $x \geq x_0$

M. Filaseta. *An elementary approach to short interval results for k-free numbers.* J. Number Theory 30 (1988), 208–225.

Remark. Earlier results are due to Halberstam and Roth. For $k = 3$ one can take $h \ll x^{7/46}$ which is more precise than the estimate b) with $k = 3$. See

O. Trifonov. *On the gaps between consecutive k-free numbers.* Math. B. 4 (1990), 50–60.

5) a) $\liminf_{x\to\infty} (Q_k(x) - [x]/\zeta(k)) \cdot x^{-1/2k} \ll -L$

b) $\limsup_{x\to\infty} (Q_k(x) - [x]/\zeta(k)) \cdot x^{-1/2k} \gg L$

H.M. Stark. *On the asymptotic density of the k-free integers.* Proc. Amer. Math. Soc. 17 (1966), 1211–1214.

6) Let $s_n = s_n(k)$ denote the n-th k-free number. Then

$$\sum_{S_n \leq x} (S_{n+1} - S_n)^\gamma \ll x$$

holds for any $\gamma < 2k - 2 + 4/(k+1)$

S.W. Graham. *Moments of gaps between k-free numbers.* J. Number Theory 44 (1993), 105–117.

§ VI.33 Number of k-free integers $\leq x$, which are relatively prime to n

Let $Q_k(x, n)$ be the number of k-free integers $\leq x$, which are relatively prime to n. Then

$$Q_k(x, n) = \frac{n^{k-1}\varphi(n)}{J_k(n)} \frac{x}{\zeta(k)} + O\left(\frac{\varphi(n)\gamma(n)}{n} \cdot x^{1/k}\right)$$

where J_k is Jordan's totient and $\gamma(n)$ denotes the number of squarefree divisors of n

D. Suryanarayana. *The number and of k-free integers $\leq x$ which are prime to n.* Indian J. Math. 11 (1969), 131–139.

§ VI.34 Schnirelmann density of the k-free integers

Let $d_k = \inf_{n \geq 1} \frac{Q_k(n)}{n}$ be the Schnirelmann density of the k-free integers. Then:

1) $d_2 = \frac{53}{88} < \frac{6}{\pi^2} = D_2$

K. Rogers. *The Schnirelmann density of the sequence of k-free integers.* Proc. Amer. Math. Soc. 15 (1964), 515–516.

2) Let $D_k = \lim_{x \to \infty} \frac{Q_k(x)}{x} = \frac{1}{\zeta(k)}$. Then, for all $k \geq 2$,

$$d_k < D_k$$

H.M. Stark. *On the asymptotic density of the k-free integers.* Proc. Amer. Soc. 17 (1966), 1211–1214.

c) $d_k < D_k < d_{k+1} < D_{k+1}$

R.L. Duncan. *The Schnirelmann density of k-free integers.* Proc. Amer. Math. Soc. 16 (1965), 1090–1091.

d) $d_k > 1 - \sum_{p \text{ prime}} 1/p^k$

R.L. Duncan. *On the density of the k-free integers.* Fib. Quart. 7 (1969), 140–142.

e) $d_k > 1 - 1/2^k - 1/3^k - 1/5^k$

f) For $k \geq 5$

$$d_k \geq 1 - \frac{1}{2^k} - \frac{1}{3^k} - \frac{1}{5^k} + \left[\frac{3^{-k} + 2.5^{-k}}{6^k - 3^k + 1}\right]$$

P.H. Diananda and M.V. Subbarao. *On the Schnirelmann density of the k-free integers.* Proc. Amer. Math. Soc. 62 (1977), 7–10.

g) $d_k = 1 - \frac{1}{2^k} - \frac{1}{3^k} - \frac{1}{5^k} + O\left(\frac{1}{9^k}\right)$

P. Erdős, G.E. Hardy and M.V. Subbarao. *On the Schnirelmann density of k-free integers.* Indian J-Math. 20 (1978), 45–56.

Note. For a lower bound for d_k see also

V. Siva Rama Prasad and M.V.S. Bhramarambica. *On the Schnirelmann density of M-free integers.* Fib. Quart. 27 (1989), 366–368.

§ VI.35 Powerfree integers represented by linear forms

1) Let $f(x) = \prod_{i=1}^{k}(a_i x + b_i)$ (a_i, b_i positive integers) and $Q_2(N, f)$ be the number of $x \leq N$ for which $f(x)$ is squarefree. Then a necessary and sufficient condition for

$$Q_2(N, f) = C \cdot N + O(N^{2/3+\varepsilon})$$

where $C > 0$, is that

$$a_i b_j - a_j b_i \neq 0$$

$(i \neq j)$ and for each prime p the number of solutions of $f(x) \equiv 0 \pmod{p^2}$ be less than p^2

G. Ricci. *Ricerche arithmetiche sui polinomi.* Rend. Circ. Mat. Palermo, 57 (1933), 433–475,

and

N.H. Shapiro. *Powerfree integers represented by linear forms.* Duke Math. J. 16 (1949), 601–607.

2) a) Let $Q_r(N, f_1, \ldots, f_k)$ be the number of integers $x \leq N$ for which $f_i = a_i x + b_i$ are simultaneously r-free. If $a_i b_j - a_j b_i \neq 0$, and $r \geq 2$, then

$$Q_r(N, f_1, \ldots, f_k) = c_r \cdot N + O(N^{2/(r+1)+\varepsilon})$$

where $c_r = c_r(f_1, \ldots, f_k) > 0$

L. Mirsky. *Note on an asymptotic formula connected with r-free integers.* Quart. J. Math. (Oxford) 18 (1947), 178–182,

and

L. Mirsky. *On a problem in the theory of numbers.* Simon Stevin 26 (1948–49), 25–27.

b) A necessary and sufficient condition that

$$Q_r(N, f_1, \ldots, f_k) = A \cdot N + O(N^{2/(r+1)+\varepsilon})$$

where $A > 0$, is that for every prime p there is an x_p such that $f_i(x_p) \not\equiv 0 \pmod{p^r}$, $i = 1, 2, \ldots, k$

(See H.N. Shapiro, where a more general theorem is also proved.)

§ VI.36 On the power-free value of a polynomial

1) Let $P(x)$ be a polynomial whose roots are all rational. Then the density of integers m for which $P(m)$ is k-free $(k \geq 2)$ exists and is positive, except if $P(x)$ has a k-fold root or if $P(x)$ is such that there exists a prime p so that, for every m,

$$P(m) \equiv 0 \pmod{p^k}$$

K. Rényi. *The distribution of numbers not divisible by the k-th power of an integer greather than one in the set of values of a polynomial having rational roots.* C.R. Premier Congres Math. Hongrois, 1950, pp. 493–506, Akadémiai Kiadó, Budapest, 1952.

2) a) Let $4 \nmid k$. Then the density of integers m such that

$$4m^3 + k$$

is squarefree, is positive.

b) Let $N(x)$ denote the number of positive integers $m \leq x$ for which $4m^3 + k$ is squarefree. Then

$$N(x) = x \cdot \prod_p (1 - \rho(p^2)/p^2) + O(x \log^{-2/3} x)$$

where $\rho(l)$ is the number of roots (mod l) of the congruence $4x^3 + k \equiv 0(\bmod l)$, $4 \nmid k$

C. Hooley. *On the squarefree values of cubic polynomials.* J. Reine Angew. Math. 229 (1968), 147–154.

3) Let A be an infinite, strictly increasing sequence of positive integers and $M(x)$ be the number of $a \in A, a \leq x$ for which $P(a)$ is k-free, where $P(x) \in \mathbb{I}[x]$ is a primitive polynomial of degree ≥ 2. If the discriminant of $P(x)$ is not zero, then

$$M(x) = c \cdot A(x) + O(A(x)/\log\log x)$$

where $c = \prod_p (1 - \rho^*(p^k)/p^{k-1}(p-1))$, and $\rho^*(k)$ denotes the number of incongruent solutions u with $(u, k) = 1$ of $P(u) \equiv 0(\bmod k)$. (Here $A(x)$ is the counting function of the set A.)

S. Uchiyama. *On the power-free value of a polynomial.* Tensor (N. S) 24 (1972), 43–48.

4) Let $P(x) \in \mathbb{I}[x]$ be an irreducible polynomial of degree g, having no fixed square divisors > 1.
Put $N_k(x) =$ card $\{n : 1 < n \leq x, f(n)$ is k-free$\}$. Then, if $g \geq 2$ and $k \geq \lambda g$, where $\lambda = \sqrt{2} - 1/2$, then

$$N_k(x) = a_k \cdot x + O(x/(\log x)^{k-1})$$

as $x \to \infty$ (a_k-constant.)

M. Nair. Power free values of polynomials. Mathematika 23 (1976), 159–183.

Remark. For some improvements, see

M.N. Huxley and M. Nair. *Power free values of polynomials.* III. Proc. London Math. Soc.(3) 41 (1980), 66–82.

5) Let $P(x) \in \mathbb{I}[x]$ be of degree n and irreducible. Let $k \geq n + 1$ be an integer with g.c.d. $(P(m), m), m \in \mathbb{I}$, k-free, and let r be the greatest integer satisfying $r(r-1) < 2n$. Then there is a constant

$$c = c(P, k)$$

such that for all sufficiently large x there is a corresponding integer $m \in (x, x+h]$ for which $P(m)$ is k-free where $h = cx^{n/(2k-n+r)}$

M. Filaseta. *Short interval results for k-free values of irreducible polynomials.* Acta Arith. 64 (1993), 249–270.

§ VI.37 Number of r-free integers $\leq x$ that are in arithmetic progression

1) Let $Q_r(x, k, l)$ denote the number of r-free integers $\leq x$ that are in the arithmetic progression $kt + l (t = 0, 1, 2, \ldots)$, where $k > l \geq 0$

a) If $(k, l) = 1$, then

$$Q_2(x, k, l) = \frac{x}{k} \sum_{p \nmid k} \left(1 - \frac{1}{p^2}\right) + O(\sqrt{x})$$

E. Landau. *Handbuch.* Leipzig, 1909 (pp. 633–636.)

b) For $r \geq 2$,

$$Q_r(x, k, l) = \frac{x}{k} \prod_{\min(\omega_p(k), r) \leq \omega_p(l)} \left(1 - p^{-\max(r - \omega_p(k), 0)}\right) + O(x^{i/r})$$

where $\omega_p(k)$ denotes the exponent of the prime p in the factorisation of k

E. Cohen and R. Robinson. *On the distribution of k-free integers in residue classes.* Acta Arith. 8 (1962–63), 283–293.

c) $Q_r(x, k, l) = A_r \cdot \frac{x}{k} + O(r^{\omega(k)}(k^{-1/r^2} x^{1/r} + k^{1/r}))$

where $A_r = \frac{1}{\zeta(r)} \prod_{p|k} (1 - p^{-r})$, and the O is uniform in k and r

K. Prachar. *Über diee kleinste quadratfrei Zahl einer arithmetische Reihe.* Monat. Math. 62 (1958), 173–176.

2) a) Let $S(x, y) = \sum_{k \leq y} \sum_{\substack{l \leq k \\ (l,k)=1}} |Q_2(x, k, l) - x \cdot F(k)|, \quad 1 \leq y \leq x,$ with

$$F(k) = \left(\sum_{d=1, (d,k)=1}^{\infty} \mu(d)/d^2 \right) / k. \text{ Then}$$

$$S(x, y) = O(xy)$$

for $x^{1/3} \cdot \log^{10/3} x \leq y \leq x$

R. Warlimont. *On squarefree numbers in arithmetic progressions.* Monatsh. Math. 73 (1969), 433–448.

b) Let $H(x, k, l)$ be the leading term in the asymptotic expansion of $Q_2(x, k, l)$ and define

$$T(x, y) = \sum_{k \le y} \sum_{l=1}^{k} (Q_2(x, k, l) - H(x, k, l))^2 \text{ for } 1 \le y \le x. \text{ Then}$$

$$T(x, y) = x^2 f(x/y) + O(x^{3/2} \cdot (\log x)^{7/2})$$

where $f(z) = c_1 \cdot z^{-3/2} + O(z^{-7/4} \cdot \exp(-c_2 \cdot (\log z)^{1/5}))$ for $z \ge 1$, $c_1, c_2 > 0$ being constants.

R. Warlimont. *Squarefree numbers in arithmetic progressions.* J. London Math. Soc.(2) 22 (1980), 21–24.

§ VI.38 Squarefree numbers as sums of two squares

1) Let $A(x, k)$ denote the number of solution of $1 \le u^2 + v^2 \le x, (u^2 + v^2)$ k-free. Then:

$$A(x, k) = A_k x + O(x^{1/k} \cdot \log x)$$

for $k = 2, 3$;

$$A(x, k) = A_k x + O(x^{\alpha})$$

for $k > 3$, where $1/4 \le \alpha < 1/3$

W. Recknagel. *Über k-freie Zahlen als Summe von zwei Quadraten.* Arch. Math. 52 (1989), 233–236.

Note. The case $k = 2$ is due to K.-H. Fischer.

K.-H. Fischer. *Über die Anzahl der Gitterpunkte auf Kreisen mit quadratfreien Radienquadraten.* Arch. Math. 33 (1979), 150–154.

2) $A(x + h, 2) - A(x, 2) \sim A \cdot h$
for $h \ge x^{12101/26852+\varepsilon}$ ($A =$ positive constant.)

E. Krätzel. *Squarefree numbers as sums of two squares.* Arch. Math. 39 (1982), 28–31.

§ VI.39 Distribution of unitary k-free integers

A positive integer n is called *unitary k-free*, if the multiplicity of each prime divisor of n is not a multiple of k.

Let $U_k(x)$ be the counting function of unitary k-free numbers. Then

$$U_k(x) = \alpha_k \cdot x + O(x^{1/k} \cdot \exp(-A \cdot \log^{3/5} x \cdot (\log \log x)^{-1/5}))$$

($\alpha_k > 0$, constant.)

D. Suryanarayana and R.S.R.C. Rao. *Distribution of unitary k-free intergers.* J. Austral Math. Soc. 20 (1975), 129–141.

§ VI.40 Counting function of the (k, r)-integers

For given integers $k, r; 1 < r < k$, define a (k, r) *-integer* to be a positive integer of the form $a^k \cdot b$ with b an r-free number.
Let $Q_{k,r}(x)$ denote the counting function of the (k, r)-integers. Then:

1) $Q_{k,r}(x) = \dfrac{x \cdot \zeta(k)}{\zeta(r)} + O(x^{1/r} \cdot \delta_r(x))$
   where the O-constant is independent of x, y, r and $\delta_r(x) = \exp(-b_r \cdot \log^{3/5} x \cdot (\log\log x)^{-1/5})$ with $b_r > 0$

2) Under R.H. (Riemann Hypothesis) we have
$$Q_{k,r}(x) = \frac{x \cdot \zeta(k)}{\zeta(r)} + O(x^{2/(2r+1)} \cdot \exp(-A \log x \cdot (\log\log x)^{-1}))$$

M.V. Subbarao and D. Suryanarayana. *On the order of the error function of the (k, r)-integers*. II. Canad. Math. Bull. 20(1977), 397–399.

§ VI.41 Asymptotic formulae for powerful numbers

Let $k \geq 2$ be an integer. The number $n > 1$ is called *k-full*, if in the prime factorization $n = \prod_{i=1}^{r} p_i^{a_i}$ we have $a_i \geq k$ for all $i = 1, 2, \ldots, r$.
Let $N_k(x)$ denote the number of k-full integers not exceeding x, and introduce
$$D_k(x) = N_k(x) - \sum_{i=k}^{2k-1} \alpha_{i,k} x^{1/i}$$
where $\alpha_{i,k} = \operatorname*{Res}_{s=1/i} F_k(s)/s$, with $F_k(s) = \sum_{n=1, n=k-\text{full}}^{\infty} n^{-s}$
Put $\lambda_k = \inf\{\rho_k : D_k(x) \ll x^{\rho_k}\}$. Then:

1) $\lambda_k \leq \dfrac{1}{k + m_k}$
   with:

   a) $m_k = 1$

P. Erdős and G. Szekeres. *Über die Anzahl Abelschen Gruppen gegebener Ordnung und über ein verwandtes zahentheoretisches Problem*. Acta. Scien. Math. Szeged 7 (1935), 95–102.

   b) $m_k = [\sqrt{2k}]$

P.T. Bateman and E. Grosswald. *On a theorem of Erdös and Szekers.* Ill. J. Math. 2 (1958), 88–98.

c) $m_k = \left[\sqrt{8k/3}\right]$
for $k \geq 5$

E. Krätzel. *Zahlen k-ter Art.* Amer. J. Math. 44 (1972), 309–328.

d) $m_k = \left[\sqrt{k \log k}\right]$
for $k \geq e^8$

E. Krätzel. *Divisor problems and powerful numbers.* Math. Nachr. 114(1983), 97–104.

2) a) $\lambda_2 \leq \frac{1}{6}$

$$\lambda_3 \leq \frac{7}{46} = 0.1521\ldots$$

(See P.T. Bateman and E. Grosswald.)

b) $\lambda_3 \leq \frac{16}{113} = 0.1415\ldots$

$$\lambda_4 \leq \frac{169}{1360} = 0.1242\ldots$$

$$\lambda_5 \leq \frac{16188}{151297} = 0.1069\ldots$$

$$\lambda_6 \leq \frac{113}{1173} = 0.0963\ldots$$

$$\lambda_7 \leq \frac{274}{3213} = 0.0852\ldots$$

(See E. Krätzel (1972).)

c) $\lambda_3 \leq \frac{655}{4643} = 0.1410\ldots$

$$\lambda_4 \leq \frac{257}{2072} = 0.1240\ldots$$

$$\lambda_5 \leq \frac{6656613}{62279970} = 0.1068\ldots$$

A. Ivić. *On the asymptotic formulas for powerful numbers.* Publ. Inst. Math. Belgrade, 23 (37)(1978), 85–94.

d) $\lambda_3 \leq \frac{577}{4176} = 0.1381\ldots$

$$\lambda_4 \leq \frac{3187}{25852} = 0.1232\ldots$$

$$\lambda_5 \leq \frac{124371}{1165874} = 0.1066\ldots$$

A. Ivić. *On the number of finite non-isomorphic abelian groups in short intervals.* Math. Nachr. 101 (1981), 257–271.

e)
$$\lambda_3 \le \frac{263}{2052} = 0.1281\ldots$$
$$\lambda_4 \le \frac{3091}{25981} = 0.1189\ldots$$
$$\lambda_5 \le \frac{1}{10}$$
$$\lambda_6 \le \frac{1}{12}$$
$$\lambda_7 \le \frac{1}{14}$$

A. Ivić and P. Shiu. *The distribution of powerful numbers.* Ill. J. Math. 26 (1982), 576–590.

f)
$$\lambda_4 \le \frac{5}{44} = 0.1136\ldots$$
$$\lambda_5 \le \frac{6}{65} = 0.0923\ldots$$
$$\lambda_6 \le \frac{13}{162} = 0.0802\ldots$$
$$\lambda_8 \le \frac{1}{16}$$

(See E. Krätzel (1983).)

g) $\lambda_3 \le \frac{1}{8}$

E. Krätzel. *Zweifache Exponentialsummen und dreidimensionale Gitterpunktprobleme.* Elementary and Analytic Theory of Numbers, Banach Center Publ. 17, PWN, Warsaw 1985, 337–369.

h) $\lambda_4 \le \frac{21}{187} = 0.1122\ldots$

E. Krätzel. *The distribution of powerful integers of type* 4. Acta Arith. 52 (1989), no. 2, 141–145.

i) $\lambda_4 \le \frac{35}{316} = 0.1107\ldots$

H. Menzer. *The distribution of powerful integers of type* 4. Monatsh. Math. 107 (1989), 69–75.

3) Let $\delta(x) = (\log x)^{3/5}(\log\log x)^{-1/5}$. Then:

a) $D_2(x) = O\left(x^{1/6} \cdot \exp(-c_1\delta(x))\right)$
where $c_1 > 0$ (constant.)

(See P.T. Bateman and E. Groswald.)

b) $D_3(x) = O\left(x^{1/8} \cdot \exp(-c_2\delta(x))\right)$
where $c_2 > 0$ (constant.)

(See E. Krätzel (1985).)

4) Let $h = x^{1/2+\theta}$. Then

$$N_2(x+h) - N_2(x) \sim \left(\zeta\left(\frac{3}{2}\right)/2\zeta(3)\right) \cdot x^{\theta}$$

for:

a) $0.1\,526 \leq \theta$

P. Shiu. *On square-full integers in a short interval.* Glasgow Math. J 25 (1984), 127–134.

b) $\frac{68}{451} \leq \theta$
where $\frac{68}{451} = 0.1507\ldots$

P.G. Schmidt. *Zur Anzahl quadratvoller Zahlen in kurzen Intervallen.* Acta Arith. 46 (1986), 159–164.

c) $0.149\ldots \leq \theta$

C.H. Jia. *The square-full integers in short intervals* (Chinese.) Acta Math. Sinica 30 (1987), 614–621.

d) $\frac{1}{7} - \frac{2}{7575} \leq \theta$

Here $\frac{1}{7} - \frac{2}{7575} = 0.1425\ldots$

P.G. Schmidt. *Über die Anzahl quadratvoller Zahlen in kurzen Intervallen und ein verwandtes Gitterpunktproblem.* Acta Arith. 50 (1988), 195–201.

e) $0.14254 \leq \theta$

H.Q. Liu. *On square-full numbers in short intervals.* Acta Math. Sinica (N.S.) 6 (1990), 148–164.

f) $\theta \geq 0.13084$

H. Liu. *The number of square-full numbers in an interval.* Acta Arith. 64 (1993), 129–149.

5) a) $D_k(x) = \Omega(x^{1/(2(k+r))}(\log x)^{1/2})$
where $k \geq 3$ and r is the least integer such that $r(r-1) \geq 2k$

R. Balasubramanian, K. Ramachandra and M.V. Subbarao. *On the error function in the asymptotic formula for the counting of k-full numbers.* Acta Arith. 50 (1988), 107–188.

b) $\lambda_k \geq \frac{r}{(r+1)(2k+r)}$

$(k \geq 5)$ where $r = \left[\frac{1+\sqrt{8k+1}}{2}\right]$

(See R. Balasubramanian, K. Ramachandra and M.V. Subbarao.)

c) Let $\alpha_2 = 1/10$ and $\alpha_k = r/((r+1)(2k+r))(k \geq 3)$, where r denotes the integral part of $(1+\sqrt{8k+1})/2$. Then

$$\int_1^T D_k^2(t)/t^{2\alpha_k+1}\mathrm{d}t \sim c_k \cdot \log T$$

$(T \to \infty)$ under the assumption of the Riemann hypothesis, if $k \geq 2$ is not of the form $\frac{m^2}{2} - \frac{m}{2}$ for any integer $m \geq 4$. For other $k \geq 5$, the integral in the above relation in unbounded.

T. Zhan. *Distribution of k-full integers.* Sci. China Ser. A 32 (1989), 20–37.

§ VI.42 Maximal k-full divisor of an integer

For a fixed integer $k \geq 2$ let $M_k(n)$ denote the maximal k-full divisor of n. If $0 < \alpha < \min\left(\frac{2k+1}{4k+4}, \frac{k-1}{x+2}\right); x \geq 2$, then

$$\sum_{n \leq x} M_k(n) = A \cdot x^{(k+1)/k} + B \cdot x^{(k+2)/(k+1)} + \Delta_k(x)$$

where

$$\Delta_k(x) \ll \begin{cases} x^{(\alpha+6)/5} & \text{if } k = 2 \\ x^{(k+3)/(k+2)} & \text{if } k \geq 3 \end{cases}$$

(A, B- constants.)

D. Suryanarayana and P. Subrahamanyam. *The maximal k-full divisor of an integer.* Indian J. Pure Appl. Math. 12 (1981), 175–190.

§ VI.43 Number of squarefull integers between successive squares

Let $f(n)$ be the number of squarefull integers q with $n^2 < q < (n+1)^2$. Then for each $m = 0, 1, 2, \ldots$, the set $\{n : f(n) = m\}$ has positive density

$$d_m = \sum_{k=0}^{+\infty} (-1)^k \binom{m+k}{m} \cdot c_{m+k}$$

where $c_0 = 1$, $c_r = \sum_{1<b_1<\cdots<b_r} \mu^2(b_1)\cdots\mu^2(b_r)/(b_1\cdots b_r)^{3/2}, r \geq 1$

P. Shiu. *On the number of square-full integers between successive squares.* Mathematika 27 (1980), 171–178.

Chapter VII

FUNCTIONS $\pi(x)$, $\psi(x)$, $\theta(x)$, AND THE SEQUENCE OF PRIME NUMBERS

§ VII. 1 Estimates on $\pi(x)$. Chebyshev's theorem. The prime number theorem

1) $\lim\limits_{x\to\infty} \dfrac{\pi(x)}{x} = 0$

L. Euler. *Variae Observationes Circa Series Infinitas.* Opera Omnia, Leipzig: B.G. Teubner, 1924, I, 14, pp. 216–244 (original 1748.)

Remark. For an exact proof, see

E. Landau. *Handbuch*, Leipzig, 1909.

2) a) There exist constants $A > 0, a > 0$, such that for all $x \geq 2$

$$a\frac{x}{\log x} < \pi(x) < A\frac{x}{\log x}$$

b) $\liminf\limits_{x\to\infty} \dfrac{\pi(x)\log x}{x} \leq 1 \leq \limsup\limits_{x\to\infty} \dfrac{\pi(x)\log x}{x}$

P. Chebyshev. *Mémoire sur les nombres premiers.* J. Math. Pures appl. 17 (1852), 366–390.

3) a) $\pi(x) \sim \dfrac{x}{\log x}$
$(x \to \infty)$

J. Hadamard. *Sur la distribution des zéros de la fonction $\zeta(s)$ et ses conséquences arithmétiques.* Bull. Soc. Math. France 24 (1896), 199–220;

and

C.J. de la Vallée Poussin. *Recherches analytiques sur la théorie des nombres* (3 parts). Ann. Soc. Sci. Bruxelles 20, Part II (1896), 183–256, 281–397.

b) $\pi(x) = \dfrac{x}{\log x} + O\left(\dfrac{x}{\log^2 x}\right)$

C.J. de la Vallée Poussin. *Sur la fonction $\zeta(s)$ de Riemann et le nombre des nombres premiers inférieures à une limite donnée.* Mém. couronnés et autres mémoires. Acad. Royal Sci. Lettres Beaux-Arts Belgique 59, 1899–1900.

Remark. The first elementary proof (without using complex variable) of a) was discovered in 1949 by P. Erdős and A. Selberg.

A. Selberg. *An elementary proof of the prime number theorem.* Ann. Math. 50 (1949), 305–313;
P. Erdős. *On a new method in elementary number theory which leads to an elementary proof of the prime number theorem.* Proc. Nat. Acad. Sci. 35 (1949), 374–384.

4) a) $\pi(x) < \frac{x}{\log x}\left(1 + \frac{3}{2\log x}\right)$

for $x > 1$, and

$\pi(x) > \frac{x}{\log x}\left(1 + \frac{1}{2\log x}\right)$

for $x \geq 59$

b) $\pi(x) < \frac{x}{\log x + 3/2}$

for $x > e^{3/2}$, and

$\pi(x) > \frac{x}{\log x - 1/2}$

for $x \geq 67$

J.B. Rosser and L. Schoenfeld. *Approximate formulas for some functions of prime numbers.* Illinois J. Math. 6 (1962), 64–94.

Remark. In 1941 Rosser proved that

$$\frac{x}{\log x + 2} < \pi(x) < \frac{x}{\log x - 4}$$

for $x \geq 55$

J.B. Rosser. *Explicit bounds for some functions of prime numbers.* Amer. J. Math. 63 (1941), 211–232.

§ VII. 2 Approximation of $\pi(x)$ by $\int_2^x \frac{dy}{\log y}$

$$\pi(x) = \int_2^x \frac{dy}{\log y} + R(x)$$

where

a) $R(x) = O(x \exp(-A \log^{1/2} x))$

C.J. de la Vallée-Poussin. *Sur la fonction $\zeta(s)$ de Riemann et le nombre des nombres premiers inférieures à une limite donnée.* Mém. couronnés et autres mém. publ. 1' Acad. roy. Sci. Lettres Beaux–Arts Belgique 59 (1899–1900), No. 1, 74 pp.

b) $R(x) = O(x \exp(-A(\log x \log\log x)^{1/2}))$

J.E. Littlewood. *Researches in the theory of Riemann ζ-function.* Proc. London Math. Soc. (2) 20 (1922), XXII–XXVIII.

c) $R(x) = O(x \exp(-A \log^{3/5} x (\log\log x)^{-1/5}))$

H.M. Korobov. *Estimates of trigonometric sums and their applications* (Russian). Uspehi Mat. Nauk. 13 (1958), 185–192.

and

I.M. Vinogradov. *A new estimate for* $\zeta(1+it)$ (Russian). Izv. Akad. Nauk. SSSR Ser. Mat. 22 (1958), 161–164.

Remark. Similar results are, clearly, valid also for $\Psi(x)$ and $\theta(x)$ (replacing the above integral with x.)

§ VII. 3 On $\pi(x) - \operatorname{li} x$. Sign changes

1) a) $\pi(x) - \operatorname{li} x$

has infinitely many sign changes.

J.E. Littlewood. *Sur la distribution des nombres premiers.* C.R. Acad. Sci. Paris 158 (1914), 1869–1872.

Remark. For $x < 10^7$ we have $\pi(x) - \operatorname{li} x < 0$

D.H. Lehmer. *List of primes from* 1 *to* 10006721, Carnegie Inst. Wash. Publ. No. 165, Wash. D.C. 1914.

Littlewood's proof was ineffective and so it could not give any explicit upper bound for the first sign change of $\pi(x) - \operatorname{li} x$.

In 1955, Skewes gives the upper bound $e_4(7.705)$, where $e_4(x)$ means the four times iterated exponential function.

S. Skewes. *On the difference* $\pi(x) - \operatorname{li} x$. II. Proc. London Math. Soc. 5 (1955), 48–70.

In 1966 this was improved to $1.6 \cdot 10^{1165}$

R. Sherman Lehman. *On the difference* $\pi(x) - \operatorname{li} x$, Acta Arith. 11 (1966), 397–410.

In 1987 H.J.J. te Riele has proved that $\pi(x) - \operatorname{li} x > 0$ for at least 10^{180} successive integers in

$[6.627\ldots \times 10^{370}, 6.687\ldots \times 10^{370}]$

H.J.J. te Riele. *On the sign of difference* $\pi(x) - \operatorname{li} x$. Math. Comp. 48 (1987), 323–328.

b) Let $V_1(T)$ denote the number of sign changes of $\pi(x) - \operatorname{li} x$ in the interval $[2, T]$.

Let θ denote the least upper bound of the real parts of the zeros of $\zeta(s)$. If there is a zero on the line $\sigma = \theta$ (where $s = \sigma + it$) then for $T > T_0$, $\pi(x) - \mathrm{li}\, x$ has a sign change in every interval of the form

$$[T, c_0 \cdot T]$$

with a constant c_0

A.E. Ingham. *A note on the distribution of primes.* Acta Atrith. 1 (2) (1936), 201–211.

Corollary. If the condition is satisfied, then

$$\liminf_{T\to\infty} V_1(T)/\log T > 0$$

c) $V_1(T) > e^{-35} \cdot \log\log\log\log T$

for $T > e_5(35)$ and $\liminf_{T\to\infty} V_1(T)/\log\log T > 0$

S. Knapowski. *On the sign changes in the remainder term in the prime number formula.* J. London Math. Soc. 36 (1961), 451–460

and

S. Knapowski. *On the sign changes of the difference* $\pi(x) - \mathrm{li}\, x$. Acta Arith. 7 (2) (1962), 107–120.

d) $\liminf_{T\to\infty} \dfrac{V_1(T)}{\log^{1/4} T(\log\log T)^{-4}} > 0$

S. Knapowski and P. Turán. *On the sign changes of* $\pi(x) - \mathrm{li}\, x$. I. Topics in Number Theory. Coll. Math. Soc. János Bolyai 13, North-Holland, 1976, 153–169.

e) There exist effectively computable positive constants c_1 and c_2 such that for $T > c_1$ the inequality

$$V_1(T) > c_2 \cdot \log\log\log T$$

holds.

S. Knapowski and P. Turán. Ibid. Monatsh. Math. 82 (1976), 163–175.

f) There exist positive effective constants c_1, c_2, c_3 such that

$$V_1(T) > c_1(\log\log T)^{c_2}$$

for $T > c_3$

J. Pintz. *Bemerkungen zur Arbeit von S. Knapowski and P. Turán.* Monatsh. Math. 82 (1976), 199–206.

g) For $T > T_1$, the interval $[T, T\exp(63\sqrt{\log T}\log\log T)]$ contains a sign-change of $\pi(x) - \mathrm{li}\, x$, where T_1 is an ineffective constant; and there exist effective constants c and T_2 such that

$$V_1(T) > c \cdot \sqrt{\log T}/\log\log T$$

for $T > T_2$

J. Pintz. *On the remainder term of the prime number formula.* III. Studia Sci. Math. Hungar. 12 (1977), 345–369.

h) $V_1(T) > 10^{-11} \cdot \log T/(\log\log T)^3$
for $T > T_3$ (ineffective.)

J. Pintz. Ibid. IV. Studia Sci. Math. Hungar. 13 (1978), 29–42.

i) Assuming the Riemann Hypothesis, one has

$$V_1(T) > \frac{1}{e_2(7.707)} \cdot \log T > 10^{-957} \cdot \log T$$

for $T > e_3(7.707)$

(Here $e_k(x)$ denotes the k-fold iterated exponential function.)

W. Dette, J. Meier and J. Pintz. *Bemerkungen zu einer Arbeit von Ingham über die Verteilung der Primzahlen.* Acta Math. Hungar. 45 (1–2) (1985), 121–132.

j) Assuming the Riemann Hypothesis,

$$\pi(x) - \operatorname{li} x$$

has at least one sign change in the interval $[T^{10^{-957}}, T]$, for $T > e_3(7.707)$

See W. Dette, J. Meier and J. Pintz.

k) $V_1(T) \geq c \cdot \log T$
($c > 0$, constant.)

J. Kaczorowski. *On sign-changes in the remainder term of the prime number formula* II. Acta Arith. 45 (1985), 65–74.

2) Let $D_1(T) = \frac{1}{T}\int_2^T |(\pi(x) - \operatorname{li} x)|dx$. Then:

a) If $\zeta(\beta_1 + i\gamma_1) = 0$, $(\beta_1 \geq 1/2, \gamma_1 > 0)$ and $T > \max(c_1, e^{\gamma_1})$ $(c_1 > 0)$, then

$$D_1(T) \geq \frac{1}{T}\int_{T\exp(-6(\log T)^{1/3}(\log\log T)^{4/3})}^T |\pi(x) - \operatorname{li} x|dx >$$

$$> T^{\beta_1} \cdot \exp(-18(\log T)^{1/3}(\log\log T)^{4/3})$$

J. Pintz. *On the remainder term of the prime number formula.* V. Effective mean value theorems. Studia Sci. Math. Hungar. 15 (1980) 215–223.

b) For $T > T_1$ (ineffective constant)

$$D_1(T) > c_2 \cdot \frac{\sqrt{T}}{\log T}$$

(see see J. Pintz. V.)

c) For $T > T_0$ (ineffective constant) we have

$$D_1(T) > \frac{1}{T}\int_{T\exp(-5\sqrt{\log T})}^T |\pi(x) = \operatorname{li} x|dx > 0.62\frac{\sqrt{T}}{\log T}$$

J. Pintz. Ibid. VI. Studia Sci. Math. Hungar. 15 (1980), 225–230.

d) If the Riemann Hypothesis is true, then for $T > c_3$

$$\frac{1}{T}\int_{10^{-3}T}^{T}(\pi(x) - \operatorname{li} x)\log x \, \mathrm{d}x < -0.62\sqrt{T}$$

e) If the Riemann Hypothesis is true, then for $T > c_4$

$$c_5\frac{\sqrt{T}}{\log T} < D_1(T) < c_6\frac{\sqrt{T}}{\log T}$$

($c_i > 0$, constants.)

(See J. Pintz. VI.)

3) Let $\operatorname{li} x = \int_0^x \frac{1}{\log t}\mathrm{d}t, (x > 2)$. Then the statement "li $(\vartheta(x)) > \pi(x)$ for sufficiently large x" is equivalent to the Riemann Hypothesis. (Here $\vartheta(x) = \sum\limits_{p\le x} \log p$).

G. Robin. *Sur la différence* li $(\vartheta(x)) - \pi(x)$. Ann. Sci. Toulouse Math. (5) 6 (1984), 257–268.

4) Without any assumption we have

$$\int_1^{\infty}(\pi(x) - \operatorname{li} x)\exp\left(-\frac{\log^2 x}{y}\right)\mathrm{d}x < -\frac{c_2}{y}\exp\left(\frac{9}{16}y\right)$$

for $y > c_1$, where c_1 and c_2 are explicitely computable positive constants.

J. Pintz. *On an assertion of Riemann concerning the distribution of prime numbers.* Acta Math. Hungar. 58 (1991), 383–387.

§ VII. 4 On $\pi(x) - \pi(x - y)$ for $y = x^{\theta}$

Let $y(x) = x^{\theta}$. Then:

1) $\pi(x) - \pi(x - y) \sim y/\log x$
for:

a) $\theta > 1 - \frac{1}{33000}$

G. Hoheisel. *Primzahlprobleme in der Analysis.* Sitzungsber. Berlin (1930), 580–588.

b) $\theta = 3/4 + \varepsilon$
($\varepsilon > 0$)

N.G. Čudakov. *On zeros of Dirichlet's L-functions.* Mat. Sb. 1 (43) (1936), 591–602.

c) $\theta = \frac{5}{8} + \varepsilon$

A.E. Ingham. *On the difference between consecutive primes.* Quart. J. Math. (Oxford) 8 (1937), 255–266.

d) $\theta = 3/5 + \varepsilon$

H.L. Montgomery. *Zeros of L-functions.* Invent. Math. 8 (1969), 346–354.

e) $\theta = \frac{7}{12} + \varepsilon$

M.N. Huxley. *On the difference between consecutive primes.* Invent. Math. 15 (1972), 164–170.

2) $\pi(x) - \pi(x - y) \gg y/\log x$
if x is large enough, where:

a) $\theta > 13/23$

H. Iwaniec and M. Jutila. *Primes in short intervals.* Ark. Mat. 17 (1979), 167–176.

b) $\theta > 11/20$

D.R. Heath-Brown and H. Iwaniec. *On the difference between consecutive primes.* Invent. Math. 55 (1979), 49–69.

3) a) $\pi(x) - \pi(x - y) > c(\theta) \cdot y/\log x$
for $x > x_0(\theta)$, where $\theta > 11/20$, and $c(\theta)$ is a constant depending on θ

D.R. Heath-Brown. *Finding primes by sieve methods.* Proc. Intern. Congress of Math. August 16–24, 1983, Warsaw, pp. 487–492.

b) If $\theta \geq 17/31 - c_1$, then

$$\pi(x) - \pi(x - y) \geq c_2 \cdot y/\log x$$

where c_1 and c_2 are explicitely calculable positive absolute constants.

H. Iwaniec. *Primes in short intervals.* Unpublished manuscript

and

J. Pintz. *On primes in short intervals.* II. Studia Sci. Math. Hungar 19 (1984), 89–96.

c) For $23/42 \leq \theta < 1$ (and $y = x^\theta$) for $x > x(\theta)$ we have

$$\pi(x) - \pi(x - y) > \frac{1}{100} \cdot \frac{y}{\log x}$$

H. Iwaniec and J. Pintz. *Primes in short intervals.* Monatsh. Math. 98 (1984), 115–143.

4) a) Let $\varepsilon(x) \to 0$ as $x \to \infty$. Then

$$\pi(x) - \pi(x - y) \sim y/\log x$$

as $x \to \infty$, uniformly for $x^{7/12-\varepsilon(x)} \leq y \leq x$

b) $\pi(x) - \pi(x - y) = \frac{y}{\log x} + O(y(\log x)^{-45/44})$
uniformly for $x^{7/12} \leq y \leq x$

D.R. Heath-Brown. *Sieve identities and gaps between primes.* Journées Arithmétiques, Metz, September 1981, pp. 61–65.

5) a) Let $\lambda > 5$ and $\varepsilon > 3$. Then we have for all $x \leq X$, apart from a set of measure $o(X)$, the upper bound

$$\frac{\pi(x+f(x)) - \pi(x)}{f(x)/\log x} \leq 4 + \varepsilon$$

J.B. Friedlander. *Sifting short intervals.* Math. Proc. Camb. Phil. Soc. 91 (1982), 9–15.

b) Let $f(x) = (\log x)^{\lambda}$, $\lambda > 1$. Then

$$\limsup_{x\to\infty} \frac{\pi(x+f(x)) - \pi(x)}{f(x)/\log x} > 1$$

and

$$\liminf_{x\to\infty} \frac{\pi(x+f(x)) - \pi(x)}{f(x)/\log x} < 1$$

For $1 < \lambda < e^{\gamma}$, the first limit is $\geq e^{\gamma}/\lambda$

H. Maier. *Primes in short intervals.* Michigan Math. J. 32 (1985), 221–225.

c) Assuming the Riemann Hypothesis, the measure of the set of $x \leq X$ for which the interval $[x, x + g(x)\log x]$ contains no prime is $o(X)$ (Here $g(x) \to \infty$ as $x \to \infty$)

D.R. Heath-Brown. Acta Arith. 41 (1982), 85–99.

6) Assuming the Riemann Hypothesis, for every $k \geq 2$ (integer) one has

$$\sum_{p\leq x} (\pi(p+h) - \pi(p))^{2k-1} \sim h^{2k-1} \cdot x(\log x)^{-2k}$$

provided that $h/f_{2k}(x) \to \infty$, where $f_{2k}(x)$ differs from $x^{(k-1)/(2k-1)}$ by an explicit log-factor.

A. Perelli and S. Salerno. *On the average of primes in short intervals.* Acta Arith. 42 (1982), 91–96.

7) Let $y = x^{1/2+\varepsilon}$ and b the unique positive solution of the equation $4b + 5\cdot\log(5-4b) = 2 + 5\cdot\log(5/2)$. Then

$$\sum_{n\leq x/y} \log\frac{x}{n}\left(\pi\left(\frac{x+y}{n}\right) - \pi\left(\frac{x}{n}\right)\right) = y\cdot\log\frac{x}{y} + O(y)$$

and

$$\sum_{x^{1-b}<n\leq x/y} \log\frac{x}{n}\left(\pi\left(\frac{x+y}{n}\right) - \pi\left(\frac{x}{n}\right)\right) < \left(\frac{1}{2} - \frac{7}{4}\varepsilon\right) y\log x$$

A. Balogh. *Numbers with a large prime factor.* Studia Sci. Math. Hungar. 15 (1980), 139–146.

8) a) $\pi(x) - \pi(x - x^{11/20+\varepsilon}) \leq \dfrac{2(1+c)x^{11/20+\varepsilon}}{\log x}$

where $c = 1/(0.92 - \varepsilon) - 1$

H. Iwaniec. *A new form of the error-term in the linear sieve.* Acta Arith. 37 (1980), 307–320.

b) The inequality form a) is valid with $c = 0.001$

H. Halberstam, S. Lou and Q. Yao. *A new upper bond in the linear sieve.* Number theory, trace formulas and discrete groups, Symp. in Honor of A. Selberg, Oslo/Norway 1987, 331–341 (1989.)

Note. According to the above paper, the last two authors have shown that

$$\pi(x) - \pi(x - x^{\theta}) > \frac{0.037x^{\theta}}{\log x}$$

for $\theta > \frac{6}{11}$

c) $$\frac{99}{100} \cdot \frac{y}{\log x} < \pi(x) - \pi(x - y) < \frac{101}{100} \cdot \frac{y}{\log x}$$

for $y = x^{11/20+\varepsilon}$ and $x \geq x(\theta)$

S. Lou and Q. Yao. *The number of primes in a short interval.* Hardy-Ramanujan J. 16 (1993), 21–43.

d) $$\frac{0.969y}{\log x} < \pi(x) - \pi(x - y) < \frac{1.031y}{\log x}$$

for $y = x^{6/11+\varepsilon}$ and $x \geq x(\theta)$

S. Lou and Q. Yao. *A Chebyshev's type of prime numbers theorem in a short interval.* II. Hardy-Ramanujan J. 15 (1992), 1–33.

§ VII. 5 On $\pi(x + y) \leq \pi(x) + \pi(y)$

1) a) The inequality

$$\pi(x + y) \leq \pi(x) + \pi(y)$$

is true for all integers $x, y \geq 2$ iff for all integers $n \geq 3$ and all integers q, $1 \leq q \leq n - 1/2$, $p_n \geq p_{n-q} + p_{q+1} - 1$ is true (where p_n is the n-th prime.)

S.L. Segal. *On* $\pi(x + y) \leq \pi(x) + \pi(y)$. Trans. Amer. Math. Soc. 104 (1962), 523–527.

b) The following two conjectures are incompatible:

(i) $\pi(x + y) \leq \pi(x) + \pi(y)$
for all $x, y \geq 2$

(ii) The prime k-tuple conjecture (due to Hardy and Littlewood.)

D. Hensley and I. Richards. *Primes in intervals.* Acta Arith. 25 (1973/74) 375–391.

See also

D. Hensley and I. Richards. *On the incompatibility of two conjectures concerning primes.* Proc. Symp. Pure Math. Amer. Math. Soc. 24 (Analytic Number theory, St. Louis, 1972), 123–127.

2) a) $\pi(2x) < 2\pi(x)$
for $x \geq x_0$

E. Landau. *Handbuch.* Band I., Leipzig, 1909.

b) $\pi(2x) < 2\pi(x)$
for all $x > 2$

J.B. Rosser and L. Schoenfeld. *Abstracts of scientific communications.* Intern. Congr. Math. Moscow 1966, Section 3, Theory of Numbers.

Note. For a simple method (based on a proof by G. Robin) see
E. Erhart. *On prime numbers.* Fib. Quart. 26 (1988), 271–274,
where it is proved that

$$\pi(2n) < 2\pi(n)$$

for $n > 10$ and $\pi(k \cdot n) < k\pi(n)$ for all $k, n \geq 2$

c) $\pi(kx) < k\pi(x)$
for $k \geq e^{1/2}$ and $x \geq 347$ and $\pi(kx) > k\pi(x)$ for $k \leq e^{-1/2}$ and $kx \geq 347$. If $a \geq e^{1/4}$ and $x \geq 347$, then either

$$\pi(ax) < a\pi(x)$$

or

$$\pi(a^2x) < a\pi(x)$$

C. Karanikolov. *On some properties of the function* $\pi(x)$. Univ. Beograd Publ. Elektr. Fac. Ser. Mat. Fiz. 1971, 357–380.

d) For $k > 1$ an sufficiently large x one has

$$\pi(kx) < k\pi(x)$$

L. Panaitopol. *Eine Eigenschaft der Funktion über die Verteilung der Primzahlen.* Bull. Math. Soc. Sci. Math. R. S. Roumanie, 23 (71) (1979), 189–194.

3) a) For any $\varepsilon > 0$ there is an $x_0(\varepsilon)$ such that

$$\limsup_{y\to\infty}(\pi(x+y) - \pi(y)) < (1+\varepsilon)e^{-\gamma} \cdot x/\log\log x$$

for $x \geq x_0(\varepsilon)$
G.H. Hardy and J.E. Littlewood. *Some problems of "partitio numerorum".* III, Acta Math. 44 (1923), 1–70.

b) There exists a positive constant $A > 0$ such that

$$\liminf_{x\to\infty}(\pi(x+y) - \pi(x)) < Ay/\log y$$

(See G. Hardy and J.E. Littlewood.)

c) $\pi(x+y) - \pi(x) < \dfrac{2y}{\log y} + o(y \log\log y/\log^2 y)$

A. Selberg. *On elementary methods in prime number theory and their limitations.* Den 11-te Skandinaviske Matematikerkongress 1952, pp. 13–22.

d) $\pi(x+y) - \pi(x) < \dfrac{2y}{\log y}$

H.L. Montgomery. *Topics in multiplicative number theory.* Springer Lecture Notes Vol. 227, 1971 (p. 34).

e) $\pi(m+n) < \pi(m) + 2\pi(n)$
for all integers $m \geq 1, n \geq 2$

H.L. Montgomery and R.C. Vaughan. *The large sieve.* Mathematika 20 (1973), 119–134.

f) For any $\varepsilon > 0$ and any $x, y \geq 17$ with $x + y \geq 1 + \exp(4(1 + 1/\varepsilon))$, one has

$$\pi(x+y) < (1+\varepsilon)\,(\pi(x) + \pi(y))$$

g) If $0 < \varepsilon \leq 1$ and $x \cdot \varepsilon \leq y \leq x$, then

$$\pi(x+y) < \pi(x) + \pi(y)$$

for x and y sufficiently large.

V. Udrescu. *Some remarks concerning the conjecture* $\pi(x+y) \leq \pi(x) + \pi(y)$. Revue Roumaine Math. Pures Appl. 20 (1975), 1201–1208.

h) If $0 < d \leq 1$ and x and y are sufficiently large with $x \geq y \geq \mathrm{d}x > 2$, then

$$\pi(x+y) < \pi(x) + \frac{\log y}{\log(x+y)} \cdot \pi(y) + O(y/\log^{n+1} y)$$

for any natural number $n \geq 2$

i) Let $0 < q \leq 1$. If m and n are sufficiently large positive integers satisfying $m \geq n \geq qm > 2$, then

$$\pi(m+n) \leq \pi(m) + 2\pi(n/2)$$

G. Giordano. *Further results on primes in small intervals.* Intern. J. Math. Sci. 12 (1989), 441–446.

j) For $0 < a_1 < a_2 \leq a_3 < a_4$ and $a_1 + a_4 = a_2 + a_3$ one has $\pi(a_1 x) + \pi(a_4 x) < \pi(a_2 x) + \pi(a_3 x)$ for sufficiently large x

L. Panaitopol. *Eine Eigenschaft der Funktion über die Verteilung der Primzahlen.* Bull Math. Soc. Sci. Math. R. S. Roumanie, 23 (71) (1979), 189–194.

§ VII. 6 On $\sum_{q \leq k \leq n} (\pi^*(k) - \pi(k))$

Let $\pi^*(n)$ denote the number of prime powers not exceeding n, $\pi^*(1) = 0$. Then

$$\sum_{k=2}^{n} (\pi^*(k) - \pi(k)) = \frac{4}{3}\frac{n^{3/2}}{\log n} + O\left(\frac{n^{3/2}}{\log^2 n}\right)$$

H. Sahu, K. Kar and B.S.K.R. Somayajulu. *On the average order of* $\pi^*(n) - \pi(n)$. Acta Cienc. Indica Math. 11 (1985), 165–168.

§ VII. 7 A sum on $\frac{1}{\pi(n)}$

$$\sum_{2\le n\le x} \frac{1}{\pi(n)} = \frac{1}{2}\log^2 x + O(\log x)$$

J.-M. de Koninck and A. Ivić. *Topics in arithmetical functions.* Notas de Matematica (72), 1980, North–Holland. Amsterdam, New York, Oxford (p. 231).

§ VII. 8 Number of primes $p \le x$ for which $p+k$ is a prime and related questions

1) a) The number of primes $p \le x$ for which $p+k$ is prime ($k \ge 2$ given integer) does not exceed

$$\frac{c_1 x}{\log^2 x}\prod_{p|k}\left(1+\frac{1}{p}\right)$$

($c_1 > 0$, constant)

Corollary. If p' runs over all the twin primes (i.e. primes p for which $p+2$ is also prime), then

$$\sum \frac{1}{p'}$$

is convergent.

V. Brun. *Le crible d'Eratosthene et le théorème de Goldbach.* Videnskapselkapets Skrifter, I, No. 3, Kristiania, 1920.

b) The number of primes $p \le x$ for which $|p+b|$ is prime (b an integer ($b > 0$ or $b < 0$)) does not exceed

$$\frac{c_2 x}{\log^2 x}\prod_{p|b}\left(1-\frac{1}{p}\right)^{-1}$$

L. Schnirelman. *Über additive Eigenschaften der Zahlen.* Izv. Donskowo Politechn. Inst. 14 (1930), 3–28 and Math. Ann. 107 (1933), 649–690.

c) The number of primes $p \le x$, for which all the numbers $p+b_1, p+b_2, \dots, p+b_s (0 < b_1 < b_2 < \cdots < b_s)$ are primes, does not exceed

$$\frac{c_3 x}{\log^{s+1} x}\prod_{p|E}\left(1-\frac{1}{p}\right)^{-(s+1-f(p))}$$

where

$$E = \prod_{1\le i\le s} b_i \cdot \prod_{1\le i\le k\le s} (b_k - b_i)$$

and $f(p)$ is the number of solutions mod p of the congruence

$$m(m+b_1)\cdots(m+b_s) \equiv O(\text{mod } p)$$

P. Erdős. *On the easier Waring problem for powers of primes.* I. Proc. Cambridge Phil. Soc. 33 (1937), 6–12.

d) Let a_i, b_i be integer numbers ($i = 1, 2, \ldots, s$) such that $a_i \neq 0$, $(a_i, b_i) = 1$, $a_i \neq \pm a_k$, $b_i \neq \pm b_k$ for $i \neq k$. Then the number of integers $n \leq x$ for which $|a_i n + b_i|$ are all primes ($i = 1, 2, \ldots, s$) does not exceed

$$c(s)\frac{x}{\log^s x} \cdot \prod_{p|E}\left(1 - \frac{1}{p}\right)^{-(s-g(p))}$$

where $E = \prod_{1\leq i\leq s} a_i \prod_{1\leq i<k\leq s}(a_i b_k - a_k b_i)$ and $g(p)$ is the number of solutions mod p of the congruence

$$(a_1 m + b_1)\cdots(a_s m + b_s) \equiv 0(\text{mod} p)$$

K. Prachar. *Primzahlverteilung.* Springer-Verlag, 1957 (p. 45.)

e) The number of prime pairs $(p, p+2k)$ with p not exceeding x is less than

$$8 \cdot \prod_{p|2k, p>2}\left(\frac{p-1}{p-2}\right)\prod_{p>2}\left(1 - \frac{1}{(p-1)^2}\right)\frac{x}{\log^2 x} + O\left(\frac{x}{\log^3 x} \cdot \log\log x\right)$$

Y. Wang. *On the representation of large integer as a sum of a prime and almost prime.* Sci. Sinica 11 (1962), 1033–1054.

f) The number of prime pairs $(p, p+2k)$ with p not exceeding x equals

$$2 \cdot \prod_{p>2}\frac{p(p-2)}{(p-1)^2} \cdot \prod_{p|k, p>2}\frac{p-1}{p-2} \cdot \int_2^x \frac{dt}{\log^2 t} + O(x(\log x)^{-c})$$

valid for all integer $2 \leq 2k \leq x(\log x)^{-c}$, excluding not more than $\ll x(\log x)^{-M-c}$ of them where $c \geq 3$ and $M > 0$ are arbitrary constants (with the constants in the symbol O being independent of k.)

A.F. Lavrik. *The number of k-twin primes lying on an interval of a given length* (Russian). Dokl. Akad. Nauk. SSSR 136 (1961), 281–283.

g) The number of twin primes $p, p+2$ with $p \leq x$ does not exceed

$$2 \cdot \left(\frac{64}{17} + \varepsilon\right)\prod_{p>2}\left(1 - \frac{1}{(p-1)^2}\right) \cdot \frac{x}{\log^2 x}$$

for all $\varepsilon > 0$ and $x \geq x_0(\varepsilon)$

É. Fouvry. *Autour du théorème de Bombieri-Vinogradov.* Acta Math. 152 (1984), 219–244.

h) The number of twin primes $p, p+2$ with $p \leq x$ does not exceed

$$2 \cdot \left(\frac{7}{2} + \varepsilon\right)\prod_{p>2}\left(1 - \frac{1}{(p-1)^2}\right) \cdot \frac{x}{\log^2 x}$$

for all $\varepsilon > 0, x \geq x_0(\varepsilon)$

E. Bombieri, J.B. Friedlander and A. Iwaniec. *Primes in arithmetic progressios to large moduli.* Acta Math. 156 (1986), 203–251.

2) Let $N(x)$ denote the number of twin primes with first elements $p \leq x$, such that $|\omega(p+1) - \log\log x| \geq (\log\log x)^{1/2+\varepsilon}$ (where $\omega(m)$ denotes the number of distinct prime factors of m). Then

$$N(x) = o(x/\log^2 x)$$

for all $\varepsilon > 0$

M.B. Barban. *On the number of divisors of "translations" of the prime number-twins* (Russian). Acta Math. Hungar. 15 (1964), 285–288.

3) For any fixed $m \geq 2$ the number of prime m-tuples $p_1 < \cdots < p_m$ such that $(p_i + p_j)/2$ is also a prime $(i, j \in \{1, 2, \ldots, m\})$ with $p_m \leq X$ is $\gg X^m (\log X)^{-m(m+1)/2}$

A. Balog. *Linear equations in primes.* Mathematika 39 (1992), 367–378.

Remark. The above result was predicted by C. Pomerance, A. Sárközy and C.L. Stewart

C. Pomerance, A. Sárközy and C.L. Stewart. Pac. J. Math. 133 (1988), 363–379.

§ VII. 9 Number of primes $p \leq x$ with $\omega(p+2) \leq 2$

Let $\pi_{1,2}(x)$ denote the number of primes $p \leq x$ such that $p+2$ has at most two prime divisors. Then:

a) $\pi_{1,2}(x) > \alpha \cdot C \cdot x/\log^2 x$
for $x \geq x_0$, where $C = 2 \cdot \prod_{p>2}(1 - 1/(p-2)^2)$ and $\alpha = 0.335$

J. Chen. Sci. Sin. 16 (1973), 157–176.

b) The same inequality holds with $\alpha = 0.71$

É. Fouvry and F. Grupp. *On the switching principle in sieve theory.* J. Reine Angew. Math. 370 (1986), 101–126.

c) The same holds true with $\alpha = 1.015$

H. Liu. *On the prime twins problem.* Sci. China, Ser. A 33, No. 3(1990), 281–298.

§ VII.10 Almost primes P_2 in intervals

1) Let P_2 denote integers with at most two prime factors. Then if

$g(x) \to \infty$ as $x \to \infty$, there exists a P_2 in the interval:

a) $(n, n + g(n)(\log n)^{7+\varepsilon}]$
for almost all n

G. Harman. *Almost-primes in short intervals.* Math. Ann. 258 (1981), 107–112.

b) $(n, n + g(n)(\log n)^5]$
for almost all n

H. Mikawa. *Almost primes in arithmetic progressions and short intervals.* Tsukuba J. Math. 13 (1989) 387–401.

2) Let q_n denote the n-th P_2 number. Then:

a) $\sum_{q_n \le x} (q_{n+1} - q_n)^2 \ll x^{1.285} \cdot (\log x)^{10}$

U. Meyer. *Eine Summe über Differenzen aufeinanderfolgender Fastprimzahlen P_2.* Arch. Math. (Basel) 42 (1984), 448–454.

b) $\sum_{q_n \le x} (q_{n+1} - q_n)^2 \ll x^{1.023}$

H. Mikawa. *The differences between consecutive almost-primes.* Tsukuba J. Math. 11 (1987), 257–264.

3) The interval

$$(x - x^\theta, x]$$

contains at least a P_2-number, ($x \ge x_0(\theta)$), where $\theta = 0.44$

J. Wu. *P_2 dans les petits intervalles.* Théorie des nombres, Sémin. Paris/Fr. 1989–1990, Prog. Math. 102 (1992), 233–267.

Remark. Iwaniec and Laborde obtained first that the assertion is valid for $\theta = 0.45$; then Halberstam and Richert improved it to $\theta = 0.4476$. More recently, Fouvry obtained $\theta = 0.4436$. The present improvement is based on Greaves' weighted sieve and the exponent pairs of Huxley and Watt.

H. Iwaniec and M. Laborde. Ann. Inst. Fourier 31 (1981), 37–56.
H. Halberstam and H.-E. Richert. Banach Cent. Publ. 17 (1985), 183–215.
É. Fouvry. Lect. Notes Math. 1434, 65–85 (1990).

§ VII.11 P_{21} in short intervals

1) a) If x is large, then the interval

$$[x, x + x^{1/2} \exp((\log x)^{0.99})]$$

contains integers which are products of exactly three primes.

Yu. V. Linnik. *A remark on products of three primes.* (Russian). Dokl. Akad. Nauk SSSR (N.S.) 72 (1950), 9–10.

b) For every integer $k \geq 2$, for large x the interval $(x, x + x^{1/k})$ contains at least one integer which has at most $2k$ prime factors.

S. Uchiyama. *On a theorem concerning the distribution of almost primes.* J. Fac. Sci. Hokkaido Univ. Ser. I 17 (1963), 152–159.

c) For all sufficiently large x, the interval

$$(x - x^{0.455}, x]$$

contains at least $(1/121) \cdot x^{0.445}/\log x$ integers with at most two prime factors.

H. Halberstam, D.R. Heath-Brown and H.E. Richert. *Almost-primes in short intervals.* Recent progress in analytic number theory, vol. I (Durham, 1979), pp. 69–101, London, 1981.

d) In c) the constant 0.455 may be replaced with 0.45

H. Iwaniec and P. Laborde. *P_2 in short intervals.* Ann. Inst. Fourier (Grenoble) 31 (1981), 37–56.

e) In d) the constant 0.45 may be replaced with $63/142 = 0.4436$

É. Fouvry. *Nombres presque premiers dans les petits intervalles.* Analytic Number Th. (Tokyo, 1988), 65–85, Lecture Notes in Math., 1434, Springer, 1990.

2) a) For infinitely many primes p the expression $(p+2)(p+6)(p+8)$ has at most 14 distinct primes.

H. Halberstam and H.E. Richert. *Sieve methods.* London, 1974.

b) 14 may be replaced with 12

S.C. Xie. *The prime 4-tuplet problem.* (Chinese). Sichuan Daxue Xuebao 26 (1989), Special Issue, 168–171.

c) If $g(y)$ is any function such that $g(y) \to \infty$ as $y \to \infty$, then in almost all intervals

$$[x, x + g(x) \log x]$$

there is a number with at most 21 prime factors.

J.B. Friedlander. *Sifting short intervals.* II. Math. Proc. Cambridge Philos. Soc. 92 (1982), 381–384.

§ VII.12 Consecutive almost primes

For $\varepsilon > 0$ define $P(\varepsilon) = \{n \in \mathbb{N} : n = pk, p \text{ prime}, k \leq n^{\varepsilon}\}$. Then there is a constant $c > 0$ such that there are infinitely many positive integers n, $n + 1$ with

$$n, n+1 \in P(c \cdot (\log\log n/ \log n)^{1/4})$$

D.R. Heath-Brown. *Consecutive almost-primes.* J. Indian Math. Soc. (N.S.) 52 (1987), 39–49.

§ VII.13 Primes in short intervals

a) Let $g(x)$ any positive function such that $g(x) \to \infty$ as $x \to \infty$. If the Riemann Hypothesis is true then almost all intervals of the form

$$[n, n + g(n) \log^2 n]$$

contain a prime.

A. Selberg. *On the normal density of primes in short intervals and the difference between consecutive primes.* Arch. Math. Naturvid 47 (1943), 87–105.

b) For almost all n (without any assumption), the interval

$$[n, n^{1/5+\varepsilon}]$$

contains a prime.

H.L. Montgomery. *Topics in multiplicative number theory.* Berlin-Heidelberg-New-York, Springer 1971 (Chapter 14.)

c) The same holds with the exponent 1/6 in place of 1/5

M.N. Huxley. *On the difference between consecutive primes.* Invent. Math. 15 (1972), 164–170.

d) For almost all n, the interval

$$[n, n + n^{1/10+\varepsilon})$$

contains $\gg n^{1/10+\varepsilon} / \log n$ primes.

G. Harman. *Primes in short intervals.* Math. Z. 180 (1982), 335–348.

§ VII.14 Primes between x and $a \cdot x$, ($a > 1$, constant). Bertrand's postulate

1) a) For every integer $n \geq 2$ there is a prime between n and $2n$, i.e. there is a prime p such that

$$n < p < 2n$$

(Bertrand's postulate, see:

J. Bertrand. *Mémoire sur le nombre de valeurs que peut prendre une fonction quand on y permute les lettres qu'elle renferme.* J. L'École Royale Polytechn. 18 (1845), 123–140.)

P. Chebychev. *Mémoire sur le nombres premiers.* J. Math. Pures Appl. 17 (1852), 366–390.

Remark. Bertrand was unable to prove his postulate, but verified it for all $n < 3\,000\,000$. The conjecture was first proved by Chebychev in 1852. For further proofs, we quote

S. Ramanujan. *A proof of Bertrand's Postulate.* J. Indian Math. Soc. 11 (1919), 181–182;

P. Erdős. *Beweis eines Satzes Tchebyschev.* Acta Litt. Sci. Regiae Univ. Hungar. Francisco-Josephinae 5 (1930–1932), 194–198;

P. Finsler. *Über die Primzahlen zwischen n und 2n.* Festschrift zum 60 Geburstag von Prof. Dr. Andreas Speiser, Zürich: Füssli, 1945, pp. 118–122.

See also

R. Archibald. *Bertrand's Postulate.* Scripta Math. 11 (1945), 109–120.

b) The interval

$$\left[x, \frac{6}{5}x\right]$$

contains a prime if $x \geq 25$

J. Nagura. *On the interval containing at least one prime number.* Proc. Japan Acad., 28 (1952), 177–181.

c) For each positive integer $n \geq 118$ there is a prime p such that

$$n < p \leq 14n/13$$

H. Rohrbach and J. Weis. *Zum finiten Fall des Bertrandschen Postulats.* J. Reine Angew. Math. 214/215 (1964), 432–440.

d) For every natural number $n > 1$ there exists a prime p such that $3n < p < 4n$

D. Hanson. *On a theorem of Sylvester and Schur.* Canad. Math. Bull. 16 (2) (1973), 195–199.

e) The interval

$$\left[x, \frac{258}{257}x\right)$$

contains a prime if $x \geq 485\,492$

N. Costa Pereira. *Elementary estimate for the Chebyshev function $\Psi(x)$ and the Möbius function $M(x)$.* Acta Arith. 52 (1989), 307–337.

Remark. For intervals of type $(x, x + x^c]$, see the results for primes in short intervals (on $\pi(x) - \pi(x - y)$)

2) Let d_k denote the least positive integer n for which

$$p_{n+1} < 2p_n - k$$

is valid (where p_n is the n-th prime). Then

a) $d_k \leq \exp\left([1 + \exp(k + 10)]^{1/2}\right)$
where [ ] denotes the integer-part function.

N.S. Udrescu. *A stronger Bertrand's postulate.* Preprint No. 34 (1974) INCREST, Bucharest (1974.)

b) $d_k \leq \dfrac{2 + M + \sqrt{M^2 + 12M + 4}}{4}$

where $M = M_k = \max\left(118, \left[\dfrac{13k}{12}\right]\right)$

C. Badea. *On a stronger Bertrand's postulate.* Anal. Univ. Al.I. Cuza. Iaşi, 32 (1986), 3–5.

c) $d_k < \max\left(\dfrac{M+4}{4c_2}, \sqrt[3]{\dfrac{M^2}{4c_3^2}}, \sqrt{\dfrac{M}{4c_4}}\right)$

where c_2, c_3, c_4 are strictly positive numbers satisfying $c_2 + c_3 + c_4 < 1$ and M is defined as in b).

Corollary. $d_k < \dfrac{M+4}{2}$

J. Sándor. *On a stronger Bertrand's postulate.* Bull. Number Theory. 11 (1987), 162–166.

Remark. This paper contains also various upper bounds for d_k. By the prime-number theorem and an elementary argument, easily follows that

$$d_k \sim k/\log k$$

and

$$d_k \leq \left[\frac{13}{12}k/(\log k - \log\log k)\right] + 1$$

for every $k \geq 4$

(See C. Pomerance. MR 88j: 11005.)

§ VII.15 On intervals containing no primes

Let $a_0(\lambda, X) = \dfrac{1}{X}$ card $\{n \in [X, 2X]$: no primes lie in the interval $[n, n + \lambda \log n]\}$. For any $\varepsilon > 0$ and sufficiently large X,

$$a_0(\lambda, X) \geq 2 \cdot \left(\frac{9}{8} - \lambda\right)^2 - \varepsilon$$

for $7/8 \leq \lambda \leq 9/8$

A.Y. Cheer and D.A. Goldston. *Longer than average intervals containing no primes.* Trans. Amer. Math. Soc. 304 (1987), 469–486.

§ VII.16 Difference between consecutive primes

a) $p_{n+1} - p_n \ll p_n^{\theta}$
where $\theta = 11/20 - \delta$ and $\delta \leq 1/384$

C.J. Mozzochi. *On the difference between consecutive primes.* J. Number theory 24 (1986), 181–187.

Remark. For earlier results see the section with $\pi(x) - \pi(x-y)$

b) The same is valid with $\theta = 6/11 + \varepsilon$, $\varepsilon > 0$.

S. Lou and Q. Yao. *A Chebyshev's type of prime numbers theorem in a short interval.* II. Hardy-Ramanujan J. 15 (1992), 1–33.

§ VII.17 Comparison of $p_1 \dots p_n$ with p_{n+1}

1) a) For all $k > 1$ there is an n_k such that

$$p_1 p_2 \dots p_n > p_{n+1}^k$$

for all $n > n_k$

L. Pósa. *Über eine Eigenschaft der Primzahlen.* (Hungarian.) Mat. Lapok 11 (1960), 124–129.

b) $p_1 p_2 \dots p_n > p_{4n}$
for $n \geq 11$ and

$p_1 p_2 \dots p_{4n-9} > p_{4n}^4$
for $n \geq 46$

S.E. Mamangakis. *Synthetic proof of some prime number inequalities.* Duke Math. J. 29 (1962), 471–473.

c) For $n \geq 3$

$$p_1 p_2 \dots p_n \geq p_1 p_2 \dots p_{n-1} + p_n + p_{p_n - 2}$$

and

$$p_1 p_2 \dots p_{n-2} p_{n-1}^2 \geq p_n + p_{p_n - 1} - 1$$

and

$$\prod_{i=1}^{n} p_i \geq \sum_{i=3}^{n} (p_i + p_{p_i - 2}) + 6$$

d) For $n \geq 24$

$$p_1 p_2 \dots p_n > p_{n+5}^2 + p_{[n/2]}^2$$

and for $n \geq 63$

$$p_1 p_2 \dots p_n > p_{n+3}^3 + p_{[n/3]}^6$$

Corollary. For $n \geq 4$ we have

$$p_1 p_2 \dots p_n > p_{n+1}^2$$

and for $n \geq 5$ we have

$$p_1 p_2 \dots p_n > p_{n+1}^3$$

(H. Bonse, See H. Rademacher and O. Toeplitz. *The enjoyment of mathematics.* Princeton Univ. Press, 1957.)
J. Sándor. *Über die Folge der Primzahlen.* Mathematica (Cluj) 30 (53)(1988), 67–74.

e) For every natural number k there exists a natural number $N(k)$ such that

$$p_1 p_2 \dots p_n > p_{n+k}^2$$

for all $n \geq N(k)$

S. Reich. *On a problem in number theory.* Math. Mag. 44 (1971), 277–278.

§ VII.18 Elementary estimates on $p_{[an]}$, p_{mn}, p_{n+1}/p_n

a) $p_m \cdot p_n > p_{mn}$

H. Ishikawa. *Über die Verteilung der Primzahlen.* Sci. Rep. Tokyo Univ. Lit. Sci. Sect. A 2 (1934), 27–40.

b) $p_{n+1}^2 \leq 2p_n^2$
for all $n > 4$

R.E. Dressler, L. Pigno and R. Young. *Sums of sequences of primes.* Nordisk Mat. Tidskrift 24 (1976), 39.

c) If $a > 1$, then

$$p_{[an]} > ap_n$$

for $n \geq n_0$

G. Giordano. *The existence of primes in small intervals.* Indian J. Math. 31 (1989), 105–110.

Remark. For estimates on $p_{n+1} - p_n$, see the general results on

$$\pi(x) - \pi(x - y)$$

§ VII.19 Sharp upper and lower bounds for p_n

a) $p_n < n \cdot \left(\log n + \log\log n - \dfrac{1}{2}\right)$

for $n \geq 20$, and

$p_n > n \cdot \left(\log n + \log\log n - \dfrac{3}{2}\right)$

for $n \geq 2$

J.B. Rosser and L. Schoenfeld. *Approximate formulas for some functions of prime numbers.* Illinois J. Math. 6 (1962), 64–94.

Remark. The inequality $p_n > n \log n$ $(n > 1)$ was first proved by Rosser.

J.B. Rosser. *The n-th prime is greater than n* log *n*. Proc. London Math. Soc. (2) 45 (1938), 21–44.

Corollary. $p_n < n \log n + n \log\log n$ for $n \geq 6$

§ VII.20 The nth composite number

Let c_n be the n-th composite number. Then

$$c_n = n \cdot \left(1 + \frac{1}{\log n} + \frac{2}{\log^2 n} + \frac{4}{\log^3 n} + \frac{19}{2}\frac{1}{\log^4 n} + \frac{181}{6}\frac{1}{\log^5 n} + o\left(\frac{1}{\log^5 n}\right)\right)$$

A.E. Bojarincev. *Asymptotic expressions for the n-th composite number*. (Russian). Ural. Gos. Univ. Mat. Zap. 6 (1967), 21–43.

§ VII.21 On infinite series involving $\sqrt{p_{n+1}} - \sqrt{p_n}$, $1/n(p_{n+1} - p_n)$ and related problems

1) a) The series

$$\sum_{n=1}^{\infty} \frac{\sqrt{p_{n+1}} - \sqrt{p_n}}{n^\theta}$$

is convergent for $\theta > 1/2$ and divergent for $\theta \le 1/2$

L. Panaitopol. Problem 141. Gaz. Mat. Ser. A (Bucureşti), 2/1974, p. 72.

b) The series

$$\sum_{n=1}^{\infty} \frac{1}{n(p_{n+1} - p_n)}$$

is divergent.

L. Panaitopol. *On the sequence of differences of consecutive prime numbers*. (Romanian). Gaz. Mat. Ser. A (Bucureşti), 6/1974, pp. 238–242.

c) Let $1 < a_1 < a_2 \cdots$ be a sequence of positive integers. Then the following series are divergent:

$$\sum_{n=1}^{\infty} \frac{a_n}{(\log a_n)^{\alpha-1}} \cdot \frac{(\log a_{n+1})^\alpha - (\log a_n)^\alpha}{p_{a_n}}$$

$$\sum_{n=1}^{\infty} \frac{\sqrt[k]{a_n^{k-1}} \cdot \left(\sqrt[k]{a_{n+1}} - \sqrt[k]{a_n}\right)}{p_{a_n}}$$

$$\sum_{n=1}^{\infty} \left(\frac{\sqrt{a_{n+1}} \cdot \log a_{n+1} - \sqrt{a_n} \log a_n}{p_{a_n}}\right) \cdot \frac{\sqrt{a_n}}{\log a_n}$$

where $\alpha \ge 1$ and $k \ge 2$. (Here p_{a_n} is the a_n-th prime.)

J. Sándor. *On certain sequences and series with applications in prime number theory*. (Romanian). Gaz. Mat. Perf. Met. Met. Mat. Inf. 6 (1985), No. 1–2, pp. 38–48.

2) a) Let $1 < a_1 < a_2 \cdots$ be a sequence of natural numbers such that $\sum_{n=1}^{\infty} 1/p_{a_n} = +\infty$. Then

$$\liminf_{n\to\infty} \left(\sqrt[k]{p_{n+1}} - \sqrt[k]{p_n}\right) \cdot (p_{a_{n+1}} - p_{a_n}) \cdot n^{\varepsilon_k} = 0$$

where $k \geq 2$ is a positive integer, and $0 \leq \varepsilon_k < \dfrac{k-1}{k}$ are constants (depending on k.)

b) $\liminf_{n\to\infty} n \cdot d_{n-1} \cdot \left(\dfrac{d_n}{p_n}\right)^2 = 0$

where $d_n = p_{n+1} - p_n$

c) $\liminf_{n\to\infty} p_n^{\alpha} \cdot (\sqrt{p_{n+1}} - \sqrt{p_n}) = 0$

where $0 < \alpha < 1/2$ and $\liminf \left(p_{n+1}^{\theta} - p_n^{\theta}\right) = 0$ for $0 < \theta < 1$

d) $\limsup_{n\to\infty} (d_{n+1} - d_n)/(\sqrt{\log(n+1)} - \sqrt{\log n}) = +\infty$

e) $\limsup_{n\to\infty} \left|\dfrac{p_{n+1}}{d_{n+1}} - \dfrac{p_n}{d_n}\right| \cdot \dfrac{p_n^2}{d_n \cdot d_{n-1}} = +\infty$

f) $\liminf_{n\to\infty} n \cdot (\sqrt[n]{n+1} - 1)^2 \cdot d_n = 0$

(See J. Sándor.)

§ VII.22 Largest gap between consecutive primes below x

1) Let $G(x)$ denote the largest gap between consecutive primes below x. Then:

a) $G(x) \geq c \cdot (e^{\gamma} + o(1)) \dfrac{\log x \cdot \log\log x}{(\log\log\log x)^2}$

where c is an explicite constant.

P. Erdős. *On the difference of consecutive primes*. Quart. J. Math. Oxford Ser. (2) 6 (1935), 124–128.

b) $G(x) \geq (e^{\gamma} + o(1)) \dfrac{\log x \cdot \log\log x \cdot \log\log\log\log x}{(\log\log\log x)^2}$

R.A. Rankin. *The difference between consecutive prime numbers*. V. Proc. Edinburgh Math. Soc. (2) 13 (1962/63), 331–332.

c) In b) the constant e^γ can be replaced by $c_0 \cdot e^\gamma$, where $c_0 = 1.31265\ldots$

H. Maier and C. Pomerance. *Unusually large gaps between consecutive primes*. Trans. Amer. Math. Soc. 322 (1990), 201–237.

2) Let $(a, q) = 1$ and $G(x; a, q)$ denote the largest gap between consecutive primes $p_n, p_{n+1} \equiv a(\mathrm{mod}\ q)$ with $p_n \leq x$. Then for any constant $C > 0$ one has

$$G(qx; q, a) \geq (e^\gamma + o(1))\, \varphi(q) \log x \cdot \log\log \frac{\log\log\log\log x}{(\log\log\log x)^2}$$

uniformly for

$$\omega(q) \leq \{C \log\log x \cdot (\log\log\log\log x)/(\log\log\log x)\}$$

A. Zaccagnini. *A note on large gaps between consecutive primes in arithmetic progressions*. J. Number Theory 42 (1992), 100–102.

§ VII.23 On $\min(d_n, d_{n+1})$ and various sums over d_n

1) a) $\limsup\limits_{n\to\infty} \min(d_n, d_{n+1}) = +\infty$

W. Sierpinski. *Remarque sur la répartition des nombres premiers*. Colloq. Math. 1 (1948), 193–194.

Remark. For a generalization of Sierpinski's theorem for an infinite sequence of k-free, well distributed integers, see

G.S.R.Ch. Murty. *A generalization of a theorem of Sierpinski*. Math. Student 46 (1978), 336–337 (1982).

b) $\limsup\limits_{n\to\infty} \dfrac{\min(d_n, d_{n+1})}{\log n} = +\infty$

P. Erdős. *Problems and results on the difference of consecutive primes*. Publ. Math. Debrecen 1 (1949), 33–37.

c) For any integer N and any $r < c_1\sqrt{\log N}$ there is a prime $p_n \leq N$ for which

$$d_{n+j} \geq c_2 \cdot \frac{\log N}{r^2}$$

$j = 0, 1, 2, \ldots, r-1$

Corollary. ($r = 2$). There exists a constant C such that for every N there is a prime $p \leq N$ for which all the numbers $p \pm k$ ($k = 1, 2, \ldots, [C \log N]$) are composite.

P. Erdős and A. Rényi. *Some problem and results on consecutive primes*. Simon Stevin 27 (1950), 115–125.

d) For any $k \in \mathbb{N}^*$

$$\limsup_{n\to\infty} \frac{\min(d_{n+1}, \ldots, d_{n+k})}{\log n \cdot \log_2 n \cdot \log_4 n / \log_3^2 n} > 0$$

where $\log_k$ denotes the iterated natural logarithm.

H. Maier. *Chains of large gaps between consecutive primes*. Adv. Math. 39 (1981), 257–269.

2) a) $c_1 \dfrac{x}{\log^2 x} < \sum_{p_n<x} \dfrac{1}{d_n} < c_2 \dfrac{x}{\log^2 x} \log\log x$

(See P. Erdős and A. Rényi.)

b) $c_3 \cdot \sum_{p_n<x} \dfrac{1}{d_n} < \sum_{p_n<x} \left| \dfrac{1}{d_n} - \dfrac{1}{d_{n+1}} \right| < c_4 \cdot \sum_{p_n<x} \dfrac{1}{d_n}$

and

$$c_5 \cdot \frac{x}{\log^2 x} < \sum_{p_n<x, d_n \ge d_{n-1}} \frac{1}{d_n} < c_6 \cdot \frac{x}{\log^2 x}$$

($c_i > 0$, constants.)

W. Knödel. *Primzahldifferenzen*. J. Reine Angew. Math. 195 (1955), 202–209.

c) $\sum_{p_n \le x, d_n \ge p_n^{1/2}} d_{n_{1/2}} \ll x^{29/30}$

D. Wolke. *Große Differenzen zwischen aufeinanderfolgenden Primzahlen*. Math. Ann. 218 (1975), 269–271.

d) Let $1/2 \le \sigma \le 1$ and $N(\sigma, T)$ be the number of zeros $\rho = \beta + i\gamma$ of ζ with $\beta \ge \sigma$, $|\gamma| \le T$. Suppose there are constants $c \ge 2$ and $h \ge 0$ such that $N(\sigma, T) = O(T^{c(1-\sigma)} \cdot \log^h T)$ uniformly for $1/2 \le \sigma \le 1$. Then there exists an absolute constant K such that for every $\varepsilon > 0$,

$$\sum_{m=1, d_m > p_m^{1-2/c+\varepsilon}}^{n} d_m = O_\varepsilon(n^{1-K\varepsilon})$$

R. Warlimont. *Über die Häufigkeit großer Differenzen konsekutiver Primzahlen*. Monatsh. Math. 83 (1977), 59–63.

Corollary. $\sum_{p_n \le x, d_n > x^{1/6}} d_n \ll x^{1+\varepsilon}$

e) $\sum_{p_n<x, d_n > p_n^{1/2}} d_n \ll x^{85/98+\varepsilon}$

for $\varepsilon > 0$

R.J. Cook. *On the occurence of large gaps between prime numbers*. Glasgow Math. J. 20 (1979), 43–48.

f) $\sum_{p_n \le x} d_n^2 \ll x^{4/3} \cdot (\log x)^{10000}$

for $x \ge 2$ and

$$\sum_{p_n \le x, d_n \ge p_n^{1/2}} d_n \ll x^{5/6} \cdot (\log x)^{10000}$$

Corollary. $p_{n+1} - p_n \ll p_{n+1}^{2/3} \cdot (\log p_n)^{5000}$

D.R. Heath-Brown. *The differences between consecutive primes*. J. London Math. Soc. (2) 18 (1978), 7–13.

g) Assuming the Lindelöf hypothesis (i.e. $\zeta(\sigma + it) \ll t^{\varepsilon}$ for $\sigma \ge 1/2, t \ge 2$ for all $\varepsilon > 0$), we have

$$\sum_{p_n \le x} d_n^2 \ll x^{7/6+\eta}$$

for any $\eta > 0$

D.R. Heath-Brown. Ibid. II. J. London Math. Soc. (2) 19 (1979), 207–220.

h) $\sum_{p_n \le x} d_n^2 \ll x^{23/18+\varepsilon}$ and $\sum_{p_n \le x, d_n \ge p_n^{1/2}} d_n \ll x^{3/4+\varepsilon}$

D.R. Heath-Brown. Ibid. III. J. London Math. Soc. (2) 20 (1979), 177–178.

i) $\sum_{p_n \le x, d_n > x^{\alpha}} d_n \ll x^{f(\alpha)+\varepsilon}$

where $f(\alpha) = 0$ for $\alpha > 11/20$

D.R. Heath-Brown and H. Iwaniec. *On the difference of consecutive primes*. Invent. Math. 55 (1979), 49–69.

j) In i)

$$f(\alpha) \le 1 - \alpha/7$$

for $7/32 \le \alpha \le 3/8$, and

$$f(\alpha) \le 11/10 - 3\alpha/5$$

for $1/6 < \alpha \le 7/32$

R.J. Cook. *An upper bound for the sum of large differences between prime numbers*. Proc. Amer. Math. Soc. 81 (1981), 33–40.

k) In i)

$$f(\alpha) \le 23/18 - \alpha$$

for $0 \le \alpha \le 1$

(see D.R. Heath-Brown III.)

l) Assuming the Riemann Hypothesis,

$$\sum_{p_n \le x, d_n > p_n^{1/2}} d_n^2 \ll \begin{cases} x^{k/2} \log^2 x & \text{for } 1 \le k < 2 \\ x \log^2 x \cdot \log\log x & \text{for } k = 2 \\ x^{k/2} \cdot \log^k x & \text{for } k > 2 \end{cases}$$

M.G. Lu. *The difference between consecutive primes*. Acta Math. Sinica (N.S.) 1 (1985), 109–118.

3) $$\sum_{X \le p_n \le 2X} d_n^2 \ge \left(\frac{193}{192} - \varepsilon\right) X \log X$$

for any $\varepsilon > 0$ and all sufficiently large X.

A.Y. Cheer and D.A. Goldston. *Longer than average intervals containing no primes.* Trans. Amer. Math. Soc. 304 (1978), 469–486.

§ VII.24 On the sign changes of $d_n - d_{n+1}$ and related theorems on primes

1) a) For infinitely many n one has $d_n > d_{n+1}$ and for an infinity of m, $d_m < d_{m+1}$. The same is true for the sequence (q_n) defined by $q_n = p_{n+1}/p_n$.

P. Erdős and P. Turán. *On some new question on the distribution of prime numbers.* Bulletin Amer. Math. Soc. 54 (1948), 371–378.

Remark. The above results can be expressed also by saying that the sequences (p_n) and $(\log p_n)$ are neither convex nor concave from some point onwards.

b) For a certain $c > 0$, both $d_{n+1} > (1+c)d_n$ and $d_{n+1} < (1-c)d_n$ have infinitely many solutions.

P. Erdős. *On the difference of consecutive primes.* Bull. Amer. Math. Soc. 54 (1948), 885–889.

c) Let $G_N = \sum_{k=2}^{n-1} \left| \arg \frac{z_{k+1} - z_k}{z_k - z_{k-1}} \right|$

where $z_k = k + i \log p_k$ and $p_n \le N$ (Thus G_N is the "total curvature" of the polygonal line with vertices z_k). Then

$$G_N \ge c \cdot \log\log\log N$$

($c > 0$, constant.)

A. Rényi. *On a theorem of Erdős and Turán.* Proc. Amer. Math. Soc. 1 (1950), 7–10.

d) $c_1 \cdot \log N < G_N < c_2 \cdot \log N$

($c_1, c_2 > 0$, constants)

P. Erdős and A. Rényi. *Some problems and results on consecutive primes.* Simon Stevin 27 (1950), 115–125.

2) a) There are infinitely many n for which

$$2p_n < p_{n-i} + p_{n+i}$$

for all positive $i < n$

b) There are infinitely many n with

$$p_n^2 > p_{n-i} \cdot p_{n+i}$$

for all positive $i < n$

c) Let $M(n) = \max\limits_{0<i<n} p_{n-i} \cdot p_{n+i}$. Then

$$\limsup_{n\to\infty} \frac{p_n^2 - M(n)}{\log^2 n} \geq 1$$

C. Pomerance. *The prime number graph.* Math. Comp. 33 (1979), 399–408.

§ VII.25 The sequence (b_n) defined by $b_n = d_n/\log p_n$

Let $b_n = \dfrac{p_{n+1} - p_n}{\log p_n}$. Then:

1) a) $\liminf\limits_{n\to\infty} b_n < 1$

P. Erdős. *The difference of consecutive primes.* Duke Math. J. 6 (1940), 438–441.

b) $\liminf\limits_{n\to\infty} b_n \leq 0.4\,665$

and

$$\liminf_{n\to\infty} (p_{n+r} - p_n)/\log p_n \leq r - \frac{1}{2}$$

E. Bombieri and H. Davenport. *Small differences between prime numbers.* Proc. Royal Soc. Ser. A, 293 (1966), 1–18.

c) $\liminf\limits_{n\to\infty} b_n \leq 0.45706$

G.Z. Pil'tjaĭ. *The value of the difference between succesive primes.* (Russian) Moskov Gos. Ped. Inst. Zap. No. 375 (1971), 133–135,

and

Studies in number theory No. 4, pp. 73–79, Saratov, 1972.

d) $\liminf\limits_{n\to\infty} b_n \leq 0.4425$ and

$$\liminf_{n\to\infty} (p_{n+r} - p_n)/\log p_n \leq \frac{2r-1}{16r}\left(4 + (4r-1)\frac{\theta_r}{\sin\theta_r}\right) = H(r)$$

where $\theta_r + \sin\theta_n = \pi/4r$

M.N. Huxley. *Small differences between consecutive primes.* II. Mathematika 24 (1977), 142–152.

e) $\liminf\limits_{n\to\infty} b_n < 0.4394$

M.N. Huxley. *An application of the Fouvry-Iwaniec Theorem.* Acta Arith. 43 (1984), 441–443.

f) $\liminf_{n\to\infty} b_n \le 0.248$ and $\liminf_{n\to\infty}(p_{n+r} - p_n)/\log p_n \le e^{-\gamma} \cdot H(r)$
where $H(r)$ is defined in d)

H. Maier. *Small differences between prime numbers.* Michigan Math. J. 35 (1988), 323–344.

Corollary. $\liminf_{n\to\infty} \frac{\max(d_n, d_{n+1})}{\log p_n} < 1$, where $d_n = p_{n+1} - p_n$

2) $\frac{b_1 + \cdots + b_n}{n} \to 1$

as $n \to \infty$ and $\sum_{n=2}^{\infty} \frac{b_n}{n(\log n)^\alpha}$ converges for $\alpha > 1$ and diverges for $\alpha \le 1$

L. Panaitopol. *On the sequence of the differences of consecutive prime numbers.* (Romanian). Gaz. Mat. Ser. A 79 (1974), 238–242.

3) Denote by $D(x, a, b)$ the number of integers k with $p_k < x$ and $a < b_k \le b$. A number u is called point of the condensation from the right of the sequence (b_n) if there exist $\gamma > 0$, $h_0 > 0$ so that for each
$0 < h < h_0$ we can find a sequence (x_n), $x_n \to \infty$ satisfying

$$D(x_n, u, u + h) > \gamma \cdot hx_n/\log x_n$$

Similarly, one can define points of condensation from the left.
Let U, V denote the set of points of condensation from the right, and from the left, respectively. Then we have:

$$\inf U < 1, \inf V < 1, \operatorname{mes} U > \frac{1}{8}, \operatorname{mes} V > \frac{1}{8}$$

G. Ricci. *Recherches sur l'allure de la suite* $((p_{n+1} - p_n)/\log p_n)$. Colloque sur la Théorie des Nombres. Bruxelles, 1955, pp. 93–106.

Remark. In 1952, K. Prachar proved that the density of integers of the form $p_{k+1} - p_k$ (p_k is the k-th prime) is positive.

K. Prachar. *Über Primzahldifferenzen.* Monatsh. Math. 56 (1952), 304–306.

4) a) For every $\varepsilon > 0$

$$b_n > 2 - \varepsilon$$

for infinitely many n

R.J. Backlund. *Über die Differenzen zwischen den Zahlen die zu den n ersten Primzahlen teilerfremd sind.* Commemoration volume in honor of E.L. Lindelöf, Helsingfors, 1929.

b) In a) $2 - \varepsilon$ can be replaced with $4 - \varepsilon$

A. Brauer and H. Zeitz. *Über eine zahlentheoretische Behauptung von Legendre.* Sitz. Berliner Math. Ges. 29 (1930), 116–125.

c) $b_n > 2 \cdot \log\log\log p_n / \log\log\log\log p_n$

for an infinity of n

E. Westzynthius. *Über die Verteilung der Zahlen die zu den n ersten Primzahlen teilerfremd sind.* Comm. Phys. Math. Soc. Sci. Fenn., Helsingfors, 5 (1931), 1–37.

Corollary. $\limsup_{n\to\infty} b_n = +\infty$

d) $b_n > c_1 \cdot \log\log\log p_n$

for an infinity of $n (c_1 > 0$, constant.)

G. Ricci. *Richerche aritmetiche sui polinomi.* II. Intorno a una proposizione non vera di Legendre Rend. Circ. Mat. Palermo 58 (1934).

e) $b_n > c_2 \cdot \log\log p_n / (\log\log\log p_n)^2$

for an infinity of $n (c_2 > 0)$

P. Erdős. *On the difference of consecutive primes.* Quart. J. Math. Oxford Ser. 6 (1935), 124–128.

f) $b_n > c_3 \cdot \dfrac{\log\log p_n \cdot \log\log\log\log p_n}{(\log\log\log p_n)^2}$

for an infinity of $n (c_3 > 0)$.

R.A. Rankin. *The difference between consecutive primes.* J. London Math. Soc. 13 (1938), 242–247.

5) For fixed integer $k \geq 1$, the sequence of k-tuples

$$(d_n / \log n, d_{n+1} / \log n, \ldots, d_{n+k} / \log n)$$

$(n = 2, 3, \ldots)$ has, for large T, a set of limit points inside $[0, T]^k$ with Lebesque measure at least $c(k) \cdot T^k$, where $c(k) > 0$ and depends only on k

A. Hildebrand and H. Maier. *Gaps between prime numbers.* Proc. Amer. Math. Soc. 104 (1988), 1–9.

Corollary. The sequence $(b_n)_{n\geq 2}$ has at least a finite limit point greater than 1

§ VII.26 Results on p_k/k

a) $c_1 \log^2 x < \sum_{p_k \leq x} \left| \dfrac{p_{k+1}}{k+1} - \dfrac{p_k}{k} \right| < c_2 \log^2 x$

where $c_1, c_2 > 0$ are constants

b) The number of primes $p_{k_i} \leq x$, for which $p_{k_i} / k_i < p_{k_{i+1}} / k_{i+1}$ (where p_{k_i}, $i = 1, 2, \ldots$ is a subsequence of the sequence of primes) equals

$$O(x / \log x)$$

P. Erdős and K. Prachar. *Sätze und Probleme über p_k/k.* Abh. Math. Sem. Univ. Hamburg 25 (1961/62), 251–256.

§ VII.27 On the sums of prime powers

a) $$\sum_{p\leq x} p = \frac{1}{2}\cdot(1+o(1))\frac{x^2}{\log x}$$

E. Landau. *Handbuch.* (Band 1, p. 226) Leipzig, 1909.

b) $$\sum_{p\leq x} p^{\alpha} \sim \frac{x^{1+\alpha}}{(1+\alpha)\log x}$$

for $\alpha \geq 0$

T. Salát and S. Znám. *On the sums of prime powers.* Acta Fac. Rer. Nat. Univ. Com. Math. 21 (1968), 21–25.

§ VII.28 Estimates on $\sum_{p\leq x}\frac{1}{p}$

1) a) $$\sum_{p\leq x}\frac{1}{p} = \log\log x + B + o(1)$$

P. Chebyshev. *Sur la fonction qui detérmine la totalité des nombres premiers inférieurs à une limite donnée.* Mém. préséntés à l'Acad. Imp. Sci. St. Pétersbourg par divers Savants, 6 (1851), 141–157.

b) $$\sum_{p\leq x}\frac{1}{p} = \log\log x + B + O\left(\frac{1}{\log x}\right)$$

F. Mertens. *Ein Beitrag zur analytischen Zahlentheorie.* Crelle's Journal 78 (1874), 46–62.

2) a) $$\sum_{p\leq x}\frac{1}{p} < \log\log x + B + \frac{1}{2\log^2 x}$$

for $x \geq 286$, and

$$\sum_{p\leq x}\frac{1}{p} > \log\log x + B - \frac{1}{2\log^2 x}$$

for $x > 1$, where $B = 0.2614972\ldots$

b) $$\sum_{p\leq x}\frac{\log p}{p} < \log x + C + \frac{1}{2\log x}$$

for $x \geq 319$, and

$$\sum_{p\leq x}\frac{\log p}{p} > \log x + C - \frac{1}{2\log x}$$

for $x > 1$

J.B. Rosser and L. Schoenfeld. *Approximate formulas for some functions of prime numbers.* Illinois J. Math. 6 (1962), 64–94.

3) There exists an absolute constant c such that for every finite set S of primes

$$\sum_{p\in S'} \frac{1}{p} - \sum_{p\in S} \frac{1}{p} \le c$$

where S' is the set of primes p such that all prime factors of $p-1$ belong to S

C. Pomerance. *On the composition of the arithmetic functions σ and φ.* Colloq. Math. 58 (1989), 11–15.

4) a) $$\sum_{n\le x} \sum_{(p-1)|2n} \frac{1}{p} = x\left(\frac{1}{2} + \sum_{p>2} \frac{2}{p(p-1)}\right) + O(\log\log x)$$

and

$$\sum_{n\le x} (-1)^n \cdot \sum_{(p-1)|2n} \frac{1}{p} = x \cdot \sum_{p\equiv 1 (\mathrm{mod} 4)} \frac{2}{p(p-1)} + O(\log\log x)$$

J. Valdez. *A new property of the Bernoulli numbers.* Math. Mag. 47 (1974), 144–145.

b) Let $f(n) = \sum_{p|n} \frac{1}{p}$. Then

$$\sum_{n\le x} f(n) = Cx + O(x^{4/7}\log^2 x)$$

X. Yu. *An estimate on the distribution of weakly composite numbers.* (Chinese.) J. Shandong Univ., Nat. Sci. Ed. 24 (1989), 1–6.

5) $$\sum_{n\le x}\left(\prod_{p|n} \frac{1}{p}\right) \sim (4\pi)^{-1/2} \cdot 2^{-1/4} \cdot \left(\frac{\log\log x}{\log x}\right)^{1/4} \cdot Q(x)$$

where $Q(x) = \min_{0<\sigma<\infty} x^\sigma \cdot \sum_{n=1}^{\infty} (\alpha(n))^{-1} \cdot n^{-\sigma}$

W. Schwarz. *Einige Anwendungen Tauberscher Sätze in der Zahlentheorie.* B.J. Reine Angew. Math. 219 (1965), 157–179.

6) Let $S = \{p \le n : n = kp + r \text{ with } p/2 < r < p,\ p \text{ prime}\}$. Then

$$\sum_{p\in S} \frac{\log p}{p} = \frac{1}{2}\log n + O((\log n)^{5/6+\delta})$$

for any fixed $\delta > 0$

J.W. Sander. *On a sum over primes.* Hardy-Ramanujan J. 17 (1994), 32–39.

§ VII.29 Estimates on $\prod_{p \leq x} \left(1 - \frac{1}{p}\right)$

1) a) $\prod_{p \leq x} \left(1 - \frac{1}{p}\right) = O\left(\frac{1}{\log x}\right)$

P. Chebychev. *Sur la fonction qui détermine la totalité des nombres premiers inférieurs à une limite donnée.* Mém. présentés à l'Acad. Imp. Sci. St. Pétersbourg par divers Savants, 6 (1851), 141–157.

b) $\prod_{p \leq x} \left(1 - \frac{1}{p}\right) = \frac{c}{\log x} \cdot \left(1 + O\left(\frac{1}{\log x}\right)\right)$

where $c > 0$ (in fact, $c = e^{-\gamma}$, where γ is Euler's constant)

F. Mertens. *Ein Beitrag zur analytischen Zahlentheorie.* Crelle's Journal 78 (1874), 46–62.

2) a) $\prod_{p \leq x} \left(1 - \frac{1}{p}\right) < \frac{e^{-\gamma}}{\log x} \cdot \left(1 + \frac{2}{\log^2 x}\right)$

for $x > 1$, and

$\prod_{p \leq x} \left(1 - \frac{1}{p}\right) > \frac{e^{-\gamma}}{\log x} \cdot \left(1 - \frac{1}{2\log^2 x}\right)$

for $x \geq 285$

b) $\prod_{p \leq x} \frac{p}{p-1} < e^{\gamma} \cdot \log x \cdot \left(1 + \frac{1}{2\log^2 x}\right)$

for $x \geq 286$

$\prod_{p \leq x} \frac{p}{p-1} > e^{\gamma} \cdot \log x \cdot \left(1 - \frac{1}{2\log^2 x}\right)$

for $x > 1$

J.B. Rosser and L. Schoenfeld. *Approximate formulas for some functions of prime numbers.* Illinois J. Math. 6 (1962), 64–94.

3) $\prod_{p \leq x} \left(1 - \frac{1}{p}\right) = \frac{e^{-\gamma}}{\log x} \cdot (1 + O(\exp(-a(\log x)^{3/5})))$

I.M. Vinogradov. *On the remainder in the Mertens' formula.* (Russian). Dokl. Akad. Nauk. SSSR 148 (1963), 262–263.

§ VII.30 Some properties of Ψ-function

1) a) There exists constants $A > 0$, $a > 0$, such that for all $x \geq 2$

$$ax < \Psi(x) < Ax$$

b) $\liminf_{x\to\infty} \frac{\psi(x)}{x} \le 1 \le \limsup_{x\to\infty} \frac{\psi(x)}{x}$

P. Chebyshev. *Mémoire sur les nombres premiers*. J. Math. Pures appl. 17 (1852), 366–390.

2) a) $\psi(x) \sim x$
$(x \to \infty)$
(This is an equivalent form of the prime number theorem, see $\pi(x)$)

J. Hadamard. *Sur la distribution des zéros de la fonction $\zeta(s)$ et ses conséquences arithmétiques*. Bull. Soc. Math. France 24 (1896), 199–220;

and

C.J. de la Vallée Poussin. *Recherches analytiques sur la théorie des nombres*. (3 parts). Ann. Soc. Sci. Bruxelles 20, Part II (1896), 183–256, 281–397.

b) If $\theta > 1/2$, then

$$\psi(x) - x = O(x^{\theta})$$

where θ is the limsup of all σ such that $\zeta(\sigma + it) = 0$

E. Grosswald. *Sur l'ordre de grandeur des différences $\Psi(x) - x$ et $\pi(x) - \mathrm{li}\ x$*. C.R. Acad. Sci. Paris 260 (1965), 3813–3816.

c) $\limsup_{x\to\infty} \frac{\psi(x) - x}{x^{1/2} \cdot \log\log\log x} \ge \frac{1}{2}$ and

$\liminf_{x\to\infty} \frac{\psi(x) - x}{x^{1/2} \cdot \log\log\log x} \le -\frac{1}{2}$

A.E. Ingham. *The distribution of prime numbers*. Cambridge, 1932.

3) a) $(1 - \delta(x)) \cdot x < \theta(x) \le \Psi(x)$
for $x \ge 2$, and

$$\theta(x) \le \psi(x) < (1 + \delta(x)) \cdot x$$

for $x \ge 1$, where

$$\delta(x) = \log^{1/2} x \cdot \exp(-\sqrt{(\log x)/A})$$

with $A = 17.51631\ldots$

b) $\psi(x) - \theta(x) < 1.42620 \cdot \sqrt{x}$
for $x > 0$, and
$\psi(x) - \theta(x) > 0.98\sqrt{x}$
for $x \ge 121$

c) $\frac{\psi(x)}{x}$ takes its maximum at $x = 113$

$\frac{\psi(x) - \theta(x)}{\sqrt{x}}$ takes its maximum at $x = 361$

d) $\psi(x) < 1.03883x$
for $x > 0$

J.B. Rosser and L. Schoenfeld. *Approximate formulas for some functions of prime numbers*. Illinois J. Math. 6 (1962), 64–94.

e) $\psi(x) < 532/531x$ for $x \geq 60299$ and
$\psi(x) > 530/531x$ for $x \geq 70841$

N. Costa Pereira. *Elementary estimates for the Chebyshev function $\Psi(x)$ and the Möbius function $M(x)$*. Acta arith. 52 (1989), 307–337.

Remarks: (i) By elementary methods, due to P. Erdős and A. Selberg, the following improvements were obtained: $\psi(x) = x + O(x/\log^m x)$, where

(a) $m = 1/10$

P. Kuhn. *Eine Verbesserung des Restgliedes beim elementaren Beweis der Primzahlsatzes*. Math. Scand. 3 (1955), 75–89.

(b) $m = 1/200$

J.G. van der Corput. *Colloques sur la Théorie des Nombres*. Liège: G. Thone, 1956.

(c) $m = 1/6$

R. Breusch. *An elementary proof of the prime number theorem with remainder term*. Pacific J. Math. 10 (1960), 487–497.

(d) $m = 3/4$

E. Wirsing. *Elementare Beweise der Primzahlsatz mit Restglied*. I.J. Reine Angew. Math. 211 (1962), 205–214.

(e) For all positive m

E. Bombieri. *Maggiorazioni del resto nel Primzahlsatz col methodo di Erdős-Selberg*. Inst. Lombardo Sienze Lettre, Rendinconti, A 96 (1962), 343–350

and

E. Wirsing. *Elementare Beweise der Primzahlsatz mit Restglied*. II. J. Reine Angew. Math. 214–215 (1964), 1–18.

(ii) By new elementary methods, the following results have been obtained:

(a) $\psi(x) = x + O(x \cdot \exp(-(\log x)^{1/7}(\log\log x)^{-2}))$

H.G. Diamond and J. Steinig. Invent. Math. 11 (1970), 199–258.

(b) $\psi(x) = x + O(x \cdot \exp(-(\log x)^{1/6}(\log\log x)^{-3}))$

A.F. Lavrik and A.S. Sobirov. Dokl. Akad. Nauk. SSSR 211 (1973), 534–536.

(c) $|\psi(x) - x| \leq x \cdot \exp\left(-\frac{1}{40}(\log x)^{1/6}\right)$

for $\log\log x \geq 40$

B.R. Srinivasan and A. Sampath. *An elementary proof of the prime number theorem with a remainder term.* J. Indian Math. Soc. (N.S.) 53 (1988), 1–50.

(iii) For a proof using large sieve type inequalities, see

A. Hildebrand. *The prime number theorem via the large sieve.* Mathematika 33 (1986), 23–30.

The first elementary proof which is not based on Selberg's formula (or equivalent assertion) was obtained by

H. Daboussi. C.R. Acad. Sci. Paris, Sér. I, Math. 298 (1984), no. 8, 161–164.

§ VII.31 Selberg's formula

1) $\Psi(x)\log x + \sum_{n\leq x} \psi\left(\frac{x}{n}\right) \cdot \Lambda(n) = 2x\log x + O(1)$

A. Selberg. *An elementary proof of the prime number theorem.* Ann. Math. 50 (1949), 305–313.

Remark. The above result is equivalent with

$$\sum_{n\leq x} \Lambda(n)\log n + \sum_{mn\leq x} \Lambda(m)\Lambda(n) = 2x\log x + O(1)$$

2) If the Riemann Hypothesis is true, then

$$\sum_{n\leq x}\sum_{m+k=n} \Lambda(m)\Lambda(k) = \frac{1}{2}x^2 + O(x^{3/2})$$

A. Fujii. *An additive problem on prime numbers.* Acta Arithm. 58 (1991), 173–179.

Remark. For a more precise, but complicated result, see

A. Fujii. *An additive problem of prime numbers.* II. Proc. Japan Acad., Ser. A 67 (1991), 248–252.

3) Let $A, B > 0$, $0 < V < N/4$ and $0 < \varepsilon < 1/2$. Then the relation

$$(*) \qquad \sum_{m,n=N,n-m=2k}^{2N} \Lambda(m)\Lambda(n) = \mathfrak{S}(2k)\cdot(N-2k) + O(N(\log N)^{-A})$$

holds true for all $V \leq k \leq V + H$ but $O(H(\log N)^{-B})$ exceptions, where k is a fixed positive integer, and

$$\mathfrak{S}(2k) = 2\prod_{p>2}\left(1 - \frac{1}{(p-1)^2}\right)\cdot \prod_{p|k,p>2} (p-1)/(p-2)$$

A. Perelli and J. Pintz. *On the exceptional set for the 2k-twin primes problem.* Compositio Math. 82 (1992), 355–372.

Remark. The classical Hardy-Littlewood conjecture states that (*) is true for any $A > 0$. For several "almost all" results, see

H.L. Montgomery. *Topics in multiplicative number theory.* Lecture Notes Math. 227, Springer Verlag 1971.

4) Let $\Lambda_N(n) = \sum_{d|n,d\leq N} \mu(d)\cdot\log(N/d)$. Then

$$\sum_{n\le M} \Lambda_N(n)\Lambda(n+k) = \mathfrak{S}(2k)\cdot M + R_N(M,K)$$

and

$$\sum_{n\le M} \Lambda_N(n)\Lambda_N(n+k) = \mathfrak{S}(2k)\cdot M + R_N^*(M,K)$$

where $R_N(M,k) = o(M)$ for $N \le M^{1/2-\varepsilon}$
and $R_N^*(M,K) = o(M)$ for $N \le M^{1/2-\varepsilon}$
(Here $\mathfrak{S}(2k)$ is defined in 3).)

D.A. Goldston. *On Bombieri and Davenport's theorem concerning small gaps between primes.* Mathematika 39 (1992), 10–17.

§ VII.32 On $\sum_{n\le x} \psi(n)$

Let $\psi_1(x) = \sum_{n\le x} \Psi(n)$. Then $\psi_1(x) \ge \alpha\cdot x^2$ for all $x \ge x_0$, where $\alpha = 0.49517\ldots$

M. Nair. *A new method in elementary number theory.* J. London Math. Soc. (2) 25 (1982), 385–391.

§ VII.33 Estimates on $\psi(x+h) - \psi(x)$

1) a) If θ is such that $\psi(n+h) - \psi(n) \sim h$ for almost all n and $h \gg x^{\theta}$, then

$$\sum_{p\le x} (\psi(p+h) - \psi(p))^k \sim h^k \cdot x/\log x$$

for any $k \in \mathbb{N}^*$

A. Perelli and A. Salerno. *On 2k-dimensional density estimates.* Studia Sci. Math. Hungar 20 (1985), 345–355.

b) If $N^{\theta+\varepsilon} < H \le N$, then

$$\psi(x+H) - \psi(x) \sim H$$

for almost all $x (0 < \theta < 1)$

A. Perelli. *Local problems with primes.* I.J. Reine Angew. Math. 401 (1989), 209–220.

2) If $0 < \delta \le 1$, $T \ge 2$ and the Riemann Hypothesis is true, then

$$\int_1^T (\psi(x+\delta x) - \psi(x) - \delta x)^2 \cdot x^{-2}\,dx \ll \delta(\log T)\left(\log\frac{2}{\delta}\right)$$

D.A. Goldston and H.L. Montgomery. *Pair correlation of zeros and primes in short intervals. Analytic number theory and Diophantine problems.* (Stillwater, OK, 1984), 183–203, Progr. Math. 70, Birkhäuser, Boston MA, 1987.

§ VII.34 On $\Delta(x) = \psi(x) - x$

Let $\Delta(x) = \psi(x) - x$. Then:

1) Let $V(T)$ denote the number of sign changes of $\Delta(x)$ in $[2, T]$. Then:

a) $\limsup\limits_{T\to\infty} \dfrac{V(T)}{\log T} > 0$

G. Pólya. *Über das Vorzeichen des Restgliedes im Primzahlsatz.* Gölt. Nachr. 1930, 19–27.

b) There exist effectively computable constants c_1 and c_2 such that for $T > c_1$ one has

$$V(T) > c_2 \frac{\sqrt{\log T}}{\log\log T}$$

J. Pintz. *On the remainder term of the prime number formula.* III. 12 (1977), 345–369.

c) For $T > T_0$ (effective constant), the interval

$$[T, T \cdot \exp(63 \cdot \sqrt{\log T} \cdot \log\log T)]$$

contains a sign-change of $\Delta(x)$

(See J. Pintz (1977).)

d) For $T > T_1$ (effective constant),

$$V(T) > 10^{-11} \frac{\log T}{(\log\log T)^3}$$

J. Pintz. Ibid. IV. Studia Sci. Math. Hungar 13 (1978), 29–42.

Corollary. $\liminf\limits_{T\to\infty} \dfrac{V(T)}{(\log T)(\log\log T)^{-3}} > 0$

e) $V(T) \geq \dfrac{\gamma_0}{4\pi} \cdot \log T$

where $\gamma_0 = 14.13\ldots$

J. Kaczorowski. *On sign-changes in the remainder term of the prime number formula.* I. Acta Arith. 44 (1984), 365–377.

f) $V(T) \geq 0.013 \log T$
for $T \geq 10^{2250}$

g) If $H \geq 501.5$ and if all nontrivial zeros $\rho = \beta + i\gamma$ of ζ with $|\gamma| < H$ have $\beta = \dfrac{1}{2}$, then for every

$T \geq \exp(0.09 \cdot \max(4400, H))$,

$$V(T) \geq \left(1 - \frac{3}{H}\right) \frac{\gamma_0}{\pi} \cdot \log T$$

where $\gamma_0 = 14.13\ldots$

B. Szydło. *Über Vorzeichenwechsel einiger arithmetischer Funktionen*. I, II. Math. Ann. 283 (1989), 139–149, 151–163.

2) a) $\Delta(x) = \Omega_\pm(x^{\theta-\varepsilon})$

where θ denotes the least upper bound of the real parts of the ζ-zeros.

E. Phragmén. *Sur une loi de symétrie relative à certaines formules asymptotiques*. Öfversigt af Kongl. Vetenskaps-Akad. Förhandlingar 58 (1901), 189–202.

b) $\Delta(x) = \Omega_\pm(\sqrt{x})$

E. Schmidt. *Über die Anzahl der Primzahlen unter gegebener Grenze*. Math. Ann. 57 (1903), 195–204.

c) Assuming the Riemann Hypothesis, we have

$$\Delta(x) = O(\sqrt{x}\log^2 x)$$

H. von Koch. *Sur la distribution des nombres premiers*. Acta Math. 24 (1901), 159–182.

d) Assuming the Riemann Hypothesis, we have

$$\Delta(x) = \Omega_\pm(\sqrt{x} \cdot \log\log\log x)$$

J.E. Littlewood. *Sur la distribution des nombres premiers*. C.R. Acad. Sci. Paris 158 (1914), 1869–1872.

e) If $\rho_0 = \beta_0 + i\gamma_0$, $\beta \geq \frac{1}{2}$, is an arbitrary nontrivial zero of $\zeta(s)$, then for $T > \max(c_0, c_1(\rho_0))$ one has

$$\max_{l \leq x \leq T} |\Delta(x)| > \frac{T^{\beta_0}}{|\rho_0|^{10\log T/(\log\log T)}} \cdot \exp\left(-c_1 \frac{\log T \cdot \log\log\log T}{\log\log T}\right)$$

where c_0, c_1 are explicitly calculable positive constants.

P. Turán. *On the remainder-term of the prime number formula*. I. Acta Math. Hungar 1 (1) (1950), 48–63.

Remark. Turán obtained the above result by his powersum method.

f) If ρ_0 is as above, then for $T > \max(c_2, \exp\exp(2|\rho_0|))$ one has

$$\max_I |\Delta(x)| > T^{\beta_0} \cdot \exp\left(-8\frac{\log T}{\log\log T}\right)$$

where $I = \left[T\exp\left(-\frac{\log T \cdot \log\log\log T}{(\log\log T)^2}\right) - 1, T\right]$

W. Staś. *Über die Abschätzung des Restgliedes in Primzahlsatz*. Acta Arith. 5 (1959), 427–434.

g) Let $0 < \varepsilon \leq 1/50$ and let us assume the existence of a

$\rho_0 = \beta_0 + i\gamma_0$, zero of $\zeta(s)$ with $\beta_0 = \frac{1}{2} + \delta_0 > \frac{1}{2} + \varepsilon$ and $\gamma_0 > \exp\exp(10^{12}/\varepsilon^3)$. Then for every H satisfying $H^{\varepsilon^4/10^7} > \max(\gamma_0, c_3)$ we have in the interval $I = [H, H^{1+\varepsilon}]$ an$x_1 \in I$ and $x_2 \in I$ for which

$$\Delta(x_1) > x_1^{\beta_0}/\gamma_0^{1+\varepsilon} \text{ and } \Delta(x_2) < -x_2^{\beta_0}/\gamma_0^{1+\varepsilon}$$

J. Pintz. *On the remainder-term of the prime number formula.* I. On a problem of Littlewood. Acta Arith. 36 (1980), 341–365.

h) For $T > \max(c_4 \cdot (\gamma_0/\varepsilon)^{14}, \exp((c_5/\varepsilon\gamma_0)^2))$ there exists an $x \in [T, T^{6\log\gamma_0+60}]$ such that

$$|\Delta(x)| > (1-\varepsilon)x^{\beta_0}/|\rho_0|$$

J. Pintz. *On the mean value of the remainder term of the prime number formula.* Banach Center Publ. 17 (1985), 411–417, PWN, Warsaw.

Remark. For similar results on certain general class of remainder terms, see Sz.Gy. Révész. *Effective oscillation theorems for a general class of real-valued remainder terms.* Acta Arith. 49 (1988), 481–505.

3) Let $D(x) = \frac{1}{x}\int_1^x |\Delta(t)|\,dt$ and $D\left(x - \frac{x}{H}, x\right) = \frac{1}{(x|H)} \cdot \int_{x-x|H}^x |\Delta(t)|\,dt$

Then:

a) Assuming the Riemann Hypothesis, one has

$$\frac{1}{x} \cdot \int_1^x \Delta^2(t)\,dt = O(x) \text{ and } D(x) = O(\sqrt{x})$$

H. Cramér. *Ein Mittelwertsatz in der Primzahltheorie.* Math. Z. 12 (1922), 147–153.

b) Assuming the Riemann Hypothesis,

$$D(x) > \sqrt{x} \cdot \exp\left(-c_1\frac{\log x}{\log\log x} \cdot \log\log\log x\right)$$

S. Knapowski. *Contributions to the theory of distribution of prime numbers in arithmetic progressions.* I. Acta Arith. 6 (1961), 415–434.

c) Without any hypothesis,

$$D(x) > \frac{\sqrt{x}}{22000}$$

for $x > 2$

J. Pintz. *On the mean value of the remainder term of the prime number formula.* Banach Center Publ. 17(1985), 411–417, PWN, Warsaw.

d) Assuming the Riemann Hypothesis,

$$D\left(x - \frac{x}{H}, x\right) \ll \sqrt{x}\log^2(H+1)$$

for $1 \le H \le x$

J. Pintz. *A note on the distribution of primes in short intervals.* Acta Math. Hungar 44 (3–4)(1984), 335–338.

4) Let $E\left(x - \frac{x}{H}, x\right) = \frac{1}{(x|H)} \cdot \int_{x-x|H}^{x} \Delta(t)\mathrm{d}t$. Then:

a) Assuming the Riemann Hypothesis,

$$E\left(x - \frac{x}{\log\log x}, x\right) = \Omega_{\pm}(\sqrt{x}\log\log x)$$

W.B. Jurkat. *On the Mertens conjecture and related Ω-theorems.* Proc. Symp. Pure Math. 24 (Providence, Rhode Island), 147–158.

b) Without any hypothesis, $E\left(x - \frac{x}{H}, x\right) = \Omega_{\pm}(\sqrt{x}\log^2(H+1)$ for

$$1 \le H \le \log\log x/(\log\log\log x)^2$$

(See J. Pintz. (1984).)

5) $\frac{1}{2u} \cdot \int_1^{\infty} \frac{\Delta(x)\log x}{x} \exp\left(-\frac{\log^2 x}{4u}\right)\mathrm{d}x = O(1)$

as $u \to \infty$

K.A. Rodosskiĭ. *On regularity in the distribution of primes.* (Russian). Uspehi Mat. Nauk 17 (1962), 189–191.

§ VII.35 Results on $\theta(x)$

1) a) There exist constants $A > 0, a > 0$, such that for all $x \ge 2$, $ax < \theta(x) < Ax$

b) $\liminf_{x\to\infty} \frac{\theta(x)}{x} \le 1 \le \limsup_{x\to\infty} \frac{\theta(x)}{x}$

P. Chebyshev. *Mémoire sur les nombres premiers.* J. Math. Pures appl. 17 (1852), 366–390.

2) a) $\theta(x) \sim x$

$(x \to \infty)$

(This is an equivalent form of the prime number theorem, see $\pi(x)$)

J. Hadamard and C.J. de la Valée Poussin.

b) $\theta(x) = x + O(x\exp(-A\log^{3/5} x(\log\log x)^{-1/5})$

H.M. Korobov and I.M. Vinogradov, see the section with $\pi(x)$

3) a) $\theta(x) < x \cdot \left(1 + \frac{1}{2\log x}\right)$
for $x > 1$, and

$$\theta(x) > x \cdot \left(1 - \frac{1}{2\log x}\right)$$

for $x \geq 563$

b) $\theta(x) < 1.01624x$
for $x > 0$

J.B. Rosser and L. Schoenfeld. *Approximate formulas for some functions of prime numbers.* Illinois J. Math. 6 (1962), 64–94.

c) $\theta(p_k) \geq k(\log k + \log\log k - 1.076868)$
for $k \geq 2$, with equality when $k = 66$

G. Robin. *Estimation de la fonction de Tchebyshev θ sur la k-ième nombre premier et grandes valeurs de la fonction $\omega(n)$ nombre de diviseurs premiers de n.* Acta Arith. 42 (1983), 367–389.

4) a) Let $\theta_k(x) = \sum_{p_1 p_2 \cdots p_k \leq x} \log p_1 \cdot \log p_2 \ldots \log p_k$. Then

$$\theta_{k+2}(x) + \frac{1}{k+1}\log x \cdot \theta_{k+1}(x) = \frac{2}{(k+1)!} x \log^{k+1} x + O(x \log^k x)$$

H.N. Shapiro. *On a theorem of Selberg and generalizations.* Ann. Math. (2) 51 (1950), 485–497.

b) Let $\theta(m_1, \ldots, m_k, x) = \sum_{p_1 \cdots p_k \leq x} \log^{m_1} p_1 \ldots \log^{m_k} p_k$
If $m_1, \ldots, m_r$ are odd, and $n_1, \ldots, n_s$ even, then

$$\theta(m_1, \ldots, m_r, n_1, \ldots, n_s, x) =$$

$$= Ax \log^{\rho-1} x + (-1)^s \prod_{i=1}^{s}(n_i - 1)! \prod_{j=1}^{r}(m_j - 1)! \theta_\rho(x) + O(x \log^{-2} x)$$

where $\rho = \sum_{i=1}^{r} m_i + \sum_{j=1}^{s} n_j$ and A is a constant depending upon the m_i and n_j

(See H.N. Shapiro.)

c) $$\sum_{p_1+\cdots+p_r=n} \log p_1 \ldots \log p_r = \frac{n^{r-1}}{(r-1)!} S_r(n) + O((n/\log n)^{r-1})$$

for $r \geq 3$, where $S_r(n)$ is a certain singular series and the constant implied by the O-symbol is independent of r

A. Walfisz. *Zur additiven Zahlentheorie.* Mitt. Akad. Wiss. Georgischen SSSR 2 (1941), 221–226.

5) Let $\Delta_3(x) = \theta(x) - x$

a) To every $\varepsilon > 0$ there exists an ineffective constant $T_0(\varepsilon)$ such that for $T > T_0(\varepsilon)$, $\Delta_3(x)$ changes its sign in the interval $[T, T^{1+\varepsilon}]$

J. Pintz. *On the remainder term of the prime number formula.* I. On a problem of Littlewood. Acta Arith. 36 (1980), 341–365.

Remark. The same is true for the functions $\Delta_1(x) = \pi(x) - x$, $\Delta_2(x) = \sum_{\nu\geq 1} \frac{1}{\nu}\pi(x^{1/\nu}) - \text{li}\, x$, $\Delta_4(x)(\text{or } \Delta(x)) = \Psi(x) - x$

b) Let $V_3(T)$ be the number of sign-changes of $\Delta_3(x)$ in $[2, T]$. Then

$$V_3(T) > 10^{-11}\frac{\log T}{(\log\log T)^3}$$

for $T > T_1$ (ineffective constant).

J. Pintz. *On the remainder term of the prime number formula.* IV. Studia Sci. Math. Hungar. 13 (1978), 29–42.

c) $V_3(T) \geq c \cdot \log T$

$(c > 0)$

J. Kaczorowski. *On sign-changes in the remainder term of the prime number formula.* I. Acta Arith. 45 (1985), 65–74.

d) $\liminf_{T\to\infty} V_3(T)/\log T \geq \gamma_0/\pi + 10^{-250}$

where $\gamma_0 = 14.137\ldots$ is the imaginary part of the first zero of $\zeta(s)$ in the supper half-plane.

J. Kaczorowski. *The k-functions in multiplicative number theory.* V: Changes of sign of some arithmetical error terms. Acta Arith. 59 (1991), 37–58.

6) a) $\max_{1\leq x\leq T} |\Delta_3(x)| > T^{\beta_0} \cdot \exp\left(-\frac{21\log T}{(\log\log T)^{1/2}}\right)$

P. Turán. *Eine neue Methode in der Analysis und deren Anwendungen.* Akadémiai Kiadó, Budapest, 1953.

b) For $T > T_0$ (ineffective constant),

$$D_3(T) > c \cdot \sqrt{T}$$

where $D_3(T) = \frac{1}{T}\int_2^T |\Delta_3(x)|\mathrm{d}x$

Remark. The same is valid for $D_4(x)$ (with $\Delta_4(x)$)

J. Pintz. Ibid. V. Studia. Sci. Math. Hungar. 15 (1980), 215–223.

c) $\frac{1}{T}\int_{T\exp(-5\sqrt{\log T})}^{T} |\Delta_3(x)|\mathrm{d}x > 0.62\sqrt{T}$

d) Assuming the Riemann Hypothesis,

$$\frac{1}{T}\int_{10^{-3}T}^{T} \Delta_3(x)\mathrm{d}x < -0.62\sqrt{T} \text{ and}$$

$$c_1\sqrt{T} < D_3(T) < c_2\sqrt{T} \text{ for } T \geq c_3 (c_i > 0 \text{ constants})$$

J. Pintz. Ibid. VI. Studia. Sci. Math. Hungar. 15 (1980), 225–230.

§ VII.36 Primes in short intervals

$$\text{Let } J(\beta, T) = \int_1^{T^\beta} \left(\sum_{x<p^k<x+\frac{x}{T}} \log p - \frac{x}{T} \right)^2 \frac{\mathrm{d}x}{x^2}$$

Assuming the Riemann Hypothesis, there are absolute constants $c_2 > c_1 > 0$ such that for each $\beta > 0$,

$$c_1 \frac{\log^2 T}{T} \leq J(\beta + 2, T) - J(\beta, T) \leq c_2 \frac{\log^2 T}{T}$$

D.A. Goldston and S.M. Gonek. *A note on the number of primes in short intervals.* Proc. Amer. Math. Soc. 108 (1990), 613–620.

§ VII.37 Estimates concerning $\Lambda(n)$ and certain generalizations. Sign-changes in the remainder

1) a) $\displaystyle\sum_{n \leq x} \frac{\Lambda(n)}{n} = \log x + O(1)$

F. Mertens. *Ein Beitrag zur analytischen Zahlentheorie.* Crelle's Journal 78 (1874), 46–62.

b) $\displaystyle\sum_{n \leq x} \frac{\Lambda(n)}{n} \cdot H\left(\frac{x}{n}\right) + H(x)\log x = O(1)$

where $\displaystyle H(x) = \sum_{n \leq x} \frac{\mu(n)}{n}$

(See F. Mertens.)

c) $\displaystyle M(x)\log^2 x - 2 \cdot \sum_{mn \leq x} \Lambda(m)\Lambda(n)M\left(\frac{x}{n}\right) = O(x \log x)$

where $\displaystyle M(x) = \sum_{n \leq x} \mu(n)$

H.N. Shapiro. *Introduction to the theory of numbers.* John Wiley and Sons, 1983 (p. 438.)

d) $$\sum_{n \le \exp x} \left(\frac{1}{n} - \frac{\Lambda(n)}{n} \right) = 2\gamma + O(x^{-\alpha})$$

for every $\alpha > 0$

E. Wirsing. *Elementare Beweise der Primzahlsatzes mit Restglied.* II. J. Reine Angew. Math. 214/215(1964), 1–18.

Remark. For results concerning $\sum_{n \le x} \Lambda(n) = \psi(x)$, see the section with the function Ψ

2) Let $V_5(T)$ denote the number of sign-changes of the function $\Delta_5(x) = \sum_{n=1}^{\infty} (\Lambda(n) - 1) \cdot e^{-n/x}$ in $(0, T]$. Then:

a) $V_5(T) \ge \frac{\gamma_0}{4} \log T$

for $T > T_0$ (effective constant), where γ_0 denotes the imaginary part of the lowest zero of ζ

b) Assuming the Riemann Hypothesis, one has

$$V_5(T) = (\gamma_0/\pi) \cdot \log T + O(1)$$

J. Kaczorowski. *On sign-changes in the remainder term of the prime-number formula.* III. Acta Arith. 48 (1987), 347–371.

c) $V_5(T) = o(\log^2 T)$

d) Let $M(T, \varepsilon) = \max_{T \le x \le (1+\varepsilon)T} |\Delta_5(x)|$ and $\theta = \sup_{\zeta(\rho)=0} \operatorname{Re} \rho (\varepsilon > 0, T > 0)$. Then for $0 > \frac{1}{2}$, one has

$$M(T, \varepsilon) \ge c_0 \cdot T^{\theta - \varepsilon}$$

for large T and for a certain constant $c_0 = c_0(\varepsilon)$

J. Kaczorowski. Ibid. IV. Acta Arith. 50 (1988), 15–21.

e) $$\max_{\delta \le y \le 1} \left| \sum_{n=1}^{\infty} (\Lambda(n) - 1) e^{-ny} \right| > \frac{1}{\sqrt{\delta}} \cdot \exp \left(\frac{-4 \log \frac{1}{\delta} \log_3 \frac{1}{\delta}}{\log_2 \frac{1}{\delta}} \right)$$

(where $\log_k$ denotes the iterated logarithm.)

S. Knapowski and W. Staś. *A note on a theorem of Hardy and Littlewood.* Acta Arith. 7 (1961/62), 161–166.

3) For a positive integer k and a real number $x \geq 2$,

let $\psi_2(x, 2k) = \sum_{2k<n\leq x} \Lambda(n) \cdot \Lambda(n-2k)$.

Let $E(x, 2k) = \Psi_2(x, 2k) - H(x, 2k)$

where $H(x, 2k) = 2\prod_p \left(1 - \frac{1}{(p-1)^2}\right) \prod_{p|k, p>2} \left(\frac{p-1}{p-2}\right) \cdot (x - 2k)$

Then, if $2x \leq y \leq x^{8/5-\varepsilon}$, the number of k such that $k \leq x$ and $E(y, 2k) \neq O(y(\log y)^{-c_1})$ is $O(x(\log x)^{-c_2})$, for any positive numbers c_1 and c_2

D. Wolke. *Über das Primzahl-Zwillingsproblem.* Math. Ann. 283 (1989), 529–537.

4) $\sum_{n\leq x}\sum_{p|n} \frac{1}{p}\Lambda\left(\frac{n}{p}\right) = cx + O(x \cdot e^{-\lambda\sqrt{\log x}})$

where $c = \sum_p 1/p^2, \lambda > 0$

X. Yu. *An estimate on the distribution of weakly composite numbers.* (Chinese.) J. Shandong Univ., Nat. Sci. Ed. 24 (1989), 1–6.

5) Let $\Lambda_{f,k}(n)$ be the arithmetical function defined by

$$f(n)\log^k n = \sum_{d|n} f(d)\Lambda_{f,k}(n/d)$$

where $f(n)$ is a non-zero arithmetical function, and k a positive integer.

a) $\sum_{n\leq x} \Lambda_{d,k}(n) = k(k+1)x\log^{k-1} x + O(x\log^{k-1} x)$

where d is the divisor function.

b) $\sum_{n\leq x} \Lambda_{\mu,k}(n) = \begin{cases} -x + O(x/\log x), & \text{for } k = 1 \\ O(x\log^{k-2} x), & \text{for } k \geq 1 \end{cases}$

where μ denotes the Möbius function.

c) $\sum_{n\leq x} \Lambda_{\varphi,k}(n) = (kx^2/2)\log^{k-1} x + O(x^2\log^{k-2} x)$

where φ is Euler's totient.

d) $\sum_{n\leq x} \Lambda_{d_m,k}(n) = m\prod_{i=1}^{k-1}(m+i)/i)x\log^{k-1} x + O(x^2\log^{k-2} x)$

where d_m is the generalized divisor function.

e) $\sum_{n\leq x} \Lambda_{r,k}(n) = kx\log^{k-1} x + O(x\log^{k-2} x)$

where r is the function counting the number of representations as a sum of two squares.

A. Ivić. *On a class of arithmetical functions connected with multiplicative functions.* Publ. Inst. Math. (Beograd) 20 (34)(1976), 131–144.

§ VII.38 A sum over $1/\Lambda(n)$

$$\sum_{n\leq x}{}' \frac{1}{\Lambda(n)} = \sum_{i=2}^{N} \frac{a_i x}{\log^i x} + O\left(\frac{x}{\log^{N+1} x}\right)$$

where $N \geq 1$ is an arbitrary fixed integer and a_i are computable constants, e.g. $a_2 = 1$. (Here the dash indicates that the sum is taken over n with $\Lambda(n) \neq 0$)

J.-M. de Koninck and A. Ivić. *Topics in arithmetical functions.* Notas de Matematica (72), 1980. North-Holland, Amsterdam, New York, Oxford (p. 232).

§ VII.39 On Chebyshev's conjecture

$$\lim_{x\to\infty} \sum_{p>2} (-1)^{(p-1)/2} \cdot p^{-\alpha} \cdot \log p \cdot \exp\left(-\frac{1}{x}\log^2 p\right) = -\infty$$

for $0 \leq \alpha \leq 1/2$

H.-J. Bentz. *Discrepancies in the distribution of prime numbers.* J. Number Theory 15 (1982), 252–274.

Remarks. 1) Chebyshev's classical conjecture (1853; still open) states that

$$\lim_{x\to\infty} \sum_{p>2} (-1)^{\frac{p-1}{2}} \cdot \exp\left(-\frac{p}{x}\right) = -\infty$$

For generalizations of Chebyshev's conjecture (which are equivalent with the Generalized Riemann Hypothesis for certain Dirichlet L-functions) see

A. Fujii. *Some generalizations of Chebyshev's conjecture.* Proc. Japan Acad. Ser. A Math. Sci. 64 (1988), 260–263.

2) S. Knapowski and P. Turán have proved that the relation of a) is equivalent with $L(s, \chi_1) \neq 0$ for $\mathrm{Re}\, s > \frac{1}{2}$, where χ_1 is a non-principal character modulo 4

S. Knapowski and P. Turán. *On the sign-changes of $\pi(x) - \mathrm{li}\, x$. I.* Topics in number theory (Proc. Colloq., Debrecen, 1974), 153–169.

§ VII.40 A sum involving primes

Let $f(n) = \sum_{p<n} \frac{1}{n-p}$, where p is prime. Then

a) $\frac{1}{x} \sum_{n \le x} f(n) \to 1$
$(x \to \infty)$

b) $\sum_{n \le x} (f(n))^2 < cx$
$(c > 0$ – constant)

c) $f(n) < a \log\log n$
with $a > 0$ (P. Erdős and P. Turán.) See

P. Erdős. *Quelques problèmes de théorie des nombres.* L'Enseign. Math. (1963), 81–135.

d) Assuming that the number of zeros of $\zeta(\sigma + it)$ in the rectangle $\alpha \le \sigma \le 1, -T \le t \le T$ is less than $c \cdot T^{2(1-\alpha)} \log^2 T$, uniformly for $\frac{1}{2} \le \alpha \le 1, T > 2$, we have

$$\sum_{p<x} (f(p) - 1)^2 = O(x(\log\log x)(\log x)^{-3/2})$$

I. Kátai. *A result on consecutive primes.* Acta Math. Hungar. 27 (1976), 153–159.

Chapter VIII

PRIMES IN ARITHMETIC PROGRESSIONS AND OTHER SEQUENCES

§ VIII. 1 Dirichlet's theorem on arithmetic progressions

1) For $k > 0$ and l, integers such that $(k, l) = 1$, the arithmetic progression $kn + l, n = 1, 2, \ldots$, contains infinitely many primes.

G.L. Dirichlet. *Beweis des Satzes daβ jede unbegrenzte arithmetische Progression deren erstes Glied und Differenz ganze Zahlen ohne gemeinschaftlichen Faktor sind unendlich viele Primzahlen enthält.* Werke, Leipzig: G. Reimer, 1889, I, pp. 313–342, (Original 1837).

Remarks. (i) An elementary proof was given by Mertens.

P. Mertens. *Wiener Sitzungsb.* 106 (1897), 254–282.

(ii) The first new "elementary proof" of Dirichlet's theorem was published by Selberg.

A. Selberg. *An elementary proof of Dirichlet's theorem about primes in arithmetic progression.* Ann. Math. 50 (2) (1947), 297–304.

See also

H.N. Shapiro. *On primes in arithmetic progression, (II).* Ann. Math. 52 (1950), 231–243.

2) If k is a power of an odd prime and l is a non-residue mod k, $(k, l) = 1$, then there exist infinitely many primes in the arithmetic progressions, (I). Ann. Math. 52 (1950), 217–230.

§ VIII. 2 Bertrand's and related problems in arithmetic progressions

1) a) For $n \geq 6$, positive integer, there is always a prime p of the form $6n + 1$, and one of the form $6n - 1$, such that

$$n < p < 2n$$

R. Breusch. *Zur Verallgemeinerung der Bertrandschen Postulates dass zwischen x und 2 x stets Primzahlen liegen.* Math. Z. 34 (1932), 505–526;

G. Ricci. *Sul teorema di Dirichlet relativo alla progresione aritmetica.* Boll. Un. Mat. Ital. 12 (1933), 304–309;

and

P. Erdős. *Über die Primzahlen gewisser arithmetischen Reihen.* Math. Zeit. 39 (1935), 473–491.

b) There is always a prime p such that

$$n < p \le \frac{8}{7}n$$

of each of the forms $3x+1, 3x-1$, provided $n \ge 199$

c) The interval

$$n < p \le \frac{4}{3}n$$

always contains a prime of each of the forms $12x+1, 12x-1, 12x+5, 12x-5$.

K. Molsen. *Zur Verallgemeinerung der Bertrandschen Postulates.* Deutsche Math. 6 (1941), 248–256.

2) If $(k,l)=1$ and k is sufficiently large then there exists a positive constant c such that for $x \ge \exp(c \log k \cdot \log\log k)$

$$\pi(2x;k,l) - \pi(x;k,l) > 0$$

T. Tatuzawa. *On Bertrand's problem in an arithmetic progression.* Proc. Japan Acad. 28 (1962), 293–294.

3) There are absolute constants $c>0, c'>0$ such that for any $0<\varepsilon\le c$, for all $k \ge k_0(\varepsilon)$ and all $x \ge k^{c'\log(c/\varepsilon)}$, there is at least one prime $p \equiv l (\mathrm{mod}\ k)$, $(l,k)=1$, in the interval

$$(x, x\cdot k^{\varepsilon})$$

E. Fogels. *On the existence of primes in short arithmetical progressions.* Acta Arith. 6 (1960/61), 295–311.

§ VIII. 3 Sums over $1/p$, $\log p/p$ when $p \le x$, $p \equiv l (\mathrm{mod}\ k)$

1) $$\sum_{p\le x, p\equiv l (\mathrm{mod}\ k)} 1/p = \frac{1}{\varphi(k)} \log\log x + A + O\left(\frac{1}{\log x}\right)$$

$(k>0)$

E. Landau. *Handbuch.* I. 1909, p. 450.

2) a) $$\sum_{p\le x, p\equiv l (\mathrm{mod}\ k)} \log p/p = \frac{1}{\varphi(k)} \log x + O(1)$$

(See E. Landau.)

Remark. For an elementary method in case $k=4, l=\pm 1$ see

R. Breusch. *An asymptotic formula for primes of the form* $4n+1$. Michigan Math. J. 11 (1964), 311–315.

b)
$$\sum_{p\le x,\, p\equiv l (\mathrm{mod}\ k)} \log^2 p + \sum_{pq\le x,\, pq\equiv l (\mathrm{mod}\ k)} \log p\cdot \log q = \frac{2}{\varphi(k)}\cdot x\log x + O(x)$$

H.N. Shapiro. *On primes in arithmetic progressions.* II. Ann. Math. (2) 52 (1950), 231–243.

c)
$$\sum_{p\le x,\, p\equiv l (\mathrm{mod}\ k)} \log^2 p/p + \sum_{pq\le x,\, pq\equiv l (\mathrm{mod}\ k)} \log p\cdot \log q/pq = \frac{1}{\varphi(k)}\cdot \log^2 x + O(x)$$

Z. Koshiba and S. Uchiyama. *On the existence of prime numbers in arithmetic progressions.* J. Math. Anal. Appl. 19 (1967), 431–443.

Note. For results on $\sum_{p_1^{\alpha_1}\cdots p_r^{\alpha_r}\le x,\, p_1^{\alpha_1}\cdots p_r^{\alpha_r}\equiv l (\mathrm{mod}\ k)} \log^{\alpha_1} p_1\cdots \log^{\alpha_r} p_r$, see

Y. Eda. *On Selberg's function.* Proc. Japan Acad. 29 (1953), 418–422.

3) a)
$$\prod_{p\le x,\, p\equiv 1 (\mathrm{mod}\ 4)}\left(1-\frac{1}{p}\right) = c_1\cdot(\log x)^{-1/2} + O((\log x)^{-3/2})$$

where $c_1 = \left(\pi\cdot e^{-\gamma}\prod_{p\equiv 1(\mathrm{mod}\ 4)}(1-p^{-2})\right)^{1/2}$

b)
$$\prod_{p\le x,\, p\equiv 3 (\mathrm{mod}\ 4)}\left(1-\frac{1}{p}\right) = c_2\cdot(\log x)^{-1/2} + O((\log x)^{-3/2})$$

where $c_2 = \left(\frac{1}{2}\pi\cdot e^{-\gamma}\prod_{p\equiv 3(\mathrm{mod}\ 4)}(1-p^{-2})\right)^{1/2}$

S. Uchiyama. *On some products involving primes.* Proc. Amer. Math. Soc. 28 (1971), 629–630.

c)
$$\sum_{p_n\le T,\, p_{n-1}\equiv p_n\equiv 1 (\mathrm{mod}\ 4)} 1 > \log^B T$$

for all $T > C$, where B and C are positive explicitly calculable constants.

S. Knapowski and P. Turán. *On prime numbers* $\equiv 1$ *or* 3(mod 4). Number theory and algebra, pp. 157–165, Academic Press, New York, 1977.

d) If $(k,l) = 1$, $k > 0$, then

$$\prod_{p\le x,\, p\equiv l (\mathrm{mod}\ k)}\left(1-\frac{1}{p}\right) \sim c_l\cdot(\log x)^{-\varphi(k)}$$

where C is a positive constant depending on l and k

K.S. Williams. *Mertens' theorem for arithmetical progressions.* J. Number Theory 6 (1974), 353–359.

Remark. This result has been rediscovered by Grosswald.

E. Grosswald. *Some number theoretical products.* Rev. Columbiana Mat. 21 (1987), 231–242.

§ VIII. 4 The n-th prime in an arithmetic progression

Let $p_n(k, l)$ denote the n-th prime in the arithmetic progression $kn + l (k > 0)$

a) Let $(k, l) = 1$. The density of integers of the form

$$p_{n+1}(k, l) - p_n(k, l)$$

is positive.

b) $\liminf_{n\to\infty} \frac{p_{n+1}(k, l) - p_n(k, l)}{\log p_n(k, l)} < \varphi(k)$

Remark. For $k = 1$ this is a result of P. Erdős.
K. Prachar. *Über Primzahldifferenzen.* II. Monatsh. Math. 56 (1952), 307–312.

c) $\liminf_{n\to\infty} \frac{p_{n+r}(k, l) - p_n(k, l)}{\varphi(k) \log p_n(k, l)} \le r - \frac{5}{8} + O\left(\frac{1}{r}\right)$
where $(k, l) = 1$, k even.
M.N. Huxley. *On the differences of primes in arithmetical progressions.* Acta Arith. 15 (1968/69), 367–392.

§ VIII. 5 Least prime in an arithmetic progression. Linnik's theorem. Various estimates on $p(k, l)$

Let $p(k, l)$ be the least prime in the arithmetic progression $l + nk$ $(n = 0, 1, 2, \ldots)$ with $(k, l) = 1$, $k \ge 2$. Then:

1) a) There exists an absolute constant L such that

$$p(k, L) < k^L$$

(L—Linnik's constant.)
Ju.V. Linnik. *On the least prime in the arithmetic progression.* I. The basic theorem. Mat. Sbornik 15 (1947), 139–178; II. The Deuring-Heilbronn phenomenon. Ibid. 347–368.

b) $p(k, l) \ll k^L$
where $L \le 550$
M. Jutila. *A new estimate for Linnik's constant.* Ann. Acad. Sci. Fenn. Ser. A, I, No. 471 (1970), 8 pp.

c) $L \le 17$
J.R. Chen. *On the least prime in an arithmetical progression and theorems concerning the zeros of Dirichlet's L-functions.* II. Sci. Sinica 22 (1979), 859–899.

d) $L \le 16$

W. Wang. *On the least prime in an arithmetical progression (Chinese).* Acta Mat Sinica 29 (1986), 826–836.

e) $p(k, l) \ll k^L$
where $L \leq 13.5$

J. Chen and J. Liu. *On the least prime in an arithmetical progression.* III, IV. Sci. China, Ser. A, 32, No. 6, 654–673, No. 7, 792–807 (1989).

f) $L \leq 5.5$

D.R. Heath-Brown. *Zero-free regions for Dirichlet L-functions and the least prime in an arithmetic progression.* Proc. London Math. Soc. III Ser. (to appear)

2) a) For almost all positive integers k

$$p(k, l) < \varphi(k) \cdot \log^A k$$

for at least $c \cdot \varphi(k)$ incongruent values of $l(\text{mod } k)$ with $(k, l) = 1$, where $A > 3$ and $0 < c < 1$

S. Uchiyama. *The maximal large sieve.* Hokkaido Math. J. 1 (1972), 117–126.

b) Let $A > 3$ be a real number and $0 < \varepsilon < A - 3$. Then for almost positive integers k

$$p(k, l) < \varphi(k) \log^A k$$

except for possibly $\varphi(k) \cdot (\log k)^{-\varepsilon}$ values of l with $(l, k) = 1$, $1 \leq l < k$.

S. Uchiyama. *An application of the large sieve.* Proc. Jap. Acad. 48 (1972), 67–69.

3) a) Let $c_1 > 0$ be any constant. Then there exists a constant c_2 depending on c_1, and infinitely many values of k such that

$$p(k, l) > (1 + c_1) \varphi(k) \log k$$

for more than $c_2\varphi(k)$ values of l

b) Let $c_3 > 0$ be any constant. Then there exists a constant $c_4 > 0$ depending on c_3 such that for $c_4\varphi(k)$ values of l

$$p(k, l) < c_3\varphi(k) \log k$$

P. Erdős. *On some Applications of Brun's method.* Acta Univ. Szeged. Sect. Sci. Math. 3 (1949), 57–63.

c) Given $l > 0$, there exists infinitely many integers k with $(k, l) = 1$ such that

$$p(k, l) > \frac{Ck \log k \cdot \log\log k \cdot \log\log\log\log k}{(\log\log\log k)^2}$$

where C is independent of k

K. Prachar. *Über die kleinste Primzahl einer arithmetischen Reihe.* J. Reine Angew. Math. 206 (1961), 3–4.

Note. See also

A. Schinzel. *Remark on a paper of K. Prachar, "Über die kleinste Primzahl einer arithmetischen Reihe."* J. Reine Angew. Math. 210 (1962), 121–122, where it is shown that C is independent of l (Thus C is an absolute constant.)

d) For any fixed l there exist infinitely many primes q such that

$$p(q,l) < c(\varepsilon)\cdot q^{\theta+\varepsilon}$$

where $\theta = 2e^{1/4}\cdot(2e^{1/4}-1)^{-1} = 1.63773\ldots$

Y. Motohashi. *A note on the least prime in an arithmetic progression with a prime difference.* Acta Arith. 17 (1970), 283–285.

e) $p(q,l) > q\log q/(\log\log q)^{1+\varepsilon}$

(q-prime) for all $l = 1, 2, \ldots$ and all $\varepsilon > 0$

S.S. Wagstaff, Jr. *The least prime in an arithmetic progression with prime difference.* J. Reine Angew. Math. 301 (1978), 114–115.

f) Let $P(k) = \max\limits_{l=1,2,3,\ldots} p(k,l)$. Then

$$\liminf_{k\to\infty} P(k)/\varphi(k)\log k \ge e^{\gamma}$$

C. Pomerance. *A note on the least prime in an arithmetic progression.* J. Number theory 12 (1980), 218–223.

4) The lower density of integers $k > 0$ for which $p(k,\ell) \le k^2/g(k)$ is 1, whenever $\log g(x) = o(\log x)$. Here l is a fixed nonzero integer.

A. Granville. *Least primes in arithmetic progressions.* Théorie des nombres (Quebec, PQ, 1987), 306–321, de Gruyter. Berlin-New-York, 1989.

The density of integers $k > 0$ for which $p(k,l) \le kf(k)$ is 1 provided $f(k) \ge k^{1-o(1)}$

A. Granville. The same paper.

§ VIII. 6 Siegel-Walfisz theorem. The Bombieri-Vinogradov theorem

1) Let $\Psi(x;q,a) = \sum\limits_{n\le x, n\equiv a(\mathrm{mod}\ q)} \Lambda(n)$, where Λ is von Mangoldt's function and $q > 0$, a are integers such that $(a,q) = 1$. Then:

a) $\Psi(x;q,a) = \dfrac{x}{\varphi(q)}\cdot\left(1+O\left(\dfrac{1}{\log^A x}\right)\right)$

for any fixed $A > 0$, uniformly in the range $q < \log^A(x)$ (Siegel-Walfisz theorem.)

C.L. Siegel. *Über die Classenzahl quadratischer Körper.* Acta Arith. 1 (1935), 83–86

and

A. Walfisz. *Zur additiven Zahlentheorie.* II. Math. Zeit. 40 (1936), 592–607.

b) For fixed $A > 0$, there exists $B = B(A) > 0$ such that

$$\max_{q \le Q (a,q)=1} \left| \Psi(x; q, a) - \frac{x}{\varphi(q)} \right| = O\left(\frac{x}{\log^A x}\right)$$

where $Q = \dfrac{\sqrt{x}}{\log^B x}$ (Bombieri-Vinogradov theorem.)

E. Bombieri. *On the large sieve.* Mathematatika 12 (1965), 201–225.

and

A.I. Vinogradov. *The density conjecture for Dirichlet L-series.* Izv. Akad. Nauk. SSSR Ser. Mat. 29 (1965), 903–934.

Remark. There are also other forms of the above theorems expressed in terms of $\pi(x; q, a)$ or $\theta(x; q, a)$. E.g.

$$\sum_{q \le Q, (q,a)=1} \left| \pi(x; q, a) - \frac{\pi(x)}{\varphi(q)} \right| = O\left(\frac{x}{\log^A x}\right)$$

See the results on $\pi(x; q, a)$

c)

$$\sum_{q \le Q} \max_{(q,a)=1} \max_{y \le x} \left| \Psi(y; q, a) - \frac{y}{\varphi(q)} \right| \le C_1(\sqrt{x} Q (\log x)^{11/8} +$$

$$+ \frac{x^{\beta_{k_0}}}{\varphi(k_0)} \cdot (\log x)^{5/4} + x \exp(-C_2 (\log x)^{1/4}))$$

where $C_1, C_2 > 0$ are effective constants.

M.M. Timofeev. *The Vinogradov-Bombieri theorem (Russian).* Mat. Zametki 38 (1985), 801–809, 956.

2) a) Let

$$E(x; b, q, a) = \sum_{bn \le x, bn \equiv a (\text{mod } q)} \Lambda(n) - \frac{1}{\varphi(q)} \cdot \frac{x}{b}$$

and let $f(b)$ be an arithmetic function satisfying $f(b) \ll d^r(b)$ (where d is the divisor function). Then given $0 < \varepsilon < 1$ and any $A > 0$, and given functions $A_1(x)$, $A_2(x)$ with $0 < A_1(x) < A_2(x) \le x^{1-\varepsilon}$ there exists

$$B \ge 2A + 2^{2r+2}(2^{2r+2} + 1) + 21$$

such that

$$\sum_{q \le \sqrt{x}/\log^B x} \max_{y \le x} \max_{(l,k)=1} \left| \sum_{\substack{A_1(y) \le b < A_2(y) \\ (b,q)=1}} f(b) \cdot E(y; b, q, a) \right| \ll x/\log^A x$$

C.D. Pan and X.X. Ding. *A new mean value theorem.* Sci. Sinica 1979, Special issue II on Math., 149–161.

b) Let $a \ne 0$, $\varepsilon > 0$ and $Q = x^{4/7-\varepsilon}$. For any well factorable function $\lambda(q)$ of level Q and any $A > 0$ we have

$$\sum_{(q,a)=1} \lambda(q) \cdot \left(\Psi(x; q, a) - \frac{x}{\varphi(q)} \right) \ll_{\varepsilon,a,A} x / \log^A x$$

E. Bombieri, J.B. Friedlander and H. Iwaniec. *Primes in arithmetic progressions to large moduli.* Acta Math. 156 (1986), No. 3–4, 203–251.

c) Let $a \neq 0$, and $x \geq y \geq 3$. Then

$$\sum_{\substack{q \leq (xy)^{1/2} \\ (q,a)=1}} \left| \Psi(x; q, a) - \frac{x}{\varphi(q)} \right| \ll x \cdot \left(\frac{\log y}{\log x} \right)^2 \cdot (\log \log x)^B$$

Here B is an absolute constant and the $\ll$ symbol depends only on a

Corollary. $\pi(x; q, a) \sim \operatorname{li} x / \varphi(q)$ for almost all q in the range

$$q \leq x^{1/2} \cdot \exp\left(\frac{\log x}{(\log \log x)^B} \right)$$

E. Bombieri, J.B. Friedlander and H. Iwaniec. Ibid. II. J. Math. Ann. 277 (1987), 361–393.

d) Let $a \in \mathbb{I} \backslash \{o\}$ and let $I \subset [1, x^\theta]$, where $\frac{1}{2} \leq \theta \leq \frac{3}{4}$
Then

$$\sum_{q \in I, (q,a)=1} \left| \pi(x; q, a) - \frac{\pi(x)}{\varphi(q)} \right| \ll x / \log^A x +$$

$$+ \left(\left(\theta - \frac{1}{2} \right)^2 \cdot \frac{x}{\log x} + \frac{x (\log \log x)^2}{(\log x)^3} \right) \sum_{q \in I, (q,a)=1} 1/\varphi(q)$$

E. Bombieri, J.B. Friedlander and H. Iwaniec. Ibid. III. J. Amer. Math. Soc. 2 (1989), No. 2, 215–224.

3) a) Fix $B > 1$. There exists arbitrarily large values of a and x for which:

(i) $$\sum_{T < q < 4T, (q,a)=1} \left| \Psi(x; q, a) - \frac{x}{\varphi(q)} \right| \gg_B \frac{x}{\log \log x}$$

where $T = x / 4 \log^B x$

(ii) $$\sum_{q < x / \log^B x, (q,a)=1} \left| \Psi(x; q, a) - \frac{x}{\varphi(q)} \right| \gg_B x$$

J. Friedlander and A. Granville. *Limitations to the equi-distribution of primes.* I. Ann. Math. 129 (1989), 363–382.

Remark. This shows that a conjecture of Elliott-Halberstam and another conjecture of Montgomery, respectively, are false.

P.D.T.A. Elliott and H. Halberstam. *A conjecture in prime number theory.* Symp. Math. IV (Rome 1968/69), 59–72;

H.L. Montgomery. *Problems concerning prime numbers.* Proc. Symp. Pure Math. 28 (1976), 307–310.

b) Fix $B > 1$. There exists arbitrarily large values of y such that for any value of Q in the range

$$y/\log^B y < Q < y/\log y \cdot \log\log y$$

there exists $x = x(Q)$ in the range $y < x < 3y$ such that

$$\sum_{Q<q<2Q} \max_{(a,q)=1} \left| \Psi(x;q,a) - \frac{x}{\varphi(q)} \right| \gg_B \sum_{Q<q<2Q} \frac{x}{\varphi(q)}$$

(See J. Friedlander and A. Granville.)

§ VIII. 7 Primes in arithmetic progressions

Let $E(x;q,a) = \left| \varphi(x;q,a) - \frac{x}{\varphi(q)} \right|$. Then:

a) For fixed $A > 0$, for $q \leq \log^A x$ and any $Q \leq x/\log^A x$, one has

$$\sum_{q \leq Q} \sum_{1 \leq a \leq q, (a,q)=1} (E(x;q,a))^2 = O(x^2 \cdot (\log x)^{5-A})$$

H. Davenport and H. Halberstam. *Primes in arithmetic progressions.* Michigan Math. J. 13 (1966), 485–489.

b) For fixed $A > 0$ and $Q \leq x$,

$$\sum_{q \leq Q} \sum_{1 \leq a \leq q, (a,q)=1} (E(x;q,a))^2 = Qx\log x + O\left(Qx \log \frac{2x}{Q}\right) +$$

$$+ O(x^2/\log^A x)$$

while for $Q \geq x$, the same sum is

$$= Qx\log x - \frac{\zeta(2)\zeta(3)}{\zeta(6)} x^2 \log \frac{Q}{x} - Qx + A_1 x^2 + O(Qx/\log^A x)$$

H.L. Montgomery. *Primes in arithmetic progressions.* Michigan Math. J. 17 (1970), 33–39.

§ VIII. 8 Bombieri's theorem in short intervals

$$(*) \quad \sum_{q \leq Q} \max_{(a,q)=1} \max_{h \leq y} \max_{\frac{1}{2}x < z \leq x} \left| \psi(z+h;q,a) - \psi(z;q,a) - \frac{h}{\varphi(q)} \right| \ll y/\log^A x$$

where $y = x^\theta$ and $Q = x^\alpha/\log^B x$ with:

a) $\alpha < \dfrac{4c\theta + 2\theta - 1 - 4c}{6 + 4c}$

Here $c = \inf\left\{\xi : \zeta\left(\frac{1}{2} + it\right) \ll t^\xi\right\}$

M. Jutila. *A statistical density theorem for L-functions with applications.* Acta Arith. 16 (1969), 207–216.

b) $\alpha \le \theta - \frac{1}{2}$

if $\frac{3}{4} \le \theta \le 1$

$\alpha \le \left(\frac{1}{5} + \left(\frac{3}{5} \cdot \left(\theta - \frac{3}{5}\right)\right)^{1/2}\right)^2 - \varepsilon$

if $\frac{29}{48} < \theta \le \frac{3}{4}$

$\alpha \le 3\theta - \frac{7}{4} - \varepsilon$

if $\frac{7}{12} < \theta \le \frac{29}{48}$

M.N. Huxley and H. Iwaniec. *Bombieri's theorem in short intervals.* Mathematika 22 (1975), 188–194.

c) $\alpha \le \min\left(\theta - \frac{1}{2}, \frac{1}{2}(5\theta - 3)\right)$

if $\theta > \frac{3}{5}$

S.J. Ricci. *Mean-value theorems for primes in short intervals.* Proc. London Math. Soc. (3) 37 (1978), 230–242.

d) The estimate (*) holds if $\alpha \le \theta - \frac{1}{2}$ and $\theta > \frac{3}{5}$

A. Perelli, J. Pintz and S. Salerno. *Bombieri's theorem in short intervals.* Ann. Scuola Normale Sup. Pisa, Serie IV, 11 (1984), 529–539.

e) The estimate (*) holds if $y = x^{\theta}$, $7/12 < \theta \le 1$ and $Q = x^{1/40}$

A. Perelli, J. Pintz and S. Salerno. Ibid. II. Invent. Math 79 (1985), 1–9.

f) In e) we can choose $Q = x^{1/38.5}$

T. Zhan. *Bombieri's theorem in short intervals.* Acta Math. Sinica 5 (1989), 37–47.

g) The estimate (*) holds if $x^{7/12+\varepsilon} \le y \le x$ with $Q = y/x^{11/20+\delta}$ and if $x^{3/5} \cdot (\log x)^{2(A+64)+1} \le y \le x$ with $Q = y/(x^{1/2} \cdot (\log x)^{A+64})$, where the constants A, ε, δ are given and the implied constants in $\ll$ depend only on A, ε, δ.

N.M. Timofeev. *Distribution of arithmetic functions in short intervals in the mean with respect to progressions (Russian).* Izv. Akad. Nauk SSSR Ser. Mat. 51 (1987), 341–362, 447.

Note. The above paper contains also similar results involving the function $\mu(n)$ and $d_k(n)$

§ VIII. 9 Prime number theorem for arithmetic progressions

Let $\pi(x, k, l) = \sum_{p \leq x, p \equiv l (\mathrm{mod}\ k)} 1$, where $k > 0$, $(k, l) = 1$ (integers). Then:

a) $\pi(x, k, l) \sim \dfrac{1}{\varphi(k)} \dfrac{x}{\log x}$
$(x \to \infty)$

C.J. de la Vallée Poussin. *Recherches analytiques sur la théorie des nombres (3 parts).* Ann. Sec. Sci. Bruxelles, 20 (1896), 183–256; 281–397.

b) $\pi(x, k, l) = \dfrac{1}{\varphi(k)} \cdot \displaystyle\int_2^x \frac{du}{\log u} + O(x \exp(-(\log x)^{1/\alpha}))$

where α is a positive constant depending on k

E. Landau. *Über die Primzahlen einer arithmetischen Progression.* Sitz. Kaiserlichen Akad. Wiss. Wien Mat.-natur. Kl. 112 (1903), 493–535.

c) $\pi(x; k, l) = \dfrac{1}{\varphi(k)} \cdot \displaystyle\int_2^x \frac{du}{\log u} + O(x \exp(-A \log^{1/2} x))$

where $A > 0$ is a constant.

E. Landau. *Handbuch.* Leipzig 1909.

d) The same formula with O replaced by

$$O(\exp(-A(\log x \cdot \log\log x)^{1/2}))$$

E. Landau. *Über die ζ-Funktion und die L-Funktionen.* Math. Z. 20 (1924), 105–125.

e) $\pi(x, k, l) = \dfrac{1}{\varphi(k)} \displaystyle\int_2^x \frac{du}{\log u} + O\left(\exp\left(-A \log^{3/5} x (\log\log x)^{-\frac{1}{5}}\right)\right)$

A. Walfisz. *Weylische Exponentialsummen in der neueren Zahlentheorie.* Berlin, 1963.

§ VIII.10 An estimate on $\pi(x; p, -1)$

Let m be a non-negative integer, and δ a real number, $0 < \delta \leq 1/2$. Then there is a number c, depending upon m but not δ, so that the inequality

$$\sum_{x^{1-\delta} < p \leq x} p^{m-1} \cdot (\pi(x; p, -1))^m \leq \delta c \left(\frac{x}{\log x}\right)^m$$

holds for sufficiently large value of x

P.D.T.A. Elliott. *Probabilistic Number Theory.* I. 1979, Springe-Verlag (p. 175).

Remark. The case $m = 2$ was established in

M.B. Barban, A.I. Vinogradov and B.V. Levin. Liet. mat. rinkinys 5 (1965), 5–8.

§ VIII.11 Assertions equivalent to the prime number theorem for arithmetic progressions. Sums over $\pi(x;k,l) - \dfrac{\operatorname{li} x}{\varphi(k)}$

1) Let $(k,l)=1$. Then the following are equivalent:

$$\pi(x;k,l) \sim \frac{x}{\varphi(k)\log x}$$

$$\theta(x;k,l) \sim \frac{x}{\varphi(k)}$$

$$\Psi(x;k,l) \sim \frac{x}{\varphi(k)}$$

$$\sum_{n=l(\mathrm{mod}\, k)} \mu(n)/n \text{ converges}$$

$$\sum_{n\le x, n=l(\mathrm{mod}\, k)} \mu(n) = o(x)$$

$$\sum_{n\le x, n=l(\mathrm{mod}\, k)} \mu(n)\log n/n = o(\log x)$$

H.N. Shapiro. *Some assertions equivalent to the prime number theorem for arithmetic progressions.* Comm. Pure Appl. Math. 2 (1949), 293–308.

2) Let
$$S(x_n, l, X, Q) = \sum_{k\le Q,(k,l)=1} \left| \sum_{n\le X, n\equiv l(\mathrm{mod}\, k)} x_n - \frac{1}{\varphi(k)} \sum_{n\le X,(n,k)=1} x_n \right|$$
$(l \in \mathbb{I}\backslash\{0\})$

Let us denote by $C(s)$ the following proposition: There exist constants $u_s > 0$ and $\theta_s \ge 1/2$ such that one has, for all A and all integer $K \ge 1$ the estimation

$$S(y_t, l, X, X^{\theta_s}) \ll_{s,A,K} X \cdot (\log X)^{-A}$$

with $y_t = \sum_{t=mn, m\le X^{u_s}} \lambda_m \cdot d_s(n)$, uniformly for $1 \le |l| \le (\log X)^a$ and for all (λ_m) satisfying $|\lambda_m| \le d_K(m)$. Then we have the following theorem:

If conjecture $C(s)$ is true for $3 \le s \le 6$, then there exists $\theta > \frac{1}{2}$ such that

$$\sum_{(k,l)=1, K\le X^{\theta}} \left| \pi(X;k,l) - \frac{\pi(X)}{\varphi(k)} \right| \ll_A X \cdot (\log X)^{-A}$$

uniformly for $|l| \le (\log X)^A$ (for all A)

É. Fouvry. *Autour du théorème de Bombieri-Vinogradov.* Acta Math. 152 (1984), 219–244.

Remark. For similar results, see also

É. Fouvry. Ibid. II. Ann. Sci. École Norm. Sup. (4) 20 (1987), 617–640.

3) $$\sum_{k\le Q,(k,l)=1} \left(\pi(x;k,l) - \frac{\operatorname{li} x}{\varphi(k)} \right) \ll_A x/\log^A x$$

uniformly for $1 < |l| \leq \log^A x$ for any $x \geq 2$, $Q \leq x^{1/2}$, and any $A > 0$

É. Fouvry. *Sur le problème des diviseurs de Titchmarsh.* J. Reine Angew. Math. 357 (1985), 51–76.

§ VIII.12 Brun-Titchmarsh theorem

a) Let $1 \leq k \leq x^\alpha$, where $0 < \alpha < 1$. Then

$$\pi(x;k,l) < c\frac{x}{\varphi(k)\log(x/k)}$$

($c > 0$) (Brun-Titchmarsh theorem.)

E.C. Titchmarsh. *A divisor problem.* Rend. Palermo 54 (1930), 414–429 and 57 (1933), 478–479.

Remark. The above result, obtained by Titchmarsh, by a method of Brun, is called as the Brun-Titchmarsh theorem.

V. Brun. *Le crible d'Eratosthène et le théorème de Goldbach.* Skrifter utgit av Videnskapsselskapet: Kristiania mat. naturvid. Kl. 1920, No. 3.

b) $$\pi(x;k,l) \leq \frac{2x}{\varphi(k)\log(x/k)} \cdot \left(1 + O\left(\frac{\log\log x}{\log x}\right)\right)$$

where $(k, l) = 1$ and $k = O(x^\alpha)$, with $0 < \alpha < 1$, fixed.

I.V. Čulanovskiĭ. *Certain estimates connected with a new method of Selberg in elementary number theory (Russian).* Dokl. Akad. Nauk SSSR (N.S.) 63 (1948), 491–494.

c) For $1 \leq k < x$, $(k, l) = 1$

$$\pi(x;k,l) < \frac{2x}{(k)\log(x/k)} \cdot \left(1 + \frac{8}{\log(x/k)}\right)$$

and

$$\pi(x;k,l) < \frac{3x}{\varphi(k)\log(x/k)}$$

J.H. van Lint and H.-E. Richert. *On primes in arithmetic progressions.* Acta Arith. 11 (1965), 209–216.

d) Let η, η_1, δ be any positive constants. Suppose that $x > x_0(\eta, \eta_1, \delta)$, $x^{4/5} \leq S \leq x^{1-\eta_1}$ and $S \leq l \leq 2S$, $(l, k) = 1$ ($k > 0$, fixed). Then

$$\pi(x;k,l) \leq (l+\eta) \cdot \frac{x}{\varphi(k)\log(x/S)}$$

holds except for at most $S/\log^\delta S$ exceptional values of l

C. Hooley. *On the Brun-Titchmarsh theorem.* J. Reine Angew. Math. 255 (1972), 60–79.

e) If k is any fixed nonzero integer, then

$$\pi(x;k,l) < (4+\varepsilon)\frac{x}{\varphi(l)\log l}$$

holds for any $\varepsilon > 0$ and almost all l satisfying $(l, k) = 1$ and $l \cdot \log^{34} l < x$

C. Hooley. Ibid. II. Proc. London Math. Soc.(3) 30 (1975), 114–128.

f) $\pi(x; k, l) \le \dfrac{2x}{\varphi(k)\log(x/k)}$

for $k < x^{3/7}$

Y. Motohashi. *On some improvements of Brun-Titchmarsh theorem.* J. Math Soc. Japan 26 (1974), 306–323.

g) $\pi(x; k, l) \le (1+\varepsilon) \cdot \dfrac{x}{\varphi(k)\log(x/k^{3/2})}$

if $x^{2/5} \le k \le x^{1/2}$ and

$$\le 2 \cdot (1+\varepsilon) \cdot \frac{x}{\varphi(k)\log(x/k^{3/8})}$$

if $1 \le k \le x^{1/3}$

Y. Motohashi. Ibid II. (Japanese). Sûrikaisekikenkyûsho Kôkyûroku No. 193 (1973), 97–109.

h) Let $x^{5/6} \le w \le x \cdot (\log x)^{-(6s+165)}$. Then for $w \le k < 2w, (k, l) = 1$, we have

$$\pi(x; k, l) \le \frac{6x}{\varphi(k)\log x} \cdot \left(1 + O\left(\frac{\log\log x}{\log x}\right)\right)$$

save for at most $\ll w \cdot (\log x)^{-s}$ exceptional values of k

Y. Motohashi. Ibid. III. J. Math. Soc. Japan 27 (1975), 444–453.

i) $\pi(x; k, l) \le (2 + O(\varepsilon))\dfrac{x}{\varphi(k)\log(x/k^{3/8})}$

for $1 \le k \le x^{24/71-\varepsilon}$ and

$$\le (1 + O(\varepsilon)) \cdot \frac{x}{\varphi(k)\log(x/k^{3/2})}$$

for $x^{\frac{2}{5}-\varepsilon} \le k \le x^{1/2}$

D.M. Goldfeld. *A further improvement of the Brun-Titchmarsh theorem.* I.J. London Math. Soc. (2) 11 (1975), 434–444.

j) $\pi(x; k, l) \le (2+\varepsilon) \cdot \dfrac{x}{\varphi(k)\log(x/k^{3/8})}$

for $k \le x^{9/20-\varepsilon}$

When k is cubefree, $x/k^{3/8}$ can be replaced by $x/k^{1/4}$.

H. Iwaniec. *On the Brun-Titchmarsh theorem.* J. Math. Soc. Japan 34 (1982), 95–123.

k) Let $x \ge 2$, $\varepsilon > 0$, $A > 0$, $|l| < \log^A x$, $Q = x^\theta$

with $1/2 \leq \theta \leq 11/20$. Then we have for all $k \in [Q, 2Q]$, with $Q/\log^A x$ possible exceptions,

$$\pi(x; k, l) < (c_1(\theta) - c_2(\theta)) \cdot \frac{x}{\varphi(k)\log x}$$

where

$$c_1(\theta) = \frac{12}{25 - 40\,\theta}$$

for $1/2 \leq \theta \leq 53/104$;

$$c_1(\theta) = \frac{48}{47 - 56\,\theta}$$

for $53/104 \leq \theta \leq 11/20$ and

$$c_2(\theta) = \log\frac{10 - 10\,\theta}{90}$$

for $1/2 \leq \theta \leq 10/19$;

$$c_2(\theta) = 0$$

in the remaining part of the interval [1/2, 11/20].

É. Fouvry. *Sur le théorème de Brun-Titchmarsh.* Acta Arith. 43 (1984), 417–424.

l) Let $A > 0$ and $l \neq 0$ be given. Then

$$0.85\frac{x}{\varphi(k)\log x} \leq \pi(x; k, l) \leq 1.48\frac{x}{\varphi(k)\log x}$$

for all $k \leq Q \leq x^{1/2+10^{-100}}$, $(k, l) = 1$, with at most $Q/\log^A x$ exceptions.

B. Rousselet. *Inégalités de type Brun-Titchmarsh en moyenne.* Groupe de travail en théorie analytique et élémentaire des nombres, 1986–1987, 91–123, Publ. Math. Orsay, 88–01, Univ. Paris XI, Orsay, 1988.

m) Let $\varepsilon > 0$, $A > 0$, $l \neq 0$ be given. Then

$$(1 - \varepsilon)\frac{x}{\varphi(k)\log x} \leq \pi(x; k, l) \leq (1 + \varepsilon)\frac{x}{\varphi(k)\log x}$$

for all $k \leq Q = x^{1/2}$ with $O(Q \cdot (\log x)^{-2+\varepsilon})$ exceptions.

E. Bombieri, J.B. Friedlander and H. Iwaniec. *Primes in arithmetic progressions in large moduli.* II. Math. Ann. 227 (1987), 361–393.

n) $\pi(x; k, l) \leq \left(1 + \left(\theta - \frac{1}{2}\right)\right)\frac{x}{\varphi(k)\log x}$

for all but a proportion $O\left(\theta - \frac{1}{2}\right)$ of the $k \leq x^{\theta}$, $(k, l) = 1$, where $\frac{1}{2} \leq \theta \leq \frac{3}{4}$

E. Bombieri, J.B. Friedlander and H. Iwaniec. Ibid. III. J. Amer. Math. Soc. 2 (1989), 215–224.

o) $\pi(x; k, l) \leq (18 + \varepsilon)\frac{x}{\varphi(k)\log(x^6/k)}$

is true for almost all reduced residue classes $l (\bmod k)$, if $x^{6/7} \leq k \leq x(\log x)^{-A}$

H. Mikawa. *On the Brun-Titchmarsh theorem.* Tsukuba J. Math. 15 (1991), 31–40.

Remark. In the proof the Rosser-Iwaniec sieve is used.

§ VIII.13 Application of the Brun-Titchmarsh theorem on lower bounds for $\pi(x; k, l)$

a) Let $1 \leq k \leq x^{\alpha}$, where $0 < \alpha < 1$. For all k, there exist more than $c_1 \varphi(k)$ distinct values of l (where $(l, k) = 1$) with

$$\pi(x; k, l) > c_2 \cdot \frac{x}{\varphi(k) \log x}$$

for suitable positive constants c_1 and c_2 (independent of x and k)

b) Let $1 \leq k \leq x^{\alpha}, 0 < \alpha < 1$. For each β with $0 < \beta < 1$, there exist an n with the following property:
for each representation $k = k_1 k_2 \cdots k_n, (k_i, k_j) = 1 (i \neq j)$ there is ak_r $(1 \leq r \leq n)$ such that

$$\pi(x; k_r, l) > c_3 \cdot \frac{x}{\varphi(k_r) \log x}$$

for at least $\beta \cdot \varphi(k_r)$ distinct residue classes $l (\bmod k_r)$, with suitable $c_3 = c_3(\beta)$

P. Erdős. *On the sum and the difference of squares of primes.* J. London Math. Soc. 12 (1937), 133–136 and 168–171.

Note. These results are based on the Brun-Titchmarsh theorem.

§ VIII.14 On $\pi(x + x^{\theta}; k \cdot l) - \pi(x; k, l)$

a) Let $(k, l) = 1, k > 0$ and $A > 0$ arbitrary. Then

$$\pi(x + x^{\theta}; k, l) - \pi(x; k, l) \sim \frac{x^{\theta}}{\varphi(k) \log x}$$

for $k < \log^A x$, where $\theta = \frac{3}{4} + \varepsilon$

N.G. Chudakov. *Sur le zéros des L-fonctions de Dirichlet.* Dokl. Akad. Sci. URSS (N.S.) 49 (1945), 89–91.

b) $\pi(x+x^{\theta};k,l)-\pi(x;k,l)\sim\dfrac{x^{\theta}}{\varphi(k)\log x}$

provided that $\log k=O((\log x)^{1-\tau})$, where $\tau>\dfrac{3}{4}$ is an absolute constant. (This holds uniformly in k.)

N.G. Chudakov. *On the limits of variation of the function $\psi(x;k,l)$ (Russian).* Izv. Akad. Nauk SSSR Ser. Mat. 12 (1948), 31–46.

§ VIII.15 Barban's theorem

For a certain positive constant $\delta>0$, and every fixed A,

$$\sum_{k\le x^{\delta}}\mu^2(k)\max_{(l,k)=1}\left|\pi(x;k,l)-\frac{\operatorname{li}x}{\varphi(k)}\right|\ll x/\log^A x$$

Remark. Any $\delta<3/23$ is permissible.

M.B. Barban. *New applications of the large sieve of Yu. V. Linnik (Russian).* Akad. Nauk Uzb. SSR, Trudy Inst. Mat. V.I. Romanov, Teor. Ver. Mat. Stat. 22 (1961), 1–20.

§ VIII.16 On generalizations of the Bombieri-Vinogradov theorem

1) Let $\pi(x;a,k,l)=\sum_{ap\le x,ap\equiv l(\operatorname{mod}k)}1$, and let $f(a)$ be a real function such that $f(a)\ll1$. Then for any given $A>0$, we have

$$\sum_{k\le\sqrt{x}/\log^B x}\max_{y\le x}\max_{(l,k)=1}\left|\sum_{a\le x^{1-\varepsilon},(a,k)=1}f(a)\left(\pi(y;a,k,l)-\frac{\pi(y;a,1,1)}{\varphi(k)}\right)\right|\ll x/\log^A x$$

where $B=\dfrac{3}{2}A+17$ and $0<\varepsilon<1$

C.D. Pan. *A new mean value theorem and its applications.* Recent progress in analytic number theory. Vol. 1 (Durham, 1979), pp. 275–287, Academic press, 1981.

Remark. This result generalizes the Bombieri-Vinogradov theorem.

2) Let $\pi(x,z;q,a)=\sum_{n\le x,n\equiv a(\operatorname{mod}q)}f_z(n)$, where $f_z(n)$ is the indicator function of the integers n having no prime factors less than z. Let $\pi(x,z;q)$ denote the sum of $\pi(x,z;q,a)$ over the integers $1\le a\le q$ with $(a,q)=1$. Let $z\le x^{1/883}$ and put $1\le|a|\le x$. Then for any $A>0$,

$$\sum_{q\le x^{11/21},(q,a)=1}\left|\pi(x,z;q,a)-\frac{\pi(x,z;q)}{\varphi(q)}\right|\ll x/\log^A x$$

where the implied constant depends only on A.

É. Fouvry and H. Iwaniec. *On a theorem of Bombieri-Vinogradov type.* Mathematika 27 (1980), 135–152.

3) Let the multiplicative function $g(n)$ satisfy $|g(p^a)\le D_1\cdot a^{D_2}$ for all prime powers p^a, and $\sum_{p\le y}|g(p)-\tau|\ll y/\log^{A_1}y(y\ge 2)$, where D_1, D_2, τ are constants, and A_1 may be fixed at any positive value. Then for each $A>0$ we can find a B such that

$$\sum_{k\le\sqrt{x}/\log^B x}\max_{(l,k)=1}\max_{y\le x}\left|\sum_{\substack{n\le y\\ n\equiv l(\text{mod } k)}}g(n)-\frac{1}{\varphi(k)}\sum_{\substack{n\le y\\ (n,k)=1}}g(n)\right|\ll x/\log^A x$$

D. Wolke. *Über die mittlere Verteilung der Werte zahlentheoretischer Funktionen auf Restklassen.* I. Math. Ann. 202 (1973), 1–25.

§ VIII.17 An upper bound for $\pi_k(y;k,l)$ = number of primes $x<p\le x+y$ with $p\equiv l(\text{mod } k)$

Let $\pi_x(y;k,l)$ be the number of primes p such that $x<p\le x+y$ and $p\equiv l(\text{mod } k)$. Then:

a) $\pi_x(y;k,l)\le\dfrac{2y}{\varphi(k)(\log(y/k)+c)}$

where $c=0.83$, provided that $c'k\le y$ for a sufficiently large constant c'

H.L. Montgomery and R.C. Vaughan. *The large sieve.* Mathematika 20 (1973), 119–134.

b) The above holds with $c=1.584$

N.I. Klimov. *The small sieve* (Russian). Mat. Zametki 27 (1980), 161–174, 317.

§ VIII.18 An analogue of the Brun-Titchmarsh inequality

Let $\pi_d(x;k,l)$ denote the number of positive integers $n\equiv l(\text{mod } k), n\le x$, having exactly d prime factors. Then, if $k\le x^{1/2d}$,

$$\pi_d(x;k,l)\le c_k\cdot d^{d-1}(k)\cdot x\cdot(\log\log x)^{d-1}/\varphi(k)\log x$$

for some constant c_k, where $d(k)$ denotes the number of divisors of k

M. Orazov. *Analogue of the Brun-Titchmarsh inequality* (Russian). Izv. Akad. Nauk Turkmen. SSR Ser. Fiz.-Tekhn. Khim. Geol. Nauk 1982, No. 2, 90–91.

§ VIII.19 On Goldbach-Vinogradov's theorem. The prime k-tuple conjecture on average

Let $\underline{a}$ and $\underline{b}$ be k-dimensional integer vectors, $x > 0$ a real number and $\pi(x; \underline{a}, \underline{b})$ the number of integers such that $1 < \underline{a}n + \underline{b} \le x$, $\underline{a}n + \underline{b}$. (The last two conditions are understood to hold in each coordinates simultaneously.) Let

$$T(x; \underline{a}, \underline{b}) = \sum_{1 < \underline{a}n + \underline{b} \le x} \frac{1}{\log(a_1 n + b_1) \cdots \log(a_k n + b_k)}$$

(where a_i and b_i are the coordinates of $\underline{a}$ and $\underline{b}$, respectively.)

a) $$\sum_{2|h} \left| \underline{\pi} - 2 \cdot \prod_{p \ne 2} \frac{p(p-2)}{(p-1)^2} \cdot \prod_{\substack{p|h \\ p \ne 2}} \frac{p-1}{p-2} \cdot \underline{T} \right| \ll x^3 / \log^A x$$

where $\underline{\pi} = \pi(x; (1,1), (0,h))$ and $\underline{T} = T(x; (1,1), (0,h))$

N.G. Chudakov. *On Goldbach-Vinogradov's theorem.* Ann. Math. 48 (1947), 515–545.

b) Let $\pi(x; \underline{a}, \underline{b}; c, d)$ be the number of integers n in the residue class $c(\text{mod } d)$ such that $1 < \underline{a}n + \underline{b} \le x$ and $\underline{a}n + \underline{b}$ are primes. Then

$$\sum_{\substack{2|h \\ (c+h,d)=1}} \left| \underline{\pi} - \frac{2}{\varphi(d)} \prod_{p \ne d} \frac{p(p-2)}{(p-1)^2} \cdot \prod_{\substack{p|hd \\ p \ne 2}} \frac{p-1}{p-2} \cdot \underline{T} \right|^2 \ll x^3 / \log^A x$$

uniformly for $d \le \log^B x$ and $(c, d) = 1$.

Here $\underline{\pi} = \pi(x; (1,1), (0,h); c, d)$ and $\underline{T} = T(x; (1,1), (0,h))$

A.F. Lavrik. *The number of k-twin primes lying on an interval of a given length.* Dokl. Akad. Nauk. SSSR 136 (1961), 281–283 (Russian).

c) With the notation of b),

$$\sum_{d \le x^\delta} \sum_{2|h} \max_{\substack{(c,d)=1 \\ (c+h,d)=1}} \left| \underline{\pi} - \frac{2}{\varphi(d)} \prod_{p \ne 2} \frac{p(p-2)}{(p-1)^2} \cdot \prod_{\substack{p|hd \\ p \ne 2}} \frac{p-1}{p-2} \cdot \underline{T} \right| \le x^2 / \log^A x$$

for any $A > 0$, where $\delta > 0$ is some small computable constant.

H. Maier and C. Pomerance. *Unusually large gaps between consecutive primes.* Trans. Amer. Math. Soc. 322 (1990), 201–237.

d) For fixed $\underline{a}$ and $\underline{b}$, let $\rho(p) = \rho(p; \underline{a}, \underline{b})$ denote the number of solutions of the congruence

$$(a_1 n + b_1) \cdots (a_k n + b_k) \equiv 0 (\text{mod } p)$$

Let $\sigma(\underline{a}, \underline{b}; c, d) = \dfrac{1}{d} \displaystyle\prod_{p|d} \frac{p}{p - \rho(p)} \cdot \prod_p \left(1 - \frac{1}{p}\right)^{-k} \cdot \left(1 - \frac{\rho(p)}{p}\right)$ if $\rho(p) < p$ for all prime p and $(\underline{ac} + \underline{b}, d) = 1$;

and $\sigma(\underline{a}, \underline{b}; c, d) = 0$ otherwise.

Let $k \geq 1$ and b_k be fixed integers and $\underline{a}$ be a fixed k-dimensional integer vector. Let x be a real number with $x \geq |b_k|$ and

Let $Z = Z(b_k; x)$ be the set of k-dimensional integer vectors $\underline{h}$ such that the last coordinate of $\underline{b}$ is b_k and the set

$\{n : 1 < \underline{a}n + \underline{b} \leq x\} \neq \oslash$.

For any $A > 0$ there is a $B = B(A) > 0$ such that for any $D \leq x^{1/3} \cdot \log^{-B} x$ we have

$$\sum_{d \leq D} \max_{c} \sum_{\underline{b} \in Z} |\pi(x; \underline{a}, \underline{b}; c, d) - \sigma(\underline{a}, \underline{b}; c, d) T(x; \underline{a}, \underline{b})| \ll x^k / \log^A x$$

Here the implied constant in $\ll$ depends at most on A and $\underline{a}$ but not on b_k

A. Balog. *The prime k-tuple conjecture on average*. Analytic Number theory, Proc. Conf. in Honor of P. T. Bateman, Birkhäuser, 1990, pp. 47–75.

e) For any fixed a and $A > 0, k \geq 2$,

$$x^{1/3} (\log x)^{-B} \text{in d)}$$

can be replaced with

$$x^{1/2} (\log x)^{-B}$$

vhere $B = B(A, k) > 0$.

K. Kawada. *The prime k-tuples in arithmetic progressions*. Tsukuba J. Math. 17 (1993), 43–57.

Remark. The proof is based on ideas initiated by H. Mikawa in

H. Mikawa. Tsukuba J. Math. 10 (1992), 377–387.

§ VIII.20 Sums over $\left(\theta(x; k, l) - \frac{x}{\varphi(k)}\right)^2$, $\left|\pi(x; k, l) - \frac{\operatorname{li} x}{\varphi(k)}\right|^2$, $\left|\psi(x; k, l) - \frac{x}{\varphi(k)}\right|^2$

1) a)

$$\sum_{k \leq Q} \sum_{l=1, (l,k)=1}^{k} \left|\pi(x; k, l) - \frac{\operatorname{li} x}{\varphi(k)}\right|^2 =$$

$$= Q \cdot \operatorname{li}(x) + O(x^2 / \log^A x) + O(Qx(\log x)^{-2} \cdot \log(x/Q))$$

for $\exp(c(\log x)^{1/2}) \leq Q \leq x$ (where A may be any positive constant, and c is an absolute constant).

b) When $Q = x$, the right-hand side of a) may be replaced by $x \operatorname{li} x + E(\operatorname{li} x)^2 + O(x^2/\log^A x)$, for a suitable constant E

M.B. Barban. *On the average error in the generalized prime number theorem.* Dokl. Akad. Nauk UzSSR, 5 (1964), 5–7.

c) Let $g(a) \ll d^r(a)\,(r \geq 0)$, where $d(a)$ denotes the number of divisors of a. Then

$$\sum_{k \leq Q} \sum_{1 \leq l \leq k, (a,k)=1} \left(\sum_{a \leq x^{1-\alpha}, (a,k)=1} g(a) \right) \cdot \left(\sum_{an \leq x, an \equiv l \pmod k} \mu(n) \right)^2 \ll Qx(\log x)^{2^{2r+2}}$$

for $x/\log^A x \leq Q \leq x$. (Here $0 \leq \alpha \leq 1$.)

D. Zhang. *An extension of Barban's theorem* (Chinese). J. Ocean Univ. Quingdao 18 (1988), 82–87.

2) a) Let $\theta(x; k, l) = \sum_{p \leq x, p \equiv l \pmod k} \log p$ and let $A > 0$ be fixed.

If $Q \leq x/\log^A x$, then

$$\sum_{k \leq Q} \sum_{l=1, (l,k)=1}^{k} \sup_{y \leq x} \left(\theta(y, k, l) - \frac{y}{\varphi(k)} \right)^2 \leq Bx^2/(\log x)^{A-3}$$

with a constant $B = B(A) > 0$

S. Uchiyama. *Prime numbers in arithmetic progressions.* Math. J. Okayama Univ. 15 (1971/72), 187–196.

b)
$$\sum_{k \leq Q} \sum_{l=1, (k,l)=1}^{k} \left(\theta(x, k, l) - \frac{x}{\varphi(k)} \right)^2 = Qx \log Q + O(Qx + x^2/\log^A x)$$

for any $A > 0$, where $1 \leq Q \leq x$

C. Hooley. *On the Barban-Davenport-Halberstam theorem.* I.J. Reine Angew. Math. 274/275(1975), 206–223.

Remark. In parts II, III, IV, VI, VII, J. London Math. Soc. 9 (1974/75), 625–636; 10 (1975), 249–256; 11 (1975), 399–407; 13 (1976), 57–64; 16 (1977), 1–8 and part V, Proc. London Math. Soc. 33 (1976), 535–548, the author obtains various extensions and other results under the assumption of the generalized Riemann Hypothesis.

3) For any $C > 0$ we have

$$\sum_{k \leq x/\log^c x} \sum_{1 \leq l \leq k, (l,k)=1} \left| \psi(x; k, l) - \frac{x}{\varphi(k)} \right|^2 \ll x^2 (\log x)^{8-C/3}$$

D.R. Heath-Brown. *The ternary Goldbach problem.* Rev. Mat. Iberoamericana 1 (1985), 45–58.

§ VIII.21 Oscillation theorems for primes in arithmetic progressions

Let $x \geq 2$, $q \geq 1$ integer, $(a, q) = 1$ and $\theta(x; q, a)$ and $\Delta(x; q, a)$ be given by

$$\theta(x;q,a) = \sum_{p\le x,\, p\equiv a (\mathrm{mod}\ q)} \log p = \frac{x}{\varphi(q)} \cdot (1 + \Delta(x;q,a))$$

a) Let $\varepsilon > 0$.
Then for all $q \ge q_0(\varepsilon)$ and all x satisfying

$$\sum_{p|q,\, p<\log q} 1 < \frac{3}{2}\log_3 q$$

$q \cdot (\log q)^{1+\varepsilon} < x \le q\exp((\log q)^{5/11-\varepsilon})$, there exist numbers $x_\pm$ with $x/2 < x_\pm \le 2x$ and integers $a_\pm$ coprime with q such that

$$\Delta(x_+;q,a_+) \ge y^{-(1+\varepsilon)\delta_1(x,y)}$$

and

$$\Delta(x_-;q,a_-) \le y^{-(1+\varepsilon)\delta_1(x,y)}$$

where $y = x/q$ and $\delta_1(x,y) = \log(\log y/\log_2 x)/\log(\log x/\log y))$
(Here $\log_k$ denotes the k-fold iterated logarithm.)

Corollary. For all sufficiently large x and

$$2x\exp(-(\log x)^{5/11-\varepsilon}) \le Q \le x/2$$

we have

$$\sum_{q\le Q} \frac{1}{\varphi(q)} \cdot \max_{x'\le x} \max_{(a,q)=1} |\Delta(x';q,a)| \gg y^{-(1+\varepsilon)\delta_1(x,y)}$$

where $y = x/Q$

b) Let $\varepsilon > 0$. Then for some $c > 0$, for all $x > x_0(\varepsilon)$, and all Q satisfying

$$x\exp(-c\sqrt{\log x}) < Q \le x(\log x)^{-(1+\varepsilon)}$$

there exist $Q_\pm$ with $Q/2 < Q_\pm \le 2Q$ and integers $a_\pm$ such that

$$\sum_{\substack{q\sim Q_+\\ (q,a_+)=1}} \frac{1}{\varphi(q)} \Delta(x;q,a_+) \ge \frac{1}{\log_2 x} \cdot y^{-(1+\varepsilon)\delta_2(x,y)}$$

$$\sum_{\substack{q\sim Q_-\\ (q,a_-)=1}} \frac{1}{\varphi(q)} \Delta(x;q,a_-) \le -\frac{1}{\log_2 x} \cdot y^{-(1+\varepsilon)\delta_2(x,y)}$$

where $y = \dfrac{x}{Q}$ and $\delta_2(x,y) = \log(\log y/\log_2 x)/\log_2 x$. Here $q \sim Q$ means that $Q < q \le 2Q$

Corollary. Let $\varepsilon > 0$, $A > 0$. For every $x > x_0(\varepsilon, A)$, there exists an integer such that for all Q with

$$Q \ge x\exp(-(1-\varepsilon)(\log_2 x)^2/\log_3 x)$$

we have

$$\sum_{q<Q,(q,a)=1} \left|\theta(x;q,a) - \frac{x}{\varphi(q)}\right| \geq \left|\sum_{\substack{q\sim Q/2\\(q,a)=1}} \left(\theta(x;q,a) - \frac{x}{\varphi(q)}\right)\right| > x/\log^A x$$

c) Let $\varepsilon > 0$. There exist $N(\varepsilon) > 0$ and $q_0 = q_0(\varepsilon) > 0$ such that for any $q > q_0$ and any x with

$$q(\log q)^{N(\varepsilon)} < x \leq q\exp((\log q)^{1/3})$$

there exist numbers $x_\pm$ with $\frac{x}{2} < x_\pm \leq 2x$ and integers $a_\pm$ coprime with q such that

$$\Delta(x_+;q,a_+) \geq \frac{1}{\log^5 x} \cdot y^{-(1+\varepsilon)\,\delta_3(x,y)}$$

$$\Delta(x_-;q,a_-) \leq \frac{-1}{\log^5 x} \cdot y^{-(1+\varepsilon)\,\delta_3(x,y)}$$

where $y = x/q$ and $\delta_3(x,y) = 3\log(\log y/\log_2 x)/\log(\log x \log y)$

J.B. Friedlander, A. Granville, A. Hildebrand and H. Maier. *Oscillation theorems for primes in arithmetic progressions and for sifting functions*. J. Amer. Math. Soc. 4 (1) (1991), 25–86.

§ VIII.22 Special results on finite sums over primes

1) Let $q, p_1, \ldots, p_n$ be distinct prime numbers. For each i let a_i be the least positive integer for which

$$q^{a_i} \equiv 1 (\bmod p_i)$$

Then

$$\sum_{1\leq i\leq n} 1/q^{a_i} < 1 - \frac{1}{q}$$

T.J. Laffy. *A problem of cyclic subgroups of finite groups*. Proc. Edinburgh Math. Soc. (2) 18 (1973), 247–249.

2) Let P be a set of primes with $\liminf_{x\to\infty} (\log x/x) \cdot \sum_{p<x, p\in P} 1 > 0$ and let Q be the set of all primes q with $q \not\equiv 1 (\bmod p)$ for all $p \in P$

Then $\sum_{q\in Q} 1/q < \infty$

H.G. Meijer. *A problem on prime numbers*. Nieuw. Arch. Wisk. (3) 22 (1974), 125–127.

§ VIII.23 Infinitely many sets of three distinct primes and an almost prime in arithmetic progressions

1) There are infinitely many sets of three distinct primes in arithmetic progression.

S.D. Chowla. *There exists an infinity of 3-combinations of primes* in A.P., Proc. Lahore Philos. Soc. 6 (1944), 15–16.

2) For $x \geq 2$ define

$$N_2(m) = \sum_{p \leq x, p+p'=m} (\log p) \cdot (\log p')$$

where p, p' run over primes.

Set $S(m) = 2 \cdot \prod_{p \geq 3} (1 - (p-1)^{-2}) \cdot \prod_{p|m, p \geq 3} \left(\frac{p-1}{p-2} \right)$, if m is even; and $S(m) = 0$ for odd m. Then for any $C > 0$ we have

$$\sum_{2x < m \leq 3x} |N_2(m) - xS(m)| \ll x^2 / \log^C x$$

D.R. Heath-Brown. *The ternary Goldbach problem.* Rev. Mat. Iberoamericana 1 (1985), 45–59.

Remark. 1) is a Corollary of 2).

3) There are infinitely many sets of three distinct primes and an almost prime in arithmetic progression.

D.R. Heath-Brown. *Three primes and an almost-prime in arithmetic progression.* J. London Math. Soc. 23 (1981), 396–414.

Remark. For a new proof of this result, see

H. Maier and C. Pomerance. *Unusually large gaps between consecutive primes.* Trans. Amer. Math. Society. 322 (1990), 201–237.

§ VIII.24 Large prime factors of integers in an arithmetic progression

1) Let $a \in \mathbb{I} \setminus \{0\}$, $k \in N^*$.

There are infinitely many natural numbers r with $(r, a) = 1$ such that for all r there exist $k+1$ distinct primes $p_0, p_1, \ldots, p_k$ with $p_i \equiv a (\bmod r)$, $\frac{p_i - a}{r}$ prime and $p_i < \lambda k r \log r \log \log r$, $0 \leq i \leq k$. (Here $\lambda > e$ is any fixed real number.)

P.T. Bateman and C. Pomerance. *Moduli r for which there are many small primes congruent to a modulo r.* H. Delange Colloquium (Orsay, 1982), 8–19, Orsay 1983.

2) Let $\alpha > 1/2$ and $(a, r) = 1$. There exists $\delta = \delta(\alpha) > 0$ such that, for all sufficiently large r, there is some $m \leq r^{1+\alpha}$ with $m \equiv a (\bmod r)$, and m having a prime factor exceeding $r^{\alpha+\delta}$

A. Balog, J. Friedlander and J. Pintz. *Large prime factors of integers in an arithmetic progression.* Studia Sci. Math. Hungar 22 (1987), 175–188.

§ VIII.25 Almost primes in arithmetic progressions

1) Let $P_2(q, a)$ denote the smallest number in the arithmetic progression $qn + a$, $(q, a) = 1, 0 < a < q$ having at most two prime factors. Then

$$P_2(k, l) \leq k^{2.3696}$$

B.V. Levin. *On the least almost prime number in an arithmetic progression and the sequence* $k^2x^2 + 1$ (Russian). Uspehi Mat. Nauk. 20 (1965), 158–162.

2) a) There exists a P_2 such that $P_2 \equiv a(\text{mod } q)$, $P_2 \leq q^{11/10}$
Y. Motohashi. J. Math. Soc. Japan 28 (1976), 363–383.

b) There exists a P_2 (number with at most two prime factors) such that

$$P_2 \equiv a(\text{mod} q),\ P_2 \leq g(q)\,q(\log q)^5$$

for almost all reduced residue classes $a(\text{mod} q)$. Here $g(x)$ denotes any positive function such that $g(x) \to \infty$ as $x \to \infty$

H. Mikawa. *Almost primes in arithmetic progressions in short intervals*. Tsukuba J. Math. 13 (1989), 387–401.

§ VIII.26 Arithmetic progressions that consist only in primes

Let $N_m(x)$ be the number of arithmetic progressions of m primes all $\leq x$. Then

$$N_3(x) = \frac{1}{2}C \cdot \frac{x^2}{\log^3 x} \cdot \left(1 + \sum_{j=1}^{n} a_j / \log^j x\right) + O(x^2/(\log x)^{n+1})$$

where C is the "twin prime constant" and $a_1, a_2, \ldots$ are explicitely given constants.
E. Grosswald. *Arithmetic progressions that consist only of primes*. J. Number theory 14 (1982), 9–31.

§ VIII.27 Number of $n \leq x$ such that there is no prime between n^2 and $(n + 1)^2$

Let $q(x)$ be the number of positive integers $n \leq x$ such that there is no prime between n^2 and $(n + 1)^2$. Then:

a) Assuming the Riemann Hypothesis,

$$q(x) = O(x^{2/3} \cdot \log^3 x)$$

H. Cramér. *Some theorems concerning prime numbers*. Arkiv. for Mat. Astr. Fys. 15 (1921), 1–33.

b) Without any assumption

$$q(x) = O\left(\frac{x}{\log x}\right)$$

L. Washington. *Class numbers of elliptic function fields and distribution of prime numbers.* Acta Arith. 28 (1975), 112.

§ VIII.28 Primes in the sequence $[n^c]$

1) Let $\pi_c(x)$ be the number of primes in the sequence $[n^c]$ not exceeding x. Then:

a) $\pi_c(x) \sim \dfrac{x^{1/c}}{\log x}$

for $1 \leq c < \dfrac{12}{11}$

I.I. Pjateckiĭ-Šapiro. *On the distribution of prime numbers in sequences of the form* $[f(n)]$ (Russian). Mat. Sbornik (N.S.) 33 (75)(1953), 559–566.

b) $\pi_c(x) \sim \dfrac{x^{1/c}}{\log x}$

for $1 < c < \dfrac{10}{9}$

G.A. Kolesnik. *The distribution of primes in sequences of the form* $[n^c]$ (Russian). Mat. Zametki 2 (1967), 117–128.

c) For any closed interval $I \subset (1, \infty)$, and not containing any integer there exists a constant $K_I \in \mathbb{R}$ such that

$$\pi_c(x) \leq K_I \cdot \frac{x^{1/c}}{\log x}$$

for all $c \in I$ and all $x > 1$

J.-M. Deshouillers. *Nombre premiers de la forme* $[n^c]$ C.R. Acad. Sci. Paris Sér. A–B 282 (1976), A131–A133.

d) The set of all c with $1 \leq c \leq 2$ for which the relation

$$\pi_c(x) \sim \frac{x^{1/c}}{\log x}$$

is false has Lebesgue measure zero.

D. Leitman and D. Wolke. *Primzahlen Der Gestalt* $[f(n)]$. Math. Z. 145 (1975), 81–92.

e) Let $c \notin \mathbb{I}$. Then for almost all $c \in (1, +\infty)$ there exist three positive constants k_c, K_c and x_c such that

$$k_c \cdot \frac{x^{1/c}}{\log x} < \pi_c(x) < K_c \cdot \frac{x^{1/c}}{\log x}$$

for all $x > x_c$

J.-M. Deshouillers. *Répartition de nombre premiers de la forme* $[n^c]$. Journées Arithmétiques (Grenoble, 1973), pp. 49–52. Bull. Soc. Math. France Mém. 37, Paris, 1974.

f) Relation a) is valid for

$$1 < c < \frac{755}{662} = 1.14\ldots$$

D.R. Heath-Brown. *The Pjateckiĭ-Šapiro prime number theorem.* J. Number theory 16 (1983), 242–266.

g) Relation a) is valid for

$$1 < c < \frac{39}{34}$$

G. Kolesnik. *Primes of the form* $[n^c]$. Pacific J. Math. 118 (1985), 437–447.

Remark. In fact the more precise relation

$$\pi_c(x) = \frac{x^{1/c}}{\log x} + O\left(\frac{c^2 \cdot x^{1/c}}{\log^2 x}\right)$$

is proved.

h) The same is valid for $1 < c < 15/13$

H. Liu and J. Rivat. *On the Pjateckiĭ-Šapiro prime number theorem.* Bull. London Math. Soc. 24 (1992), 143–147.

2) If $1 < c < 2$, then the number of natural numbers $n \leq x$ for which $[n^c]$ and $[n^c] + 2$ are primes equals $O(x/\log^2 x)$.

G.J. Rieger. *Über ein additives Problem mit Primzahlen.* Arch. Math. (Basel) 21 (1970), 54–58.

§ VIII.29 Number of primes $p \leq x$ for which $[p^c]$ is prime

For a positive constant c, let $\pi_c(x)$ be the number of primes $p \leq x$, for which $[p^c]$ is prime. Then:

a) $\pi_c(x) \sim \frac{1}{c} \cdot x/\log^2 x$

for $0 < c < \frac{5}{6}$

b) $\limsup_{x\to\infty} \Pi_c(x)/(c\log^2 x)^{-1} \cdot x \geq 1$

for almost all positive c (in the sense of Lebesgue measure.)

A. Balog. *On a variant of the Pjateckiĭ-Šapiro prime number theorem Groupe de travail* en Th. Analy. et Elém. des Nombres, 1987–1988, 3–11, Publ. Math. Orsay, 89–01, Univ. Paris XI, Orsay, 1989.

§ VIII.30 Almost primes in (n^2+1) and related sequences

1) a) The sequence $(n^2+1)_{n=1,2,\ldots,N}$ contains at least

$$\frac{a\cdot N}{\log N}+O\left(\frac{N\log\log N}{(\log N)^{3/2}}\right)$$

members each having at most five prime factors. The prime factors exceed $N^{1/2.91}$ and $\underline{a}$ is a certain positive constant.

B.V. Levin. *Estimates from below for the number of nearly-prime integers belonging to some general sequences* (Russian). Vestnik Leningrad, Univ. 15 (1960), 48–65.

b) Let $J(N)$ be the the number of integers in the sequence $(n^2+1), n\le N$, which have at most three prime factors. Then

$$J(N)>A\cdot\frac{N}{\log N}+O\left(\frac{N(\log\log N)^2}{(\log N)^{3/2}}\right)$$

where

$$A=\frac{2}{\pi}\cdot\prod_{p\equiv 1(\bmod 4)}\left(1-\frac{1}{(p-1)^2}\right)\cdot\prod_{p\equiv 3(\bmod 4)}\left(1-\frac{1}{p^2}\right)^{-1}$$

B.V. Levin. *The weak Landau problem and its generalization* (Russian). Uspehi Mat. Nauk. 16 (1961), 123–125.

c) There are infinitely many integers n such that n^2+1 is the product of at most two primes.

H. Iwaniec. *Almost-primes reprezentated by quadratic polynomials.* Invent. Math. 47 (1978), 171–188.

2) Let $p_0=2$ and $p_k(k>0)$ be the k-th prime $\equiv 1(\bmod 4)$. Define $P(t)=\prod_{k=0}^{t}p_k$ $(t=0,1,2,\ldots)$. Then for each sufficiently large integer t there exists a sequence C_t of consecutive integers n such that

(i) $(n^2+1,P(t))>1$
for all $n\in C_t$

(ii) card $C_p\ge[(1-\varepsilon)\cdot\lambda\cdot p_t]$
$0<\varepsilon<1$ for a certain positive constant λ

(iii) $p_t<n<P(t)$
for all $n\in C_t$

B. Garrison. *Consecutive integers for which n^2+1 is composite.* Pacific J. Math. 97 (1981), 93–96.

3) a) If $1 < y \leq x$, then the number of integers n satisfying

$$x - y < n \leq x$$

with both $n^2 + 1$ and $n^2 + 3$ prime is

$$\leq 32 \prod_{p>2} (1 - (p-1)^{-2}) \cdot \prod_{p \equiv \pm 1 (\mathrm{mod}\ 12)} \left(1 - \frac{2\chi(p)}{p-2}\right) \cdot \frac{y}{\log^2 y} \cdot \left(1 + O\left(\frac{\log\log 3y}{\log y}\right)\right)$$

b) If $1 < y \leq x$, then the number of integers n satisfying

$$x - y < n \leq x$$

with both $n^2 - 2n + 2$ and $n^2 + 2n + 2$ prime is

$$\leq 16 \cdot \prod_{p>2} (1 - (p-1)^{-2}) \cdot \prod_{p>2} \left(1 - \frac{2(-1)^{(p-1)/2}}{p-2}\right) \cdot \frac{y}{\log^2 y} \cdot \left(1 + O\left(\frac{\log\log 3y}{\log y}\right)\right)$$

c) Let $a \neq -b^2 (a, b$ integers). If $1 < y \leq x$ then the number of integers n satisfying

$$x - y < n \leq x^2$$

with $n^2 + a$ prime is

$$\leq 2 \cdot \prod_{2<p|a} \left(1 - \frac{(-a/p)}{p-1}\right) \frac{y}{\log y} \cdot \left(1 + O_a\left(\frac{\log\log 3y}{\log y}\right)\right)$$

H. Halberstam and H.-E. Richert. *Sieve methods.* Academic Press, 1974.

4) a) If $1 < y \leq x$ then the number of integers n satisfying

$$x - y < n \leq x$$

with $n^4 + 1$ prime is

$$\leq 2 \cdot \prod_{p>2} \left(1 - \frac{1 + 2\varepsilon(p)}{p-1}\right) \frac{y}{\log y} \cdot \left(1 + O\left(\frac{\log\log 3y}{\log y}\right)\right)$$

where

$$\varepsilon(p) = \begin{cases} 1 & \text{if } p \equiv 1 (\mathrm{mod}\ 8) \\ -1 & \text{if } p \not\equiv 1 (\mathrm{mod}\ 8) \end{cases}$$

(See H. Halberstam and H.-E. Richert.)

b) The number of primes $\leq n$ and of the form $x^4 + 1$ is

$$\sim n^{1/4} \cdot (\log n)^{-1} \cdot \prod_{p \geq 3} (1 - (-1/p)/(p-1)) \cdot \prod_{p \equiv 1 (\mathrm{mod}\ 4)} (1 - (-1/p)_4 \cdot 2/(p-2))$$

M. Dos Reis. *On conjectured asymptotic formulas concerning the distribution of prime numbers.* Gaz. Mat., Lisboa, 12 (1951), 83–90.

5) a) The number of primes $p \leq x$ with $p^2 + 4$ prime is

$$\leq 8 \cdot \prod_{p>2} (1 - (p-1)^{-2}) \cdot \prod_{p>2} \left(1 - \frac{(-1)^{(p-1)/2}}{p-2}\right) \cdot \frac{x}{\log^2 x} \cdot \left(1 + O\left(\frac{\log\log 3x}{\log x}\right)\right)$$

b) The number of primes $p \leq x$ with $p^2 + p + 1$ prime is

$$\leq 8 \cdot \prod_{p>2}(1-(p-1)^{-2}) \cdot \prod_{p>3}\left(1 - \frac{\chi(p)}{p-2}\right) \cdot \frac{x}{\log^2 x} \cdot \left(1 + O\left(\frac{\log\log 3x}{\log x}\right)\right)$$

where $\chi(p) = 1$ if $p \equiv 1 \pmod 3$ and -1 if $p \equiv 2 \pmod 3$

c) The number of primes p with $x - y < p \leq x$, $p + 2$ prime, $p^2 + 4$ prime is

$$\leq 2^7 \cdot 3 \prod_{p>2}(1-(p-1)^{-2})^2 \cdot \prod_{p>3}(1-(p-2)^{-2}) \cdot \prod_{p>3}\left(1 - \frac{(-1)^{(p-1)/2}}{p-3}\right) \cdot \frac{y}{\log^3 y} \cdot \left(1 + O\left(\frac{\log\log 3y}{\log y}\right)\right)$$

where the constant implied by the O-symbol is absolute.

6) For infinitely many integers n, $n^3 + 2$ has at most 4 prime factors.

Y. Wang. *On sieve methods and some of their applications.* Sci. Record (N.S.) 1 (1957), 1–5.

§ VIII.31 Primes $p \leq N$ of the form $p = [cn]$

1) Let $\pi_{2,c}(N)$ be the number of primes $p \leq N$ of the form

$$p = [cn] \ (c \in \mathbb{R})$$

Then:

a) If $c \in Q$, $c = \frac{a}{q}$, $(a, q) = 1$, $q > 0$, then

$$\pi_{2,c}(N) \sim \psi(a, q)\pi(N)$$

where $\varphi(a)\Psi(a, q)$ equals the number of all integers b with $1 \leq b \leq q$ and $([ab/q], a) = 1$

b) If $c \notin Q$, then

$$\pi_{2,c}(N) \sim \frac{1}{c} \cdot \pi(N)$$

$(N \to \infty)$

D. Leitman and D. Wolke. *Primzahlen der Gestalt* $[f(n)]$. Math. Z. 145 (1975), 81–92.

Note. See also

D. Leitman. *The distribution of prime numbers in sequences of the form* $[f(n)]$. Proc. London Math. Soc. (3) 35 (1977), 448–462.

2) For $c > 0$ let $\pi_{3,c}(x)$ be the number of $n \leq x$ such that $[cn^2]$ is prime. Then:

a) $$\int_{\alpha}^{\beta} \pi_{3,c}(x)\mathrm{d}c = \frac{1}{2}(\beta - \alpha)\frac{x}{\log x} + O\left(\frac{x}{\log^2 x}\right)$$

b) For almost all c there exists an $x_c \in \mathbb{R}$ such that, for all $x \geq x_c$, the inequality

$$\pi_{3,c}(x) \leq A \cdot x/\log x$$

is true, with A absolute constant.

c) For almost all c

$$\pi_{3,c}(x) \to \infty$$

as $x \to \infty$

D. Nordon. *Nombre premiers de la forme* $[un^2]$. Arch. Math. (Basel) 28 (1977), 395–400.

§ VIII.32 Primes of the form $n \cdot 2^n + 1$ or $p \cdot 2^p + 1$ or $2^p \pm p$

1) a) Let $T(x) = \text{card}\{n < x : n \cdot 2^n + 1 = \text{prime}\}$ and $S(x) = \text{card}\{p < x : p \text{ prime and } p \cdot 2^p + 1 = \text{prime}\}$. Then

$$T(x) = o(x)$$

C. Hooley. *Applications of sieve methods to the theory of numbers.* Cambridge, 1976.

b) $T(x) = O(x/\log x)$ and $S(x) = O(x/\log^2 x)$

E. Heppner. *Über Primzahlen der Form* $n \cdot 2^n + 1$ *bzw.* $p \cdot 2^p + 1$. Monatsh. Math. 85 (1978), 99–103.

2) Let $N(x) = \text{card}\{k \leq x : k \text{ odd}, k \cdot 2^n + 1 = \text{prime for some } n \geq 1\}$. Then $\left(\frac{1}{2} - c_1\right)x \geq N(x) \geq c_2(x)$ for $x \geq 1$, where $c_1, c_2 > 0$ are computable constants.

P. Erdős and A.M. Odlyzko. *On the density of integers of the form* $(p-1)\cdot 2^{-n}$ *and related questions.* J. Number Theory 11 (1979), 257–263.

Remark. Earlier, Sierpiński proved that $N(x) \to \infty$ as $x \to \infty$

W. Sierpiński. *Sur un problème concernant les nombres* $k \cdot 2^n + 1$. Elem. Math. 15 (1960), 73–74.

3) The number of primes of the form $2^p \pm p (p \leq x, \text{prime})$ is $o(x/\log x)$.

A. Mil'uolo. *On primes in sparse sequences* (Russian). Vestnik Moskov Univ. Ser. I Mat. Mekh. 1987, no. 2, 75–77, 104.

4) Let A be the set of all natural numbers N such that $N - 2^K$, where $1 \leq k \leq \log N/\log 2$ are primes. Then

$$A(x) = O\left(x \exp\left(-c\frac{\log x \log\log x}{\log\log\log x}\right)\right)$$

($c > 0$), where $A(x)$ denotes the counting function of A

R.C. Vaughan. *Some applications of Montgomery's sieve.* J. Number theory 5 (1973), 64–79.

Remark. P. Erdős conjectured that $A = \{7, 15, 21, 45, 75, 105\}$

P. Erdős. *On integers of the form* $2^k + p$ *and some related questions.* Summa Bras. Math. 2 (1950), 113–123.

§ VIII.33 Primes of the form $x^2 + y^2 + 1$

a) There exist infinitely many primes of the form $x^2 + y^2 + 1$

B.M. Bredihin. *Binary additive problems of indeterminate type II. Analogue of the problem of Hardy and Littlewood* (Russian). Izv. Akad. Nauk. SSSR. Ser. Mat. 27 (1963), 577–612

and

Y. Motohashi. *On the distribution of prime numbers which are of the form* $x^2 + y^2 + 1$. Acta Arith. 16 (1969/70), 351–363.

b) Let $I(N)$ be the number of primes $p \leq N$ of the form

$$x^2 + y^2 + 1$$

Then

$$I(N) \leq c \cdot N/(\log N)^{3/2}$$

($c > 0$, constant.)

Y. Motohashi. Ibid. II. Acta Math. Hungar. 22 (1971/72), 207–210.

§ VIII.34 On a sum on $\frac{\log p}{p}$ when $p \in L \neq$ arithmetic progression

Let $m \geq 1, r \geq 1, q \geq 1$ be integers. Let

$$L = \{n \in \mathbb{N} : m^s \cdot r + qm \cdot \frac{m^s - 1}{m - 1} \leq n \leq m^s \cdot (r + 1) + qm \cdot \frac{m^s - 1}{m - 1} - 1$$

with any $s \geq 1$ (integer) $\}$. Then:

a) $$\sum_{p \leq x, p \in L} \frac{\log p}{p} = \frac{1}{\log m} \cdot \log\left(1 + 1/\left(r + \frac{qm}{m-1}\right)\right) \cdot$$
$$\cdot \log x + o(\log x)$$

Corollary. L contains an infinity of primes.

b) $$\sum_{\substack{p\leq x\\ p=l(\text{mod }k),\,p\in L}} \frac{\log p}{p} = \frac{1}{\varphi(k)\log m}\cdot\log\left(1+1/\left(r+\frac{qm}{m-1}\right)\right)\cdot$$
$$\cdot\log x + o(\log x)$$

where $(l,k)=1, k>0$

A.O. Gelfond. *On an arithmetic equivalent of the analyticity of the Dirichlet L-series on the straight line* $Rs=1$. Izv. Akad. Nauk., Ser. Mat. 20 (1956), 145–166.

§ VIII.35 Recurrent sequences of primes

Let (a_n) be a recurrent sequence of natural numbers, i.e. satisfying

$$a_n = c_1 a_{n-1} + \cdots + c_k a_{n-k}$$

$(c_k \neq 0.)$ If (a_n) consists only of prime powers, then there exist natural numbers M and d_l, $1\leq l\leq M$ with $a_{l+jM} = a_{l+j_0M}\cdot d_l^{j-j_0}$, for all $j\geq j_0$ and all $1\leq l\leq M$

Corollary. A recurrent sequence of primes must be periodic.

P. Erdős, Th. Maxsein and P.R. Smith. *Primzahlpotenzen in rekurrenten Folgen.* Analysis 10 (1990), 71–83.

§ VIII.36 Composite values of exponential and related sequences

1) There is a number $A>1, A\notin\mathbb{N}$, such that $[A^{3^n}]$ is prime for all $n=1,2,3,\ldots$

W.H. Mills. *A prime representing function.* Bull. Amer. Math. Soc. 53 (1947), 604.

Note. See also

U. Dudley. *History of a formula of primes.* Amer. Math. Monthly 76 (1969), 23–28.

2) a) Let $\theta>1$ be a root of polynomial of the form $x^s - x^{s-r} - 1$, where $s>r\geq 1$. Then $[\theta^n]$ is composite infinitely often (in fact $[\theta^n]$ is even infinitely often.)

b) Let $\theta>1$ be a root of the polynomial

$$x^n + x^{n-1} + \cdots + x^{l+1} - x^l - x^{l-1} - \cdots - 1$$

where $n>l>\frac{1}{2}(n+1)$. Then $[\theta^n]$ is composite infinitely often.

c) Let $\theta>1$ be a root of the polynomial

$$x^s - 3x^{s-r} - 3$$

where $s > r \geq 1$. Then $[\theta^n]$ is composite infinitely often.

Remark. If θ is a root of the above polynomial, and which is not the root of largest absolute value, then $[\theta^n]$ is infinitely often divisible by 2 or 3

H.N. Shapiro and G.N. Sparer. *Composite values of exponential and related sequences.* Comm. Pure Appl. Math. 25 (1972), 569–615.

3) a) Infinitely many integers of the form $\left[\left(\frac{4}{3}\right)^n\right]$ are composite.

b) The same for $\left[\left(\frac{3}{2}\right)^n\right]$

W. Forman and H.N. Shapiro. *An arithmetic property of certain rational powers.* Comm. Pure Appl. Math. 20 (1967), 561–573.

4) If f is any non-constant polynomial with positive leading coefficient, then for any M there are infinitely many $c \geq 0$ such that the sequence $[f(n)] + c\ (n = 0, 1, 2, \dots)$ contains at least M primes.

R. Forman. *Sequences with many primes.* Amer. Math. Monthly 99 (1992), 548–557.

§ VIII.37 Primes in partial sums of n^n

The number of primes of the form $\sum_{1 \leq r \leq n} r^r$ not exceeding x has an upper bound $\ll (\log x)/(\log\log x)^2$.

K. Soundararajan. *Primes in a sparse sequence.* J. Number Theory 43 (1993), 220–227.

§ VIII.38 Beurling's generalized integers

a) Let $\wp = (p_i)_{i=1,2,\dots}$ be a sequence of real numbers satisfying $1 < p_1 \leq p_2 \leq \dots, p_j \to \infty$. Let N be the semigroup generated by $\wp$ under multiplication. Assume that the elements of N are arranged in ascending order and are denoted by $(n_i)_{i=1,2,\dots}$. Define $N(x) = \sum_{n_i \leq x} 1$, $\pi(x) = \sum_{p_i \leq x} 1$. Then, if $\pi(x) \ll x/\log x$ and $\limsup_{s \to 1+} \left(\int_1^\infty x^{-s} d\pi(x) - \log(1/(s-1))\right) < \infty$, then

$$N(x) \ll x$$

W.-B. Zhang. *Density and O-density of Beurling generalized integers.* J. Number Theory 30 (1988), 120–139.

b) Let $\theta(x) = \sum_{p \in \wp, p \leq x} \log p$. If $\theta(p_k) \leq k(\log k + \log\log k + O(1))$ for $k \geq 2$, and if there exists an increasing function g satisfying $g > 0$, $g(x) = O(x \log x)$, $g(x) \nearrow \infty$ and $p_k \geq k \log k / g(k)$, then

$$p_k \leq ck(\log k) \cdot (\log g(k))$$

for a suitable constant $c > 0$ (This result is best possible.)

G. Robin. *Comportement asymptotique du produit des k premiers nombres premiers généralisés.* Colloq. Math. 54 (1987), 333–338.

c) $\theta(p_k) \geq k(\log k + \log\log k - 1.076868)$
for $k \geq 2$, with equality when $k = 66$

G. Robin. *Estimation de la fonction de Tchebyschev θ sur le k-iéme nombre premier et grandes valeurs de la fonction ω(n), nombre de diviseurs premiers de n.* Acta Arithm. 42 (1983), 367–389.

d) Let $\Psi(x) = \sum_{p_i^k \leq x} \log p_i$ and let $0 < \rho_1 < \rho_2 < \cdots < \rho_n$ and $A_1, A_2, \ldots, A_n$ be arbitrary real numbers.

If $\int_1^\infty x^{-1} \left\{ \sup_{x \leq y} y^{-1} \cdot \left| N(y) - y \sum_{i=1}^n A_i \log^{\rho_i - 1} y \right| \right\} dx < \infty$ holds with $\rho_n \geq 1$ and $A_n > 0$, then there exist numbers $\alpha > 0$ and $\beta < \infty$ for which $0 < \alpha \leq \liminf_{x\to\infty} \frac{\Psi(x)}{x}$ and $\limsup_{x\to\infty} \frac{\Psi(x)}{x} \leq \beta < \infty$ hold.

W. Zhang. *Chebyshev type estimates for Beurling generalized prime numbers.* II. Trans. Amer. Math. Soc. 337 (1993), 651–675.

Remark. If $N(x) = Ax + O(x \log^{-\gamma} x)$ for some positive constant $A > 0$ and $\gamma > \frac{3}{2}$, then A. Beurling has proved that $\Psi(x) \sim x$ holds. If $\gamma > 1$ then H.G. Diamond proved some estimates of Chebyshev type.

A. Beurling. Acta Math. 68 (1937), 255–291.
H.G. Diamond. Illinois J. Math. 14 (1970), 29–34.

§ VIII.39 Accumulation theorems for primes in arithmetic progressions

Let $(l_1, q) = (l_2, q) = 1, l_1 \not\equiv l_2 (\text{mod } q)$ and introduce the notation

$$\varepsilon(n, q, l_1, l_2) = \begin{cases} 1 \text{ if } n \equiv l_1 (\bmod q) \\ -1 \text{ if } n \equiv l_2 (\bmod q) \\ 0 \text{ otherwise} \end{cases}$$

Let us suppose that $L(s, X, q) = 0$ for $0 < \sigma < 1, |t| < A(q)$ (where $0 < A(q) \leq 1$) i.e. the "Haselgrove-condition", and that $L(s, X, q) \neq 0$ for $\sigma > 1/2 \quad |t| \leq D$, with D satisfying $D \geq c_1 q^{15}$ and $D \geq \frac{12\pi}{A(q)}$, i.e. "finite Riemann-Piltz conjecture". Then:

a) If $q > c_2$ and $Y \geq \exp(D^8 \cdot \exp(q^{10}))$, there exist x and k with $x \in \left[Y^{1-\frac{2}{D}}, Y\right], k \in \left[\frac{2\log Y}{D^2}, \frac{2\log Y}{D^2} + \sqrt{\log Y}\right]$ such that

$$\sum_n \varepsilon(n, q, l_1, l_2)\Lambda(n) \exp\left(-\frac{\log^2(n/x)}{4k}\right) > Y^{\frac{1}{2}-\frac{8}{D}}$$

If we additionally assume that l_1 and l_2 are both quadratic non-residues, then also

$$\sum \varepsilon(p, q, l_1, l_2) \log p \cdot \exp\left(-\frac{\log^2(p/x)}{4k}\right) > Y^{\frac{1}{2}-\frac{8}{D}}$$

S. Knapowski and P. Turán. *Further developments in the comparative prime-number theory.* IV. Acta Arith. 11 (1965), 147–161.

b) If $q > c_3, l_1 \not\equiv l_2 \pmod q$ are both quadratic residues, then for $Y \geq \exp(\exp(D^6))$ we have x and k as above, satisfying the last inequality in a).

S. Knapowski and P. Turán. Ibid. VI. Acta Arith. 12 (1966), 85–96.

c) Let us suppose for q the truth of the Haselgrove condition, the finite Riemann-Piltz conjecture, further $\lambda \geq 20D_0$, where $D_0 = c_0 q^2 \log^6 q$, with c_0 a sufficiently large absolute constant. Then for every $T > c$ with

$$L = \log T > \max\left(\frac{cq^2}{A(q)} \cdot \lambda \log^3 L, c\lambda^2\right)$$

there exist $x = e^\mu$, k with

$$x \in \left[T \exp\left(-\frac{cq^2}{A(q)}\lambda \log^3 \lambda\right), T\right]$$

$$k \in \left[\frac{\mu}{\lambda^2} - 1, \frac{\mu}{\lambda^2} + 1\right]$$

and

$$\sum_n \varepsilon(n, q, l_1, l_2)\Lambda(n) \exp\left(-\frac{\log^2(n/x)}{4k}\right) >$$

$$> \sqrt{x} \exp\left(-\frac{LD^2}{\lambda^2} - \frac{cq^2}{A(q)}\lambda \log^3 L\right) > x^{0.49}$$

If we additionally assume that l_1 and l_2 are both quadratic non-residues or if they are both quadratic residues, then

$$\sum_p \varepsilon(p, q, l_1, l_2) \log p \cdot \exp\left(-\frac{\log^2(p/x)}{4k}\right) >$$

$$> \sqrt{x} \exp\left(-\frac{LD^2}{\lambda^2} - \frac{cq^2}{A(q)}\lambda \log^3 L\right) > x^{0.49}$$

J. Pintz and S. Salerno. *Accumulation theorems for primes in arithmetic progressions.* Acta Math. Hung. 46 (1–2) (1985), 151–172.

§ VIII.40 About the Shanks-Rényi race problem

Assuming the generalized Riemann Hypothesis, there exist infinitely many integers m with

$$\pi(m; q, 1) > \max_{a \not\equiv 1 \pmod q} \pi(m; q, a)$$

and the same with "<" and "min". The set of m in either case has positive lower density.

J. Kaczorowski. *A contribution to the Shanks-Rényi race problem.* Quart. J. Math. Oxford II. Ser. 44 (1993), 451–458.

Chapter IX

ADDITIVE AND DIOPHANTINE PROBLEMS INVOLVING PRIMES

§ IX. 1 Schnirelman's theorem. Vinogradov's theorem

1) a) Every sufficiently large positive integer is representable as the sum of not more than c_0 primes, where c_0 is an absolute positive constant.

L.G. Schnirelmann. *On the additive properties of numbers.* Rostov n/D, Izv. Dopets. Politekhn. in-ta 14 (1930), 3–28.

Remarks. (i) Schnirelmann obtained $c_0 = 8 \cdot 10^5$. In 1951 Shapiro and Warga obtained $c_0 \leq 20$.

H.N. Shapiro and J. Warga. *On the representation of large integers as sums of primes.* I. Comm. Pure Appl. Math. 3 (1950), 153–176.

(ii) If Q is a set that contains a positive proportion of the primes, then there is a number c_0 such that every sufficiently large integer is the sum of at most c_0 primes belonging to Q.

This result is also attributed to Schnirelmann (1930); for a short proof, using Selberg's sieve, see

M.B. Nathanson. *A generalization of the Goldbach-Schnirelmann theorem.* Amer. Math. Monthly 94 (1987), 768–771.

b) $c_0 \leq 4$

for all sufficiently large numbers.

I.M. Vinogradov (1937). See *Selected Works (Izbrannye trudy).* Akad. Nauk. SSSR. Moskow (1952).

2) Every even number is the sum of 18 or fewer primes.

H. Reisel and R.C. Vaughan. *On sums of primes.* Ark. Mat. 21 (1983), 46–74.

Remark. Since Vinogradov's theorem (see 1) b)) assures that 4 primes suffice for all sufficiently large numbers, the interest here is that all numbers are covered.

§ IX. 2 Number of representations of N in the form $p_1^n + \cdots + p_k^n$. Vinogradov's three primes theorem

1) a) There is a large $N_0 > 0$ such that each odd number $n \geq N_0$ can be expressed as a sum of 3 primes.

I.M. Vinogradov (1937). See *Selected works (Izbrannye trudy)*. Akad. Nauk. SSSR. Moskow (1952).

Remark. In fact, Vinogradov shows that the number of solutions of $p_1 + p_2 + p_3 = x$ is

$$\frac{1}{2}A(x)x^2/\log^3 x + O(x^2 \cdot \log\log x/\log^4 x)$$

where

$$A(x) = \prod_p (1 + (p-1)^{-3}) \prod_{p|x} (1 - 1/(p^2 - 3p + 3))$$

b) $N_0 \leq \exp(\exp(16.038))$

K.G. Borozdkin. *On the question of I.M. Vinogradov's constant.* Tr. 3-rd Vses. Mat. S'ezda Moscow, vol 1, p. 3 (1956).

c) $N_0 \leq \exp(\exp(11.503))$

J. Chen and T. Wang. *On the odd Goldbach problem* (Chinese.) Acta Math. Sin. 32 (1989), 702–718.

2) Let $1 < c < 2, \quad \gamma = 1/c$.
Suppose that $n \geq 2, \quad k \geq k_0$, where
$k_0 = 2^n + 2$ if $2 \leq n \leq 10$ and
$k_0 = 2[n^2(2\log n + \log\log n + 5)]$ if $n > 10$, $s = 3n + 3$,
$K = 2[s^2(2\log s + \log\log s + 4)]$ and $\eta = \omega(n-1)(12K^2)^{-1}$

a) Let N be odd number, let $I_1(N)$ denote the number of representations of N as the sum of three primes $p_i (1 \leq i \leq 3)$ with $p_i \in [(2t)^c, (2t+1)^c)$, where t is an integer, and let $I(N)$ denote the number of representations of N as the sum of three arbitrary primes. Then

$$I_1(N) = \frac{I(N)}{8} + O(N^{2-(1-\gamma)/6} \cdot \log^5 N)$$

b) Let $J_1(N)$ denote the number of representations of N in the form $N = p_1^n + \cdots + p_k^n$ (N positive number), where $p_i \in [(2t)^c, (2t+1)^c), (1 \leq i \leq k)$ are primes and $J(N)$ the number of representations of N as the sum of k arbitrary primes; then

$$J_1(N) = 2^{-k} J(N) + O\left(N^{(k-n-\eta)/n+\varepsilon} \cdot (c+1)^{-((n-1)/(4K^2))}\right)$$

S.A. Gritsenko. *The ternary Goldbach problem and the Goldbach-Waring problem with prime numbers lying in intervals of special type* (Russian.) Uspekhi Mat. Nauk. 43 (1988), No. 4, 203–204; translation in Russian Math. Surveys 43 (1988), No. 4, 209–210.

Remark. For the number of solutions of $a_1p_1 + a_2p_2 + a_3p_3 = b$, where $(a_1, a_2, a_3) = 1$, see

K.M. Tsang. *Small prime solutions of linear equations and the exceptional set in Goldbach's problem.* Number theory and its application in China, 153–158, Contemp. Math., 77, Amer. Math. Society Providence, RI, 1988.

3) a) If $63/64 < \theta < 1$, then there exists n_0 such that each odd number $N > N_0$ is representable as the sum of three primes each of which differs from $\frac{N}{3}$ by no more than N^{θ}.

C.B. Haselgrove. *Some theorems in analytic theory of numbers.* J. London Math. Soc. 26 (1951), 273–277.

b) Every sufficiently large odd number N can be represented as sum of three primes satisfying

$$\left| p - \frac{1}{3}N \right| \le N^{279/308+\varepsilon}$$

V. Statulevičius. *On the representation of odd numbers as the sum of three almost equal prime numbers.* Vilniaus Valst. Univ. Mosklo Darbai Mat. Fiz. Chem. Mosklu Ser. 3 (1955), 5–23 (Lithuanian).

c) Every sufficiently large odd number N can be expressed as $N = p_1 + p_2 + p_3$, where p_i $(1 \le i \le 3)$ are prime satisfying

$$\frac{N}{3} - N^{\frac{13}{17}+\varepsilon} < p_i \le \frac{N}{3} + N^{\frac{13}{17}+\varepsilon}$$

$(1 \le i \le 3)$

C. Jia. *Three primes theorem in short intervals* (Chinese). Acta Math. Sin. 32 (1989), 464–473.

d) Every large odd integer N can be represented as $N = p_1 + p_2 + p_3$, where p_i are primes such that $p_i = \frac{N}{3} + O(N^{5/8}(\log N)^c)$ with $c > 0$ a constant.

Remark. The proof is based on the Hardy-Littlewood method, and the key result is an estimate for short exponential sums over primes. The weaker result with 2/3 in place of 5/8 was discovered by Pan and Pan.

C. Pan and C. Pan. Chin. Ann. Math. Ser. B 11 (1990), 138–147.

e) If the generalized Riemann Hypothesis is true, then the same is true with

$$p_i = \frac{N}{3} + O(N^{1/2}(\log N)^{7+\varepsilon})$$

D. Wolke. *Über Goldbach-Zerlegungen mit nahe zu gleichen Summanden.* J. Number Theory 39 (1991), 237–244.

f) Without any hypothesis, the result holds with

$$p_i = \frac{N}{3} + O(N^{a+\varepsilon})$$

where $a = \frac{23}{39} \left(< \frac{5}{8} \right)$

C. Jia. *Three primes theorem in short intervals* (Chinese). Acta Math. Sin. 34 (1991), 832–850.

§ IX. 3 Rényi's theorem. Chen's theorem

a) All large even integers are sums of two integers having at most four prime factors.

A.A. Buchstab. *Sur la décomposition des nombres pairs en somme de deux composantes dont chacune est formée d'un nombre borné de facteurs premiers.* C.R. Acad. Sci. URSS (N.S.) 29 (1940), 544–548.

b) For all sufficiently large N there exists a representation of the form $2N = p + N'$ (p prime), where the prime factors of N' exceed $(\log N)^\lambda$ for any λ. The number of such representations is

$$\geq c \cdot N/(\log N \cdot \log\log N)$$

A.A. Buchstab. *On an additive representation of integers* (Russian). Mat. Sb. (N.S.) 10 (52) (1942), 87–91.

Remark. By the same method may be proved that there are infinitely many primes p such that the prime factors of $p + 2$ exceed $(\log p)^\lambda$ for any λ.

c) Each even integer is a sum of a prime and an almost prime (almost prime, in the sense: there exists an absolute constant K such that the number has at most K prime factors).

A. Rényi. *On the representation of an even number as the sum of a single prime and single almost-prime number* (Russian). Dokl. Akad. Nauk SSSR (N.S.) 56 (1948), 455–458.

d) Let $f(n)$ be the number of representations of an even integer n as the sum of two primes. Then for all sufficiently large x, there are more than $\exp(c_1(\log x)^{1/2})$ even integers $n \leq x$ for which

$$f(n) > c_2 x \log\log x / \log^2 x$$

($c_1, c_2 > 0$, constants.)

K. Prachar. *On integers n having many representations as sum of two primes.* J. London Math. Soc. 29 (1954), 347–350.

e) Every sufficiently large even number is a sum of a prime and a product of at most two primes.

J.-R. Chen. *On the representation of large even integer as the sum of a prime and the product of at most two primes.* Kexue Tongbao 17 (1966), 385–386.

§ IX. 4 Improvements on Chen's theorem

1) a) Let x be a large even integer. Let $G_2(x)$ be the number of primes $p \le x$ such that $x - p$ has at most two prime factors. Then

$$G_2(x) \ge \frac{0.67x \cdot C_x}{(\log x)^2}$$

where $C_x = \prod_{\substack{p|x \\ p>2}} \frac{p-1}{p-2} \prod_{p>2} \left(1 - \frac{1}{(p-1)^2}\right)$

J.-R. Chen. *On the representation of a larger even integer as a sum of a prime and the product of at most two primes.* Sci. Sinica 16 (1973), 157–176.

b) 0.67 can be replaced by 0.689

H. Halberstam. *A proof of Chen's theorem.* Bordeaux Conference on Number Theory, 1974.

c) Every sufficiently large even integer x can be written as a sum of a prime and a natural number which has at most one prime factor less than $x^{1089/2089}$

A. Fujii. *Some remarks on Goldbach's problem.* Acta Arithm. 32 (1977), 27–35.

d) 0.689 in b) can be replaced with 0.7544

J.-R. Chen. *On the representation of a larger even integer as a sum of a prime and the product of at most two primes.* Sci. Sinica 21 (1978), 421–430.

e) 0.7544 can be replaced with 0.81

J.-R. Chen. *The exceptional set of Goldbach numbers.* I. Sci. Sinica 23 (1980), 416–430.

2) Let N be a large positive even number, and p a prime.
Then:

a) There exists N_1 such that for all $N \ge N_1$,

$$\text{card}\,\{p : p < N,\ N - p = p_1 p_2 \text{ or } N - p = p_1 p_2 p_3\} >$$
$$> 0.003 C_N \cdot \frac{N}{\log^2 N}$$

b) $\text{card}\,\{p : p < N,\ p + 2 = p_1 p_2 \text{ or } p + 2 = p_1 p_2 p_3\} >$

$$> 0.003c \frac{N}{\log^2 N}$$

where C_N and c are explicitely known positive constants.

E.K.-S. Ng. *On the sequences $N-p, p+2$ and the parity problem.* Arch. Math. (Basel) 42 (1984), 430–438.

3) For a large even integer x let $P_x(1,1)$ denote the number of primes p such that $x-p$ is a prime.

a) $$P_x(1,1) \leq 7.8342 \prod_{p>2}\left(1-\frac{1}{(p-1)^2}\right)\cdot\left(\prod_{2<p|x}\frac{p-1}{p-2}\right)\cdot\frac{x}{\log^2 x}$$

J.-R. Chen. Sci. Sinica 21 (1978), No. 6, 701–739.

b) The same is true with the constant 7.81565

D.H. Wu. *An improvement of J.-R. Chen's theorem* (Chinese). Shanghai Keji Daxue Xuebao 1987, 94–99.

Remark. For a Chen-type theorem for algebraic number fields, see

J.G. Hinz. *Chen's theorem in totally real algebraic fields.* Acta Arith. 58 (1991), 335–361.

§ IX. 5 On number of writings of N as $\varepsilon_1\ldots\varepsilon_s + p_1\ldots p_r$ or $\varepsilon_1\ldots\varepsilon_s + p_1\ldots p_{r+1}$. A common generalization of Chen's and Linnik's theorems

1) Let $f(N;r,s)$ denote the number of ways of writing N in either the form $q_1\ldots q_s + p_1\ldots p_r$ or $q_1\ldots q_s + p_1\ldots p_{r+1}$, where q_i, p_j are primes.

a) $f(N;1,1) > 0.81\cdot C_N\cdot N/\log^2 N$

where

$$C_N = \prod_{p>2}(1-(p-1)^{-2})\cdot\prod_{2<p|N}\left(\frac{p-1}{p-2}\right)$$

for all large even N

J.R. Chen. *The exceptional set of Goldbach numbers.* I. Sci. Sinica 23 (1980), 416–430.

b) If $r+s\geq 3$ and the primes concerned exceed $\exp(\log N)^{\varepsilon}$, where $0<\varepsilon<1$ is a constant, then

$$f(N;s,r) \gg C_N\cdot N(\log\log N)^{s+r-3} + \log^2 N$$

M. Zhang. *Some new applications of the mean-value theorems* (Chinese). J. China Univ. Sci. Technol. 19 (1989), 38–50.

2) Let $b_1, b_2 > 0$ and b_3 be pairwise coprime integers such that $2|b_1b_2b_3$. Then there are infinitely many solutions of

$$b_1p - b_2\cdot P_2 = b_3$$

with p prime and P_2 a product of at most two prime factors.

There exists an effective constant c such that the first solution satisfies

$$p\cdot P_2 \ll (2+b_1b_2)^c\cdot|b_3|$$

M.D. Coleman. *On the equation $b_1 p - b_2 P_2 = b_3$*. J. Reine Angew. Math. 403 (1990), 1–66.

Remark. The theorem contains for $b_1 = b_2 = 1$ and for $b_1 = 1$ theorems by J. Chen and Yu.V. Linnik, respectively.

3) Let ζ, δ be fixed numbers with $0 < \zeta < 1, 0 < \delta < 1$. For every sufficiently large even integers N and any $s \geq 1, r \geq 2$,

$$\begin{aligned} \text{card}\,\{a : a = N - q_1 \cdots q_s, (a, N) = 1, a = p_1 \cdots p_{r-1} \text{ or } p_1 \cdots p_r, \\ \exp(\log^\zeta N) < q_1 < \ldots < q_s, \\ \exp(\log^\delta N) < p_1 < \ldots < p_r\} > \\ > \frac{0.77(1-\zeta)^{s-1}(1-\delta)}{(s-1)!(r-2)!} C_N \log^{-2} N \cdot (\log\log N)^{s+r-3} \end{aligned}$$

where p_i, q_j are primes, and

$$C_N = \prod_{p>2}(1 - (p-1)^{-2}) \cdot \prod_{2<p|N} (p-1)|(p-2) \; (p \text{ prime.})$$

J. Kan. *On the problem of Goldbach's type.* Math. Annalen (to appear).

Remark. For more special results, see

J. Kan. *On the representation of an even number as a sum of a prime and almost primes* (Chinese). J. Math., Wahan Univ. 11 (1991), 196–204,

where it is proved e.g. that

$$\frac{3.1245}{(r-3)!} \frac{C_N N(\log\log N)^{r-3}}{\log^2 N} <$$

$$< \text{cart}\,\{N : N = p + P_r, p \text{ prime}, \omega(P_r) \leq r\} \frac{C_N N(\log\log N)^{r-1}}{\log^2 N}$$

where the left-side inequality holds for $r \geq 4$, and the right-side one for $r \geq 1$

4) Let $(a_1, a_2, a_3) = (b, a_i, a_j) = 1$ for $1 \leq i < j \leq 3$, $b \equiv (a_1 + a_2 + a_3) \pmod 2$ and write $B = \max(3, |a_1|, |a_2|, |a_3|)$

a) Suppose $a_i > 0$ for $1 \leq i \leq 3$. Then there is an effective constant $A > 0$ such that the equation

(*) $\quad a_1 p_1 + a_2 p_2 + a_3 p_3 = b$

has a solution for all $b > B^A$ (in primes p_i)

b) Suppose $a_i \neq 0$ are not all of the same sign. Then for all b there is a solution of (*) with $\max(p_1, p_2, p_3) < 3|b| + B^A$

M.-C. Liu and K.-M. Tsang. *Small prime solutions of linear equations.* Théorie des nombres, C.R. Conf. Int., Québec, Canada, 1987, 595–624.

5) Let $S_{2,3}(\alpha, \beta)$ denote the number of solutions in primes p, p_i of the conditions $p+2 = p_1p_2$ or $p_1p_2p_3, q(p+2) > x^{\alpha}$, $P(p+2) < x^{\beta}$, where $q(n)$, $P(n)$ are the least and greatest prime factors of n. Then

$$S_{2,3}(1/6, 5/6) < c_1 x/\log^2 x \text{ and } S_{2,3}(0, 4/5) < c_2 x/\log^2 x$$

with c_1, c_2 positive constants.

S. Salerno and A. Vitolo. *$p+2$ with few and bounded prim factors.* Analysis 11 (1991), 129–148.

§ IX. 6 On $p_1^k + p_2^k = N$. Estimates on the number of solutions. Binary Hardy-Littlewood problem

1) a) Let $B(u)$ be the number of solutions of $p_1^k + p_2^k = N$, $0 < p_1, p_2 \leq u^{1/k}$($p_1$, p_2 primes). Then

$$\liminf_{u\to\infty} \frac{B(u)}{u^{2/k}/\log^2 u} > 0 \text{ and } \limsup_{u\to\infty} \frac{B(u)}{u^{2/k}/\log^2 u} < \infty$$

G.J. Rieger. *Über die Summe von zwei n-ten Primzahlpotenzen.* Math. Z. 84 (1964), 137–142.

b) The number of integers $n \leq x$ which are of the form $p^k + q^m (p, q$ primes), $k, m \geq 2$ (fixed) is

$$\sim c_{k,m} \cdot x^{\frac{1}{k}+\frac{1}{m}} \cdot \log^{-2} x, \text{ where}$$

$$c_{k,m} = \Gamma\left(\frac{1}{k}\right)\Gamma\left(\frac{1}{m}\right)\Gamma\left(\frac{1}{r}\right), \text{ with } r = \frac{1}{k} + \frac{1}{m} + 1$$

(and Γ is Euler's Gamma function).

M. Orazov. *Some applications of the Romanov-Erdős inequality* (Russian). Izv. Akad. Nauk Turkmen. SSR Ser. Fiz.-Tekhn. Him. Geol. Nauk 1978, 3–9.

2) Let $f(x)$ be a polynomial with positive highest coefficient which assumes integral values for integral x. The density of integers of the form $p_1 + f(p_2)$ (p_1, p_2 primes) is positive.

E. Wirsing. *Eine Erweiterung des ersten Romanovschen Satzes.* Arch. Math. 9 (1958), 407–409.

3) Let $k \geq 2$ and $E_2(N)$ and $E_{1+1/k}(N)$ denote the set of all $n \leq N$ for which $2n = p_1 + p_2 (p_1, p_2$ primes) and $n = p + x^k (p$ prime) respectively, do not hold.

a) For all $N \geq N_0$, $E_{3/2}(N) < N^{1-\delta}$ and $E_{4/3}(N) <^{1-\delta}$, where $\delta > 0$ is an effective constant.

b) Assuming the Riemann Hypothesis, $E_2(N) < N^{1/2+\varepsilon}$,

$$E_{3/2}(N) < N^{2/3+\varepsilon},\ E_{4/3}(N) < N^{5/6+\varepsilon} (\varepsilon > 0 \text{ fixed.})$$

A.I. Vinogradov. *The binary Hardy-Littlewood problem* (Russian). Acta Arith. 46 (1985), 33–56.

c) For every fixed integer $k \geq 2$ there exist positive constants $C(k)$ and $\delta(k) < 1$ such that

$$E_{1+1/k}(N) < C(k) \cdot N^{1-\delta(k)}$$

A. Zaccagnini. *On the exceptional set for the sum of a prime and a k-th power.* Mathematika 39 (1992), 400–421.

Remark. The proof is based on methods discovered by R. Brünner, A. Perelli and J. Pintz

in Acta Math. Hung. 53 (1989), 347–365.

and H.L. Montgomery and R.C. Vaughan

in Acta Arith. 27 (1975), 353–370.

d) Let $k \geq 2$ be a natural number and $K = \prod p$, the product being extended over those primes p for which $\varphi(p)|k$. The number of positive integers $N_k \leq X$ with $(N-1, K) = 1$, not representable in the form $N = p_1 + p_2^k$ with primes p_1 and p_2, is of order X^γ, where $\gamma < 1$ always and $\gamma < 1 - 1/137k^3 \log k$ if k is sufficiently large.

V.A. Plaksin. *On a question of Hua Lookeng.* Math. Notes. 47 (1990), 278–286 (translation from Mat. Zametki 47 (1990), 78–90.

4) a) For infinitely many n the number of solutions of $n = p^2 + q^2$ (p, q primes) is greater than

$$\exp(c \log n / \log\log n) \qquad (c > 0)$$

P. Erdős. *On the sum and difference of square of primes.* I, II. J. London Math. Soc. 12 (1937), 133–136, 168–171.

b) The number of positive integers $n \leq x$, with at least one representation $n = p^2 + q^2$ (p, q odd primes) is

$$\frac{1}{2}\pi x \cdot (\log x)^{-2} \cdot (1 + O((\log\log x)^{2/3} \cdot (\log x)^{-2/3}))$$

G.J. Rieger. *Über die Summe aus einem Quadrat und einem Primzahlquadrat.* J. Reine Angew. Math. 231 (1968), 89–100.

§ IX. 7 Number of Goldbach numbers and related problems

1) a) Let $3/5 < \lambda < 1$, and A any large constant. The number of even numbers that can be represented as a sum of two odd prime ("Goldbach numbers") in the interval $[x, x + x^\lambda]$ is

$$\frac{x^{\lambda}}{2} + O(x^{\lambda}/\log^A x)$$

where the implied constant depends only on λ and A

Corollary. If $x \geq x_0(\lambda)$, then there exists a prime $p \in [x, x + x^{\lambda}]$ such that $p - 1$ and $p + 1$ are Goldbach numbers.

K. Ramachandra. *On the number of Goldbach numbers in small intervals.* J. Indian Math. Soc. (N.S.) 37 (1973), 157–170.

b) For all large x, the interval $(x, x + x^{7/72+\varepsilon})$ contains a sum of two prime numbers. Assuming the Riemann Hypothesis, this can be replaced by the interval $(x, x + c\log^2 x)$ ($c > 0$ constant.)

H.L. Montgomery and R.C. Vaughan. *The exceptional set in Goldbach's problem.* Acta Arith. 27 (1975), 353–370.

c) For all sufficiently large N there are primes p_1, p_2 with $|p_1 + p_1 - N| < (\log N)^A$, for a certain constant A

K. Prachar. *Über eine Anwendung einer Methode von Linnik.* Acta Arith. 29 (1976), 367–376.

d) If $\theta > 7/72, x > x_0(\theta)$, then the number of natural numbers in the interval $[x, x + x^{\theta}]$ which equals the sum of two odd prime numbers exceeds $c \cdot x^{\theta}$ ($c > 0$, absolute constant.)

K. Ramachandra. *Two remarks in prime number theory.* Bull. Soc. Math. France 105 (1977), 433–437.

2) Given $\theta > \frac{3}{4}$, there exist n_0 and δ (which depend at most on θ) such that each $n > n_0$ is representable in the form $n = p + ab$, p prime and $a, b \in [1, n^{1/2-\delta}]; a \cdot b \leq n^{\theta}$

D.R. Heath-Brown. *Representation of an integer of a prime plus a product of two small factors.* Math. Proc. Cambridge Phil 89 (1981), 29–33.

3) Assuming the Extended Riemann Hypothesis, for a given natural number $N > 1$ and for a given $q \ll N/\log^2 N$ (q natural number) there is a Goldbach number $N + mq$ with $m \leq \log^2 N$

K. Prachar. *Bemerkungen über Primzahlen in kurzen Reihen.* Acta Arith. 44 (1984), 175–180.

4) Let q, N be natural numbers, $q \leq \dfrac{N}{c(\log N)^2}$ and N is even if q is even (c = constant.) Under the Generalized Riemann Hypothesis, the equation $N = p_1 + p_2 + hq$ has always a solution in primes p_1, p_2 and integer h satisfying $0 \leq h \leq c \cdot (\log N)^2$, when N is sufficiently large.

Y. Wang and Z. Shan. *A conditional result on Goldbach's problem.* Acta Math. Sinica (N.S.) 1 (1985), 72–78.

§ IX. 8 The exceptional set in Goldbach's problem

1) For any $C > 0$ there are at most $O(x/\log^C x)$ even integers $\le x$ which arc not the sum of two primes.

J.G. van der Corput. *Sur l'hypothèse de Goldbach pour presque tous les nombres pairs.* Acta Arith. 2 (1937), 266–290;
N.G. Čudakov. *On the density of the set of even numbers which are not representable as a sum of two odd primes.* Izv. Akad. Nauk. SSSR Ser. Mat. 2 (1938), 25–39;
T. Estermann. *On Goldbach's problem: Proof that almost even positive integers are sums of two primes.* Proc. London Math. Soc. (2) 44 (1938), 307–314.

and

H. Heilbronn. Zentralblatt, 16 (1937), 291–292.

2) Let $E(x)$ be the number of even integers $\le x$ and > 4, that are not the sum of two odd primes. Then:

a) $E(x) = O(x \exp(-(\log x)^{1/2}))$

R.C. Vaughan. *On Goldbach's problem.* Acta Arith. 22 (1972), 21–48.

b) $E(x) < x^{1-\delta}$
where $\delta > 0$ is an effective constant.

H.L. Montgomery and R.C. Vaughan. *The exceptional set in Goldbach's problem.* Acta Arith. 27 (1975), 353–370.

c) Relation b) holds with $\delta = \dfrac{1}{100}$

J.-R. Chen. *The exceptional set of Goldbach numbers.* I. Sci. Sinica 23 (1980), 416–430.

d) Let $g_1 < g_2 < \cdots$ be the sequence of Goldbach numbers (even numbers which are sum of two primes). Then

$$\sum_{g_n \le x} (g_{n+1} - g_n)^3 \ll x(\log x)^{300}$$

H. Mikawa. *On the intervals between consecutive numbers that are sums of two primes.* Tsukuba J. Math. 17 (1993), 443–453.

3) Let $E_\theta(x)$ be the number of even integers $2n \le x$ that cannot be represented in the form $p_1 + p_2$ with $|n - p_i| < n^\theta$.
There exist effectively computable constants $\theta < 1$ and $\delta > 0$ such that

$$E_\theta(x) \le x^{1-\delta}$$

É. Fouvry. *Un résultat nouveau en Théorie additive des nombres premiers.* Séminaire de Théorie des Nombres, 1975–76 (Univ. Bordeaux 1, Talence), Talence, 1976.

4) Let E denote the set of those even numbers which cannot be written as a sum of two primes, and let $E_1 = E \cap N_1$, with

$N_1 = \{n = 2ab : a < \log n, p|b \Rightarrow p \geq b^{1/3}\}$ $(a, b, n \in \mathbb{N}, p$ prime.)

Then there exists a set N_2 of even numbers with $N_2(x) \ll_\varepsilon x^{2/3+\varepsilon}$ for any $\varepsilon > 0$ such that

$$\sum_{n=p_1+p_2} 1 = c(n) \cdot \frac{n}{\log^2 n} + O\left(\frac{n}{\log^{11/5} n}\right)$$

if $n \in N_1 \backslash N_2$

(Thus $E_1(x) \ll x^{2/3+\varepsilon}$). Here $E_1(x)$ denotes the counting function of the set E_1

J. Pintz. *A note on the exceptional set in Goldbach's problem.* Colloque de Théorie Analytique des Nombres ≪Jean Coquet≫ (Marseille, 1985), 101–115. Publ. Math. Orsay, 88–02, Univ. Paris XI, Orsay, 1988.

5) a) $E(x + H) - E(x) \ll_A H \cdot (\log x)^{-A}$

if $x^{\frac{7}{36}+\varepsilon} \leq H \leq N$ and A is any positive number.

A. Perelli and J. Pintz. J. London Math. Soc. II. Ser. 47 (1993), 41–19.

b) The above result is true for $x^{\frac{7}{48}+\varepsilon} \leq H \leq N$

H. Mikawa. *On the exceptional set in Golbach's problem.* Tsukuba J. Math. 16 (1992), 513–543.

c) Assume that the generalized Riemann Hypothesis is true. If $H/\log^6 x \to \infty$, then all even integers in the interval $(x, x + H)$ with at most $\ll H^{1/2} \log^3 x$ exceptions, are sums of two primes.

J. Kaczorowski, A. Perelli and J. Pintz. *A note on the exceptional set for Goldbach's problem in short intervals.* Monatsh. Math. 116 (1993), 275–282.

§ IX. 9 Partitions into primes

1) a) 6 is the largest integer which is not representable as a sum of distinct primes.

H.-E. Richert. *Über Zerfällungen in ungleiche Primzahlen.* Math. Z. 52 (1949), 342–343.

b) 17163 is the largest integer which is not representable as a sum of distinct squares of primes.

R.E. Dressler, L. Pigno and R. Young. *Sums of squares of primes.* Nordisk Mat. Tidskrift 24 (1976), 39–40.

2) Let $y(n) = 3 + \sum_{k=4}^{n} p_k$ (where p_k is the k-th prime) and put

$U(y, c) = \sqrt{2y} \cdot \sqrt{\log \sqrt{2y} + c \log\log \sqrt{2y}}$. Then:

a) For $0 < \varepsilon \leq 1, n$ sufficiently large and $y(n-1) < y \leq y(n), y$ can be partitioned into distinct primes p satisfying

$$p < U\left(y, \frac{1}{2} + \varepsilon\right)$$

J. Riddell. *Partitions into distinct small primes.* Acta Arith. 41 (1982), 71–84.

3) a) Let $P(n)$ denote the number of partitions of the integer n into primes, repetitions being allowed. Then

$$\lim_{n\to\infty} (P(n+1) - P(n)) = +\infty$$

P.T. Baleman and P. Erdős. *Monotony of partition functions.* Mathematika 3 (1956), 1–14.

b) $P(n+1) \geq P(n)$
for all $n = 1, 2, 3, \ldots$

P.T. Bateman and P. Erdős. *Partitions into primes.* Publ. Math. Debrecen 4 (1956), 198–200.

c) $P(n+1) > P(n)$
for $n \geq 8$

J. Browkin. *Sur la décomposition des nombres naturels en sommes de nombres premiers.* Colloq. Math. 5 (1957), 205–207.

4) a) Every integer greater than 55, 121, 161, 205 is a sum of distinct primes of the form $4k-1, 4k+1, 6k-1, 6k+1$, respectively. Furthermore, these lower bounds are best possible.

A. Mąkowski. *Partitions into unequal primes.* Bull. Acad. Polon. Sci. Sér. Math. Astronom. Phys. 8 (1960), 125–126 (Russian.)

b) Every integer greater than 1969, 1349, 1387, 1475 is a sum of distinct primes of the form $12k+1$, $12k+5$, $12k+7$,
$12k+11$, respectively. Furthermore, these lower bounds are best possible.

R.E. Dressler, A. Mąkowski and T. Parker. *Sums of distinct primes from congruence classes modulo* 12. Math Comp. 28 (1974), 651–652.

5) Let $f(n)$ be the smallest integer so that every $m > f(n)$ is the sum of n distinct primes or squares of primes where a prime and its square are not both used. Then

$$f(n) < \sum_{i=2}^{n+1} p_i + C$$

where p_i is the i-th prime and C an absolute constant independent of n

P. Erdős. *On a problem of Sierpiński.* Acta Arith. 11 (1965), 189–192.

6) Let $g(n)$ be the number of ways of representing n as the sum of one or more consecutive primes. Then

$$\lim_{x\to\infty} \frac{1}{x} \cdot \sum_{n\leq x} g(n) = \log 2$$

L. Moser. *Notes on number theory,* III. On the sum of consecutive primes. Canad. Math. Bull. 6 (1963), 159–161.

§ IX.10 Partitions of n into parts, or distinct parts in a set A

1) a) Let $P^d(n)$ be the number of partitions of n into distinct primes. Then

$$\log P^d(n) = \pi\left(\frac{2}{3}\cdot\frac{n}{\log n}\right)^{1/2}\cdot\left(1+O\left(\frac{\log\log n}{\log n}\right)\right)$$

K.F. Roth and G. Szekeres. *Some asymptotic formulae in the theory of partitions.* Quart. J. Math. Oxford Ser. (2) 5 (1954), 241–259.

b) Let $P_A(n)$ and $P_A^d(n)$ be the number of partitions of n into parts, and distinct parts in A, respectively, where A is a given set of positive primes. Then

$$\log P_A(n) \sim 2\pi\sqrt{n/(3\log n)}$$

and assuming that $P_A^d(n)$ is monotonic increasing for $n > n_0$,

$$\log P_A^d(n) \sim \pi\left(\frac{2}{3}\cdot\frac{n}{\log n}\right)^{1/2}$$

S.M. Kerawala. *On the asymptotic values of* log $P_A(n)$ *and* log $P_A^d(n)$ *with A as a set of primes.* J. Natur. Sci. Math. 9 (1969), 209–216.

2) Let $T(n; m, k)$ be the number of partitions of n into k powers of prime numbers not exceeding m, where $m \le n^{1/k}(k \ge 1)$.
Then

$$T(n; m, k) \sim (2\pi A_2)^{-1/2}\cdot\exp(n\alpha + A_1)$$

as $n, m \to \infty$, where α is the root of the equation

$$n = \sum_{p\le m} p^k\cdot(\exp(\alpha p^k) - 1)^{-1}$$

and $A_1, A_2 > 0$, constants. (Here $n\alpha \to \infty$ as $m, n \to \infty$.)

T. Mitsui. *On the partitions of a number into the powers of prime numbers.* J. Math. Soc. Japan 9 (1957), 428–447.

3) Let $Q(x)$ be the number of integers $\le x$ of the form $p_1^{\alpha_1}\dots p_r^{\alpha_r}$ where $\alpha_1 \ge \dots \ge \alpha_r$ and p_i is the i-th prime. Then

$$\log Q(x) = \frac{2\pi}{\sqrt{3}}\cdot\left(\frac{\log x}{\log\log x}\right)^{1/2}\cdot(1-(2\log\pi+12B_1/\pi^2-2)/(2\cdot\log\log x) -$$
$$-(\log 3 - \log\log\log x)/(2\log\log x) + O(((\log\log\log x)/(\log\log x))^2)$$

where

$$B_1 = -\int_0^{+\infty}\log(1-e^{-y})\log y\, dy$$

L.B. Richmond. *Asymptotic results for partitions.* I. The distribution of certain integers. J. Number Theory 8 (1976), 372–389.

Note. For other results of this type see also

P. Erdős and L.B. Richmond. *Concerning periodicity in the asymptotic behavior of partition functions.* J. Austral. Math. Soc. A21 (1976), 447–456.

§ IX.11 Representations in the form $k = ap_1 + \cdots + a_r p_r$ (p_i primes) with restricted primes p_i

1) Let $A(N, k)$ be the number of representations of an integer k in the form $k = a_1 p_1 + \ldots a_r p_r$ by means of primes subject to the inequalities $p_1 \leq \beta_1 \cdot N, \ldots, p_r \leq \beta_r \cdot N$ (Here $a_i \geq 0, r \geq 3, \beta_i > 0$.) Then

$$A(N, k) = R(N, k) \cdot S(k) + (\rho/N) \cdot (N/\log N)^r$$

where ρ tends to zero (uniformly in k as $N \to \infty$) more rapidly than any negative power of $\log N$, and

$$S(k) = \sum_{q=1}^{\infty} \frac{\varphi(q) \cdot \mu(q/(q,k))}{\varphi(q/(q,k))} \cdot \prod_{\upsilon=1}^{r} \frac{\mu(q/(q,a_\upsilon))}{q/(q,a_\upsilon))}$$

$$R(N, x) = \int_{-\infty}^{+\infty} e^{-2\pi i x z} \prod_{\upsilon=1}^{z} \int_{\sqrt{N}}^{\beta_\upsilon \cdot N} e^{2\pi i a_\upsilon z t} \cdot (\log t)^{-1}\, dt\, dz$$

2) Let $n \geq 3$, $K \geq 1$ be given and let $b_1, \ldots, b_m$ be relatively prime to k (with fixed $m, K, b_1, \ldots, b_m$). Let $N(n)$ be the number of solutions in odd primes $p_r \equiv b_r \pmod K$ of the equation $n = p_1 + \cdots + p_m$. Then there is a singular series $S(n)$ such that for all large $n \equiv m \pmod 2$ such that $n \equiv b_1 + \cdots + b_m \pmod K$, we have

$$N(n) = \frac{n^{m-1}}{(\log n)^m} \cdot \frac{S(n)}{(m-1)!(\varphi(K))^m} \cdot (1 + O(\log\log n/\log n))$$

and $0 < c < S(n)/K < c'$ for suitable absolute constants c and c'

A. Zulauf. *Beweis einer Erweiterung des Satzes von Goldbach-Vinogradov.* J. Reine Angew. Math. 190 (1952), 169–198.

§ IX.12 Representations in the form $N = p + n$, p prime, with certain restrictions on n

Let k and l we positive integers; let $a_1, \ldots, a_k$ be distinct non-zero integers; let $a_{k+1}, \ldots, a_{k+l}$ be distinct integers.

Put $\Phi(\alpha_1, \beta_1) = \dfrac{1}{\sqrt{2\pi}} \displaystyle\int_{\alpha_i}^{\beta_i} \exp(-x^2/2)\, dx$ for $\alpha_i < \beta_i$ $(i = 1, \ldots k+1)$.

If $A(N) = A(n; a_1, \ldots, a_{k+l}; \alpha_1, \beta_1, \ldots, \alpha_{k+l}, \beta_{k+l})$ denotes the number of representations of N as the sum of the form $N = p + n$, where p is prime, and n is a positive integer such that

$$\log\log N + \alpha_i \sqrt{\log\log N} < \omega(p + a_i) < \log\log N + \beta_i \sqrt{\log\log N}$$

for $i = k+1, \ldots, k+l$ simultaneously. Then

$$A(N) \sim \frac{N}{\log N} \cdot \prod_{i=1}^{k+l} \Phi(\alpha_i, \beta_i)$$

M. Tanaka. *Some results on additive number theory.* V. Tokyo J. Math. 12 (1989), 457–473.

§ IX.13 On integers of the form $p + a^k$ (p prime, $a > 1$) or $p^2 + a^k$ or $p + q!$ (q prime), etc.

1) a) The density of numbers of the form $p + a^k$ (p prime, $a > 1$, integer) is positive.

N.P. Romanoff. *Ueber einige Satze der additiven Zahlentheorie.* Math. Ann. 109 (1934), 668–678.

b) Let $a_1 < a_2 < \ldots$ be a sequence of positive integers with the property that $a_k \mid a_{k+1}$ for all k. The density of numbers of the form $p + a_k$ (p prime) is positive if there exists an absolute constant c such that

$$a_k < c^k$$

and

$$\sum_{d \mid a_k} (1/d) < c$$

$k = 1, 2, \ldots$

P. Erdős. *On integers of the form $2^k + p$ and some related problems.* Summa Brasil. Math. 2 (1950), 113–123.

c) Let $N(x)$ denote the number of odd positive integers $n \leq x$ which are not of the form $n = 2^k + p$ (p prime). Then

$$\limsup_{x\to\infty} \frac{N(x)}{x} < 1$$

(See N.P. Romanoff.)

d) $\liminf_{x\to\infty} \frac{N(x)}{x} > 0$

J.G. van der Corput. *On the Polignac's conjecture* (Dutch). Simon Stevin 27 (1950), 99–105.

e) The number of $n \leq x$, that can be represented in the form $n = 2^k + p^2$ (p prime) is

$$\frac{2}{\log 2} \cdot x^{1/2} + o(x^{1/2} \cdot \log\log x / \log x)$$

M. Orazov. *Squares of prime numbers and powers of two* (Russian). Taškent Gos. Univ. Naučn. Trudy No. 548 Voprosy Mat. (1977), 67–68, 143.

2) Let $f(n)$ denote the number of solutions of

$$p + a^k = n \ (p \text{ prime}, \alpha > 1, \text{ integer})$$

For infinitely many n one has

$$f(n) > c \log\log n$$

(See P. Erdős.)

3) The number of natural numbers $n \leq N$ such that the equation $n = p + a^2$ (p prime, $a \geq 1$) has no solution is:

a) $O(N/\log^A N)$

R.J. Miech. *On the equation $n = p + x^2$*. Trans. Amer. Math. Soc. 130 (1968), 494–512.

b) $O(N \exp(-a(\log N)^{1/2}))$
and

$$\sum_p \sum_{\substack{m \\ n=p+m^2}} \log p \sim n^{1/2} \cdot \prod_p \left(1 - \left(\frac{n}{p}\right) /(p-1)\right)$$

for those numbers $n \leq N$ not included in an exceptional set of this order of magnitude ($a > 0$ constant).

I.V. Polyakov. *On the exceptional set of the sum of prime and the square of an integer* (Russian). Izv. Akad. Nauk. SSSR. Ser. Mat. 45 (1981), 1391–1423.

c) $\ll N^{1-\gamma}$
where $\gamma > 0$ and the comnstant implied by $\ll$ are effectively calculable.

I.V. Polyakov. *Sum of a prime and a square.* Math. Notes 47 (1990), 373–380 (translation from Mat. Zametki 47 (1990), 90–99).

4) Let $A(x)$ denote the number of $n \leq x$ with $n = pa^2$ (p prime). Then

$$A(x) = \frac{\pi^2}{6} \cdot \frac{x}{\log x} + O\left(\frac{x}{\log^2 x}\right)$$

for $x \geq 2$

E. Cohen. *Arithmetical notes. IX. On the set of integers representable as a product of a prime and a square.* Acta Arith. 7 (1961/62), 417–420.

5) A positive integer N is called a Hardy-Littlewood (H-L) number if it is a sum of a prime and a square.
There is an H-L number

$$N \equiv a \,(\mathrm{mod}\, D), \quad N \leq D^{3/2+\varepsilon}$$

for any $1 \leq a < D$

Z.Kh. Rakhmonov. *The distribution of Hardy-Littlewood numbers in arithmetic progressions* (Russian). Izv. Akad. Nauk. SSSR. Ser. Mat. 53 (1989), 211–224; translation in Math. USSR-Izv. 34 (1990), 213–228.

6) The number of integers $n \le x$, which have the form $n = p + q!\,(p, q$ primes) is

$$x/\log\log^2 x + 2x \log\log\log x/\log\log^3 x + O(x \log\log^{-3} x)$$

M. Orazov. *Some applications of the Romanov-Erdős inequality* (Russian). Izv. Akad. Nauk Turkmen. SSR Ser. Fiz.-Tekhn. Knim. Geol. Nauk 1978, 3–9.

7) There are infinitely many odd integers not of the form

$$n = 2^k + 2^l + p,$$

where p is prime.

R. Crocker. *On the sum of a prime and of two powers of two.* Pacific. J. Math. 36 (1971), 103–107.

§ IX.14 Linnik's theorem (on the Hardy-Littlewood problem)

1) a) For all sufficiently large n,

$$n = p + a^2 + b^2$$

(p prime)

Yu.V. Linnik. *Hardy-Littlewood problem on the representation as the sum of a prime and two squares* (Russian). Dokl. Akad. Nauk SSSR, 124 (1959), 29–30.

b) Let $Q(n)$ be the number of representations of n as a sum of a prime and two squares.

$$Q(n)=\frac{\pi n}{\log n}\cdot\prod_p\left(1+\frac{\chi_4(p)}{p(p-1)}\right)\cdot\prod_{p|n}\frac{(p-1)(p-\chi_4(p))}{p^2-p-\chi_4(p)}+O(n/(\log n)^{1.028}),$$

where χ_4 is the non-principal character modulo 4.

Yu.V. Linnik. *An asymptotic formula in an additive problem of Hardy and Littlewood* (Russian). Izv. Akad. Nauk. SSSR. Ser. Mat. 24 (1960), 629–706.

c) For $0 < a < b \le 2\pi$ let $Q'(n)$ be the number of solutions (p, x, y) of $n = p + x^2 + y^2$ with $a < \tan^{-1} x/y < b$. Then for some constant c:

$$Q'(n) = \frac{1}{2}\pi^{-1}(b-a)Q(n)(1 + O((\log\log n)^c/(b-a)(\log n)^{1-2/\pi})),$$

where $Q(n)$ is defined in b).

M.B. Barban and B.V. Levin. *Multiplicative functions on shifted prime numbers* (Russian). Dokl. Akad. Nauk SSSR, 181 (1968), 778–780.

2) Let $Q_1(n)$ be the number of solutions of $p - x^2 - y^2 = a$, where x, y run over integers with $0 < x^2 + y^2 \le n$ and a is a fixed integer (p runs over primes). Then

$$Q_1(n) = \frac{\pi n}{\log n} \cdot \prod_{p>2}\left(1 + \frac{\chi_4(p)}{p^2 - p}\right) \cdot \prod_{p|a,\, p\equiv 1 (\mathrm{mod}\, 4)} (p-1)^2/(p^2-p+1) \cdot$$
$$\prod_{p|a,\, p\equiv 3(\mathrm{mod}\, 4)} (p^2-1)/(p^2-p-1) + O(n/(\log n)^{-1.042}),$$

where χ_4 is the non-principal character modulo 4.

B.M. Bredihin. *Binary additive problems of indeterminate type.* II. Analogue of the problem of Hardy and Littlewood. (Russian). Izv. Akad. Nauk. SSSR. Ser. Mat. 27 (1963), 577–612.

3) The number of solutions of $n = p + x_1^2 + x_2^2 + x_3^2$ (x_i integers, p prime) is

$$\frac{2(\pi n)^{3/2}}{3\Gamma(3/2)\log n} \cdot \prod_{\substack{p \nmid n \\ p \geq 3}} \left(1 - \frac{(-n/p)}{p(p-1)}\right) + O\left(\frac{n^{3/2} \cdot \log\log n}{\log^2 n}\right)$$

M.A. Subhankulov. *Additive proprieties of certain sequences of numbers* (Russian). Issled. po mat. analizu i mehanike v Uzbekistane, pp. 220–241. Taškent, 1960.

4) a) The number of solutions of $n = x^2 + y^2 + p_1 p_2$, where p_1, p_2 run through all primes such that $p_i > \exp(\log\log n)^2$ ($i = 1, 2$) is

$$\sim \pi n \frac{\log\log n}{\log n} \cdot \prod_{p|n} \frac{(p-1)(p-\chi_4(p))}{p^2 - p + \chi_4(p)} \prod_p \left(1 + \frac{\chi_4(p)}{p^2 - p}\right)$$

b) The number of solutions of $n = x^2 + y^2 + N$, where all the prime divisor of N are greater than $n^\alpha (\alpha > 0)$, in

$$\ll \frac{n}{\log n} \cdot \prod_{p|n} \frac{(p-1)(p-\chi_4(p))}{p^2 - p + \chi_4(p)}$$

Yu.V. Linnik. *The dispersion method in binary additive problems.* Amer. Math. Society, 1963.

5) Let $K_\sigma, a_\sigma, b_\tau$ be fixed positive integers with $(K_\sigma, a_\sigma) = 1$ for $\sigma = 1, 2, \ldots, s$ and $\tau = 1, 2, \ldots, t$. Let $s \geq 1, t > 0$ and $w = s + \frac{t}{2} \geq 2$. Then the number $N_{s,t}(h)$ of solutions in odd primes $p_\sigma \equiv a_\sigma (\mathrm{mod}\, k_\sigma)$ and integers g_τ of the equation

$$n = p_1 + \ldots + p_s + b_1 g_1^2 + \ldots + b_t g_t^2$$

is given by

$$N_{s,t}(n) = \frac{n^{w-1}}{(\log n)^s} \cdot \sigma_{s,t}(n) + O\left(\frac{n^{w-1} \cdot \log\log n}{(\log n)^{s+1}}\right)$$

A. Zuhlauf. *Zur additiven Zerfällung natürlicher Zahlen in Primzahlen un Quadrate.* Arch. Math. 3 (1952), 327–333.

Note. For related problems, see also

A. Zuhlauf. *On sums and differences of primes and squares.* Compositio. Math. 13 (1958), 103–112.

6) The number $f(n, k)$ of representations of an integer n as a sum of two primes and k powers of 2 satisfies

$$f(N,k) > N\left(\frac{\log N}{\log 2}\right)^{k-1} \cdot (c_1 - c_2 \cdot (1-\eta)^{k-2}),$$

where c_1, c_2 and $\eta < 1$ are absolute constants and $k \geq 3$.

Yu.V. Linnik. *Prime numbers and powers of* 2 (Russian). Trudy. Mat. Inst. Steklov., V.38 Izdat. Akad. Nauk. SSSR Moskow (1951), 152–169.

§ IX.15 Representations in the form $p_1^3 + p_2^3 + p_3^3 + x^3$ (p_i primes), etc.

1) a) Almost all positive integers n are representable in the form $n = p_1^3 + p_2^3 + p_3^3 + x^3$ (p_i-primes); $x = 1, 2, \ldots$

K.F. Roth. *On Waring's problem for cubes.* Proc. London Math. Soc. (2) 53 (1951), 268–279.

b) Every sufficiently large positive integer n may be represented in the form $n = p_1^3 + \cdots + p_7^3 + x^3$ (p_i-primes); $x = 1, 2, \ldots$ (See K.F. Roth).

2) Let $S(n)$ be the number of pairs (p, q) of odd primes, $p, q \leq \sqrt{n/2}$ such that $n - p^2 - q^2$ is representable as $x^2 + y^2 (x, y$ integers). Then, for $n \not\equiv 0,1,5 (\text{mod } 8)$,

$$S(n) > a \cdot n(\log n)^{-5/2} \cdot (1 + O(\log\log n \cdot (\log n)^{-1/10})$$

where $\alpha > 0$ is a constant (explicitely given).

G. Greaves. *On the representation of a number in the form $x^2 + y^2 + p^2 + q^2$ (p, q-odd primes).* Acta Arith. 29 (1976), 257–274.

§ IX.16 Number of solutions of $n = p + xy$ (p prime; $x, y \geq 1$)

a) The number of solutions of $n = p + xy$ (p prime, $x, y = 1, 2, 3\ldots$) is

$$\frac{315\zeta(3)}{2\pi^4} \cdot \prod_{p|n} \frac{(p-1)^2}{p^2 - p + 1} \cdot n + O\left(\frac{n}{\log^{1-\varepsilon} n}\right)$$

B.M. Bredihin. *Applications of the dispersion method in binary additive problems* (Russian). Dokl. Akad. Nauk SSSR, 149 (1963), 9–11.

b) The number of solutions of $l = p_1 p_2 \ldots p_k - xy$, p_j prime, $p_j \leq n^{\alpha_j}$ $(j = 1,2,\ldots k)$, $0 < \alpha_1 < \alpha_2 \leq \ldots \leq \alpha_k \leq 1/7$ and $\alpha_1 + \alpha_2 + \cdots + \alpha_k = 1$, $xy \leq n$ is

$$315\pi^4 \cdot \frac{1}{\alpha_1\alpha_2\ldots\alpha_k} \cdot \zeta(3) \cdot \sum_{p|\ell} \frac{(p-1)^2}{p^2 - p + 1} \cdot \frac{n}{\log^{k-1} n} + O(n/\log^k n)$$

A.K. Karšiev. *The generalized problem of Titchmarsh divisors.* Izv. Akad. Nauk Uz. SSR Ser. Fiz.-Mat. Nauk 13 (1969), 69–70.

§ IX.17 Representations of primes by quadratic forms

a) There exist imfinitely many primes p for which there are integers k and m with

$$p = k^2 + m^2, m = O(p^{25/64})$$

I.P. Kubilius. *On some problems of the geometry of prime numbers.* Mat. Sb. 31 (1952), 507–542.

b) Assuming the Extended Riemann Hypothesis on certain Hecke series, there exist infinitely many primes p for which there are integers k and m with

$$p = k^2 + m^2, m = O(\log p)$$

N.C. Ankeny. *Representations of primes by quadratic forms.* Amer. J. Math. 74 (1952), 913–919.

c) There exist infinitely many pairs of primes p_1, p_2 for which there are integers k and m with

$$p_1 p_2 = k^2 + m^2, m = O(\log p_1 p_2)$$

I.P. Kubilius and Yu.V. Linnik. *An elementary theorem on the theory of prime numbers* (Russian). Uspehi Mat. Nauk. (N.S.) 11 (1956), 191–192.

§ IX.18 Number of solutions of $m = p_1 + v^a$, $n = p_2 + v^a$, $(m < x, n < x, p_i$ primes)

a) Let $A_1(x, a)$ denote the number of solutions of the equations $m = p_1 + v^a$, $n = p_2 + v^a$, with $m < x, n < x; p_1$ and p_2 primes, a and v positive integers with $a > 1$. Then there exist constants $c_1(a)$ and $c_2(a)$ such that for $x > c_1(a)$

$$A_1(x, a) > c_2(a) \cdot x^2$$

b) Let $A_2(x, a)$ be the number of solutions of the equations $m = p_1 + a^v, n = p_2 + a^v$, with $m < x, n < x; p_1$ and p_2 primes, a and v positive integers with $a > 1$. Then there exist constants $c_1'(a)$ and $c_2'(a)$ such that for $x > c_1'(a)$

$$A_2(x, a) > c_2'(a) x^2 \log x$$

G.J. Rieger. *On linked binary representations of pairs of integers: some theorems of the Romanov type.* Bull. Amer. Math. Soc., 69 (1963), 558–563.

§ IX.19 Number of representations of n as the sum of the square of a prime and an r-free integer

Let $Q_k(n)$ be the number of representations of n as the sum of the square of a prime and a k-free positive integer ($k \geq 2$). Let $C_k(n) = 1$ or 2 for $k = 2$, $n \equiv 1(\text{mod}4)$ or $k \geq 3, n \equiv 1(\text{mod}8)$, respectively, and put $C_k(n) = 0$, otherwise. Then

$$Q_k(n) = A_k(n) \cdot li\sqrt{n} + O(\sqrt{n}/\log^H n)$$

for all $H > 0$, where

$$A_k(n) = (1 - C_k(n)/2^{k-2}) \prod_{p \nmid 2n} \left(1 - \frac{1 + n/p}{p^{k-1}(p-1)}\right)$$

with the O-constant depending only on k and H

S. Uchiyama. *On the number of representations of an integer as the sum of the square of a prime and an r-free integer.* J. Fac. Sci. Shinsu Univ. 5 (1970), 141–146.

§ IX.20 Distinct integers $\leq x$ which can be expressed as $p + a^{k_i}$, where (k_i) is a certain sequence

1) Let $k_0, k_1, k_2, \ldots$ be a strictly increasing sequence of non-negative integers, $a > 1$, a fixed positive integer, $k(x)$ the number of $a^{k_i} \leq x$. Then the number of distinct integers $\leq x$ which can be expressed in the form

$$p + a^{k_i}$$

is greater than $cx\, k(x)/\log x$, $x > 2a^{k_0}$ (where $c = c(a)$ is a constant).

K. Prachar. *Über einen Satz der additiven Zahlentheorie.* Monatsh. Math. 56 (1952), 101–104.

2) Let $(b)_j$, $j = 1, 2, \ldots$ be a finit or infinit sequence of integers satisfying $3 \leq b_1 < b_2 < b_3 \ldots$ and $\sum 1/b_j < \infty$. Let $1 = d_1 < d_2 < d_3 < \ldots$ be the sequence of all integers d_i ($i = 1, 2, 3, \ldots$) which are not divisible by any b_j. Let $B(x) = \sum_{b_j \leq x} 1$ and suppose that $B(x) = o(x/\log x \cdot \log\log x)$. Then the number of solutions for any fixed $n \geq n_0$ of the equation $n = p + d_j$ (p-prime) is $\gg \dfrac{n}{\log n}$, and in particular ≥ 1.

P. Erdős, G.J. Babu and K. Ramachandra. *An asymptotic formula in additive number theory.* II. J. Indian Math. Soc. 41 (1977), 281–291.

§ IX.21 Waring-Goldbach-type problems for the function $f(x) = x^c$, $c > 12$. Hybrid of theorems by Vinogradov and Pjateckiĭ-Šapiro

1) If $1 < \sigma < \frac{12}{11}$, then there exist $c_1, c_2 > 0$ (constants depending at most on σ) such that for any natural number $N > c_1$ there exist at least $c_2 \cdot \frac{N^{1/\sigma}}{\log N}$ primes of the form $N - [a^\sigma]$, with natural numbers a.

Corollary. Every sufficiently large integer N can be written in the form $N = p + [a^\sigma]$ for some prime p and natural number a.

G.J. Rieger. *Über ein additives Problem mit Primzahlen.* Arch. Math. (Basel) 21 (1970), 54–58.

2) For $c > 12$, $c \in \mathbb{R}\backslash\mathbb{I}$, $s, N \in \mathbb{I}$ let $R_S(N)$ be the number of ways of writing N as $N = [p_1^c] + \cdots + [p_S^c]$, p_i primes.
Let $G'(c)$ be the least integer s such that $R_S(N) > 0$ if $N > N_0(s, c)$.

a) $$R_s(N) \sim \frac{\left(\Gamma\left(\frac{1}{c}\right)\right)^s}{\Gamma\left(\frac{s}{c}\right)} \cdot \frac{N^{\frac{1}{c}-1}}{\log^s N}$$

if $s > 1650c^3 \log c$

b) $$\limsup_{c \to \infty} \frac{G'(c)}{c \log c} \le 4$$

Ph. Toffin. *Probleme de Waring-Goldbach pour la fonction $f(x) = x^c$ lorsque $c > 12$, c non entier.* C.R. Acad. Sci. Paris Sér. A–B, 280 (1975), Ai, A755–A757.

Remark. Relation b) improves a result of Pjateckiĭ-Šapiro (where is 8 in place of 4)

I.I. Pjateckiĭ-Šapiro. *On a variant of the waring-Goldbach problem* (Russian). Mat. Sb. (N. S.) 30 (72) (1952), 105–120.

c) Corresponding to each $c \in \left(1, \frac{15}{14}\right)$, and each $N > N_0(c)$, the inequality

$$|p_1^c + p_2^c + p_3^c - N| < N^{-\alpha} \cdot \log^\beta N$$

is soluble in primes p_1, p_2, p_3, where $\alpha = (15/14 - c)/c$ and $\beta = 9$

D.I. Tolev. *On a diophantine inequality involving prime numbers.* Acta Arith. 61 (1992), 289–306.

3) a) For any fixed $\frac{20}{21} < \gamma \leq 1$ the primes p of the form $[n^{1/\gamma}]$ have the property that every sufficiently large odd integer can be written as the sum of three of them.

b) For any fixed $\frac{8}{9} < \gamma \leq 1$ every sufficiently large odd integer can be written as the sum of three primes one of which is the form $[n^{1/\gamma}]$.

c) For any fixed $\frac{22}{25} < \gamma \leq 1$ the set of primes p satisfying $\{p^\gamma\} < p^{-2(1-\gamma)/3}$ has the property that every sufficiently large odd integer is the sum of three of them.

A. Balog and J.B. Friedlander. *A hybrid of theorems of Vinogradov and Pjateckiĭ-Šapiro. Pacific J. Math.* 156 (1992), 45–62.

§ IX.22 Integers not representable in the form $p + [n^c]$ $(c > 1)$

Let $c > 1$ be a constant and $E(x)$ the number natural numbers $N \leq x$ which cannot be represented in the form $N = p + [n^c]$ (p prime). Then

$$E(x) \ll x^{1-\delta}$$

with $\delta = \delta(c) > 0$

K. Buriev. *On an exceptional set in the Hardy-Littlewood problem for non-integral powers* (Russiaan). Mat. Zametki 46 (1989), 127–128.

Remarks: (i) In fact, $\delta \leq \frac{1}{c} \cdot 2^{-[c+1]}$ for $c < 100$ and $\delta \leq 1/\{10^3 \cdot c^3\}$ for $c > 100$

(ii) In an earlier paper, Buriev proved that the same estimate holds for any $\delta < 1/(2c^3 \cdot (\log c + 14))$

K. Buriev. *An additive problem with prime numbers* (Russian). Dokl. Akad. Nauk Tadzhik SSR, 30 (1987), 686–688.

§ IX.23 On the maximal distance between integers composed of small primes

1) Let $n_1^{(k)} < n_2^{(k)} < \ldots$ denote those positive integers that have at least one prime factor $p > k$ and let

$$f(k) = \max_i \left(n_{i+1}^{(k)} - n_i^{(k)}\right)$$

Then:

a) $f(k) \leq c \cdot \pi(k)$
for some constant $c > 1$

P. Erdős. *On consecutive integers.* Nieuw Arch. Wisk. (3) 3 (1958), 124–128.

b) $f(k) < \pi(k)$
except perhaps for a finite number of k

R. Tijdeman. *On the maximal distance of numbers with a large prime factor.* J. London Math. Soc. (2) 5 (1972), 313–320.

c) $f(k) \ll \dfrac{k \log\log\log k}{\log k \cdot \log\log k}$

T.N. Shorey. *On gaps between numbers with a large prime factor.* II. Acta Arith. 25 (1973/74), 365–373.

2) If $0 < n_1 < n_2 < \ldots$ is a sequence of integers composed of primes all $\leq p$, then there is an effectively computable constant $C = C(p)$ such that

$$n_{i+1} - n_i > n_i/(\log n_i)^C$$

$i = 1, 2, \ldots$

R. Tijdeman. *On integers with many small prime divisors.* Compositio Math. 26 (1973), 319–330.

3) For $r \geq 2$ let $p_1, p_2, \ldots, p_r$ be fixed primes, and let $1 = n_1 < n_2 < \ldots$ be the increasing sequence of all positive integers composed of these and only these primes. Then

$$r_{i+1} - n_i < n_i/(\log n_i)^B$$

for $n_i \geq N$, where $B > 0$ and N are effective constants.

R. Tijdeman. *On the maximal distance between integers composed of small primes.* Compositio Math. 28 (1974), 159–162.

§ IX.24 On the representation of N as $N = a + b$ or $N = a + b + c$ with restrictions on $P(ab)$ or $P(abc)$

1) a) If $N > N_0$ then N can be written in the form

$$N = a + b + c$$

where $P(abc) < N^{\varepsilon}$. (Here $P(n)$ denotes the greatest prime factor of n)

A. Fujii. *An additive problem in theory of numbers.* Acta Arith. 40 (1981), 41–49.

b) In a) we can take $P(abc) \leq \exp(3(\log N \cdot \log\log N)^{1/2})$

A. Balog and A. Sárközy. *On sums of integers having small prime factors.* I. Studia Sci. Math. Hung. 19 (1984), 35–47.

c) There exist absolute constants $M_0, c_1\ (>0)$ such that if $M > M_0$ and $\exp(5(\log M \cdot \log\log M)^{1/2}) \le y \le M^{1/3}$, then

$$n = a + b, \quad P(ab) \le y$$

can be solved for all but $c_1 \cdot \frac{M}{y} \exp(10\frac{\log M}{\log y} \cdot \log\log M)$ (in fact $< \frac{M}{\sqrt{y}}$) integers $n \le M$

d) If $N > N_1$ then N can be written in the form

$$a + b = N$$

where $P(ab) \le 2N^{2/5}$

A. Balog and A. Sárközy. Ibid. II. Studia Sci. Math. Hung. 19 (1984), 81–88.

e) Let $\varepsilon > 0$ be fixed number. There exists an $N_0 = N_0(\varepsilon)$ such that every integer $N > N_0$ can be expressed in the form

$$N = a + b$$

with $a > 1, b > 1$ and $P(ab) < N^{\alpha+\varepsilon}$, where $\alpha = 4/9\sqrt{e}$

A. Balog. *On additive representation of integers.* Acta Math. Hung. 54 (1989), 297–301.

2) a) Let A and B be sets of positive integers and suppose that for $x > x_0$, $A(x)B(x) \ll x^{24/13} \cdot \log^{42/13} x$ (where $A(x)$ is the counting function of A). Then there are integers $a \in A, b \in B, a \le x, b \le x$ and prime p such that

$$p^2 | (a + b)$$

$$p^2 \gg \frac{(A(x)B(x))^{5/2}}{x^4 \log^7 x}$$

A. Balog and A. Sárközy. *On sums of sequences of integers.* III. Acta Math. Hung. 44 (1984), 339–349.

Corollary. The "small" multiples of prime squares form a sum-intersective set H with $H(x) = O(\alpha(x)x^{1/2}\log^6 x)$ for any positive function $\alpha(x) \to \infty (x \to \infty)$. (The set H is called *sum-intersective* if H intersects $A + A$ for any set having positive upper density.)

b) The set

$$H = \left\{ mp^k : \underset{(p \text{ prime})}{m \le p}, \ k \ge 0.2(\log p)^{1/3} \cdot (\log\log p)^{-2/3} \right\}$$

is sum-intersective, and we have

$$H(x) = O(\exp(10(\log x)^{3/4} \cdot (\log\log x)^{1/2}))$$

A. Balog. *On sum-intersective sets.* Acta Math. Hung. 55 (1–2) (1990), 143–148.

§ IX.25 On the maximal length of two sequences of consecutive integers with the same prime divisors

1) Let $F_m(x)$ denote the number of vectors $(a_1, \dots, a_m)$ with positive integral components, $1 \le a_1 \le \dots \le a_m \le x$ and $\mathrm{Supp}(a_1) = \dots = \mathrm{Supp}(a_m)$, where $\mathrm{Supp}(n)$, $n \ge 2$ integer, denotes the set of prime factors of n. Then

$$F_m(x) = c_m \cdot x + O(x^{m/(m+1)+\varepsilon})$$

for any $\varepsilon > 0$, c_m a constant depending on m

H.N. Shapiro. *Integer vectors with interprimed components*. Math. Comp. 27 (1973), 455–462.

Note. For $m = 2$ this settles a problem of Erdős and Motzkin.

P. Erdős and T. Motzkin. *Advanced problem No. 5735*. Amer. Math. Monthly 77 (1970), 532.

2) Suppose that $\mathrm{Supp}(x+i) = \mathrm{Supp}(y+i)$ for $1 \le i \le k$, where x, y, k are positive integers. Then:

a) $\log k \le c_1(\log x \cdot \log\log x)^{1/2}$
for $x \ge 3$

b) $y - x > \exp(c_2 k(\log k)^2 \cdot (\log\log k)^{-1})$
for $k \ge 3$

c) $y - x > (k \log\log y)^D$
for $y \ge 27$, where $D = c_3 k(\log\log y) \cdot (\log\log\log y)^{-1}$. (Here c_1, c_2, c_3 are effective positive constants.)

R. Balasubramanian, T.N. Shorey and M. Waldschmidt. *On the maximal length of two sequences of consecutive integers with the same prime divisors*. Acta Math. Hungar. 54 (1989), 225–236.

§ IX.26 Representation of n as $n = \dfrac{p+1}{q+1}$ (p, q primes)

1) a) For every nonzero rational number r there is a constant $K = K(r)$ such that for infinitely many natural numbers a and b with at most K prime factors, we have

$$r = \frac{a+1}{b+1}$$

C. Badea. *Note on a conjecture of P.D.T.A. Elliot*. Arch. Math. (Brno) 23 (1987), 89–94.

Remark. Assuming the H Hypothesis of Schinzel, the above result holds without any condition on a and b.

A. Schinzel and W. Sierpinski. *Sur certaines hypothèses concernant les nombres premiers.* Acta Arith. 4 (1958), 185–208 (See, especially, pp. 191–192.)

b) If n is sufficiently large, then there is a prime q and a number $p \leq n^{357/200}$ with at most three prime factors such that

$$n = \frac{p+1}{q+1}$$

(See C. Badea.)

2) Let n be a positive integer in an interval $1 < n < x^{4/5}$ and denote by $E(n)$ the number of solutions to the equation

$$n = \frac{p+1}{q+1}$$

in primes $p \leq x$ and q. Then:

a) $E(n) \leq \frac{c_1}{n} \cdot \prod_{\substack{p|n(n-1) \\ p>2}} \left(\frac{p-1}{p-2}\right) \frac{x}{\log^2 x}$

for all large $x (c_1 > 0$, absolute constant)

b) $\sum_{n \leq x^{4/5}} n \cdot E(n)^2 \leq c_2 \cdot \frac{x^2}{\log^3 x}$

$(c_2 > 0)$

c) $\sum_{n \leq x^{4/5}} E(n)^2 > c \cdot \frac{x}{\log x}$

for all large x

P.D.T.A. Elliott. *A conjecture of Kátai.* Acta Arith. 26 (1974), 11–20.

Note. See also

P.D.T.A. Elliott. *Arithmetic functions and integer products.* Springer-Verlag, 1985, p. 403.

3) The positive integers which are representable in the form $\frac{p+1}{q+1}$ with primes p, q, have a positive upper density.

(See P.D.T.A. Elliott. (1974); (1985)).

4) Every positive integer n has a representation

$$n^v = \frac{(p_1+1)(p_2+1)}{(p_3+1)(p_4+1)}$$

with primes $p_j (j = 1, 2, 3, 4)$ and a positive integer $v \leq 8$

J. Meyer. *Représentation multiplicative des entiers à l'aide de l'ensemble* $P+1$. II. Astérisque 94 (1982), 133–142.

Remark. Meyer announces that the above is true with $v \leq 4$.

§ IX.27 An additive property of squares and primes

1) a) Let P be the set of primes and $P' = P \cup \{0, 1\}$.
If $\sigma(A) = \inf A(n)/n$ denotes the Schnirelmann density of the set A of integers, then $\sigma(A + P') \gg K^{-2/3}$, where $\sigma(A) = 1/K$

H. Plünnecke. *Eine zahlentheoretische Anwendung der Graphentheorie*. J. Reine Angew. Math. 243 (1970), 171–183.

Note. This result is not stated explicitly in the above paper, but the used method easily implies the theorem. See also

H. Plünnecke. *Über die Dichte der Summe zweier Mengen, deren eine die Dichte null hat*. Ibid. 205 (1960/61), 1–20.

b) Let A be any set of integers of Schnirelmann density $\sigma(A) = 1/K$, with $K > 2$. Then

$$\sigma(A + P') \geq c/\log K$$

($c > 0$ absolute constant.)

c) Let A be a set with $\underline{d}(A) = 1/K$, $K > 3$, where $\underline{d}(A)$ is the lower asymptotic density of the set A. Then

$$\underline{d}(A + P) \geq c_1/\log\log K$$

($c_1 > 0$)
There is a constant c_2 such that for every $K > 3$ one can find a set A with $\underline{d}(A) = 1/K$ and $\underline{d}(A + P) \leq c_2/\log\log K$

I.Z. Ruzsa. *On an additive property of squares and primes*. Acta Arithm. 49 (1988), 281–289.

2) a) Let Q_k and P_k denote the set of k-th powers of nonnegative integers and primes, respectively, and let $d(A)$ denote the (asymptotic) density of the set A. Then, if A is a set of positive integers with a positive lower density, and $K = 1/d(A)$, then

$$d(A + Q_k) \geq d(A + P_k) \geq K^{-c/\log\log K}$$

with a positive constant c depending on k

b) For every $K > 3$ there is a set A such that

$$d(A) = 1/K$$

$$d(A + P_k) \leq d(A + Q_k) \leq K^{-C/\log\log K}$$

where C is another constant depending on k

I.Z. Ruzsa. *An additive problem for powers of primes*. J. Number Theory 33 (1989), 71–82.

§ IX.28 On the distribution of $\{\sqrt{p}\}$ and $\{p^\theta\}$, $\frac{1}{2} \le \theta \le 1$

1) a) For $1/2 < \theta < 1$, the number of primes $p \le x$ such that $\{p^\theta\} < \delta$ is

$$\delta\pi(x) + O(x^{(1+\theta)/2} \cdot \omega^2 \cdot \log^8 x + \delta x/\omega \log x)$$

uniformly in $0 \le \delta \le 1$ and $1 \le \omega \le x^{1/25}$

A. Balog. *On the fractional part of p^β*. Arch. Math. (Basel) 40 (1983), 434–440.

b) Let $D(N) = \sup\limits_{0\le\delta\le 1} \left| \sum\limits_{p\le N, \{p^\theta\}\le\delta} 1 - \pi(N)\cdot\delta \right|$, where $\theta > 1$, $\theta \notin \mathbb{N}$.

(i) $D(N) < N^{1-\gamma}$
for $N > C_1(\theta)$, where $\gamma = (15000\,\theta^2)^{-1}$

(ii) $D(N) < N^{157/168+\varepsilon}$
for $\theta = \dfrac{3}{2}$ and $N > C_2(\varepsilon)$

R.C. Baker and G.A. Kolesnik. *On the distribution of p modulo one*. J. Reine Angew. Math. 356 (1985), 174–193.

c) For $X \ge 2, \Delta > 0$ and $0 < \beta \le 1$ we define $S(X, \Delta, \beta)$ as the number of primes $p \le X$ such that $\{p^{1/2} - \beta\} < \Delta \cdot X^{-1/2}$. Then $S(X, \Delta, \beta) \gg_\varepsilon \Delta \cdot X^{1/2} \cdot \log^{-1} X$ if $\Delta \ge \log^\varepsilon X$ and β is an algebraic number.

M. Nair and A. Perelli. *On the distribution of $p^{1/2}$ modulo one.* Number Theory. Vol. I. Elementary and analytic, Proc. Conf., Budapest/Hung. 1987, Colloq. Math. Soc. János Bolyai 51 (1990), 393–435.

2) a) There are infinitely many primes p such that

$$\{\sqrt{p}\} < p^{-\sqrt{15}/(2(8+\sqrt{15}))} \cdot \log^{14} p$$

Assuming the Riemann Hypothesis, the right side may be replaced by $p^{-1/4} \cdot \log^5 p$. Here {} denotes the fractional part.

R.M. Kaufman. *The distribution of $\{\sqrt{p}\}$* (Russian). Mat. Zametki 26 (1979), 497–504; 653; Correction. Mat. Zametki 29 (1981), 636.

b) For every real θ and every $\varepsilon > 0$ there are infinitely many solutions to $\{p^{1/2} - \theta\} < p^{-1/8+\varepsilon}$ with $[p^{1/2}]$ and p both primes. (Here { } and [ ] denote fractional and integral parts, respectively.)

G. Harman. *Fractional and integral parts of p^λ*. Acta Arith. (to appear).

3) Let T_σ be the number of primes $p \leq N$ for which

$$\{b \cdot p^{3/2}\} < \sigma$$

($b > 0$). Then

$$T_\sigma = \sigma\pi(N) + O(N^{1+\varepsilon-1/56})$$

for all $\varepsilon > 0$

E.P. Golubeva and O.M. Fomenko. *On the distribution of the sequence $b \cdot p^{3/2}$ modulo* 1 (Russian). Zap. Naučn. Sem. Leningrad, Otdel. Mat. Inst. Steklov (LOMI) 91 (1979), 31–39, 180.

§ IX.29 Diophantine approximations by almost primes

1) a) Let α be irrational. There are infinitely many solutions of

$$\|\alpha \cdot q\| < q^{-1/3} \cdot (\log q)^{18}$$

where q is a number with precisely two prime factors.

S.W. Graham. *Diophantine approximation by almost primes*. Unpublished.

b) 18 may be replaced with 4/3

G. Harman. *Trigonometric sums over primes*. II. Glasgow Math. J. 24 (1983), 23–37.

2) Suppose f is a real polynomial of degree $k \geq 2$ with an irrational leading coefficient. Then, for $\varepsilon > 0$, there are infinitely many solutions of

$$\|f(q)\| < q^{-\sigma+\varepsilon}$$

where $\sigma = 1/(2^k + 2)$(q is a number with two prime factors.)

(See G. Harman).

§ IX.30 Number of solutions of $\|f(p)\| < p^{-\tau+\varepsilon}$ (p prime)

Let $k \geq 4$ an integer and $f(x)$ a real polynomial in x with irrational leading coefficient. Then, for a given $\varepsilon > 0$, there are infinitely many solutions of the inequality

$$\|f(p)\| < p^{-\tau+\varepsilon}$$

(p prime.) Here, for $k \leq 11$, $\tau = (2T + (2^{k+1} - 1 - 2k)/k)^{-1}$, where T is defined by the following table.

k	4	5	6	7	8	9	10	11
T	46	110	240	414	672	1080	1770	3000

For $k \geq 2$, we have

$$\tau = \left(12\left[k^2\left(\log k + \frac{1}{2}\log\log k + 1.3\right)\right]\right)^{-1}$$

G. Harman. *Trigonometric sums over primes*. II. Glasgow Math. J. 24 (1983), 23–37.

§ IX.31 A sum involving $\|p\,\alpha\|$ (p prime)

$$\sum_{p\le n}\min\left(y, \frac{1}{\|p\alpha\|}\right) < c\cdot\frac{N\log y\cdot\log\log y}{\log N}$$

where $3 \le y \le N^{1/4-\varepsilon}$, $\frac{1}{N} \le \alpha \le 1 - \frac{1}{N}$ and $c > 0$ is an absolute effective constant. (Here $\|x\|$ denotes the distance of x from the nearest integer.)

A. Sárközy and C.L. Stewart. *On exponential sums over prime numbers*. J. Austral. Math. Soc. Ser. A 46 (1989), 423–437.

§ IX.32 On the distribution of $\alpha\,p$ modulo one

1) Suppose that α is an irrational number and β is a real number. Then there exist infinitely many primes p such that

$$\|\alpha\,p - \beta\| < p^{-c}$$

where:

a) $c = \frac{1}{5}$

A.I. Vinogradov. *The method of trigonometrical sums in the theory of numbers* (translated, revised and annotated A. Davenport and K.F. Roth.) Interscience, New York, 1954.

b) $c = \frac{1}{4}$

R.C. Vaughan. *On the distribution of $\alpha\,p$ modulo one*. Mathematika 24 (1977), 135–141.

c) $c = \frac{3}{10}$

G. Harman. *On the distribution of $\alpha\,p$ modulo one*. J. London Math. Soc. (2) 27 (1983), 9–18.

d) $c = \frac{4}{13}$

C. Jia. *On the distribution of α_p modulo one*. J. Number Theory 45 (1993), 241–253.

2) Suppose that α is an irrational number and $k \ge 2$ is integer. Then, for any real β and any $\varepsilon > 0$, there are infinitely many primes p satisfying

$$\|\alpha p^k + \beta\| < p^{-\lambda+\varepsilon}$$

where $\lambda = \dfrac{3}{20}$ if $k = 2$, $\lambda = (3 \cdot 2^{k-1})^{-1}$ if $k \geq 3$

R.C. Baker and G. Harman. *On the distribution of $\alpha\, p^k$ modulo one*. Mathematika 38 (1991), 170–184.

§ IX.33 Simultaneous diophantine approximation with primes

A set of real numbers $\{\alpha_1, \ldots, \alpha_s\}$ is called *compatible* if $\mathbb{C}\sum_{j=1}^{s} n_j\alpha_j \in Q$ implies that $\sum_{j=1}^{s} n_j\alpha_j \in \mathbb{I}$ whenever $n_1, \ldots, n_s$ are integers.

a) Let $\{\alpha_1, \ldots, \alpha_s\}$ be a compatible set of real algebraic numbers, lying in a field of degree d.
Then, for any $A < 1/(3ds + d - s - 1)$, there are infinitely many solutions in primes p to

$$\max_{1\leq j\leq s} \|p\,\alpha_j\| < p^{-A}$$

(Here $\|x\|$ denotes the smallest distance of x from an integer.)

A. Balog and J. Friedlander. *Simultaneous Diophantine approximation using primes*. Bull. London Math. Soc. 20 (1988), 289–292.

b) Let $\{\alpha_1, \ldots, \alpha_s\}$ be a compatible set of real algebraic numbers, such that $1, \alpha_1, \ldots, \alpha_s$ span a vector space of dimension $d \geq 2$ over $\boldsymbol{Q}$. Then, for any $A < 1/(2d(d-1))$, there are infinitely many solutions in primes p to

$$\max_{1\leq j\leq s} \|p\,\alpha_j\| < p^{-A}$$

Moreover, if $d = 2$, one can take $A < 3/10$

G. Harman. *Simultaneous diophantine approximation with primes*. J. London Math. Soc. (2) 39 (1989), 405–413.

c) Let $\{\alpha_1, \ldots, \alpha_s\}$ be a compatible set of real numbers, contained in a vector space of dimension d over $\boldsymbol{Q}$ spanned by $1, \alpha_1, \ldots, \alpha_{d-1}$ Write

$$r = \mathrm{up}\{\gamma : \lim_{N\to\infty} N^{\gamma} \cdot \mathop{\mathrm{in}}_{|j|\leq N, \sum j>0} \left\| \sum_{j=1}^{-1} jj \right\| = 0\}$$

Let $\varepsilon > 0$ be given. Then there are infinitely many solutions in primes p to

$$\max_{1\leq j\leq s} \|p\,\alpha_j\| < p^{-c(d,r)+\varepsilon}$$

where

$$c(d, r) = \begin{cases} 1/(2(d-1)(r+1)) \text{ if } & d \geq 3 \\ 3/10 & \text{if } \; d = 2 \end{cases}$$

(See G. Harman.)

§ IX.34 Diophantine approximation by prime numbers

Let λ_1, λ_2 be non-zero real numbers, not both of the same sign and with λ_1/λ_2 irrational, and let λ_0 be an arbitrary real number.

1) There are infinitely many solutions in positive integers n_1 and primes p such that

$$|\lambda_0 + \lambda_1 n_1 + \lambda_2 p| < n_1^{-0.3}$$

G. Harman. *On the distribution of* αp *modulo one*. J. London Math. Soc. (2) 27 (1983), 9–18.

2) a) There are infinitely many solutions of

$$|\lambda_0 + \lambda_1 p + \lambda_2 P_4| < p^{-\tau}$$

where $\tau = 10^{-5}/6$ (Here p denotes a prime and P_r a number with at most r prime factors.)

R.C. Vaughan. *Diophantine approximation by prime numbers*. III. Proc. London Math. Soc. (3) 33 (1976), 177–192.

b) There are infinitely many solutions of

$$|\lambda_0 + \lambda_1 p + \lambda_2 P_3| < p^{-1/300}$$

G. Harman. *Diophantine approximation with a prime and an almost prime*. J. London Math. Soc. (2) 29 (1984), 13–22.

Notes: (i) The method of the above paper can be adapted to change P_3 with P_r for $r > 3$ and $p^{-1/300}$ to $p^{-\tau_r}$, where τ_r increases with r (for example $\tau_4 > \frac{1}{12}$)

(ii) The methods of the above paper combined with results on trigonometric sums give results for

$$|\lambda_0 + \lambda_1 p^k + \lambda_2 P_r| < p^{-\tau_r}$$

where $\tau_r > 0$ and r is approximately $(k+1) \cdot 2^{k+1}$ for small k and $\frac{25}{2} k^3 \log k$ for large k

3) For any $\varepsilon > 0$, there are infinitely many ordered triples of primes p_1, p_2, p_3 for which

$$|\lambda_0 + \lambda_1 p_1 + \lambda_2 p_2 + \lambda_3 p_3| < (\max p_j)^{-1/5+\varepsilon}$$

(Here $\lambda_1, \lambda_2, \lambda_3$ are non-zero real numbers not all of the same sign, λ_1/λ_2 irrational.)

G. Harman. *Diophantine approximation by prime numbers.* J. London Math. Soc. II. Ser. 44, No. 2, (1991), 218–226.

§ IX.35 Metric diophantine approximation with two restricted prime variables

a) Let $f(n)$ be a non-increasing positive function and suppose that $\sum_p f(p)/\log p < \infty$. Then there are only finitely many solutions to the inequality

$$|\alpha p \pm q| < f(p)$$

p, q primes, for almost all α

G. Harman. *Metric Diophantine approximation with two restricted variables.* I. Math. Proc. Cambridge Philos. Soc. 103 (1988), 197–206.

b) Let F_1 be the set of real functions f of an integer variable such that $0 \leq f(n) < \frac{1}{2}$ for all positive integer n and for each f there exist constants σ_1, σ_2, N such that

$$0 < \sigma_1 \leq \frac{f(m)}{f(n)} \leq \sigma_2$$

for all m with $N \leq n < m < 2n$

Also, let F_2 be the set of all non-increasing positive functions of a real variable.

Suppose $f \in F_1 \bigcup F_2$ and that $\sum_p f(p)/\log p$ diverges. Then, for almost all real α, there are infinitely many solutions of the inequality

$$|\alpha p \pm q| < f(p)$$

p, q primes.

G. Harman. *Metric Diophantine approximation with two restricted variables.* III. J. Number Theory 29 (1988), 364–375.

c) Let $\beta \in \mathbb{R}$, a and b two coprime integers, and $\varepsilon > 0$ be given. Suppose $f \in F_1$ (see b)) and write $\theta(\alpha, N)$ for the number of solutions with $p \leq N$ of

$$|\alpha p - m + \beta| < f(p)$$

$p \equiv a (\text{mod } b)$, M squarefree.

Then, for almost all α, we have, as $N \to \infty$, that

$$\theta(\alpha, N) = \frac{12}{\pi^2} \cdot F(N) + O(F(N)^{3/4} \cdot (\log F(N))^{3/2+\varepsilon})$$

where

$$F(N) = \sum_{p \le N, p \equiv a (\mathrm{mod}\, b)} f(p)$$

Note. It is possible to replace 3/4 with 1/2, when $F(N) \ll \log N$ for all $N \geq 2$

G. Harman. Ibid. II. Mathematika 35 (1988), 59–68.

§ IX.36 The uniform distributed sequences (αp) and (p^α), where $0 < \alpha < 1$, and (p^γ), $\gamma > 1$, $\gamma \neq$ integer

1) Let $\alpha \in (0, 1)$ be irrational. Then the sequence (αp) (p prime) is uniformly distributed.

I.M. Vinogradov (1937). See: *Selected Works (Izbrannye Trudy)*. Akad. Nauk SSSR, Moskow (1952).

2) a) Let $0 < \gamma < 1$. Then the sequence (p^γ) (p prime) is unifformly distributed modulo one.

I. Stux. *On the uniform distribution of prime powers*. Comm. Pure Appl. Math. 27 (1974), 729–740

and

D. Wolke. *Zur Gleichverteilung einiger Zahlenfolgen*. Math. Z. 142 (1975), 181–184.

b) Let $\gamma > 1, \gamma \neq$ integer. Then the sequence (p^γ) (p prime) is uniformly distributed.

D. Leitman. *On the uniform distribution of some sequences*. J. London Math. Soc. (2) 14 (1976), 430–432.

Chapter X

EXPONENTIAL SUMS

§ X. 1 Basic estimates on $\sum_{n \leq x} e(m\,\alpha)$

1) For arbitrary real numbers α, x we have

$$\left| \sum_{1 \leq m \leq x} e(m\,\alpha) \right| \leq \min\left(x, \frac{1}{2\|\alpha\|}\right)$$

H. Weyl. *Über die Gleichverteilung der Zahlen mod. Eins*. Math. Ann. 77 (1916), 313–352.

2) If M is a positive integer, α a real number, then we have

$$\left| \sum_{n=0}^{M-1} e(n\,\alpha) - M \right| < 4M^2|\alpha|$$

A. Balog and A. Sárközy. *On sums of integers having small prime factors*, I. Studia Sci. Math. Hung 19 (1984), 35–47.

§ X. 2 Weyl's method

1) For $k \geq 1$, $P \geq 1$, P and Q integers, let $f(x) = \alpha x^k + \alpha_1 x^{k-1} + \cdots + \alpha_k \in \mathbb{R}[x]$ and denote

$$S = \sum_{Q < n \leq Q+P} e(f(n))$$

If $K = 2^{k-1}$, then

$$|S|^K < 4^K \left(P^{K-1} + P^{K-k} \sum_{1 \leq n_1, \ldots, n_{k-1} \leq P} \cdots \sum \min\left(P, \|\alpha k! n_1 \ldots n_{k-1}\|^{-1}\right) \right)$$

(Here $\|\lambda\| = \min(\lambda - [\lambda], 1 - \lambda + [\lambda])$ and $\min(P, 0^{-1}) = P$ by convention.) (Weyl's method.)

H. Weyl. *Zur Abschätzung von $\zeta(1 + it)$*. Math. Z. 10 (1921), 88–101;
E. Landau. *Über die ζ-Funktion und die L-Funktionen*. Math. Z. 20 (1924), 105–125.

Remark. For the formulation

$$\left|\sum_{n=0}^{m-1} e(f(n))\right|^K < 4^K (m^{K-1} + m^{K-k} \sum_{(h)=1}^{m} \min\{m, |\operatorname{cosec}(\pi\,\alpha\, k!\, H)|\})$$

where $(h) = (h_1, \ldots, h_{k-1})$, $H = h_1, \ldots, h_{k-1}$, see
A. Walfisz. *Gitterpunkte in mehrdimensionalen Kugeln.* Warsawa 1957 (pp. 54–56).

2) Let $S(\alpha) = \sum_{x=1}^{P} e(\alpha x^k)$

a) If $|\alpha - a/q| \le 1/q^2$, with $(a, q) = 1$, then

$$S(\alpha) \ll_\varepsilon P^{1+\varepsilon}(q^{-1} + P^{-1} + qP^{-k})^{2^{1-k}}$$

(Weyl's inequality.)

b) If $|\alpha - a/q| \le 1/q^2$, with $(a, q) = 1$, and $k \ge 6$, then

$$S(\alpha) \ll_\varepsilon P^{1+\varepsilon}(Pq^{-1} + P^{-2} + qP^{1-k})^{(4/3)\,\cdot\,2^{-k}}$$

D.R. Heath-Brown. *Weyl's inequality, Waring's problem and Diophantine approximation. Number theory and related topics.* Tata Institute of Fundamental Research, Bombay (1988), 41–46.

3) Let S as in 1) and suppose that α is irrational. Then

a) There exists Q independent of P such that
$|S| \le ((k-1)!)^{1/2(k-1)} \cdot P^{1-\rho}$ for each $0 < \rho < 1/2$ and $k \ge k_0(\rho)$

b) If $D(n)$ denotes the number of Q in $1, 2, \ldots, n$ for which a) holds, then

$$\lim_{n\to\infty} \frac{D(n)}{n} \ge C(k, \rho) > 0, \text{ where } \lim_{k\to\infty} C(k, \rho) = 1$$

L.D. Pustil'nikov. *New estimates of Weyl sums and the remainder term in the law of distribution of the fractional part of a polynomial.* Ergodic Theory Dyn. Syst. 11 (1991), 515–534.

Remark. The proof is based on ergodic theory.

§ X. 3 Van der Corput's method

1) a) Let $a < b$, and $f \colon [a, b] \to \mathbb{R}$ differentiable with $f'(x)$ monotonic (i.e. increasing or decreasing) and $|f'(x)| \le \frac{1}{2}$ on $[a, b]$. Then

$$\sum_{a \le m \le b} e(f(m)) = \int_a^b e(f(m))\,\mathrm{d}x + O(1)$$

J.G. van der Corput. *Zahlentheoretische Abschätzangen.* Math. Ann. 84 (1921), 53–79.

b) Let $f\colon [a,b] \to \mathbb{R}$ twofold differentiable and suppose $f''(x) \geq \varepsilon$, where ε is independent of x. Then

$$\int_a^b e(f(x))\mathrm{d}x \ll \varepsilon^{-1/2}$$

(See J.G. van der Corput.)

c) Let f as above, with $f''(x) \geq \varepsilon$ for all $x \in [a,b]$ or $f''(x) \leq -\varepsilon$ for all $x \in [a,b]$, where ε is independent of x. Then

$$\sum_{a \leq m \leq b} e(f(m)) \ll (|f'(b) - f'(a)| + 1) \cdot \varepsilon^{-1/2}$$

(See J.G. van der Corput.)

2) a) Let $f\colon [a,b] \to \mathbb{R}$ and H an integer with $1 \leq H \leq b-a$. Then

$$\sum_{a<n\leq b} e(f(n)) \ll \frac{(b-a)}{\sqrt{H}} + \left\{ \frac{b-a}{H} \sum_{h=1}^{H-1} \left| \sum_{a<n\leq b-h} e(f(n+h) - f(n)) \right| \right\}^{1/2}$$

H. Weyl. *Über die Gleichverteilung der Zahlen mod Eins.* Math. Ann. 77 (1916), 313–352.

b) Let $f(t)$ be a real function with continuous derivatives up to the k-th order in $[a,b]$. Let $k \geq 2$, $K = 2^k$ and $0 < \lambda_k \leq |f^{(k)}(t)| \ll \lambda_k$. Then

$$\sum_{a<n\leq b} e(f(n)) \ll (b-a) \cdot \lambda_k^{1/(K-2)} + (b-a)^{1-4/K} \cdot \lambda_k^{-1/(K-2)}$$

E.C. Titchmarsh. *The theory of Riemann zeta-function*, Oxford, 1951.

c) Let u be a fixed constant and $1 \leq a < b \leq a \cdot u$. Let $f(t)$ be a real function in $[a,b]$ with continuous derivatives up to the k-th order. Let $k \geq 2$, $K = 2^k$ and $\lambda/a^\nu \ll |f^{(\nu)}(t)| \ll \lambda/a^\nu (\nu = 2, 3, \ldots, k.)$ If $\lambda \gg a$, then

$$\sum_{a<n\leq b} e(f(n)) \ll a^{1-k/(K-2)} \cdot \lambda^{1/(K-2)}$$

E. Krätzel. *Lattice Points.* Berlin 1988 (p. 35).

3) a) Let $f(t)$ be a real function with continuous derivatives up to the third order in $[a,b]$. Let $\lambda_2 \ll |f''(t)| \ll \lambda_2, |f'''(t)| \ll \lambda_3$ throughout the interval. Let $\varphi(t)$ be defined by $f'(\varphi) = t$.
If $\alpha = \min f'(t)$, $\beta = \max f'(t)$ and $\varepsilon = \exp(\pi i/4)$ for $f''(t) > 0$; $\varepsilon = \exp(-\pi i/4)$ for $f''(t) < 0$, then

$$\sum_{a<n\leq b} e(f(n)) = \varepsilon \cdot \sum_{\alpha<\nu\leq\beta} \frac{1}{\sqrt{|f''(\varphi(\nu))|}} \cdot e(f(\varphi(\nu)) - \nu\varphi(\nu)) +$$
$$+ O\left(\frac{1}{\sqrt{\lambda_2}}\right) + O(\log((b-a)\lambda_2 + 2)) + O(R),$$

where $R = (b-a) \cdot (\lambda_2\lambda_3)^{1/5}$

J.G. van der Corput. *Verschärfung der Abschätzung beim Teilerproblem.* Math. Ann. 87 (1922), 39–65.

b) Moreover (in a)), if $f(t)$ possesses continuous derivatives up to the fourth order with $|f^4(t)| \ll \lambda_4$ and $\lambda_2\lambda_4 \ll \lambda_3^2 \ll \lambda_2\lambda_4$ then the same estimate holds with

$$R = (b-a) \cdot \lambda_3^{1/3}$$

E. Phillips. *The zeta function of Riemann; Further developments of van der Corput's method.* Quart. J. Math. (Oxford) 4 (1933), 209–225.

4) Let $f(t)$ be a real functions with continuous derivatives up to the third order in $[a, b]$. Let $\lambda_2 \ll |f''(t)| \ll \lambda_2, 0 < |f'''(t)| \ll \lambda_3$ throughout the interval. Suppose that for each $c \in [a, b]$ the function

$$F(t, c; f) = (f'(t) - f'(c))^6 - 8f''(c)f''^2(t)(f(t) - f(c) - f'(c)(t-c))^3$$

only has a bounded number of points of zero. Let $\varphi(t)$ be defined by $f'(\varphi) = t$. Let $g(t)$ be a real function with a continuous and monotonical derivative in $[a, b]$ and $|g(t)| \leq G, |g'(t)| \leq G_1$. If $\alpha = \min f'(t)$, $\beta = \max f'(t)$, $T(z) = 0$ for $f'(z) \in \mathbb{I}$;

$$T(z) = \min\left(\frac{1}{\|f'(z)\|}, \frac{1}{\sqrt{\lambda_2}}\right) \text{ for } f'(z) \in \mathbb{I} \text{ ($\mathbb{I}$ is the set of integers)},$$

ε as in 3), and

$$\Delta = (b-a)^2 \cdot \frac{\lambda_3^2}{\lambda_2^2} + (b-a) \cdot \frac{\lambda_3^2}{\lambda_2^3} + (b-a) \cdot \frac{\lambda_3}{\lambda_2} + \frac{\lambda_3}{\lambda_2^2} + \log((b-a)\lambda_2 + 2) \text{ then}$$

$$\sum_{a<n\leq b} g(n) \cdot e(f(n)) = \varepsilon \sum_{\alpha\leq\nu\leq\beta} \frac{g(\varphi(\nu))}{\sqrt{|f''(\varphi(\nu))|}} e(f(\varphi(\nu)) - \nu\varphi(\nu)) +$$
$$+ G \cdot (O(T(a)) + O(T(b)) + O(\Delta)) + G_1 \cdot \left(O(b-a) + O\left(\frac{1}{\lambda_2}\right)\right)$$

G. Kolesnik. *On the estimation of multiple exponential sums.* Recent progresses in analytic number theory, vol. 1 (Durham 1979), 231–246, Academic Press, London 1981

and

E. Krätzel. *Zweifache Exponentialsummen und dreidimensionale Gitterpunktprobleme*, Elementary and Analytic Theory of Numbers, Banach Center Publications 17, PW — Polish Scientific Publishers, Warsaw, 1985, 337–369.

§ X. 4 Vinogradov's method

1) Let $k \geq 12$. Let f be a real function defined on $Q \leq x \leq Q + P$.
If either:

(i) $\lambda \leq \pm \frac{1}{(k+1)!} f^{k+1}(x) \leq 2^{k+1} x$
throughout the interval, where λ is a constant satisfying $P \leq \lambda^{-1} \leq P^3$, or

(ii) f is a polynomial of degree $\ll k$ whose k-th coefficients is rational with denominator d satisfying $P \leq d \leq P^{k-1}$, then

$$\max_{M \leq P} \left| \sum_{n=Q}^{Q+M} e(f(n)) \right| \ll B_k \cdot P^{1-\varepsilon_k}$$

where $B_k = \exp(c_1 k \log^2 k)$ and $\varepsilon_k = (c_2 k^2 \log k)^{-1}$

I.M. Vinogradov. *The upper bound of the modulus of a trigonometric sum.* Izvestia A. N. SSSR 14(1950), 119–214;

I.M. Vinogradov. *General theorems on the upper bound of the modulus of a trigonometric sum.* Izvestia A. N. SSSR 15(1951), 109–130.

2) Let T and $P > 0$ be integers and suppose that $f: \mathbb{R} \to \mathbb{R}$ has continuous derivative of order $n+1$ in $T \leq x \leq T + P$. Let there exist absolute positive constants c_0, c_1, c_2, c_3, c_4 with $c_0 < 1, c_2 + c_4 < c_1$, an integer r, $c_0 \cdot n \leq r \leq n$; integers $s_j \geq 2 (j = 1, 2, \ldots, r), s_j \leq n$ and such that for $T \leq x \leq T + P$,

$$|f^{n+1}(x)|/(n+1)! < P^{-c_1(n+1)}$$

$$P^{-c_2 j} \leq |f^{s_j}(x)|/s_j! \leq P^{-c_3 s_j}$$

$(j = 1, 2, \ldots, r)$
Then for any integer $0 < P_1 \leq P$,

$$\left| \sum_{x=T}^{T+P_1-1} e(f(x)) \right| \leq A \cdot P^{1-c/n^2}$$

with some positive absolute constants A and c

A.A. Karacuba. *Estimates of trigonometric sums by the method of I.M. Vinogradov, and their application* (Russian). Trudy Mat. Inst. Steklov 112 (1971), 241–255, 388.

§ X. 5 Theory of exponent pairs

A pair (k, l) of real numbers is called an *exponent pair* if $0 \leq k \leq 1/2 \leq l \leq 1$, and if, corresponding to every positive number s, there exist two numbers r and c

depending only on s (r an integer greater than 4 and $0 < c < 1/2$) such that the inequality

$$\sum_{a<n\leq b} e(f(n)) \ll z^k a^l$$

holds with respect to s and u when the following conditions are satisfied:

$$u > 0,\, 1 \leq a < b < a \cdot u,\, y > 0,\, z = ya^{-s} > 1$$

$f(t)$ being any real function with differential coefficients of the first r orders in $[a, b]$ and

$$\left| f^{(\nu+1)}(t) - y\frac{\mathrm{d}^\nu}{\mathrm{d}t^\nu}t^{-s} \right| < (-1)^\nu cy \frac{\mathrm{d}^\nu}{\mathrm{d}t^\nu}t^{-s}$$

for $a \leq t \leq b$ and $\nu = 0, 1, \ldots, r-1$

1) a) If (θ, λ) is an exponent pair, then so is

$$(k, l) = A(\theta, \lambda) = \left(\frac{\theta}{2(\theta+1)}, \frac{\lambda}{2(\theta+1)} + \frac{1}{2}\right)$$

("A-process")

J.G. van der Corput. *Verschärfung der Abschätzung beim Teilerproblem.* Math. Ann. 87 (1922), 39–65.

b) If (θ, λ) is an exponent pair with $\theta + 2\lambda \geq 3/2$, then so is

$$(k, l) = B(\theta, \lambda) = \left(\lambda - \frac{1}{2}, \theta + \frac{1}{2}\right)$$

("B-process")

(See J.G. van der Corput.) See also

E. Phillips. *The zeta function of Riemann; Further developments of van der Corput's method.* Quart. J. Math. (Oxford) 4 (1933), 209–225 and

R.A. Rankin. *van der Corput's method and the theory of exponent pairs.* Quart. J. Math. (2) 6 (1955), 147–153 and

E.C. Titchmarsh. *On van der Corput's method and the zeta-function of Riemann* (I). Quart. J. Math. (Oxford) 2 (1931), 161–173; (II), ibid. 2 (1931), 313–320; (III). ibid. 3 (1932), 133–141; (IV) ibid. 5 (1934), 98–105; (V) ibid. 5 (1934), 195–210; (VI) ibid. 6 (1935), 106–112.

2) The following are exponent pairs:

a) $\left(\frac{11}{82}, \frac{57}{82}\right)$

L.W. Nieland. *Zum Kreisproblem.* Math. Ann. 98 (1928), 717–736, (and E.C. Titchmarsh).

b) $\left(\frac{97}{696}, \frac{480}{696}\right)$

(See E. Phillips.)

c) $\left(\frac{141841}{1019718}, \frac{703527}{1019718}\right)$

(See R.A. Rankin.)

d) $\left(\frac{9}{56}+\varepsilon, \frac{37}{56}+\varepsilon\right)$

for all $\varepsilon \geq 0$

M.N. Huxley and N. Watt. *Exponential sums and the Riemann zeta-function.* Proc. London Math. Soc. III. Ser. 57, (1988), 1–24.

Remark. The result is based on an important method created by Bombieri and Iwaniec.

E. Bombieri. and H. Iwaniec. *Some mean-value theorems for exponential sums.* Ann. Scuola Norm. Sup. Pisa Cl. Sci. (4) 13 (1986), 473–486.

e) $\left(\frac{89}{560}+\varepsilon, \frac{369}{560}+\varepsilon\right)$

N. Watt. *Exponential sums and the Riemann zeta-function.* II. J. London Math. Soc. II. Ser. 39, No. 3(1989), 385–404.

f) $\left(\frac{2}{13}+\varepsilon, \frac{35}{52}+\varepsilon\right)$

M.N. Huxley and N. Watt. *The Hardy-Littlewood method for exponential sums.* Number theory, vol. I (Budapest, 1987), 173–191, Colloq. Math. Soc. János Bolyai, 51, North-Holland, Amsterdam-New-York, 1990.

§ X. 6 Multiple trigonometric sums

1) Let $f(x_1, \ldots, x_n)$ be any form of degree d in n variables with integral coefficients which is expressible as a sum of n d-th powers of linear forms with real or complex coefficients and nonzero coefficients. Let $\Phi(x_1, \ldots, x_n)$ be any real polynomial of degree $\leq d$. Let

$$S_n - \sum_{x_1=1}^{P_1} \cdots \sum_{x_n=1}^{P_n} e(af(x_1, \ldots, x_n) + \Phi(x_1, \ldots, x_n))$$

$0 < P_j \leq P$, $|a - h/q| < 1/q^2$, $(h, q) = 1$ and $P \leq q \leq P^{d-1}$

Then

$$|S_n|^K \ll P^{\varepsilon} \cdot (P^{K-1} + P^K \cdot q^{-1} + P^{K-d} \cdot q)^n$$

where $\varepsilon > 0$ and $K = 2^{d-1}$

B.J. Birch and H. Davenport. *Note on Weyl's inequality.* Acta Arith. 7 (1961/62), 273–277.

2) a) Let $f(x, y) = \sum_{r=0}^{n} \sum_{s=0}^{m} \alpha_{rs} \cdot x^r y^s (\alpha_{rs} \in \mathbb{R})$ and put

$S(P_1, P_2) = \sum_{x=1}^{P_1} \sum_{y=1}^{P_2} e(f(x, y))$. Let $N_{k,r}(P)$ denote the number of integral solutions of the system $\sum_{i=1}^{k} x_i^{\nu} = \sum_{i=1}^{k} y_i^{\nu}$

($\nu = 1, 2, \ldots, n$) such that $1 \leq x_i, y_i \leq P$.

For fixed integers $\lambda_1, \ldots, \lambda_n$, put $\beta_s = \sum_{r=0}^{n} \alpha_{rs} \cdot \lambda_r$.

Then, for all $k_1, k_2 = 1, 2, 3, \ldots$

$$|S(P_1, P_2)|^{4k_1k_2} \leq (2k_2)^m \cdot P_1^{2k_1(2k_2-1)} \cdot P_2^{2k_2(2k_1-1)} \cdot N_{k_1,n}(P_1) \cdot N_{k_2,m}(P_2) \cdot V$$

where

$$V = \sum_{\lambda_1,\ldots,\lambda_n, |\lambda_\nu| < k_1 P_1^{\nu}} \prod_{s=1}^{m} \min\left(P_2^s, \frac{1}{\|\beta_s\|}\right)$$

N.M. Korobov. *Double trigonometric sums and their applications to the estimation of rational sums* (Russian). Mat. Zametki 6 (1969), 25–34.

b) Let p be a prime number, $p > n \geq 2, q = p^{m-1} = P^{\theta n}, m \geq 2n$.
Let $S = \sum_{x,y \leq P} e(f(x, y))$,

where $f(x, y) = \sum_{s=0}^{n} \sum_{t=0}^{n} (a(s, t)/p^{m-s-t}) x^s y^t$,

with $(a(s, t), p) = 1$ and $s + t \geq 1$. If $1 < \theta \leq \frac{3}{2}$, then

$|S| \ll P^{2-c(\theta)/n^3}$, where $c(\theta)$ is a certain positive constant.

M.Kh. Kamilov. *Estimates for double trigonometric sums* (Russian). Dokl. Akad. Nauk Tadzhik SSR, 30 (1987), 471–474.

§ X. 7 Estimates on $\int_c^b g(t) \cdot e^{if(t)} \mathrm{d}t$

Let $a < c < b$ and let $f(t)$ be a real function in $[a, c]$ and $[c, b]$, respectively, with continuous derivatives up to the third order. Let $f'(c) = 0$ and $\lambda_2 \ll |f''(t)| \ll \lambda_2, 0 < |f'''(t)| \ll \lambda_3$.

Suppose that the function $f'^6(t) - 8f''(c)f''^2(t)(f(t) - f(c))^3$ only has a bounded number of points of zero.

Let $g(t)$ be a real function in $[a, c]$ and $[c, b]$, respectively, with a continuous and monotonical derivative and $|g(t)| \leq G, |g'(t)| \leq G_1$. Let $\varepsilon = \exp(\pi i/4)$ for $f''(t) > 0$; $\varepsilon = \exp(-\pi i/4)$ for $f''(t) < 0$. Then

$$\int_c^b g(t)e^{if(t)}\mathrm{d}t = \varepsilon g(c)\cdot\left(\frac{\pi}{2|f''(c)|}\right)^{1/2} e^{if(c)} + O\left(\frac{G_1}{\lambda_2}\right)+$$

$$+G\cdot\left(O\left((b-c)\frac{\lambda_3^2}{\lambda_2^3}\right)+O\left(\frac{\lambda_3}{\lambda_2^2}\right)+O\left(\min\left(\frac{1}{|f'(b)|},\frac{1}{\sqrt{\lambda_2}}\right)\right)\right)$$

and

$$\int_a^c g(t)e^{if(t)}\mathrm{d}t = \varepsilon g(c)\cdot\left(\frac{\pi}{2|f''(c)|}\right)^{1/2} e^{if(c)} + O\left(\frac{G_1}{\lambda_2}\right)+$$

$$+G\cdot\left(O\left((c-a)\frac{\lambda_3^2}{\lambda_2^3}\right)+O\left(\frac{\lambda_3}{\lambda_2^2}\right)+O\left(\min\left(\frac{1}{|f'(a)|},\frac{1}{\sqrt{\lambda_2}}\right)\right)\right)$$

I.M. Vinogradov. *Izbrannye trudy.* Izdat. Akad. Nauk. SSSR, Moskow 1952;

See also

I.M. Vinogradov. *Special variants of the method of trigonometric sums* (Russian), Moskow, 1976;

and

E. Krätzel. *Lattice points.* Berlin 1988 (p. 44).

§ X. 8 Estimates of type $\iint_D e^{if(x,y)}\mathrm{d}x\mathrm{d}y$ or $\sum_{(n,m)\in D} e(f(n,m))$ where D is a plane domain

In what follows, the following conditions are always assumed to be true.

(A) Let D be a bounded plane domain with an area $|D|$, where the number of lattice points are of order $|D|$

(B) Suppose that D is a subset of the rectangle $D' = \{(a,b) : a_1 \leq a \leq b_1, a_2 \leq b \leq b_2\}$ with $c_1 = b_1 - a_1 \geq 1$, $c_2 = b_2 - a_2 \geq 1$, $|D'| = c_1c_2$

(C) Any straight line parallel to any of the coordinate axes intersects D in a bounded number of line segments.

(D) Let $f(x,y)$ be a real function in D' with continuous partial derivatives of as many orders as may be required. Suppose that the functions $f_x(x,y)$, $f_y(x,y)$ are monotonic in x and y, respectively.

(E) Intersections of D with domains of the type $f_x(x,y) \leq c$, $f_y(x,y) \leq c$ or $f_x(x,y) \geq c$, $f_y(x,y) \geq c$ are to satisfy condition (C) as well.

(F) The boundary of D can be divided into a bounded number of parts. In each part the curve of boundary is given by $y =$ constant or a function $x = \rho(y)$, which is continuous in the closed intervals described above.

The Hessian of $f(x, y)$ is denoted by $H(f) = f_{xx}f_{yy} - f_{xy}^2$.

1) a) Suppose that $\lambda_1 \leq f_{xx}(x, y) \ll \lambda_1, \lambda_2 \leq f_{yy}(x, y) \ll \lambda_2$, $|f_{xy}(x, y)| \leq \sqrt{\lambda_1\lambda_2}, |H(f)| \gg \lambda_1\lambda_2$ throughout the rectangle D'. For all parts of the curve of boundary, let $y =$ constant or $x = \rho(y)$, where $\rho(t)$ is partly twice differentiable and $|\rho''(t)| \ll r$. Then

$$\iint_D e^{if(x,y)}dx\ dy \ll \frac{1 + \log|D'| + |\log\lambda_1| + |\log\lambda_2|}{\sqrt{\lambda_1\lambda_2}} + \frac{c_2 r}{\lambda_2}$$

E.C. Titchmarsh. *On Epstein's zeta-function.* Proc. London Math. Soc. (2) 36 (1934), 485–500;
E.C. Titchmarsh. *The lattice-points in a circle.* Proc. London Math. (2) 38 (1934), 96–115; Corrigendum 555 (for rectangles);
E. Krätzel. *Zweifache Exponentialsummen und dreidimensionale Gitterpunktprobleme.* Banach Center Publ. 17-PWN Warsaw 1985, 337–369.

b) Suppose that $\alpha_1 \leq f_x(x, y) \leq \beta_1, \gamma_1 = \beta_1 - \alpha_1$

$$\lambda_1 \leq |f_{xy}(x, y)| \ll \lambda_1, |H(f)| \geq \lambda$$

Moreover, with the notation

$$u = f_x, v = f_y - f_x\frac{f_{xy}}{f_{xx}},\ \text{let}\ \left|\frac{\partial(u, v)}{\partial(x, y)}\right| \gg \lambda$$

For all parts of the curve of boundary let $y =$ constant or $x = \rho(y)$, where $\rho(t)$ is partly twice differentiable and
$|\rho''(t)| \ll r_0$. If on all parts $x = \rho(y)$ the condition

$$|f_{xx}f_x\rho''| \leq \theta|H(f)|$$

is satisfied with some $\theta, 0 < \theta < 1$, put $r = 0$. Otherwise put $r = r_0$. Then

$$\iint_D e^{if(x,y)}dx\ dy \ll \frac{1 + |\log\gamma_1| + |\log\gamma_1|}{\sqrt{\lambda}} + \frac{c_2 r\lambda_1}{\lambda}$$

E.C. Titchmarsh. *The lattice points in a circle.* Proc. London Math. Soc (2) 38(1934), 96–115; Corrigendum: 555,

and

E. Krätzel. *Zur Anwendung der Methode von Titchmarsh auf dreidimensionale Gitterpunktprobleme.* Forscherungsergebnisse FSU Jena N/83/28(1983), Math. Nachr. 123 (1985), 197–204.

2) a) Suppose that

$$\lambda_1 \ll |f_{xx}(x, y)| \ll \lambda_1, \lambda_2 \ll |f_{yy}(x, y)| \ll \lambda_2,$$

$$|f_{xy}(x, y)| \ll \sqrt{\lambda_1\lambda_2}, |H(f)| \gg \lambda_1\lambda_2$$

throughout the rectangle D'. For all parts of the curve of boundary let $y =$ constant or $x = \rho(y)$, where $\rho(t)$ is partly twice differentiable and $|\rho''(t)| \ll r$. If R is defined by

$$R = 1 + \log |D'| + |\log \lambda_1| + |\log \lambda_2| + c_2 r\sqrt{\lambda_1/\lambda_2}$$

then

$$\sum_{(n,m)\in D} e(f(n,m)) \ll (c_1\lambda_1 + c_2\sqrt{\lambda_1\lambda_2} + 1)(c_2\lambda_2 + c_1\sqrt{\lambda_1\lambda_2} + 1) \cdot \frac{R}{\sqrt{\lambda_1\lambda_2}}$$

(See the References from 1) a))

b) Suppose that

$\alpha_1 \le f_x(x,y) \le \beta_1; \alpha_2 \le f_y(x,y) \le \beta_2, \gamma_1 = \beta_1 - \alpha_1,$
$\gamma_2 = \beta_2 - \alpha_2, \alpha_1 \cdot \beta_1 > 0, \delta_1 = \beta_1/\alpha_1,$
$\lambda_1 \le |f_{xx}(x,y)| \ll \lambda_1, \lambda \ll |H(f)| \ll \lambda,$
$c_1 \cdot (\gamma_1 + 1) \ll |D| \cdot \sqrt{\lambda}$. Moreover, with the notation $u = f_x$, $w = f_{xy}/f_{xx}$ let

$$\left| f_x \frac{\partial(u,v)}{\partial(x,y)} \right| \le \theta \cdot |H(f)|$$

where $0 < \theta < 1$ is a suitable constant. For all parts of the curve of boundary let $y =$ constant or $x = \rho(y)$, where $\rho(t)$ is partly twice differentiable. Suppose that

$$|f_x f_{xx} \rho''| \le \theta |H(f)| (0 < \theta < 1)$$

at the bound. Let

$R_1 = 1 + |\log \gamma_1| + |\log \gamma_2| + |\log \lambda_1|,$
$R_2 = R_1 \cdot (1 + \log \delta_1),$

Then

$$\sum_{(n,m)\in D} e(f(n,m)) \ll \left(|D|\sqrt{\lambda} + \frac{\gamma_1 + \gamma_2 + 1}{\sqrt{\lambda}} + c_2 \right) \cdot R_2$$

(See the References of 1) b)).

§ X. 9 Vinogradov's mean-value theorem

1) Let P and T be integers and $P \ge 2$, $f(x) = \alpha_k x^k + \cdots + \alpha_1 x$ $(k \ge 2)$ a polynomial with real coefficients, and $C_k = C_k(P) = \sum_{T < x \le T+P} e(f(x))$ (where x takes integer values).

Then, when $s \ge \frac{1}{4}k(k+1) + lk$, we have

$$\int_0^1 \cdots \int_0^1 |C_k(P)|^{2s} d\alpha_1 \ldots d\alpha_k \le (7s)^{4sl} \cdot (\log P)^{2l} \cdot P^{2s-\frac{1}{2}k(k+1)+\delta}$$

where

$$\delta = \frac{1}{2}k(k+1)\left(1-\frac{1}{k}\right)^l$$

(Vinogradov's mean-value theorem)

I.M. Vinogradov. *On Weyl's sums.* Mat. Sbornik 42 (1935), 521–530;

I.M. Vinogradov. *The method of trigonometrical sums in the theory of numbers.* Trav. Inst. Math. Stekloff, 23, 1947;

and

L.-K. Hua. *An improvement of Vinogradov's mean-value theorem and several applications.* Quart. J. Math. (Oxford) 20 (1949), 48–61.

2) Let P, k, s, l be integers; $P \ge 1, k \ge 2, s \ge k(l+1), l \ge 0$. Then

$$\int_0^1 \cdots \int_0^1 |C_k(P)|^{2s} d\alpha_1 \ldots d\alpha_k \le k!s^{2kl} \cdot k^{6k^2u} \cdot 2^{2k^2l} \cdot P^{2s-\Delta}$$

where

$$u = \min(k+1, l) \text{ and } \Delta = \frac{1}{2}k(k+1) - \frac{1}{2}k^2 \cdot \left(1-\frac{1}{k}\right)^{l+1}$$

G.I. Arhipov. *The mean value of H. Weyl sums* (Russian). Mat. Zametki 23 (1978), 785–788.

§ X.10 Exponential sums containing primes

1) Suppose that $a, b \ll N \le a, b\,(0 < a < b)$; $f(x)$ is real in $[a, b]$; $f'(x) \approx A, f''(x) \ne 0, (xf'(x)) \approx B, (xf(x))'' \ne 0$, for $a \le x \le b, x \ge x_0$, where $A, B > 0$ and $N^{-1} \ll A \ll N^{-3/4}, B \ll 1$. ($X \approx Y$ means that $X \ll Y \ll X$). Then

$$\sum_{a \le p \le b} e^{if(p)} \ll \left(\left(\frac{N}{A}\right)^{1/2} + \left(\frac{N}{B}\right)^{1/2}\right) \log N$$

(p prime).

N.G. Čudakov. *On certain trigonometric sums containing prime numbers.* Dokl. Akad. Nauk SSSR (N. S.) 58 (1947), 1291–1294.

2) a) For $N < ct, t > 2$, we have

$$\sum_{n<N} n^{it} \ll N \exp\left(-\frac{c\log^3 N}{\log^2 t}\right)$$

(I.M. Vinogradov). See A.A. Karacuba, *Foundations of analytic number theory.* Nauka, Moskow 1975 (Russian).

b) Let $F(T)$ be an increasing function such that $F(T) \to \infty$ $(T \to \infty)$, $F(T) < \log T$. Also, let $U = \sqrt{T}F(T) \cdot \log T$,

$F(T) \leq K \leq T^{1/6} \cdot \log^2 T$, $P_0 = \sqrt{T/2\pi}$. Then, for sufficiently large T there exists a number $t \in [T, T+U]$ such that

$$\left| \sum_{\exp(-1/K) \cdot P_0 < n < P_0} n^{it} \right| > \frac{1}{2}\sqrt{P_0/K}$$

Y. Moser. *An Ω-theorem for a short trigonometric sum* (Russian). Acta Arith. 42 (1983), 153–161.

c) If $-\frac{1}{2} \leq \gamma \leq 1 (\gamma \neq 0)$, $c \leq t^{10} \leq \frac{1}{2}N \leq N' < N'' \leq N$, $t > 0$, then

$$\left| \sum_{N' \leq p \leq N''} e^{it \log^{1+\gamma} p} \right| < C \cdot \left(\frac{N \log^8 N}{(|\gamma| t)^{1/2}} + \frac{N \log^8 N}{|\gamma| t} \right)$$

where p denotes a prime, and c, C are numerical constants.

P. Turán. *On certain exponential sums.* Nederl. Akad. Wetensch. Proc. 51, 343–352.

3) For $T \geq 1$ we have

$$\int_{-T}^{T} \left| \sum_{n=1}^{\infty} a_n \cdot n^{it} \right|^2 dt \ll T^2 \cdot \int_0^{+\infty} \left| \sum_{y \leq n < y \cdot \exp(1/T)} a_n \right|^2 \frac{dy}{y}$$

provided the right side converges ($a_n \in \mathbb{C}$ for $n = 1, 2, 3, \ldots$)

P.X. Gallagher. *A large sieve density estimate near* $\sigma = 1$. Invent. Math. 11 (1970), 329–339.

4) For fixed non-zero real numbers α and β, and for $T, M, N \geq 1$, one has

$$\int_0^T \left| \sum_{N < n \leq 2N} a_n \cdot n^{i\alpha t} \right|^2 \cdot \left| \sum_{M < m \leq 2M} b_m \cdot m^{i\beta t} \right|^2 dt \ll (T + MN) MN \log 2MNT$$

where $|a_n| \leq 1$, $|b_m| \leq 1$, $(a_n, b_m \in \mathbb{C}$ for $n, m = 1, 2, \ldots)$

A. Balog and G. Harman. *On mean values of Dirichlet polynomials.* Arch. Math. 57 (1991), 581–587.

§ X.11 Exponential sums of type $\sum_{M' \leq m \leq M} (m+w)^{ti}$

1) Let $R = 2^{r-1}$, $R_1 = R \cdot (r+1)$, $0 \leq w \leq 1$, $t \geq 2^{r+3}$ and $t^{1/(r+2)} \leq M \leq M' \leq 2M \leq 2 \cdot t^{2/(r+3)}$. Then

$$\sum_{m=M}^{M'} e\left(\frac{t}{m+w}\right) \ll M^{1-\frac{1}{R}-\frac{1}{R_1}} \cdot t^{\frac{1}{R_1}} \cdot \log t$$

A. Walfisz. *Über Gitterpunkte in mehrdimensionalen Ellipsoiden.* Vierte Abhandl. Math. Z. 35 (1932), 212–229.

See also

A. Walfisz. *Weylsche Exponentialsummen in der neuren Zahlentheorie.* Berlin 1963 (p. 22).

2) Let R, R_1, w as above and $t \geq 2^{r+1}$ and

$t^{1/(r+1)} \leq M \leq M' \leq 2M \leq 2t^2$.

Then

$$\sum_{m=M}^{M'} (m+w)^{ti} \ll \left(M^{1-\frac{1}{R}} \cdot t^{\frac{1}{R_1}} + Mt^{-\frac{1}{R_1}}\right) \cdot \log t$$

E. Landau. *Zum Waringschen Problem.* Math. Z. 12 (1922), 219–247.

See also

A. Walfisz, p. 28.

§ X.12 Complete trigonometric sums

1) a) Let p be a prime and $f(x) = a_k x^k + \cdots + a_1 x + a_0$, $P \nmid (a_k, \ldots, a_1)$. Then

$$\left|\sum_{j \leq p} e(f(j)/p)\right| \leq kp^{1/2}$$

A. Weil. *On some exponential sums.* Proc. Nat. Acad. Sci. USA 34 (1948), 204–207.

b) Let $f(x) = a_k x^k + \cdots + a_1 x$ with k and a_j positive integers. If $1 \leq k \leq p-1$ (p prime), then

$$\max_{(a_1,\ldots,a_k) \neq (0,\ldots,0)} \left|\sum_{j \leq p-1} e(f(j)/p)\right| > \left((k!)^2 \binom{p}{k} - p^k\right)^{1/2k}$$

D.R. Anderson and J.J. Stiffler. *Lower bounds for the maximum moduli of certain classes of trigonometric sums.* Duke Math. J. 30 (1963), 171–176.

c) Let $k \geq 3$ be an integer, p be a prime and $a_1, \ldots, a_k$ be integers not all congruent to 0 (modulo p).
If $S(p, f(x)) = \sum_{j \leq p} e(f(j)/p)$, where $f(x) = a_k x^k + \cdots + a_1$ and

$$\left|S_{\max}\right| = \max_{a_1,\ldots,a_k; a_k \not\equiv 0 (\text{mod } p)} |S(p, f(x))|, \text{ then}$$

$$\liminf_{k \to \infty} \liminf_{p \to \infty} \frac{|S_{\max}|}{\sqrt{kp}} \geq \frac{16}{9\sqrt{e}}$$

N.M. Korobov and D.A. Mit'kin. *Lower bounds of complete trigonometric sums* (Russian). Vestnik Moskov Univ. Ser. I Mat. Meh. 1977, 54–57, no. 5.

d) Let $f(x) = a_k x^k + \cdots + a_1 x + a_0$, with $p \nmid a_k$.
Then for $p = k + 1$ (p prime) and $p \geq 3$,

$$\left|\sum_{j=0}^{p-1} e(f(j)/p)\right| \leq \sqrt{p^2 - 4(p-1)\sin^2 \frac{\pi}{p}}$$

for $k + 2 \leq p \leq 2k + 1$ and $p \geq 5$,

$$\left|\sum_{j=0}^{p-1} e(f(j)/p)\right| \le p - 4(p-k-1)^2 \sin^2 \frac{\pi}{p}$$

M.Y. Zhang and Y. Hong. *On the maximum modulus of complete trigonometric sums.* Acta Math. Sinica (N.S.) 3 (1987), 341–350.

2) a) Let $f(x) = a_k x^k + \cdots + a_1 x \in \mathbb{I}[x]$ and $q > 1$ an integer such that $(a_1, \ldots, a_k, q) = 1$. Then

$$\left|\sum_{j \le q} e(f(j)/q)\right| \le \sqrt[2k]{2k \cdot k!} \cdot q^{1-\frac{1}{k}}$$

I.M. Vinogradov. *The method of trigonometrical sums in number theory* (Russian). Moskva, 1971 (p. 35).

Remark. For a given k this is the best possible estimation.

b) Let $f(x) = a_k x^k + \cdots + a_0 \in \mathbb{I}[x]$ and $q = 1, 2, \ldots$ with $(a_1, \ldots, a_k, q) = 1$, $a_0 = 0$. Put $S(q, f(x)) = \sum_{j \le q} e(f(j)/q)$. Then

$$|S(q, f(x))| \le B_k \cdot q^{1-1/k}$$

where $B_k \le \exp(k + O(k/\log k))$

S.B. Stečkin. *An estimate of a complete rational trigonometrical sum* (Russian). Trudy. Mat. Inst. Steklov 143 (1977), 188–207, 211.

c) Under the same assumptions,

$$|S(q, f(x))| \le \exp(1.85k) \cdot q^{1-\frac{1}{k}}$$

M.G. Lu. *Estimate of a complete trigonometric sum.* Sci. Sinica Ser. A 28 (1985), 561–578.

d) Let $q = 1, 2, 3, \ldots$ and $f(x) = a_k x^k + \cdots + a_0$ be a polynomial of degree $k \ge 3$ with integer cofficients satisfying $(a_1, \ldots, a_k, q) = 1$. Then

$$|S(q, f(x))| \le (k-1) \cdot q^{1-1/k}$$

if $q = p^l$ is a power of a prime p, where

$$S(q, f(x)) = \sum_{j \le q} e(f(j)/q)$$

M. Lu. *A note on complete trigonometric sum for prime power.* J. Sichuan Univ. Nat. Sci. Ed. 26, Spec. Issue (1989), 156–159.

e) Let p be a prime and $f(x) = a_k x^k + \cdots + a_1 x + a_0 \in \mathbb{I}[x]$ with $(a_1, \ldots, a_k, p) = 1$. For any integer $n \ge 1$ let

$$S(p^n; f) = \sum_{x=0}^{p^n - 1} e(f(x)/p^n)$$

Define t satisfying $p^t \| (ka_k, \ldots, 2a_2, a_1)$.

Let $\mu_1, \dots, \mu_r$ be the different zeros modulo p of the congruence $p^{-1}f'(x) \equiv 0 (\text{mod } p)$, $0 \le x < p$ and let $m_1, \dots, m_p$ be their multiplicities. Put $\max_{1 \le i \le r} m_i = M = M(f)$, $m_1 + \cdots + m_r = m = m(f)$. Then

$$|S(p^n, f)| \le mk^{1/2} \cdot p^{t/(M+1)} \cdot p^{n(1-1(M+1))}$$

P. Ding. *An improvement of Chalk's estimation of exponential sums.* Acta Arith. (to appear).

§ X.13 Nearly complete and supercomplete rational trigonometric sums

1) Let $n \ge 2$, $1 \le P < q$ and $f(x) = a_n x^n + \cdots + a_1 x \in \mathbb{I}[x]$ with $(a_n, \dots, a_2, q) = d$. Then:

a) $$\left|\sum_{x=1}^{P} e(f(x)/q)\right| \ll q^{1-\frac{1}{n}} \cdot d^{1-\frac{1}{n}+\varepsilon}$$

L.-K. Hua. *Additive number theory*, 1953, Acad. Sinica Press.

b) $$\left|\sum_{x=1}^{P} e(f(x)/q)\right| \ll q^{1-\frac{1}{n}} \cdot d^{\frac{1}{n}}$$

D.A. Mit'kin. *Estimates and asymptotic formulas for rational trigonometric sums that are nearly complete* (Russian). Mat. Sb. (N.S.) 122(164) (1983), 527–545.

2) Let $S_p(q, f(x)) = \sum_{x=1}^{P} e(f(x)/q)$, where $f(x)$ is integral-valued polynomial of degree k. Let d denote the least common denominator of the coefficients of $f(x)$.

a) $S_P(q, f(x)) = O(q^{1-1/k+\varepsilon})$
if $m = q \prod_{p|q,\, p^t \| q} p^t$ and q is squarefree.

L.-K. Hua. *Additive theory of prime numbers.* Amer. Math. Soc., Providence, RI, 1965.

b) The same result holds with $O(q^{1/2+\varepsilon})$

P.Z. Shao. *Estimations for a class of supercomplete trigonometric sums.* Kexue Tongbao (English Ed.) 33 (1988), 1319–1321.

3) If $f(x) \in \mathbb{I}[x]$ and
$f(x) = a_k x^k + \cdots + a_1 x + a_0 = a_k(x - \xi_1)^{e_1} \cdots (x - \xi_m)^{e_m}$, where ξ_i are distinct algebraic numbers, the semidiscriminant of f is defined by

$$\Delta(f) = a_k^{2k-2} \prod_{i \neq j} (\xi_i - \xi_j)^{e_i e_j}$$

and the exponent of f is defined by $e(f) = \max\{e_1, \dots, e_m\}$. Then if $f \in \mathbb{I}[X]$ has degree $n + 1 \geq 3$ and the derivative f' has exponent e and semidiscriminant Δ, we have

$$|S(f, q)| \leq q^{1-1/2e} \cdot (\Delta, q)^{1/2e} \cdot d_n(q)$$

where $d_n(q)$ is the generalized divisor function.

J.H. Loxton and R.A. Smith. *On Hua's estimate for exponential sums.* J. London Math. Soc. (2) 26 (1982), 15–20.

§ X.14 Hua's estimate

1) For an integer $k \geq 2$ and $(h, q) = 1$, let $S_q = \sum_{x=1}^{q} e(hx^k/q)$. If $\varepsilon > 0$ is fixed, then

$$\sum_{x=1}^{m} e(ax^k/q) = \frac{m}{q} \cdot S_q + O_{k,\varepsilon}\left(q^{\frac{1}{2}+\varepsilon}\right)$$

L.-K. Hua. *On exponential sums.* Sci. Record (N.S.) 1 (1957), 1–4.

Remark. The above result improves a theorem of the author in 1940.

L.-K. Hua. *On an exponential sum.* J. Chinese Math. Soc. 2 (1940), 301–312.

2) Let $S(\alpha) = \sum_{x=1}^{P} e(\alpha x^k)$. Then:

a) $\int_0^1 |S(\alpha)|^{2^k} \mathrm{d}\alpha \ll_\varepsilon P^{2^k - k + \varepsilon}$

for any $\varepsilon > 0$.

L.-K. Hua. *On Waring's problem.* Quart. J. Math. Oxford Ser. 9 (1938), 199–202.

Remark. For a generalization of Hua's inequality, see R.J. Cook. A note on a lemma of Hua. Quart. J. Oxford Ser. (2) 23 (1972), 287–288.

b) Let $k \geq 6$ and $\varepsilon > 0$. Then

$$\int_0^1 |S(\alpha)|^{(7/8)2^k} \mathrm{d}\alpha \ll P^{(7/8)2^k - k + \varepsilon}$$

D.R. Heath-Brown. *Weyl's inequality, Waring's problem and Diophantine approximation.* Number theory and related topics. Tata Institute of Fund. Research, Bombay (1988), 41–46.

§ X.15 Gaussian sums

1) a) For $n, q = 1, 2, 3, \ldots$ and a integer, let

$$S_n(a, q) = \sum_{x=0}^{q-1} e(k^n \cdot a/q)$$

If $n \geq 3$ and $(a, q) = 1$, then there exists an absolute constant $c > 0$ such that

$$|S_n(a, q)| \leq \exp\left(c \cdot \left(\frac{n}{\varphi(n)}\right)^2\right) \cdot q^{1-1/n}$$

S.B. Stečkin. *An estimate for Gaussian sums.* (Russian). Mat. Zametki 17 (1975), 579–588.

b) For $q = p (p$ prime), one has

$$\max_{(a,p)=1} |S_n(a, p)| \leq 2m^{7/12} \cdot p^{2/3}$$

J.E. Sharplinskij. *On estimates of Gaussian sums.* (Russian). Mat. Zametki 50 (1991), 122–130.

Remark. The above result can be used to show that

$$\lim_{n\to\infty} \max_{q\geq 1} |S_n(a, q)| \cdot q^{-1+1/n} = 1$$

for $(a, q) = 1$, which was conjectured by Stechkin. (See the reference from a).)

2) Let $S_N(m) = \sum_{j=0}^{m-1} e(j^2/N)$, where $m < N (N = 1, 2, \ldots)$

If $N = 4k + 3$, then $|S_N(m)| \leq \frac{1}{2}\sqrt{N+1}$

This bound is achieved for $m = 2k + 1$

D.H. Lehmer. *Incomplete Gauss sums.* Mathematika 23 (1976), 125–135.

§ X.16 Estimates by Linnik and Vinogradov

1) a) Let $S = \sum_{x=1}^{P} e(\alpha f(x))$, where $\alpha = \frac{a}{q} + \frac{\theta}{q^2}, (a, q) = 1, |\theta| \leq 1$, $P \leq q \leq P^{n-1}$. If $f(x) = a_0 x^n + \cdots + a_n$, then

$$|S| \leq P^{1-s}$$

where $s = 1/(22400\, n^2 \log n)$

Yu.V. Linnik. *On Weyl's sum.* Dokl. Akad. Sci. URSS (N.S.) 34 (1942), 184–186.

b) Let $f(x) = a_{n+1}x^{n+1} + \cdots + a_1 x, \quad n+1 \geq 2; \quad a_i \in \mathbb{R}; \quad m, P > 0$ and $S = \sum_{x=1}^{P} e(mf(x))$

Let $r \in \{2, 3, \ldots, n+1\}$ and suppose that $a_r = \frac{a}{q} + \frac{\theta}{q^2}$, with $a, q \in \mathbb{I}$, $1 < q < P^r$, $(a, q) = 1$ and $|\theta| \leq 1$.

Put $\tau = (\log q)/(\log r)$ if $1 < q < P; \tau = 1$ if $P \leq q \leq P^{r-1}$ and $\tau = r - (\log q)/(\log P)$ if $P^{r-1} < q < P^r$. Then

$$|S| < 8n^{\frac{1}{2}nl} \cdot m^{2\rho/\tau} \cdot P^{1-\rho}$$

where $l = \log((12n(n+1))/\tau)$ and $\rho = \pi/(3n^2 l)$.

I.M. Vinogradov. *The upper bound of the modulus of a trigonometric sum.* Izv. Akad. Nauk. SSSR. Ser. Mat. 14 (1950), 199–214.

§ X.17 Sums of type $\sum_{p \leq N} (\log p) \cdot e(ap^k/q)$ (p prime) and $\sum_{p \leq N} e(p\alpha)$ where $\left|\alpha - \frac{a}{q}\right| \leq \frac{1}{q^2}$ for $(a, q) = 1$

1) a) Let $\alpha \in \mathbb{R}$, and suppose that there are integers a, q which satisfy $\left|\alpha - \frac{a}{q}\right| < \frac{1}{q^2}$, $(a, q) = 1$. Then, for any $\eta > 0$, $N \geq 2$, we have

$$\sum_{\frac{N}{2} < p \leq N} e(p\alpha) \ll N^{1+\eta} \cdot (N^{-1/5} + q^{-1/2} + N^{-1/2} \cdot q^{1/2})$$

(p prime.)

I.M. Vinogradov. *The method of trigonometric sums in the theory of numbers* (translated, revised and annotated by A. Davenport and K.F. Roth). Interscience, New York, 1954.

b) Suppose $\left|\alpha - \frac{a}{q}\right| \leq \frac{1}{q^2}$, $(a, q) = 1$, $1 \leq q \leq N$. Then

$$\sum_{p \leq N} e(p\alpha) \ll N \cdot \left(\frac{1}{q} + q(\log q)/N + (\log q) \cdot \exp(-\sqrt{\log N})\right)^{1/2} \cdot r^{3/4} \cdot \log r$$

where $r = \log N$

J.-R. Chen. *On the estimation of some trigonometrical sums and their applications.* Sci. Sinica Ser. A 28 (1985), 449–458.

2) a) If a, q are integers, $(a,q)=1$ and p prime,

$$\sum_{p\le N}(\log p)e\left(\frac{ap^k}{q}\right)\ll$$

$$\ll(\log N)^{7/2}\cdot q^{\varepsilon}\cdot(N^{1/2}\cdot q^{1/2}+N\cdot q^{-1/2}+N^{3/4}\cdot q^{1/8})$$

for $q\le N^{2/9}$

I.M. Vinogradov. *On the estimation of simplest trigonometric sums involving primes.* Izv. Akad. Nauk. SSSR. Ser. Mat. 2 (1939), 371–395.

b) $\sum_{p\le N}(\log p)\cdot e\left(\frac{ap^k}{q}\right)\ll N^{\frac{2}{3}+\varepsilon}\cdot q^{1/3}$

for $q\ge N^{5/8}$

J.-R. Chen. *Estimates for trigonometric sums.* Chinese Math. 6 (1965), 163–167.

c) The estimate from a) is valid (without the assumption $q\le N^{2/9}$).

R.C. Vaughan. *Mean value theorems in prime number theory.* J. London Math. Soc. (2) 10 (1975), 153–162.

3) a) Suppose $\varepsilon>0$ is given. Let $f(x)$ be a real valued polynomial in x of degree $k\ge 2$. Put $\gamma=4^{1-k}$. Suppose α is the leading coefficient of f and there are integers a, q such that $|q\alpha-a|<1/q$ with $(a,q)=1$. Then we have

$$\left|\sum_{p\le N}(\log p)\cdot e(f(p))\right|\ll N^{1+\varepsilon}\cdot\left(\frac{1}{q}+\frac{1}{N^{1/2}}+\frac{q}{N^k}\right)^{\gamma}$$

G. Harman. *Trigonometric sums over primes.* I. Mathematika 28 (1981), 249–254.

b) Let $k\ge 3$ be an integer, and $\varepsilon>0$. Suppose

$$N^{1-1/k}\le q\le N^{k/2},\ |\alpha q-a|<N^{-k/2},\ (a,q)=1$$

Then

$$\left|\sum_{p\le N}(\log p)\cdot e(\alpha p^k)\right|\ll N^{1+\varepsilon-\delta}$$

where $\delta=1/k\cdot 2^k$

G. Harman. Ibid. II. Glasgow Math. J. 24 (1983), 23–37.

Remark. The proofs of the above results are based on the following estimate. For $\log q\ll\log N$, f a real value function, we have

$$\sum_{n=1}^{N}\Lambda(n)e(f(n))=O(N^{1/3})+S_1-S_2-S_3$$

where

$$S_1=\sum_{d\le N^{1/3}}{}'\mu(d)\sum_{l\le N/d}{}'(\log l)\cdot e(f(dl))$$

$$S_2=\sum_{r\le N^{2/3}}{}'f_1(r)\sum_{m\le N/r}{}'e(f(rm))$$

$$S_3 = \sum_{N^{1/3} \le m \le N^{2/3}} f_2(m) \sum\nolimits'_{N^{1/3} < n < N/m} \Lambda(n) e(f(mn))$$

and for any $\delta > 0$, $|f_1(m)| = o(m^\delta)$, $f_2(m) = o(m^\delta)$. Here $\sum'$ indicates that the variable summed over takes values coprime to the number q

R.C. Vaughan. *Sommes trigonométriques sur le nombre premiers.* C.R. Acad. Sci. Paris Sér. A, 285 (1977), 981–983.

and

R.C. Vaughan. *An elementary method in prime number theory.* Acta Arith. 37 (1980), 111–115.

4) If $2 \le N \le X$ then we have

$$\left| \sum_{p \le N} e(X/p) \right| < c_1 (N^{1 - c_2((\log N)/(\log X))^2} + N^{3/2} X^{-1/2}) \cdot \exp(c_3 (\log\log N)^2)$$

with c_1, c_2, c_3 positive absolute constants.

M. Jutilla. *On numbers with a large prime factor.* II. J. Indian Math. Soc. 38 (1974), 125–130.

§ X.18 Estimates of trigonometric sums over primes in short intervals

1) a)

$$\sum_{x - A < n \le x} \Lambda(n) e(na/q) \ll$$
$$\ll (x^{1/2} \cdot q^{1/2} + Aq^{-1/2} + x^{3/10} \cdot A^{1/2})(\log x)^{100}$$

where $1 \le a < q$, $(a, q) = 1$ and $1 \le A \le x$

A. Balog and A. Perelli. *Exponential sums over primes in short intervals.* Acta Math. Hungar. 48 (1986), 223–228.

b) For a given $c > 0$, there exists a constant c_1, such that

$$\sum_{x - A < n \le x} \Lambda(n) e(n\alpha) \ll A(\log x)^{-c}$$

for $x^{91/96} \le A \le x$ and $A^{-1} \cdot (\log x)^c \le |\alpha| \le (\log x)^{-c_1}$

C.D. Pan and C.B. Pan. *On estimates on trigonometric sums over primes in short intervals.* I. Sci. China Ser. A 32 (1989), 408–416.

c) Let $S(\alpha; x, A) = \sum\nolimits'_{x - A < n \le x} \Lambda(n) \cdot e(n\alpha)$, where $\Lambda(n)$ is the von Mangoldt function. Then if $91/96 < \theta < 1$ and $c > 0$,

$$S\left(\frac{a}{q} + \lambda; x, A\right) \ll A/\log^c x$$

for $x^\theta < A < x$ provided that
$(a, q) = 1$, $1 \le q \le \log^{c_1} x$ and
$\log^{c_2} x/A < |\lambda| \le 1/(q \log^{c_1} x)$ with $c_1, c_2 > 0$ (constants).

C.D. Pan and C.B. Pan. *On estimates on trigonometric sums over primes in short intervals.* II. Sci. China Ser. A 32, No. 6, 1989, 641–653.

2) a) Let $x \in \mathbb{R}$, r, s natural numbers such that $(r, s) = 1$, $\left|x - \frac{r}{s}\right| < s^{-2}$. Then for any $N = 1, 2, \ldots$, given $\varepsilon > 0$, we have

$$\sum_{n \le N} \Lambda(n) \cdot e(xn^2) \ll N^{1+\varepsilon} \cdot \left(\frac{1}{s} + \frac{1}{\sqrt{N}} + \frac{s}{n^2}\right)^{1/4}$$

where the implied constant depends at most on ε

A. Ghosh. *The distribution of αp^2 modulo* 1. Proc. London Math. Soc. (3) 42 (1981), 252–269.

b) Suppose that $\left|x - \frac{r}{s}\right| < s^{-2}$ and $d^2 \le N$. Then

$$\sum_{k \le M} \left| \sum_{n \le N, n \equiv h (\text{mod } d)} \Lambda(n) e(xkn^2) \right| \ll$$

$$\ll MN^{1+\varepsilon} d^{-1} (m^3 s^{-1} + dN^{-1/2} + sdM^{-1}N^{-2}m^{-3})^{1/2}$$

where $m = (s, d)$ and the implied constant depends only on $\varepsilon > 0$.

M.I. Israilov, I.A. Allakov. *Estimation of trigonometric sums over the square of prime numbers in an arithmetic progression.* (Russian). Izv. Akad. Nauk Uz. SSR Ser. Fiz.-Mat. Nauk 1990, no. 5, 3–10, 94.

3) a) Let $S(x) = \sum_{n \le N} \Lambda(n) e(xn)$. There is $C > 0$ such that, for all $N \ge 2$

$$\int_0^1 |S(x)| \mathrm{d}x \ge C \cdot \sqrt{N}$$

and

$$\liminf_{N \to \infty} \frac{1}{\sqrt{N}} \cdot \int_0^1 |S(x)| \mathrm{d}x \ge C_1$$

where $C_1 = (12\sqrt{2}/\pi^2\sqrt{\theta}) \cdot \int_0^\theta (|\sin \pi x|/\pi x) \mathrm{d}x$ and θ satisfies the equation

$$\int_0^\theta (|\sin \pi x|/\pi x) \mathrm{d}x = \theta \cdot \int_0^\infty (|\sin \pi x|/\pi x^2) \mathrm{d}x$$

R.C. Vaughan. *The L^1 mean of exponential sums over primes.* Bull. London Math. Soc. 20 (1988), 121–123.

b) Let $(a, q) = 1$, $|x - a/q| \le 1/q^2$, $N \ge 2$, $L = \log N$. Then

$$|S(x)| \ll L^4 \cdot (Nq^{-1/2} + N^{4/5} + N^{1/2} \cdot q^{1/2})$$

R.C. Vaughan. *The Hardy-Littlewood method.* Cambridge Tracts in Math. 80 (Cambridge Univ. Press, 1981).

c) For $A > 0$, $1 \le Q \le x^{1/4}$ and $\theta = \min(Q^{-4}, (\log x)^{-8 \cdot (A+21)})$, we have

$$\sum_{q \le Q} \max_{(a,q)=1} \max_{y \le x} \max_{|\beta| \le \theta} \left| \sum_{n \le y} \Lambda(n) \cdot e\left[n\left(\frac{a}{q} + \beta\right)\right] - \frac{\mu(q)}{\varphi(q)} \sum_{n \le y} e(n\beta) \right| \ll x / \log^A x$$

D. Wolke. *Some applications to zero density theorems for L functions.* Acta Math. Hung. 61 (1993), 241–258.

4) Let $S(\alpha) = \sum_{n \le N, n \equiv f (\mathrm{mod}\ d)} \Lambda(n) \cdot e(n\alpha)$, where $N \geq 1$, $1 \leq f \leq d$, $(f, d) = 1$

a) If $\left|\alpha - \frac{a}{q}\right| \leq \frac{2}{N}$, $(a, q) = 1$ and $h = (q, d)$, then

$$S(\alpha) \ll (hN/dq^{1/2} + q^{1/2} \cdot N^{1/2} + (h/d)^{2/7} \cdot q^{3/14} \cdot N^{5/7}) \log^{18} N$$

A.F. Lavrik. *An analytic method of estimating trigonometric sums over primes in an arithmetic progression.* Dokl. Akad. Nauk SSSR, 248 (1979), no. 5; English transl., Soviet Dokl. Math. 20 (1979), 1121–1124.

b) Let $\left|\alpha - \frac{a}{q}\right| \leq \frac{2}{N}$, $(a, q) = 1$ and $h = (q, d)$. Then

$$S(\alpha) \ll \left(\frac{hN}{dq^{1/2}} + \frac{q^{1/2} \cdot N^{1/2}}{h^{1/2}} + \frac{N^{4/5}}{d^{2/5}}\right) \log^3 N$$

A. Balog and A. Perelli. *Exponential sums over primes in an arithmetic progression.* Proc. Amer. Math. Soc. 93 (1985), 578–582.

§ X.19 A short exponential rational trigonometric sum

a) Let $g \geq 2$ be an integer, p prime and $\lambda > 0$. Let h be an unbounded, positive, integer-valued function on the primes satisfying $h(p) \leq \frac{\log p}{\log g}$. Let $N_p(\lambda)$ be the number of integers $a = 1, 2, \ldots, p-1$ such that $\left|\sum_{x=0}^{h(p)} e\left(\frac{ag^x}{p}\right)\right| < \lambda\sqrt{h(p)}$. Then $\lim_{p\to\infty} \frac{N_p(\lambda)}{p} = 1 - e^{-\lambda^2}$.

A.G. Postnikov. *On a very short exponential trigonometric sum.* Dokl. Akad. Nauk SSSR, 133 (1960), 1298–1299.

b) If in a) p is of the form $p = g^\tau - 1 (\tau \geq 2$, integer), and

$$\left|\sum_{x=0}^{h(p)} e\left(\frac{ag^x}{p}\right)\right| < \lambda\sqrt{(2c-1)h(p) - c(c-1)\tau}$$

where $(c-1)\tau < h(p) \leq c\tau$ and $c\tau \to \infty$, then

$$\lim_{p\to\infty} \frac{N_p(\lambda)}{p} = 1 - e^{-\lambda^2}$$

L.P. Usol'cev. *On an exponential rational trigonometric sum of special type* (Russian). Dokl. Akad. Nauk SSSR, 151 (1963), 62–64.

§ X.20 Estimates on sums over $e(uh/k)$, when $f(u) \equiv 0 (\mathrm{mod}\ k)$, $0 < u \leq k$ and $k \leq x$

1) a) If D is an integer, which is not a perfect square, then for every integer $h \neq 0$ there exists a constant $A(h)$ such that

$$\left|\sum_{k \leq x} \sum_{\substack{n^2 \equiv D (\mathrm{mod}\ k) \\ 0 < n \leq k}} e(nh/k)\right| \leq A(h) \cdot x^{3/4} \log^2 x$$

C. Hooley. *On the number of divisors of quadratic polynomials*, Acta Math. 110 (1963), 97–114.

b) Let $f(u)$ be an irreducible primitive polynomial of degree $n > 1$ with integer coefficients. Then, for every integer $h \neq 0$ there exists a constant $C(h)$ such that

$$\left|\sum_{k \leq x} \sum_{\substack{f(u) \equiv 0 (\mathrm{mod}\ k) \\ 0 < u \leq k}} e(uh/k)\right| \leq C(h) \cdot x \cdot (\log\log x)^{\frac{1}{2}(n^2-1)} / \log^{\delta_n} x$$

where $\delta_n = \dfrac{n - \sqrt{n}}{n!}$

C. Hooley. *On the distribution of the roots of polynomial congruences.* Mathematika 11 (1964), 39–49.

§ X.21 Exponential sums formed with the Möbius function

1) Let $S(x, \theta) = \sum_{n \leq x} \mu(n) \cdot e(n\theta); \theta \in (0, 1)$. Then:

a) $\sup_{\theta} |S(x, \theta)| = O(x/\log^A x)$
for any $A > 0$

H. Davenport. Quart. J. Math. Oxford Ser. 8 (1937), 313–320.

b) Assuming the Generalized Riemann Hypothesis,

$$\sup_{\theta} |S(x, \theta)| = O(x^{5/6+\varepsilon})$$

D. Hajela and B. Smith. *On the maximum of an exponential sum of the Möbius function.* Lect. Notes Math. 1240, (1987), 145–164.

c) 5/6 in b) can be reduced to 3/4.

R.C. Baker and G. Harman. *Exponential sums formed with the Möbius function.* J. London Math. Soc. II. Ser. (to appear).

d) For sufficiently small $c > 0$

$$\sup_{\theta} |S(x, \theta)| \ll x \cdot \exp(-(c \log x)^{1/2})$$

provided that none of the Dirichlet series $L(s,\chi)$ have Siegel zeros.

(See D. Hajela and B. Smith.)

2) $$\sum_{x\le n\le x+y,\, y\ge x^{2/3+\varepsilon}} \mu(n)e(n\theta) = O(y/\log^A y) \text{ for any } A > 0, \text{ uniformly in } \theta$$

T. Zhan. *Davenport's theorem in short intervals.* Chin. Ann. Math. Ser. B 12, No. 4 (1990), 421–431.

§ X.22 On $\sum_{n\le x} \mu^2(n)e(\alpha n^3)$

Let $f(\alpha) = \sum_{n\le x} \mu^2(n)e(\alpha n^3)$,

where $\alpha = \frac{a}{q} + \beta$ with $(a,q) = 1$ and $|\beta| \le \frac{1}{q^2}$

a) Let $0 < \nu \le 1/7$ and suppose that $|\beta| \le \frac{1}{q^2}\cdot x^{5\nu-3}$ with $q \le x^{3-5\nu}$. Then for $q \ge x^{5\nu}$, we have

$$|f(\alpha)| \ll x^{1-\nu+\varepsilon}$$

b) If $S(q,b) = \sum_{r=1}^{q} e\left(\frac{br^3}{q}\right)$ and the conditions of a) are valid, for $q \le x^{5\nu}$, we have

$$f(\alpha) = \sum_{d=1}^{\infty} \frac{\mu(d)S(q,ad^6)}{qd^2} \int_0^x e(\beta\gamma^3)d\gamma + O(x^{1-\sigma+\varepsilon})$$

where $\sigma = \frac{1}{2} - \frac{5\nu}{4}$

c) Let $s(q)$ be a multiplicative function defined on prime powers by $s(p^{3j+k}) = p^{-j-k/2} (0 \le k \le 2)$. Then

$$\sum_{d=1}^{\infty} \frac{\mu(d)S(q,ad^6)}{d^2 q} \ll q^{\varepsilon}\cdot s(q)$$

R.C. Baker and J. Brüdern. *Sums of cubes of squarefree numbers.* Monatsh. Math. 111 (1991), 1–21.

§ X.23 The sum of $e(\alpha n)$, when $\Omega(n) = k$

Let $\pi_k(x, \alpha) = \sum_{n \le x, \Omega(n)=k} e(\alpha n)$, where $\Omega(n)$ is the total number of prime factors of n. Then, uniformly for $x \ge 1$, $k \ge 1$ and all real numbers α,

$$\pi_k(x, \alpha) = \pi_k(x, 0) \cdot (E(x, \alpha)) + O(\delta_k(x)))$$

where $E(x, \alpha) = \frac{1}{x} \sum_{n \le x} e(\alpha n)$, $\delta_k(n) = \frac{1}{\sqrt{\log \log x}} + \frac{|k - \log \log x|}{k + \log \log x}$

G. Tenenbaum. *Facteurs premiers de sommes d'entiers.* Proc. Amer. Math. Soc. 106 (1989), 287–296.

§ X.24 Exponential sums involving the Ramanujan function

1) a) $\sum_{n \le x} \tau(n) \cdot e(n\alpha) \ll x^6 \cdot \log x$, for $x > 2$ and $\alpha \in \mathbb{R}$, where $\tau(n)$ denotes Ramanujan's arithmetic function.

J.R. Wilton. Math. Proc. Cambridge Philos. Soc. 25 (1929), 121–129.

b) $\sum_{n \le x} \tau(n) \cdot e(n\alpha) \ll x^6$

M. Jutila. *On exponential sums involving the Ramanujan function.* Proc. Indian Acad. Sci. Math. Sci. 97 (1987), 157–166.

Remark. For the methods of proof, see

M. Jutila. *Lectures on a method in the theory of exponential sums.* Tata Instit. Fund. Res., Bombay (1987).

2) a) $\frac{1}{m} \cdot \sum_{n \le m} \tau(n)(m - n) \cdot e(nx) \le A \cdot m^6$

and

b) $\left(\frac{1}{m} \cdot \sum_{n \le m} \left| \sum_{k \le n} \tau(k) \cdot e(kx) \right|^2 \right)^{1/2} \le Am^6$

where A is independent of x

L.A. Parson and M. Sheingorn. *Exponential sums connected with Ramanujan's function $\tau(n)$.* Mathematika 29 (1982), 270–277.

3) Let $\rho(s; p, q) = \sum_{n=1}^{\infty} \tau(n) \cdot e(np/q) \cdot n^{-s}$, where p, q are integers with $0 \le p < q$.

Then $\rho(s; p, q)$ converges for $Re\ s > 6 - \frac{1}{6}$ and

$\rho(6; p, q) \ll \log q$

(See L.A. Parson and M. Sheingorn).

§ X.25 An exponential sum involving $r(n)$ (number of representations of n as a sum of two squares)

Let $Q(N) = \sum_{n=1}^{N} r(n) \cdot e(\sqrt{xn})$, where $x > \varepsilon_0 > 0$ is a real parameter and $r(n)$ is the number of representations of n as a sum of two squares. Then

$$Q(N) \leq c(x) \cdot \sqrt{N} \cdot \log N$$

where $c(x) \ll x$

V.V. Potockiĭ. *A sharpening of the estimate of a certain trigonometric sum* (Russian). Izv. Vysš. Učebn. Zaved. Matematika 1969, 3(82), 42–51.

§ X.26 Exponential sums on integers having small prime factors

1) Let $S(x, y) = \{n \leq x\colon P(n) \leq y\}$, where $P(n)$ denotes the greatest prime factor of n (with the convention $P(1) = 1$).

   a) Let $A > 0$. Assuming
   $x \geq 3, \exp(c_1(\log\log x)^2) \leq y \leq x, 2 \leq q \leq (\log x)^A$,
   $(a, q) = 1$, we have uniformly

   $$E\left(x, y; \frac{a}{q}\right) = V_q(x, y) + O(\psi(x, y) \cdot \exp(-c_2\sqrt{\log y}))$$

   where $E(x, y, \theta) := \sum_{n \in S(s,y)} e(\theta n)$

   and

   $$V_q(x, y) = x \int_{-\infty}^{+\infty} \rho(u - v) dS_q(y^v)$$

   for $x \notin \mathbb{I}^+$ and $V_q(x, y) = V_q(x + 0, y)$ for $x \in \mathbb{I}^+$, with ρ the Dickmann function and

   $$S_q(t) = \frac{1}{t} \sum_{n \leq t} \frac{\mu(q/(n, q))}{\varphi(q/(n, q))}$$

   $(t > 0)$

   ($\rho(v)$ is the solution of the differential equation
   $v\rho'(v) + \rho(v - 1) = 0$ for $v > 1$ and $\rho(v) = 0$ for $-\infty < v < 0$,
   $\rho(v) = 0$ for $0 \leq v \leq 1$).

b) Let $\delta > 0$, $A > 0$. There exists a constant $B = B(\delta, A)$ such that, for $Q = x/\log^B x$ and with the conditions

$$x \geq 3,\ x^{\delta \log\log\log x/\log\log x} \leq y \leq x,$$

$$2 \leq q \leq Q,\ (a, q) = 1,\ \left|\theta - \frac{a}{q}\right| \leq \frac{1}{qQ},$$

we have uniformly

$$E(x, y; \theta) \ll \psi(x, y) \cdot \left(\frac{2^{\omega(q)} \cdot \log q}{\varphi(q)} \cdot \frac{\log(u+1)}{\log y} + \left(\frac{1}{\log y}\right)^A\right)$$

where $u = \dfrac{\log x}{\log y}$, $\psi(x, y) = \operatorname{card} S(x, y)$ and $\omega(q) =$ number of distinct prime factors of q

É. Fouvry and G. Tenenbaum. *Entiers sans grand facteur premier en progressions arithmétiques*. Proc. London Math. Soc. III. Ser. 63, No. 3 (1991), 449–494.

2) Let $\theta \in (0, 1)$ and put $m(\theta) = \begin{cases} \dfrac{\Lambda(q)}{\varphi(q)} & \text{if } \theta = \dfrac{a}{q},\ (a, q) = 1 \\ 0 & \text{if } \theta \in \mathbb{R}\backslash Q \end{cases}$

(where Λ is von Mangoldt function). Then

$$\lim_{y \to \infty} \sum_{\substack{n=1 \\ P(n) \leq y}}^{\infty} \frac{e(\theta n)}{n} = \log\left(\frac{1}{1 - e(\theta)}\right) + m(\theta)$$

(See É. Fouvry and G. Tenenbaum.)

§ X.27 A result on $\sum_{n \leq N} e(x\sqrt{n})$

Let N be a positive integer and $x > 0$. Denote

$$E_N(x) = \sum_{n \leq N} e(x\sqrt{n})$$

Then $E_N(x) = \displaystyle\int_1^N e(x\sqrt{t})\mathrm{d}t + \frac{1}{\sqrt{2}} x A(x) + O(\sqrt{x+1})$, where

$0 < x < \sqrt{cN}$, $0 < c < 4$ and $A(x) = \displaystyle\sum_{m=1}^{\infty} m^{-3/2} \cdot \exp\left(i\left(\frac{\pi x^2}{2m} - \frac{\pi}{4}\right)\right)$

S. Kanemitsu. *On an exponential sum*. I. Prospects of mathematical science (Tokio, 1986), 63–72, World Sci. Publishing, Singapore, 1988.

Remark. $\limsup_{x \to \infty} |A(x)| > \dfrac{1}{\sqrt{2}} \cdot \zeta\left(\dfrac{3}{2}\right)$

§ X.28 Kloosterman sums. Salié's and Weil's estimates

1) Let p be a prime and write

$$S(a,b;p)=\sum_{h=1}^{p-1} e\left(\frac{ah+b\overline{h}}{p}\right)$$

where $\overline{h}h \equiv 1 (\bmod p)$. Then, if $p \nmid b$,

$$|S(a,b;p)| \le 2p^{1/2}$$

(Weil's estimate for the Kloosterman sum).

A. Weil. *On some exponential sums.* Proc. Nat. Acad. Sci. USA 34 (1948), 204–207.

2) Let $S(a,b;k)=\sum_{\substack{0<h\le k\\(h,k=1)}} e\left(\frac{ah+b\overline{h}}{k}\right)$ where $\bar{h}\cdot h \equiv 1 (\bmod k)$

a) If $k=p^{\alpha}$ ($\alpha \ge 2$), then

$$S(a,b;p^{\alpha}) \le 3p^{\alpha/2}$$

if $(b,p)=1$, (p prime.)

H. Salié. *Über die Kloostermanschen Summen $S(u,v;q)$.* Math. Z. 34 (1931), 91–109.

Remark. From the multiplicative property of Kloosterman sums (see G. H. Hardy and E. M. Wright. *An introduction to the theory of numbers*, Oxford, 1960) and from inequalities 1) and 2) a) it follows that

$$|S(a,b;k)| \le k^{1/2}\cdot d(k)\cdot (b,k)^{1/2}$$

See also

C. Hooley. Proc. London Math. Soc. (3) 7 (1957), 396–413.

b) $$\sum_{k\le x} S(a,b;k) = O\left(x^{3/2}\cdot \sigma^2_{-\frac{1}{2}}(b)\log^{\sqrt{2}-1}x\cdot(\log\log x)^c\right)$$

where c is a positive constant and $\sigma_{-\frac{1}{2}}(b)=\sum_{d|b} d^{-1/2}$

C. Hooley. *On the distribution of the roots of polynomial congruences.* Mathematika 11 (1964), 39–49.

3) $$\sum_{\nu_1\le r\le \nu_2} e\left(\frac{b\overline{r}}{q}\right) \ll q^{1/2+\varepsilon}\cdot (b,q)^{1/2} + \frac{(\nu_2-\nu_1)}{q}(b,q)$$

C. Hooley. *On the Brun-Titchmarsh theorem.* J. Reine Angew. Math. 255 (1972), 60–79.

4) For any $\varepsilon>0$ and a,b positive integers with $(a,b)=1$ we have

$$\sum_{\substack{k\le n\\ k\overline{k}\equiv 1(\text{mod b})}} \mu(k)\chi_b(k)e\left(\frac{\overline{k}a}{b}\right) \ll_{\varepsilon} nb^{\varepsilon}\cdot\left(\frac{(\log n)^{5/2}}{b^{1/2}}+\frac{(\log n)^{11/5}\cdot b^{3/10}}{n^{1/5}}\right)$$

$k\bar{k} \equiv 1 (\mathrm{mod}\ b)$

D. Hajela, A. Pollington and B. Smith. *On Kloosterman sums with oscillating coefficients.* Canad. Math. Bull. 31 (1988), 32–36.

5) Let $\varepsilon > 0$, a complex number w with $\mathrm{Re}\, w = \theta > 0$ and positive integers h_1, h_2, d, M, l be given. Suppose that $r_1, r_2 \in \{-1, 1\}$
Then

$$\sum_{\substack{1 \le m_1, m_2 \le M, m_j \equiv r_j (\mathrm{mod}\ 4) \\ (m_1 h_1, m_2 h_2) = d}} (m_1 m_2)^w \cdot e\left(l \frac{\overline{m_1 h_1 / d}}{m_2 h_2 / d} \right) \ll$$

$$\ll (1 + |w|) M^{2\theta + 1} \cdot d(l) (l, h_2)^{1/2} \cdot (M h_2)^{1/2 + \varepsilon}$$

Here $^-$ denotes inverse modulo the denominator.

G. Harman. *Sums of two squares in short intervals.* Proc. London Math. Soc. (3) 62 (1991), 225–241.

Note. For related results, see

H. Iwaniec. *On mean values for Dirichlet polynomials and the Riemann zeta-function.* J. London Math. Soc. (2) 22 (1980), 39–45.

6) Let $U, V \ge 1$ be real, and t, w be integers with $(t, w) = 1$, $V \le 3w$. Then for any $\varepsilon > 0$, we have

$$\sum_{1 \le u \le U} \left| \sum_{\substack{1 \le v \le V \\ (v, w) = 1}} e(\bar{v} u t / w) \right| \ll U \cdot w^{1/2 + \varepsilon}$$

(Here $\bar{v}$ is a solution of $xv \equiv 1 (\mathrm{mod}\ w)$)

C. Hooley. *Application of Sieve Methods.* Cambridge, 1976. (See chap. 2, Lemmas 6 and 7).

7) Let p be a prime, $(p, a) = 1$, and χ a character modulo p which is not a Legendre symbol. Then

$$\left| \sum_{x=1}^{p-1} \chi(x) \cdot e(ax + \bar{x}) p \right| \le 2\sqrt{p}$$

S. Chowla. *On Kloosterman's sum.* Norske Vid. Selsk. Forh. (Trondheim) 40 (1967), 70–72.

§ X.29 Exponential sums connected with the distribution of $\alpha p (\mathrm{mod}\ 1)$ and with diophantine approximation with primes or almost primes

a) Suppose that $\varepsilon > 0$, $N > R, J, M \ge 1$, $1 < q \le N$, $\log |a| \ll \log N$, $(a, q) = 1$. Then

$$\sum_{R \le r < 2R} \sum_{J \le j < 2J} \sum_{M \le m < 2M} \left| \sum_{n \le N/m} e\left(\frac{ajmn}{rq} \right) \right| \ll$$

$$\ll (\log N) \cdot (JM)^{\varepsilon/3} \cdot \sum_{R \le r < 2R} \left(\frac{JN \cdot (r, a)}{rq} + JM + qR \right)$$

R.C. Vaughan. *On the distribution of* αp modulo 1. Mathematika 24 (1977), 135–141.

b) Suppose that $\varepsilon > 0, N \geq R, L, M \geq 1, 1 < q \leq N, (a, q) = 1$ and $q \ll a \ll q, \max\left(\frac{JM}{qR}, \frac{qM}{N}\right) < 1, |a_n|, |b_m| \ll N^{\varepsilon}$. Then

$$\sum_{R \leq r < 2R} \sum_{L \leq l < 2L} \left| \sum_{M \leq m < 2M} b_m \cdot \sum_{n \leq N/m} a_n e\left(\frac{lmna}{qr}\right) \right| \ll$$

$$\ll N^{3\varepsilon} \cdot NR\left(L + \frac{R}{M}\right) \cdot \left(\frac{M}{N} + \frac{1}{RM(L + R/M)}\right)^{1/2}$$

c) Suppose that $N, R, L \geq 1, q \ll a \ll q, (a, q) = 1, \varepsilon > 0$ and $TN^{1/3}/R < q < N^{2/3}, T = \max(L, R)$. Then we have

$$\sum_{R \leq r < 2R} \sum_{L \leq \ell < 2L} \left| \sum_{n \leq N} \Lambda(n) e\left(\frac{na\ell}{qr}\right) \right| \ll N^{\varepsilon} \cdot (N^{2/3} \cdot T \cdot R + N^{11/12} \cdot (TR)^{1/2})$$

Remark. The exponent $11/12$ can be reduced to $9/10$.

G. Harman. *Diophantine approximations with a prime and an almost prime.* J. London Math. Soc. (2) 29 (1984), 13–22.

§ X.30 On $e(\alpha x^3)$

1) Let $S(q, b) = \sum_{r=1}^{q} e\left(\frac{br^3}{q}\right)$

a) Write $q = uv^3$ where u is cube-free, and define $\psi(q) = u^{-1/2} \cdot v^{-1}$. If $(b, r) = 1$ we have

$$r^{-1} \cdot S(r, b) \ll \Psi(r) \cdot r^{\varepsilon}$$

R.C. Vaughan. *The Hardy-Littlewood method.* Cambridge, 1981 (Lemma 3, 4.3–4.5).

b) For real α and rational a/q with $(a, q) = 1$ let $\beta = \alpha - a/q$. Then

$$\sum_{x=1}^{X} e(\alpha x^3) - \frac{1}{q} S(q, a) \cdot \int_0^X e(\beta x^3) dx \ll q^{\frac{1}{2}+\varepsilon}(1 + X^3 \cdot |\beta|)^{1/2}$$

R.C. Vaughan. *Some remarks on Weyl sums.* Colloq. Math. Soc. János Bolyai 34 (1585–1602). Topics in Classical Number Theory, Budapest 1981. Elsevier, 1984.

2) Let $S(\alpha) = \sum_{P \leq x \leq 2P} e(\alpha x^3)$ and $T(\alpha) = \sum_{P^{4/5} \leq x \leq 2P^{4/5}} e(\alpha x^3)$

Then

$$\int_0^1 |S(\alpha)|^2 \cdot |T(\alpha)|^4 d\alpha = O(P^{13/5+\varepsilon})$$

(Davenport's mean value theorem.)

H. Davenport. *On Waring's problem for cubes.* Acta Math. 71 (1939), 123–143.

3) Let $f_d(X, \alpha) = \sum_{X \le x \le 2X, (x,d)=1} e(\alpha x^3)$

If d is an integer with $\ll 1$ divisors, and

$s_d(q, c) = q^{-1} \cdot \sum_{b|d} \mu(b) \cdot \frac{S(q, b^3 c)}{b}$, then for $\lambda = 1, 2, 3, \ldots$ we have

$$f_d(X, \lambda\alpha) = s_d(q, \lambda a) \cdot J\left(\lambda\left(\alpha - \frac{a}{q}\right), X\right) +$$

$$+ O\left(q^{\frac{1}{2}+\varepsilon} \cdot \left(1 + \lambda X^3 \cdot \left|\alpha - \frac{a}{q}\right|\right)^{\frac{1}{2}}\right)$$

for any $X > 0$, real α and rational number a/q. Here

$$J(\beta, A) = \int_A^{2A} e(\beta x^3) \mathrm{d}x$$

R.C. Baker. *Diagonal cubic equations.* II. Acta Arith. 53 (1989), 217–250.

§ X.31 Exponential sums and the logarithmic uniform distribution of $(\alpha n + \beta \log n)$

1) A function f: $[1, \infty) \to \mathbb{R}$ is said to be of class H if there are real numbers $1 = x_0 < x_1 < \cdots < x_H$ such that f is monotonic in each of the intervals $[x_{j-1}, x_j] (j = 1, \ldots, H)$ and $[x_H, \infty)$.
Let f be a real valued twice differentiable function on $[1, \infty)$. Suppose that there are positive constants c, K, δ, H with the following properties:

(i) $x(f'(x) - \lambda)$ is of class H for every real λ

(ii) either $\frac{c}{x^2} \le f''(x) \le \frac{K}{x^{1+\delta}}$ or $\frac{c}{x^2} \le -f''(x) \le \frac{K}{x^{1+\delta}}$ for $x \ge 1$

a) If h and ν are integers with $h \ge 1$; and $B > A \ge 1$, then

$$\left| \int_A^B \frac{e(hf(x) - \nu x)}{x} \mathrm{d}x \right| < C_1(c, H) h^{-1/2}$$

(where $C_1(c, H)$ is a constant depending on c and H)

b) For any natural numbers $h, A, B (\ge 1)$ we have

$$\left| \sum_{n=A}^{B} \frac{e(hf(n))}{n} \right| < C_2(c, K, \delta, H) \cdot (h^{1/2} \cdot A^{-\delta} + h^{-1/2})$$

R.C. Baker and G. Harman. *Sequences with bounded logarithmic discrepancy.* Math. Proc. Camb. Phil. Soc. 107 (1990), 213–225.

Corollary. ($\delta = H = 1, c = K = |\beta|$, where $\alpha, \beta \neq 0$).

$$\left|\sum_{n=A}^{B} \frac{e(h(\alpha n + \beta \log n))}{n}\right| < C_3(\beta) \cdot (h^{1/2} \cdot A^{-1} + h^{-1/2})$$

R. Tichy and G. Turnwald. *Logarithmic uniform distribution of* $(\alpha n + \beta \log n)$. Tsukuba Math. J. 10 (1986), 351–366.

2) Let p be a prime, and denote

$$S(\gamma, \theta, A, B) = \sum_{A \leq P \leq B} \frac{\log p}{p} \cdot e(\gamma p + \theta \log p)$$

a) Let $\gamma \in \mathbb{R}$, and suppose that h, A, B are positive integers with $h^2 \leq A < B$. Then, for all non-zero real β, we have

$$S(\gamma, \theta, A, B) \ll (\log h)^{-100}$$

b) Let the hypothesis of a) be given, and suppose that a, q are integers with $|\gamma - \frac{a}{q}| < q^{-2}$ and $(a, q) = 1, q \geq 1$. Then

$$S(\gamma, h\beta, A, B) \ll h((\log B)^5 q^{-1/2} + (\log A)^4 \cdot (A^{-1/5} + A^{1/2} \cdot q^{1/2}))$$

c) Under the same hypothesis as in b), we have

$$S(\gamma, h\beta, A, B) \ll (\log B)^5 \cdot (h^{-1/2} + q^{-1/2}) + (\log A)^4 \cdot (h^{1/2} \cdot A^{-2/3} + A^{-1/6} + A^{-1/2}.$$

d) Let the hypothesis of b) given, and suppose also that

$$q \leq (\log A)^{1000}, \gamma - a/q = \theta, B|\theta| \leq (\log B)^{2000}$$

Then

$$S(\gamma, h\beta, A, B) = \frac{\mu(q)}{\varphi(q)} \cdot \sum_{n=A}^{B} \frac{1}{n} e(\theta n + h\beta \log n) + O(\exp(-c(\log A)^{1/2}))$$

($c > 0$, a constant and φ is Euler's function).

(See R.C. Baker and G. Harman.)

§ X.32 Exponential sums with multiplicative coefficients

1) Let $f: \mathbb{N} \to \mathbb{C}$ be a multiplicative function satisfying

$$\sum_{n \leq x} |f(n)|^2 = O(x)$$

Then

$$\lim_{x \to \infty} \frac{1}{x} \cdot \sum_{n \leq x} f(n) \cdot e(\alpha n) = 0$$

for α irrational.

H. Daboussi and H. Delange. *On multiplicative arithmetical functions whose modulus does not exceed one.* J. London Math. Soc. (2) 26 (1982), 245–264.

2) Let N be a positive integer, let $2 \leq R \leq \sqrt{N}$, and suppose that α is a real number satisfying $|\alpha - r/s| \leq R/sN$, where $(r, s) = 1$, $R \leq s \leq N/R$.
Then, for any multiplicative function f satisfying $|f| \leq 1$, we have

$$\left|\sum_{n \leq N} f(n)e(\alpha n)\right| \ll \frac{N}{\log N} + \frac{N \log R}{\sqrt{R}}$$

where the implied constant is absolute.

H. Montgomery and R.C. Vaughan. *Exponential sums with multiplicative coefficients.* Invent. Math. 43 (1977), 69–82.

§ X.33 On $\sum \Phi(u)\Psi(v)e(f(uv))$

a) Suppose $\varepsilon > 0$ is given. Let $f(x)$ be a real valued polynomial in x of degree $k \geq 2$.
Put $\gamma = 4^{1-k}$. Suppose α is the leading coefficient of f and there are integers a, q such that
$|q\alpha - a| < 1/q, (a, q) = 1$.
Let $\Phi(u)$, $\Psi(v)$ be real functions and denote $T = \max |\Psi(v)|$,

$$F = \left(\frac{1}{w}\left(\sum_{u \leq w} \Phi^2(u)\right)\right)^{1/2}.$$

If $T = o(x^\delta)$ and $F = o(x^\delta)$ for every $\delta > 0$, then

$$S = \sum_{u=1}^{w} \sum_{\substack{v=1 \\ uv \leq M}}^{x} \Phi(u)\psi(v)e(f(uv)) \ll (xw)^{1+\varepsilon}(x^{-R} + w^{-1} + \\ + q^{-1} + (xw)^{-k}q)^\gamma$$

where $R = 2^{k-1}(M, w, x$ positive integers.)

G. Harman. *Trigonometric sums over primes.* I. Mathematika 28 (1981), 249–254.

b) Suppose the conditions of a) are satisfied, but either $\Phi(x) = 1$ for all x, or $\Phi(x) = \log x$ for all x. Then

$$S \ll (xw)^{1+\varepsilon} \cdot x^{(k-1)/R} \cdot (q^{-1} + q(wx)^{-k}w^{-1})^{1/R}$$

(See G. Harman.)

§ X.34 Exponential sums involving quadratic polynomials and sequences

1) Let (a_n) be a sequence of non-negative reals and (β_n) a sequence of real numbers. Then, if

$$\sum_{l=1}^{L}\left|\sum_{n=1}^{N} a_n e(l\beta_n)\right| \leq \frac{1}{6}\sum_{n=1}^{N} a_n$$

we have $\max\limits_{0\leq\alpha<1}\min\limits_{1\leq n\leq N}\|\beta_n+\alpha\| \leq \frac{1}{L}$ (L, N are positive integers.)

R.C. Baker. *Diophantine Inequalities.* Clarendon Press, Oxford, 1986. (Theorem 2.2).

2) Let $h(y) = \alpha y^2 + by$ be a quadratic polynomial with real coefficients, α irrational. Then

$$\sum_{m=1}^{M}|\sum_{n=1}^{N} c_n e(h(n+m))|^2 \leq J\cdot\sum_{n=1}^{N}|c_n|^2$$

with $J = \min\left(M + \sum_{j=1}^{N}\frac{1}{\|2j\alpha\|}, N + \sum_{j=1}^{M}\frac{1}{\|2j\alpha\|}\right)$, where $c_n \in \mathbb{C}$; $M, N \in \mathbb{I}_*^+$

P.D.T.A. Elliott. *Arithmetic functions and integer products.* Springer Verlag, 1985, pp. 317–318.

§ X.35 The large sieve as an estimate for exponential sums

1) Let $S(\alpha) = \sum\limits_{n=M+1}^{M+N} a_n\cdot e(n\alpha)$, where the a_n are any complex numbers. Then, if

$Z = \sum\limits_{n=M+1}^{M+N}|a_n|^2$,

a) $\sum\limits_{q\leq Q}\sum\limits_{a=1,(a,q)=1}^{q}\left|S\left(\frac{a}{q}\right)\right|^2 \ll (Q^2+N)\cdot Z$

E. Bombieri. *On the large sieve.* Mathematika 12 (1965), 201–255.

Corollary. Denote by $Z(a, p)$ the number of integers in an arbitrary set of Z, integers of length N, which are $\equiv a (\text{mod}\, p)$. Then

$$\sum_{p\leq Q} p\cdot\sum_{a=1}^{p}\left(Z(a,p)-\frac{Z}{p}\right)^2 \ll (Q^2+N)Z$$

(p-prime)

Remark. This inequality of "large-sieve" type improves the earlier results (large sieves) by Linnik, Rényi and Roth.

Yu.V. Linnik. *The large sieve.* Dokl. Akad. Nauk SSSR, 30 (1941), 292–294;
A. Rényi. *On the large sieve of Yu. V. Linnik.* Compositio Math. 8 (1950), 68–75;
K.F. Roth. *On the large sieve of Linnik and Rényi.* Mathematika 12 (1965), 1–9.

b) Let $\alpha_1, \ldots, \alpha_R$ be an arbitrary sequence of real numbers with $\|\alpha_j - \alpha_k\| \geq \delta$ for $j \neq k$. Then

$$\sum_{r=1}^{R} |S(\alpha_r)|^2 \leq 2.2 \max\left(\frac{1}{\delta}, N\right) \cdot Z = \Delta \cdot Z$$

H. Davenport and H. Halberstam. *The values of a trigonometric polynomial at well spaced points.* Mathematika 13 (1966), 91–96.

c) The result from b) is valid with $\Delta = \frac{1}{\delta} + \pi \cdot N$

P.X. Gallagher. *The large sieve.* Mathematika 14 (1967), 14–20.

d) When $N\delta \leq 1$, the optimal Δ satisfies

$$\frac{1}{\delta} \cdot \left(1 + \frac{1}{12}(N\delta)^3\right) \leq \Delta(N, \delta) \leq \frac{1}{\delta} \cdot (1 + 270(N\delta)^3)$$

E. Bombieri and H. Davenport. *Some inequalities involving trigonometrical polynomials.* Ann. Scuola Norm. Sup. Pisa Cl. Sci. 23 (1969), 223–241.

Remark. It is shown also that in b) one can replace 2.2 with 2.

e) $\Delta \leq N + \frac{2}{\delta}$

E. Bombieri. *A note on the large sieve.* Acta Arith. 18 (1971), 401–404.

f) The theorem from b) ("The large sieve") is valid with

$$\Delta = N - 1 + \frac{1}{\delta}$$

(A. Selberg.)

See

H.L. Montgomery. *The analytic principle of the large sieve.* Bull. Amer. Math. Soc. 84 (1978), no. 4–6, 547–567).

g) Let $\alpha_1, \ldots, \alpha_R \in \mathbb{R}$ be distinct modulo 1 and $\delta = \min_{r \neq s} \|\alpha_r - \alpha_s\|$ Then

$$\sum_{r=1}^{R} |S(\alpha_r)|^2 \leq \left(N + \frac{1}{\delta}\right) \cdot Z$$

and

$$\sum_{r=1}^{R} \left(N + \frac{3}{2} \cdot \frac{1}{\delta_r}\right)^{-1} \cdot |S(\alpha_r)|^2 \leq Z$$

where $\delta_r = \min_{s, r \neq s} \|\alpha_r - \alpha_s\|$

H.L. Montgomery and R.C. Vaughan. *The large sieve.* Mathematika 20 (1973), 119–134.

h) Let $N - \delta(\alpha)$ be the number of r for which $\|\alpha - \alpha_r\| < \delta$.
Then

$$\sum_r N_\delta(\alpha_r)^{-1}|S(\alpha_r)|^2 \leq \left(\pi N + \frac{1}{\delta}\right) \cdot \sum |a_n|^2$$

H.L. Montgomery. *Topics in multiplicative number theory.* Lectures Notes No. 227, Springer, 1971.

i) If $N \cdot \delta \leq \frac{1}{4}$, then

$$\sum_{r=1}^{R} |S(\alpha_r)|^2 < \frac{1}{\delta} \cdot (1 + 22N^3\delta^3) \cdot Z$$

M.G. Lu. *An inequality involving trigonometrical polynomials.* Kexue Tongbao 27 (1982), 1151–1156.

2) a) Let $Q \geq 10$ and $N \leq Q^{1+\theta}$, where $0 < \theta < 1$. Then

$$\sum_{p \leq Q} \sum_{k=1}^{p-1} \left| S\left(\frac{k}{p}\right) \right|^2 \leq C \cdot (1-\theta)^{-1} \cdot Q^2 \cdot \log\log Q \cdot (\log Q)^{-1} \cdot Z$$

where p is a prime and C an absolute constant.

D. Wolke. *Farey fractions with primne denominator and the large sieve.* Colloq. de Théorie des Nombres (Bordeaux, 1969), pp. 183–188, Soc. Math. France, Paris, 1971.

b) $$\sum_{p \leq Q} \sum_{k=1}^{p-1} \left| S\left(\frac{k}{p}\right) \right|^2 \ll ((\log\log Q)/(2\nu - 1)\log Q) \cdot Q^2 \cdot Z$$

where $Q = N^\nu$, $\frac{1}{2} < \nu < 1$

D. Wolke. *On the large sieve with primes.* Acta Math. Hungar. 22 (1971/72), 239–247.

3) Let P denote a set of primes $\leq Q$. Define $\Delta(N, Q) =$ the least number for which the inequality

$$\sum_{p \in P} \sum_{k=1}^{p-1} \left| S\left(\frac{k}{p}\right) \right|^2 \leq \Delta(N, Q) \cdot \sum_{n=1}^{N} |a_n|^2$$

is satisfied ($a_i \in \mathbb{C}$). Then

$$\Delta(N, Q) \geq \max\left(N, \sum_{p \in P}(p-1)\right)$$

and

$$\Delta(N, Q) = \sum_{p \in P} p + O(N^2/\log N)$$

P.D.T.A. Elliott. *On inequalities of large sieve type.* Acta Arith. 18 (1971), 405–422.

§ X.36 An estimate for the derivative of a trigonometric polynomial

Let $S(x) = \sum_{n=-N}^{N} a_n \cdot e(nx)$ be a real valued trigonometric polynomial, and let the zeros of $S(x)$ in $0 \leq x \leq 1$ be at $y_1, \ldots, y_m$. Then

$$\sum_{m=1}^{M} |S'(y_m)|^2 \ll N^3 \cdot \sum_{n=-N}^{N} |a_n|^2$$

H. Davenport. *The zeros of trigonometric polynomials.* Mathematika 19 (1972), 88–89.

§ X.37 Weighted exponential sums and discrepancy

Let $P = (p_n)_{n=1,2,\ldots}$ with $p_n \geq 0$, $p_1 > 0$ and $S_N = \sum_{n=1}^{N} p_n \to \infty$ as $N \to \infty$

For $0 \leq x_1, \ldots, x_N < 1$ let

$$D_N^*(P) = D_N^*(P; x_1, \ldots, x_N) = \sup_{0<t\leq 1} \left| \frac{1}{S_N} \cdot \sum_{n \leq N, x_n < t} p_n - t \right|$$

If $C_N(P)$ denotes the least constant such that

$$\Big| \sum_{n \leq N} p_n e(x_n) \Big| \leq C_N(P) \cdot D_N^*(P) S_N$$

for all such sequences $x_1, \ldots, x_N$, then

$$4 - (12\pi m_N / S_N)^{2/3} \leq C_N(P) \leq 4 - (\sqrt{3} m_N / 2S_N)^2$$

Here $m_N = \max\{p_1, \ldots, p_N\}$

J. Horbowicz and H. Niederreiter. *Weighted exponential sums and discrepancy.* Acta Math. Hung. 54 (1989), 89–97.

§ X.38 Deligne's estimates

Let $g = g(x_1, \ldots, x_s)$ be a polynomial of degree d over a finite field $GF(q)$ of characteristic p, with $d \not\equiv 0 (\mathrm{mod}\, p)$.

Assume that g is nonsingular at infinity, i.e. the maximal degree homogeneous part of g is nonsingular as a form over the algebraic closure of $GF(q)$.

Let T denote the trace map from $GF(q)$ to $GF(p)$.

Then

$$\left| \sum_{x_1,\ldots,x_s \in GF(q)} e(T(g)/p) \right| \le (d-1)^s \cdot q^{s/2}$$

P. Deligne. *La conjecture de Weil*, I. Inst. Haute Etudes Sci. Publ. Math. No. 43 (1974), 273–307.

§ X.39 On fourth moments of exponential sums

For large N, fixed α with $\frac{1}{2} < \alpha < 1$, and every $\varepsilon > 0$, one has

$$\int_0^1 \left| \sum_{N \le k \le N+N^\alpha} e^{2\pi i k^2 x} \right|^4 \mathrm{d}x = 2N^{2\alpha} + O(N^{3\alpha-1+\varepsilon})$$

J. Cillernelo and A. Cordoba. *Trigonometric polynomials and lattice points.* Proc. Amer. Math. Soc. 115 (1992), 899–905.

§ X.40 Biquadratic Weyl sums

Let $S_\varepsilon = \sum_{P-\frac{1}{2}\varepsilon < x \le 2P-\frac{1}{2}\varepsilon} e^{2\pi i\alpha\cdot(2x+\varepsilon)^4}$

where $\varepsilon \in \{0, 1\}$ and assume that $P > 10^{80}, 0 < a < q, (a, q) = 1,$ $4 \cdot 10^6 P < q \le \frac{1}{974} P^3, \alpha \in \mathbb{R}$ and $\left|\alpha - \frac{a}{q}\right| < \frac{975}{q} \cdot P^{-3}.$

Then

$$|S_\varepsilon| \le 15.7 P^{0.884} (\log P)^{0.25}$$

J.-M. Deshouillers. *Sur la majoration des sommes de Weyl biquadratiques.* Ann. Sci. Norm. Sup. Pisa, Cl. Sci. IV, Ser. 19, No. 2 (1992), 291–304.

Chapter XI

CHARACTER SUMS

§ XI. 1 Pólya-Vinogradov inequality and a generalization. Character sums modulo a prime power. Burgess' estimate

1) For χ any nonprincipal character modulo p (prime) and any positive integer x

a) $\left|\sum_{a=1}^{x} \chi(a)\right| \leq c\sqrt{p}\log p$

G. Pólya. *Über die Verteilung der quadratische Reste und Nichtreste.* Göttingen Nachrichten, 1918, 21–29

and

I.M. Vinogradov. *On the distribution of residues and non-residues of powers.* Journal of the Physico-Mathematical Society of Perm. 1 (1918), 94–96.

Remark. Actually, one can establish the above inequality with the constant $c = 1$

b) $\left|\sum_{a=N+1}^{N+x} \chi(a)\right| \ll x^{1-1/r} \cdot p^{(r+1)/4r^2} \cdot \log p$

where x and r are arbitrary positive integers and N is any integer.

D.A. Burgess. *On character sums and L-series.* II. Proc. London Math. Soc. (3), 13 (1963), 524–536.

2) Let χ denote a primitive character modulo k. Write

$$S(N,H) = \sum_{n=N+1}^{N+H} \chi(n)$$

a) If $r = 1$ or 2 then, for every $\varepsilon > 0$,

$$S(N, H) \ll H^{1-\frac{1}{r}} \cdot k^{(r+1)/4r^2+\varepsilon}$$

b) For any integer $r > 0$, if k has non-trivial cubic factor then the estimate from a) holds.

D.A. Burgess. *On character sums and L-series.* II. Proc. London Math. Soc. (3), 13 (1963), 524–536.

c) Let $r = 3$. Let $\varepsilon > 0$. For every prime power $k = p^\alpha$, of every prime $p > 3$, the estimate from a) holds (with the implied constant independent of both p and α).

D.A. Burgess. *Estimation of character sums modulo a power of a prime.* Proc. London Math. Soc. (3), 52 (1986), 215–235.

d) Let $r = 3$. Then the estimate from a) holds.

D.A. Burgess. *The character sum estimate with* $r = 3$. J. London Math. (2), 33 (1986), 219–226.

3) Let χ be a nonprincipal primitive even character modulo k. Then

$$\left|\sum_{a=1}^{x}(x-a)^{n-1}\cdot\chi(a) - \frac{1}{n}\left((B_\chi - x)^n - B_\chi^n\right)\right| \leq k^{n-\frac{1}{2}}\cdot\frac{1}{n}\cdot|B_n(x/k) - B_n(0)|$$

for even n, where B_n is the Bernoulli polynomial, and

$$B_\chi^n = k^{n-1}\cdot\sum_{a=1}^{k}\chi(a)\cdot B_n(a/k)$$

is a generalized Bernoulli number. (Generalization of the Pólya-Vinogradov inequality.)

S. Kanemitsu and K. Shiratani. *An application of the Bernoulli functions to character sums.* Mem. Fac. Sci. Kyushu Univ. Ser. A. 30 (1976), 65–73.

§ XI. 2 On the constant in the Pólya-Vinogradov inequality. Large values of character sums

1) Let $S(\chi, x) = \sum_{n\leq x}\chi(n)$, where χ is a nonprincipal character mod q. Then:

a) $S(\chi, x) \ll \sqrt{q}\log q$
uniformly in χ and x

G. Pólya. *Über die Verteilung der quadratischen Reste und Nichtreste.* Nachr. Königl. Ges. Wiss. Göttingen (1918), 21–29

and

I.M. Vinogradov. *Über die Verteilung der quadratischen Reste und Nichtreste.* J. Soc. Phys. Math. Univ. Permi. 2 (1919), 1–14.

Remarks: (i) For a generalization of the Pólya-Vinogradov inequality to arbitrary algebraic number fields of finite degree over the rationals, see

J.G. Hinz. *Character sums in algebric number fields.* J. Number theory 17 (1983), 52–70.

(ii) Pólya prowed the following sharper result: let χ be a nonprincipal character mod q with conductor f. Then

$$\left|\sum_{n\leq x}\chi(n)\right| \leq \left(\frac{1}{\pi} + o(1)\right) f^{1/2} d\left(\frac{q}{f}\right)\log f$$

where the $o(1)$ term is to be interpreted as $f \to \infty$ (and d is the divisor function).

(See G. Pólya (1918).)

b) Let χ be a primitive character mod q. Then

$$\left|\sum_{n=N+1}^{M} \frac{\chi(n)}{n}\right| < \frac{2}{N+1} \cdot \sqrt{q} \log q$$

T.M. Apostol. *Introduction to analytic number theory*. Springer-Verlag, 1976 (See p. 176)

c) For primitive nonprincipal characters χ mod q,

$$\sup_{X \geq 1} |S(\chi, x)| \leq \begin{cases} (c_+ + o(1))\sqrt{q} \log q & \text{if } \chi(-1) = 1 \\ (c_- + o(1))\sqrt{q} \log q & \text{if } \chi(-1) = -1 \end{cases}$$

with $c_+ = 2/(3\pi^2) = 0.0675\ldots$ and $c_- = 1/(3\pi) = 0.106\ldots$ where the term $o(1)$ tends to zero, as $q \to \infty$, uniformly in χ mod p

A. Hildebrand. *On the constant in the Pólya-Vinogradov inequality*. Canad. Math. Bull. 31 (1988), 347–352.

Remark. The previously best known bounds are $c_+ + 1/\pi^2$ and $c_- = 1/(2\pi)$ as constants, due to Landau and Bateman.

E. Landau. *Abschätzungen von Charactersummen, Einheiten und Klassenzahlen*. Nachr. Königl. Ges. Wiss. Göttingen (1918), 79–97.

P.T. Bateman. Unpublished manuscript.

See also

A. Hildebrand. *Large values of character sums*. J. Number Theory 29 (1988), 271–296

where the constant in b) for the case $\chi(-1) = 1$ is reduced by a factor of $\pi/(2\sqrt{3}) = 0.906\ldots$

2) Let $S_0(\chi, x) = S(\chi, x) - \dfrac{\tau(\chi)\varepsilon(\chi)}{\pi i} L(1, \chi)$

where $\tau(\chi)$ is the Gaussian sum for the character χ,

$$\varepsilon(\chi) = \begin{cases} 1 & \text{if } \chi(-1) = -1 \\ 0 & \text{if } \chi(-1) = 1 \end{cases}$$

and

$$L(1, \chi) = \sum_{n \geq 1} \frac{\chi(n)}{n}$$

a) Let χ be a primitive character mod q ($q \geq 3$) and let $(\log q)^{-1/21} \leq \varepsilon \leq 1$. Then the set of real numbers $\alpha \in [0, 1]$, for which

$$|S_0(\chi, \alpha q)| \geq \varepsilon\sqrt{q} \log q$$

holds, has Lebesgue measure $\ll q^{-c\varepsilon}$, where $c > 0$ is an absolute constant.

b) Let χ be as in a) satisfying $\chi(-1) = 1$, and let $0 < \varepsilon \leq 1$. Then we have

$$|S_0(\chi, \varepsilon q)| \ll \sqrt{q} \log q \left(\sqrt{\varepsilon} \log\log \left(\frac{1}{\varepsilon} + 2 \right) + (\log q)^{-1/20} \right)$$

c) Let $S_0(\chi) = \sup_{x \ge 1} |S_0(\chi, x)|$ and define the constants

$$c_+^{(0)} = 1/(3\pi\sqrt{3}) = 0.0612\ldots, c_-^{(0)} = 1/(3\pi) = 0.106\ldots$$

and set

$$c_\pm^{(1)} = 0.75c_\pm^{(0)}, c_\pm^{(2)} = (3/2 - 3/2\sqrt{e})c_\pm^{(0)} = 0.590\ldots c_\pm^{(0)}$$

Then the inequality

$$S_0(\chi) \le \begin{cases} (c_+ + o(1))\sqrt{q} \log q & \text{if } \chi(-1) = 1 \\ (c_- + o(1))\sqrt{q} \log q & \text{if } \chi(-1) = -1 \end{cases}$$

holds with $c_\pm = c_\pm^{(0)}$ for arbitrary primitive characters χ mod q $(q \ge 3)$, $c_\pm = c_\pm^{(1)}$ for primitive characters to cubefree moduli $q (q \ge 3)$, and $c_\pm = c_\pm^{(2)}$ for real-valued primitive characters χ mod q $(q \ge 3)$

A. Hildebrand. *Large value of character sums*. J. Number Theory 29 (1988), 271–296.

3) Let $T(\chi, x) = \sum_{n \le x} \frac{\chi(n)}{n}$ and $T(\chi) = \sup_{X \ge 1} |T(\chi, x)|$

Then, if χ is a nonprincipal character mod q,

$$T(\chi) \le (c + o(1)) \cdot \prod_{p \le \log q} \left| \frac{p-1}{p - \chi(p)} \right| \log q$$

with $c = \frac{1}{4}$ if q is cubefree, and $c = \frac{1}{3}$ otherwise. If, in addition, χ is real-valued, $c = \frac{1}{2} - \frac{1}{2\sqrt{e}}$ may be chosen.

(See A. Hildebrand, (1988).)

4) If χ is a nonprincipal character mod q, then

$$\left| \sum_{n=M+1}^{M+N} \chi(n) \right| < 4\pi^{-2}\sqrt{q} \log q + 0.38q^{1/2} + 0.608q^{-1/2} + 0.116(N, q)^2 q^{-3/2}$$

for primitive character χ (mod q); and

$$\left| \sum_{n=M+1}^{M+N} \chi(n) \right| < (8\sqrt{6}/3\pi^2)\sqrt{q} \log q + 0.63\sqrt{q} + q^{-1/2} + 0.2(N, q)^2 q^{-3/2}$$

for a nonprimitive character χ (mod q)

Z. Qiu. *On an inequality of Vinogradov for character sums*. (Chinese.) J. Shandong Univ., Nat. Sci. Ed. 26 (1991), 125–128.

Remark. For a generalization of the Pólya-Vinogradov inequality to algebraic number fields, see

P. Söhne. *The Pólya-Vinogradov inequality for totally real algebraic number fields.* Acta Arith. 65 (1993), 197–212.

§ XI. 3 Burgess' character sum estimate

1) Let χ be a nonprincipal character modulo a prime p. Then

$$\left|\sum_{n \le N} \chi(n)\right| \le \varepsilon N$$

$N \ge N_0(\varepsilon, p)$ holds with:

a) $N_0(\varepsilon, p) = (\log p) \cdot \sqrt{p}/\varepsilon$

G. Pólya. *Über die Verteilung der quadratische Reste und Nichtreste.* Göttingen Nachrichten, 1918, 21–29

and

I.M. Vinogradov. *On the distribution of residues and non-residues of powers.* Journal of the Physico-Mathematical Society of Perm 1 (1918), 94–96.

b) $N_0(\varepsilon, p) = p^{1/4+\delta}$
for any fixed $\varepsilon, \delta > 0$ and $p \ge p_0(\varepsilon, \delta)$

D.A. Burgess. *On character sums and primitive roots.* Proc. London Math. Soc.(3) 12 (1962), 179–192.

See also

D.A. Burgess. *The distribution of quadratic residues and non-residues.* Mathematika 4 (1957), 106–112.

2) For given $\varepsilon > 0$ there exists $\delta = \delta(\varepsilon) > 0$ and $p_0(\varepsilon) \ge 2$ such that for any non-principal character χ modulo a prime $p \ge p_0(\varepsilon)$, the estimate of 1) holds with $N_0(\varepsilon, p) = p^{1/4-\delta}$. A possible choice for $\delta(\varepsilon)$ is $\delta(\varepsilon) = \exp\left(-c\left(\frac{1}{\varepsilon^2} + 1\right)\right)$ with a sufficiently large absolute constant c

A. Hildebrand. *A note an Burgess' character sum estimate.* C.R. Math. Rep. Acad. Sci. Canada 8 (1986), 35–37.

§ XI. 4 A character sum estimate for nonprincipal character $\chi \pmod q$

Let γ be a non-negative constant with the property that for any $\varepsilon > 0$ there is an $\omega = \omega(\varepsilon) > 0$ such that

$$\sum_{l \le L} \chi(l) \ll L/q^{\omega}$$

for all non-principal characters $\chi \pmod q$ and all $L \ge q^{\gamma+\varepsilon}$.

Then, for any q we have $\gamma = \frac{3}{8}$ and for cube-free q we have $\gamma = \frac{1}{4}$.

D.A. Burgess. *On character sums and L-series*. I–II. Proc. London Math. Soc. 13 (1962), 193–206 and 13 (1963), 524–536.

§ XI. 5 A sum on $\chi(u+v)$, on sets with no two integers of which are congruent

Let S and T be any two sets of integers, such that modulo a given prime p, no two integers of S are congruent, and no two integers of T are congruent. Denote by $N(S)$, $N(T)$ the number of integers in S and T respectively. We have:
For any non-principal character modulo p,

$$\left|\sum_{u\in S, v\in T} \chi(u+v)\right| \le p^{1/2}\cdot\sqrt{N(S)\cdot N(T)}$$

P. Erdős and H.N. Shapiro. On the least primitive root of a prime. Pacific J. Math. 7 (1957), 861–865.

§ XI. 6 A lower bound on a character sum estimate arising in a problem concerning the distribution of sequences of integers in arithmetic progressions

Let χ be a primitive character mod q. Then there exists a positive integer x, such that

$$\left|\sum_{n=x}^{x+[q/2]} \chi(n)\right| > \sqrt{1-8(\log q)/q}\cdot\frac{1}{2\sqrt{2}}\cdot\sqrt{q}$$

A.V. Sokolovskiĭ. *On a theorem of Sárközy*. (Russian.) Acta Arith. 41 (1982), 27–31.

Remark. In case of $q=p$ (prime), Sárközy proved the existence of x with

$$\left|\sum_{n=x}^{x+(p-3)/2} \chi(n)\right| \ge \pi^{-1}\cdot\left(\sqrt{p}-\frac{1}{\sqrt{p}}\right)$$

A. Sárközy. *Some remarks concerning irregularities of distribution of sequences of integers in arithmetic progressions*. IV. Acta Math. Acad. Sci. Hungar. 30 (1977), 155–162.

§ XI. 7 Powers of character sums

1) Let $k>1$ be a positive integer. For a nonprincipal character mod k,

$$\left|\sum_{a=0}^{A}\sum_{n=-a}^{a} \chi(n)\right| \le (A^*+1)\sqrt{k}$$

where A^* is the least positive integer satisfying $A^*\equiv A(\mathrm{mod}\ k)$

L.-K. Hua. *On character sums*. Acad. Sinica Sci. Record 1 (1942), 21–23.

2) Let k and χ be as in 1). Then for any positive integer h,

$$\sum_{n=1}^{k} \left| \sum_{m=1}^{h} \chi(n+m) \right|^2 < k \cdot h$$

D.A. Burgess. *On a conjecture of Norton*. Acta Arith. 27 (1975), 265–267.

Remark. This was conjectured by Norton, who obtained the weaker upper bound $\frac{9}{8}kh$

K.K. Norton. *On character sums and power residues*. Trans. Amer. Math. Soc. 167 (1972), 203–226.

3) a) Let p be a prime. We have

$$\sum_{\chi \neq \chi_0} \sum_{n=1}^{p} \left| \sum_{m=1}^{h} \chi(n+m) \right|^4 \leq 6p^2h^2$$

the sum being over nonprincipal characters χ modulo p.

D.A. Burgess. *Mean values of character sums*. Mathematika 33 (1986), 1–5.

b) For any integer $k > 1$ we have

$$\sum_{(k)} \sum_{n=1}^{k} \left| \sum_{m=1}^{h} \chi(n+m) \right|^4 \leq 8(d(k))^7 \cdot k^2 \cdot h^2$$

where $\sum_{(k)}$ denotes summation over all the primitive characters mod k

D.A. Burgess. *Mean values of character sums*. II. Mathematika 34 (1987), 1–7.

4) Let χ denote a primitive character mod k. Let r be a positive integer, and let h be an integer satisfying $0 < h \leq k^{1/2r}$.

Write $T_r = \sum_{n=1}^{k} \left| \sum_{m=1}^{h} \chi(n+m) \right|^{2r}$. Then:

a) If k is a prime power then

$$T_3 \ll kh^3 \cdot (\log k)^5$$

D.A. Burgess. *Estimation of character sums modulo a power of a prime*. Proc. London Math. Soc. 52 (1986), 215–235.

b) For any k

$$T_3 \ll k^{1+\varepsilon} \cdot h^3$$

D.A. Burgess. *The character sum estimate with $r = 3$*. J. London Math. Soc. (2) 33 (1986), 219–226.

c) Let $k = p^\alpha$ be a power of the prime p. Let r be a positive integer and let h be an integer satisfying $0 < h \leq k^{1/2r}$. Then

$$T_t \ll k \cdot h^r$$

if $(r, \alpha) = (4,4); (4,5); (4,8)$ or $(5,5)$.

D.A. Burgess. *On a set of congruences related to character sums.* J. London Math. Soc. (2) 37 (1988), 385–394.

5) a) Let $k = p^\alpha$ (p prime), let χ be a primitive character mod k and let Ψ run through the additive characters mod k. Write

$$T = \sum_{\psi} \sum_{n=1}^{k} \left| \sum_{m=1}^{k} \psi(n+m)\chi(n+m) \right|^{2r}$$

Then, for $0 < h \leq \dfrac{k^{1/2(r-1)}}{2r}$ we have

$$T \ll k^2 h^r$$

if $(r, \alpha) = (4,3); (4,4); (4,5); (4,6); (5,4); (5,5); (5,8)$ or $(6,5)$.

D.A. Burgess. *On a set of congruences related to character sums.* II. Bull. London Math. Soc. 22 (1990), 333–338.

b) The result is true also for $\alpha = 7$.

D.A. Burgess. Idem, III. J. London Math. Soc. II. Ser. 45 (1992), 201–214.

§ XI. 8 Sums of characters with primes. Vinogradov's theorem

1) Let $S = \sum_{p \leq N} \chi(p+k)$ and $S' = \sum_{p \leq N} \chi(p(p+k))$, where χ is a nonprincipal character mod q with q an odd prime and $(k,q) = 1, N > 1$ (p prime). Then:

a) $S \ll N^{1+\varepsilon} \cdot G; S' \ll N^{1+\varepsilon} \cdot G$, where

$$G = \left(\frac{1}{q} + \frac{q}{N} \right)^{1/2} + N^{-1/6}$$

I.M. Vinogradov. *An improvement of the estimation of sums with primes.* Bull. Acad. Sci. URSS Ser. Mat. 7 (1943), 17–34. (Russian.)

b) If $cq^{3/4} \leq N \leq c'q^{5/4}$, then

$$S \ll N^{1+\varepsilon}(q^{1/4} \cdot N^{-1/3} + N^{-1/10})$$

I.M. Vinogradov. *Improvement of an estimate for the sum of the values* $\chi(p+k)$ (Russian.) Izv. Akad. Nauk SSSR Ser. Mat. 17 (1953), 285–290.

c) If q is sufficiently large (prime);

$0 < \omega \leq \dfrac{1}{4}$ and $q^{\frac{1}{2}+\omega} \leq N < q$, then

$$S \ll N \cdot q^{-\omega^2/1024}$$

A.A. Karacuba. *Sums of characters with prime numbers.* (Russian.) Izv. Akad. Nauk SSSR Ser. Mat. 34 (1970), 299–321.

Note. See also

A.A. Karacuba. *Sums of characters with prime numbers.* (Russian.) Dokl. Akad. Nauk SSSR 190 (1970), 517–518.

2) Let $\varepsilon > 0$. Let $q \geq c_0$ be a prime, where c_0 is sufficiently large; let l be a natural number and w runs over all products of l different prime factors. Let k be a fixed integer, $(k, q) = 1, 0 < |k| < c_0$. Let χ be a nonprincipal character mod q. If $q^{3/4} \ll N \ll q^{5/4}$, then

$$\sum_{w \leq N} \chi(w + k) \ll N^{1+\varepsilon} \cdot \left(\left(\frac{q^{3/4}}{N} \right)^{1/3} + N^{-0.1} \right)$$

3) If $S(x, \chi) = \sum_{p \leq x} \chi\ (p)$ (p-prime), then for $N \geq N_0, X = \sqrt{N}/\log N$ and $M \geq X$, we have

$$\sum_{q \leq X} \log \frac{X}{q} \sum_{\chi}{}^{*} |S(M+N, \chi) - S(M, \chi)|^2 \leq (\sqrt{N} + X)^2 (\pi(M+N) - \pi(M))$$

where in $\sum^*$ the sum is over primitive characters mod q

E. Bombieri and H. Davenport. *On the large sieve method.* Number theory and analysis (Papers in Honor of E. Landau), pp. 9–22, Plenum New-York, 1969.

§ XI. 9 Distribution of pairs of residues and nonresidues of special form

Let $S = \sum_{\substack{p \leq N \\ p \equiv l (\mathrm{mod}\ k)}} \left(\frac{(p+a)(p+b)}{q} \right)$, where q is an odd prime, $(k, q) = 1$, $(l, k) = 1$, and a, b are integers satisfying $a \not\equiv b (\mathrm{mod}\ q)$. If $A > 1, B > 1$ and $0 < \delta < 0.1$ are given constants, then

$$S \ll \frac{N}{k} \cdot q^{-0.003\,\delta^2}$$

uniformly for $1 \leq k \leq q^A, k^3 q^{0.75+\delta} \leq N \leq k^3 q^B$

A.A. Karatsuba. *Distribution of pairs of residues and nonresidues of special form.* (Russian.) Izv. Akad. Nauk SSSR Ser. Mat. 51 (1987), 994–1009; translated in Math. USSR-Izv. 31 (1988), no. 2, 307–323.

§ XI.10 A character sum estimate involving $\mu(n)$ and $\Lambda(n)$

a) $\sum_{n \leq x} \frac{\chi(n)\mu(n)}{n^{1+i\theta}} = O(\log^7(2 + |\theta|))$

where χ is any character of the modulus $D > 1$ and θ is arbitrary.

A.O. Gelfond. *On an arithmetic equivalent of the analyticity of the Dirichlet L-series on the straight line Rs* = 1. Izv. Akad. Nauk, ser. Mat. 20 (1956), 145–166.

b) $$\sum_{n\le x}\frac{\chi(n)\Lambda(n)}{n^{1+i\theta}} = \delta(\chi,\theta)\log x + O(1)$$
where $\delta(\chi,\theta) = 0$ for $\alpha \neq 0$;

$$\delta(\chi,\theta) = -1$$

for $\alpha = 0$, where $\alpha = \sum_{n=1}^{\infty}\frac{\chi(n)}{n^{1+i\theta}}$ and $\theta \neq 0$ (and $\chi \neq \chi_0$)

(See A.O. Gelfond.)

§ XI.11 An upper bound for a character sum involving $\Lambda(n)$

If χ is the nonprincipal real character mod 4 (i.e. $\chi(n) = 0$ if n is even; 1 if $n \equiv 1 \pmod 4$; -1 if $n \equiv -1 \pmod 4$)), then for all integers $x \geq e \cdot 10^4$, we have

$$\sum_{n\le x}\frac{\Lambda(n)\chi(n)}{n} \leq 0.277$$

R.M. Pollack and H.N. Shapiro. *The next to last case of a factorial diophantine equation*. Comm. Pure Appl. Math. 26 (1973), 313–325.

§ XI.12 Half Gauss sums

Let χ be a real primitive character modulo k, where $k > 1$ is an odd number. Then, for $0 < \alpha < 1$

a) $$\sum_{n=1}^{k-1}\chi(n)\cos 2\pi\alpha n/k \begin{cases} < 0 & \text{if } \chi \text{ is even and } \alpha < \frac{1}{2} \\ = 0 & \text{if } \chi \text{ is even and } \alpha = \frac{1}{2} \\ > 0 & \text{otherwise} \end{cases}$$

b) $$\sum_{n=1}^{k-1}\chi(n)\sin 2\pi\alpha n/k \begin{cases} > 0 & \text{if } \chi \text{ is odd and } \alpha > \frac{1}{2} \\ = 0 & \text{if } \chi \text{ is odd and } \alpha = \frac{1}{2} \\ < 0 & \text{otherwise} \end{cases}$$

B.C. Berndt and R.J. Evans. *Half Gauss sums*. Math. Ann. 249 (1980), 115–125.

§ XI.13 Exponential sums with characters. A large-sieve density estimate

1) Let χ be a Dirichlet character mod q, t a real number, and $\tau = |t| + 2$. Then:

a) $$\sum_{n \le N} \chi(n) \cdot n^{it} = \varepsilon(\chi) \cdot (\varphi(q) \cdot N^{1+it}/q(1+it)) + O((q\tau)^{1/2} \cdot \log q\tau)$$

b) $$\sum_{n \le N} \left(1 - \left(\frac{n}{N}\right)\right) \cdot \chi(n) \cdot n^{it} = \varepsilon(\chi)(\varphi(q) \cdot N^{1+it}/q(1+it)(2+it)) + \\ + O((q\tau)^{1/2} \cdot \log q\tau)$$

where $\varepsilon(\chi) = 1$ or 0 according to whether χ is principal or not.

A. Fujii, P.X. Gallagher and H.L. Montgomery. *Some hybrid bounds for character sums and Dirichlet L-series.* Topics in number theory (Proc. Colloq. Debrecen, 1974), pp. 41–57. Colloq. Math. Soc. János Bolyai, vol. 13, North-Holland, 1976.

2) a) Let χ be a primitive character to modulus $q = p^r > 1$ (p odd prime.) Then for $N \le N' \le 2N$ and $N^z = q(|t| + 2)$, with $z \ge \frac{1}{2}$,

$$\sum_{n=N}^{N'} \chi(n) \cdot n^{it} \ll_p B(z) \cdot N^{1-\varepsilon(z)}$$

with $B(z) = \exp(C_1 z \log^2 4z)$, and $\varepsilon(z) = (C_2 z^2 \log 4z)^{-1}$

P.X. Gallagher. *Primes in Progressions to prime-power modulus.* Invent. Math. 16 (1972), 191–201.

b) $$\left| \sum_{N<n \le N'} \chi(n) \cdot n^{it} \right| \le AN \exp\left(-\frac{1}{2976} \cdot \frac{(\log ND^{-1})^3}{\log^2 t}\right)$$

where χ is a Dirichlet character mod $q, q \le N/2, N' \le 2N \le t, t \ge t_0$, $A > 0$ is an absolute constant.

E.I. Panteleeva. *On a problem of Dirichlet divisors in number fields.* (Russian.) Mat. Zametki 44 (1988), 494–505, 557.

3) Suppose that $\sum_{n=1}^{\infty} a_n \cdot n^{it}$ is absolutely convergent and denote

$$S(\chi, t) = \sum_{n=1}^{\infty} a_n \cdot \chi(n) \cdot n^{it}$$

(a_n are complex numbers), with χ a Dirichlet character mod q. Then:

a) For $T \ge 1$, we have

$$\sum_{\chi} \int_{-T}^{T} |S(\chi, t)|^2 \mathrm{d}t \ll \sum_{n=1}^{\infty} (qT + n) \cdot |a_n|^2$$

b) For $T \geq 1$, we have

$$\sum_{q \leq Q} \sum_{\chi}{}^{*} \int_{-T}^{T} |S(\chi, t)|^2 \mathrm{d}t \ll \sum_{n=1}^{\infty} (Q^2 T + n) |a_n|^2$$

c) Assume $a_n = 0$ if n has any prime factor $\leq Q$. Then for $T \geq 1$, we have

$$\sum_{q \leq Q} \log \frac{Q}{q} \sum_{\chi}{}^{*} \int_{-T}^{T} |S(\chi, t)|^2 \mathrm{d}t \ll \sum_{n=1}^{\infty} (Q^2 T + n) |a_n|^2$$

P.X. Gallagher. *A large sieve density estimate near* $\sigma = 1$. Invent. Math. 11 (1970), 329–339.

4) a) For any $T \geq 2$, $M \geq 1/2$ and complex numbers a_n we have

$$\sum_{\chi (\bmod q)} \int_{-T}^{T} \left| \sum_{n \leq M} a_n \cdot \chi(n) \cdot n^{-it-1/2} \right|^4 \mathrm{d}t \ll (qT + M) \cdot \sum_{n \leq M} \frac{|a_n|^2}{n}$$

H.L. Montgomery. *Topics in multiplicative number theory*. Springer Lecture Notes 227 (1971) (See Theorem 6.4.)

b) For any $T \geq 2$ and $M \geq 1/2$ we have

$$\sum_{\chi (\bmod q)} \int_{-T}^{T} \left| \sum_{M \leq n \leq 2M} \chi(n) \cdot n^{-it-1/2} \right|^4 \mathrm{d}t \ll qT \log^8 qNT$$

A. Balog and A. Perelli. *Exponential sums over primes in short intervales.* Acta Math. Hung. 48 (1–2)(1986), 223–228.

§ XI.14 On $\sum_{k=1}^{q-1} \chi(n) \cdot k^n$

Let χ be a primitive character mod q, $q \geq 2$

a) If $\chi(-1) = 1$ and $n \geq 2$, then

$$\left| \sum_{a=1}^{q-1} \chi(a) \cdot a^n \right| < C_1(n) \cdot q^{n+\frac{1}{2}}$$

b) If $\chi(-1) = -1$ and $n \geq 3$, then

$$\left| \sum_{a=1}^{q-1} \chi(a) \cdot a^n \right| < (C_2(n) + |L(1, \chi)|/\pi) \cdot q^{n+\frac{1}{2}}$$

Here

$$C_1(n) = \frac{2\zeta(2)n!}{(2\pi)^{n+1}} \cdot \sum_{1\le m\le n/2} \frac{(2\pi)^{n+1-2m}}{(n+1-2m)!}$$

and

$$C_2(n) = \frac{2\zeta(3)n!}{(2\pi)^{n+1}} \cdot \sum_{1\le m\le (n-1)/2} \frac{(2\pi)^{n-2m}}{(n-2m)!}$$

M. Toyoizumi. *On certain character sums*. Acta Arith. 55 (1990), 229–232.

§ XI.15 Estimates on $\sum_{x=N+1}^{M} \chi(x)\cdot e(ax/p)$

Let $S_a(N, M) = \sum_{x=N+1}^{M} \chi(x)\,e(ax/p)$, where χ is a nonprincipal character modulo the prime p. Then:

a) $S_0(0, H) \ll H^{1/r} \cdot p^{(2r^2-3r+1)/4r^2} \cdot \log p$

A.I. Vinogradov. *On the symmetry property of sums with Dirichlet characters*. (Russian.) Izv. Akad. Nauk. UzSSR Ser. Fiz.-Mat. Nauk 9 (1965), 21–27.

b) $S_a(N, N+H) \ll H^{1-1/r} \cdot p^{1/4(r-1)} \cdot \log^2 p$
for $0 < H < p$ and any integers $r \ge 2$, a, and N

c) $S_a(N, N+H) \ll H^{1/r} \cdot p^{(2r^2-5r+4)/4r(r-1)} \cdot \log^3 p$

D.A. Burgess. *Partial Gaussian sums*. Bull. London Math. Soc. 20 (1988), 589–592.

d) $S_a(N, N+H) \ll H^{2/3} \cdot k^{1/8+\varepsilon}$
if $S_a(N, N+H)$ is defined as before except that p is replaced by k (composite) and χ is a nonprincipal character (mod k)

D.A. Burgess. Ibid. II. Bull. London Math. Soc. 21 (1989), 153–158.

e) For $r = 4$ the estimate from b) is valid also for prime powers $p^\alpha (p > 3)$ in place of p.

D.A. Burgess. Ibid. III. Glasgow Math. J. 34 (1992), 253–261.

f) The result from e) is valid for characters modulo q (not only for prime powers $q = p^\alpha$).

C. Liu. *On incomplete Gaussian sums*. Adv. Math., Beijing 22 (1993), 370–372.

§ XI.16 An infinite series of characters with application to zero density estimates for $\mathcal{L}$ functions

Let χ be a nonprincipal character mod k. If $1 \le K \le 2q$, $h = \log^2 q$, $K^* = \dfrac{q \log^6 q}{K}$ then

$$\left|\sum_{l=1}^{\infty} \chi(l) \cdot \exp(-(l/K)^h)\right| \ll 1 + d(q) \cdot \int_{-h^2}^{h^2} \left|\sum_{l \le K^*} \frac{\chi(l) \cdot l^{it}}{(l/K)^{1/2}}\right| dt$$

M. Jutila. *Zero density estimates for $\mathcal{L}$ functions.* Acta Arith. 32 (1977), 55–62.

§ XI.17 Character sums of polynomials

1) Let $\chi_1, \ldots, \chi_r$ be nonprincipal multiplicative characters mod $p (p \ge 5$, prime), and $f_1(x), \ldots, f_r(x)$ different normalized polynomials of degree $k_1, \ldots, k_r$, each irreducible mod p. Put $k = k_1 + \cdots + k_r$. Then:

a) $$S_{p-1} = \left|\sum_{x=0}^{p-1} \chi_1(f_1(x)) \ldots \chi_r(f_r(x))\right| \ll (k-1)p^{1-\theta_k}$$

where $\theta_2 = \dfrac{1}{2}$, $\theta_3 = \dfrac{1}{4}$, $\theta_k = \dfrac{3}{2k+8}$ if $k \ge 4$; and θ_k can be replaced by θ_{k-1} if

$$\chi_1^{k_1} \cdots \chi_r^{k_r} = \chi_0$$

H. Davenport. *On character sums in finite fields.* Acta Math. 71 (1939), 99–121.

b) Replacing $p-1$ with $1 \le q \le p-1$ $(0 \le x \le q)$ and θ_q with $\rho_k = \dfrac{3}{8k+16}$, one has

$$S_q < 2(r^2 - r + 2k)^{1/2} \cdot p^{1-\rho_k} \cdot \log p$$

B. Segal. *Character sums and their application.* (Russian.) Bull. Acad. Sci. URSS, Ser. Math. 5 (1941), 401–410.

2) Let $q = p_1^{a_1} \cdots p_t^{a_t}$ be the canonical factorization of q into prime powers and let $\chi = \chi_1 \cdots \chi_t$ be the corresponding decomposition of χ into primitive characters χ_j mod $p_j^{a_j}$, $\chi_j(f(x)) \ne$ constant, $j = 1, 2, \ldots, t$, where $f \in \mathbb{I}[X]$, $\deg f \ge 2$. Moreover, let g_j be integers satisfying $0 \le g_j \le [\log n / \log p_j]$ and let m denote

the maximal multiplicity of the roots of all congruence $p_j^{-g_j} \cdot f'(x) \equiv 0 (\text{mod } p_j)$. Then

$$\left| \sum_{x=1}^{q} \chi(f(x)) \right| \ll q^{1-\frac{1}{m+1}+\varepsilon}$$

for every $\varepsilon > 0$, where the implied constant depends on χ and $\deg f$

D. Ismoilov. *An estimate for the sum of characters of polynomials.* (Russian.) Dokl. Akad. Nauk Tadzhik. SSR 29 (1986), 567–571.

§ XI.18 Quadratic character of a polynomial

1) For any nonprincipal character χ mod q,

$$\sum_{n \le \sqrt{x}} \chi(n^2 - a) \ll \sqrt{q} \log q$$

for any $1 \le a < q$

A. Weil. *On some exponential sums.* Proc. Nat. Acad. Sci. USA 34 (1948), 204–207.

2) Let $f(x_1, x_2) = x_1^2 + ax_1x_2 + bx_2^2$ be a binary quadratic form that is not congruent to a perfect square (mod p). Let χ be a nonprincipal character to the odd prime p and let B denote the set of points (x_1, x_2) satisfying

$$N_i < x_i \le N_i + H$$

$(i = 1, 2)$

a) For each $\varepsilon > 0$ there exists $\delta > 0$ such that

$$S = \sum_{B} \chi(f(x_1, x_2)) = O(H^2 \cdot p^{-\delta})$$

for all $H > p^{\frac{1}{3}+\varepsilon}$

D.A. Burgess. *On the quadratic character of a polynomial.* J. London Math. Soc. 42 (1967), 73–80.

b) If f is a form of degree n in n variables that factorizes mod p into a product of n linearly independent factors, then

$$\sum_{x_1=N_1+1}^{N_1+H} \cdots \sum_{x_n=N_n+1}^{N_n+H} \chi(f(x_1, \ldots, x_n)) = O(H^n p^{-\delta})$$

provided that $H > p^{\frac{n}{2n+1}+\varepsilon}$

D.A. Burgess. *A note on character sums of binary quadratic forms.* J. London Math. Soc. 43 (1968), 271–274.

§ XI.19 Distribution of values of characters in sparse sequences

Supose that $A > 1$, $B > 1$ and $0 < w < 1$. Let q be a sufficiently large prime, p a prime and k a natural number with $(k, q) = 1$ and $1 \leq k \leq q^A$.
Define $f(x) = (x + a)(x + b)$. Also, let χ be a non-principal character to the modulus q, $n_r = p_1 p_2 \cdots p_r$, where $r \geq 2$, and $p_i (1 \leq i \leq r)$ are distinct primes.
If $k^3 \cdot q^{0.75+\varepsilon} < N < k^3 q^B$ then

$$\sum_{n_r \leq N; n_r \equiv l (\bmod k)} \chi(f(n_r) \ll k^{-1} \cdot N \cdot q^{-0.003\, w^2}$$

G.D. Negmatova. *The distribution of values of non-principal characters in sparse sequences.* Russ. Math. Surv. 44 (1989), 214–215; translation from Usp. Mat. Nauk 44 (1989), 177–178.

Remark. This generalizes an earlier result of:
A.A. Karatsuba. *Distribution of pairs of residues and nonresidues of special form.* (Russian.) Izv. Akad. Nauk SSSR, Ser. Mat. 51 (1987), 994–1009.

§ XI.20 Estimation of character sums modulo a power of a prime

For vectors $\underline{m} = (m_1, m_2, m_3, m_4, m_5, m_6)$ with components satisfying $0 < m_i \leq h$, put

$$f_1(X) = (X + m_1)(X + m_2)(X + m_3),\ f_2(X) = (X + m_4)(X + m_5)(X + m_6)$$

(where X denotes an indeterminate).

a) For every prime power $k = p^\alpha$, of every prime $p > 3$, and every primitive character $\chi \,(\bmod k)$ we have

$$\sum_m \left| \sum_{x \in A_1} \left(\frac{f_1}{f_2}(x) \right) \right| \ll kh^3 \log^5 k$$

for $h \leq k^{1/6}$ (with the implied constant independent of both p and α), where

$$A_1 = \{x : 0 \leq x < k,\ p \nmid f_1(x) f_2(x)\}$$

D.A. Burgess. *On character sums and L-series.* II. Proc. London Math. Soc. (3) 13 (1963), 524–536.

b) For $p > 3$ prime let $k = p^\alpha$, $\alpha \geq 2$ and write $\beta = \alpha - \left[\frac{\alpha}{2}\right]$. Put

$$F(X) = f_1'(X) f_2(X) - f_2'(X) f_1(X)$$

If $0 \leq \delta \leq \frac{1}{2}\alpha - 1$ and $A_2 = \{x \in A_1 : p^\delta \| F(x)\}$, then

$$\sum_{x \in A_2} \chi \left(\frac{f_1}{f_2}(x) \right) = 0$$

c) Let $\delta = \left[\frac{\alpha}{2}\right]$ and write $A_3 = \{x \in A_1 : p^{\delta} | F(x), p \nmid F'(x)\}$
Then

$$\sum_{x \in A_3} \chi\left(\frac{f_1}{f_2}(x)\right) \ll p^{\alpha/2}$$

d) Suppose α is odd. Let $\delta = \frac{1}{2}(\alpha - 1)$. Write

$$A_4 = \{x \in A_1 : p^{\delta} \| F(x), p | F'(x)\}$$

Then

$$\sum_{x \in A_4} \chi\left(\frac{f_1}{f_2}(x)\right) = 0$$

e) Let $1 \leq \gamma < \frac{\alpha}{3}$. Write

$$A_5 = \{x \in A_1 : p^{\beta} | F(x), p^{\gamma} \| F'(x)\}$$

Then

$$\sum_{x \in A_5} \chi\left(\frac{f_1}{f_2}(x)\right) \ll \gamma \cdot p^{\frac{1}{2}\alpha + \frac{1}{2}\gamma}$$

f) Let $\gamma = \alpha - \left[\frac{2\alpha}{3}\right]$. Write

$$A_6 = \{x \in A_1 : p^{\beta} | F'(x), p^{\gamma} | F(x), p \nmid F''(x)\}$$

Then

$$\sum_{x \in A_6} \chi\left(\frac{f_1}{f_2}(x)\right) \ll p^{\alpha - \gamma}$$

g) $\gamma = \alpha - \left[\frac{2\alpha}{3}\right]$ and $1 \leq \rho \leq \frac{1}{2}\gamma$. Write

$$A_7 = \{x \in A_1 : p^{\beta} | F(x), p^{\gamma} | F'(x), p^{\rho} \| F''(x)\}$$

Then we have

$$\sum_{x \in A_7} \chi\left(\frac{f_1}{f_2}(x)\right) \ll \begin{cases} p^{\alpha - \gamma + \frac{1}{2}\rho + \frac{1}{2}} & \text{if } 3\gamma - \alpha = 2, \rho = 1, \\ p^{\alpha - \gamma + \frac{1}{2}\rho} & \text{otherwise} \end{cases}$$

D.A. Burgess. *Estimation of character sums modulo a power of a prime*. Proc. London Math. Soc. (3) 52 (1986), 215–235.

§ XI.21 Mean values of character sums

1) Let χ be a nonprincipal Dirichlet character mod q and set

$$M(\chi) = \max_N \left|\sum_{n=1}^{N} \chi(n)\right|$$

Then, for all $\lambda > 0$,

$$\sum_{\chi} (m(\chi))^{2\lambda} \ll_\lambda \varphi(q) \cdot q^\lambda$$

Corollary. $M(\chi) \ll q^{1/2}$ holds for some characters χ mod q.

H.L. Montgomery and R.C. Vaughan. *Mean values of characters sums.* Canad. J. Math. 31 (1979), 476–487.

2) a) Let S be a set of nonprincipal characters modulo the prime p. Let $0 \leq \delta < \frac{1}{4}$. Let 2^v be the largest power of 2 that is less than p. Then we have

$$\sum_{\chi \in S} \max_{N,H} H^{-\frac{1}{4}+\delta} \cdot \left|\sum_{n=N+1}^{N+H} \chi(n)\right| \leq 4p^{\frac{1}{2}} \cdot (\text{card } S)^{\frac{3}{4}} \cdot \sum_{\mu=0}^{v} 2^{\delta\mu}$$

Corollary. Let $0 < \theta < 1$ and $0 < \delta < \frac{1}{4}$. There exist constants $c(\theta)$ and $c'(\theta, \delta)$ such that:

(i) for at least θp of the nonprincipal characters χ mod p

$$\left|\sum_{n=N+1}^{N+H} \chi(n)\right| \leq c(\theta) p^{\frac{1}{4}} \cdot H^{\frac{1}{4}} \cdot \log p$$

for all N and all $H > 0$;

(ii) for at least θp of the nonprincipal characters χ mod p

$$\left|\sum_{n=N+1}^{N+H} \chi(n)\right| \leq c'(\theta, \delta) p^{\frac{1}{4}+\delta} \cdot H^{\frac{1}{4}-\delta}$$

for all N and all $H > 0$.

D.A. Burgess. *Mean values of character sums.* Mathematika 33 (1986), 1–5.

b) For any $\theta \in (0, 1)$ and $\varepsilon > 0$ there exists a constant $c(\theta, \varepsilon)$ such that for at least θ. card {primitive characters modulo k} ($k > 1$) of the primitive characters χ mod k we have both

$$\left|\sum_{n=N+1}^{N+H} \chi(n)\right| \leq c(\theta, \varepsilon) k^{\frac{1}{4}+\varepsilon} \cdot H^{\frac{1}{4}}$$

and

$$\left|\sum_{n=1}^{H} \chi(n)\right| \leq c(\theta, \varepsilon) k^{\varepsilon} \cdot H^{\frac{3}{4}}$$

D.A. Burgess. *Mean values of character sums*. II. Mathematika 34 (1987), 1–7.

§ XI.22 On $\sum_{n \in S(x,y)} \chi(n)$, with $S(x, y) = \{n \leq x : P(n) \leq y\}$

Let $A > 0$. Assume that $x \geq 3$ and $\exp\ (c_0(\log\log x)^2) \leq y \leq x$. If χ is a nonprincipal Dirichlet character mod q, then:

a) For $1 < q \leq (\log x)^A$,

$$\sum_{n \in S(x,y)} \chi(n) \ll \psi(x, y) \cdot Y^{-c_1}$$

where $S(x, y) = \{n \leq x : P(n) \leq y\}$, $\Psi(x, y) = \text{card } S(x, y)$ and $Y = \exp(\sqrt{\log y})$

b) For $(\log x)^A < q \leq y^{c_2/\log\log x}$,

$$\sum_{n \in S(x,y)} \chi(n) \ll \psi(x, y) \cdot \left(Y^{-c_3} + y^{-c_3/\log q} + \theta(\chi)\frac{\log q}{\log y} H(u)^{-c_3}\right)$$

where $H(u) := \exp\left(\dfrac{u}{\log^2(u+1)}\right)$ with $u = \dfrac{\log x}{\log y} (x \geq y \geq 2)$, and $\theta(\chi) = 1$ for at most one exceptional real character, and $\theta(\chi) = 0$ for the remaining $\varphi(q) - 2$ nonprincipal characters.

É. Fouvry and G. Tenenbaum. *Entiers sans grand facteur premier en progressions arithmétiques*. Proc. London Math. Soc. III. Ser. 63 No. 3 (1991), 449–494.

§ XI.23 Large sieve-type inequalities via character sum estimates

1) Let A denote a finite set of positive integers, and put $M = \max_{q \in A} q$, $D = \max_{q \in A} d(q)$, where d is the divisor function.
For any character χ mod q let $\tau(\chi)$ denote the Gaussian sum

$$\tau(\chi) = \sum_{a=1}^{q} \chi(a) \cdot e(a/q)$$

Then, for any complex numbers a_n,

$$\sum_{q \in A} \frac{1}{\varphi(q)} \sum_{\chi} |\tau(\chi)|^2 \cdot \left|\sum_{Y < n \leq Z} \chi(n) \cdot a_n\right|^2 \leq 7D \max(Z - Y, M^2) \sum_{Y < n \leq Z} d(n) \cdot |a_n|^2$$

E. Bombieri. *On the large sieve*. Mathematika 12 (1965), 201–225.

2) a) For all complex numbers a_n and Dirichlet characters χ mod q,

$$\frac{q}{\varphi(q)}\cdot\sum_{\chi}{}^{*}\left|\sum_{n=M+1}^{M+N}a_n\chi(n)\right|^2\ll\sum_{\substack{a=1\\(a,q)=1}}^{q}\left|\sum_{n=M+1}^{M+N}a_n e\left(\frac{na}{q}\right)\right|^2$$

where $\sum^*$ indicates that the sum is over primitive characters.

P.X. Gallagher. *The large sieve*. Mathematika 14 (1967), 14–20.

b) $$\sum_{q\le X}\frac{q}{\varphi(q)}\sum_{\chi}{}^{*}\left|\sum_{n=M+1}^{M+N}a_n\chi(n)\right|^2\ll(X^2+N)\cdot\sum_{n=M+1}^{M+N}|a_n|^2$$

P.X. Gallagher. *Bombieri's mean value theorem*. Mathematika 15 (1968), 1–6.

c) Assume $a_n=0$ if n has any prime factor $\le X$. Then

$$\sum_{q\le X}\log\frac{X}{q}\sum_{\chi}{}^{*}\left|\sum_{n=M+1}^{M+N}a_n\chi(n)\right|^2\ll(X^2+N)\cdot\sum_{n=M+1}^{M+N}|a_n|^2$$

P.X. Gallagher. *A large sieve density estimate near* $\sigma=1$. Invent. math. 11 (1970), 329–339.

d) If $a_n=0$ whenever $(n,q)>1$ for some $q\le X$, then the same result as in c) is valid.

E. Bombieri and H. Davenport. *On the large sieve method.* Number theory and analysis (Papers in Honor of E. Landau) pp. 9–22, New York, 1969.

See also

E. Bombieri. *Le grand crible dans la théorie analytique des nombers*. Soc. Math. France, Astérisque No. 18, 1974.

e) Let D be a positive integer, M, N, integers, $N\ge1$ and real $X\ge1$. Then

$$\sum_{q\le X}\frac{Dq}{\varphi(Dq)}\sum_{\chi(\text{mod }Dq)}{}^{*}\left|\sum_{n=M}^{M+N}a_n\chi(n)\right|^2\ll\left(N+DX^2\right)\cdot\sum_{n=M}^{M+N}|a_n|^2$$

for all complex numbers a_n

P.D.T.A. Elliott. *Arithmetic functions and integer products*. Springer-Verlag, 1985. (See p. 111.)

3) a) For any complex numbers a_n we have

$$\frac{1}{\varphi(q)}\sum_{\chi(\text{mod }q)}\left|\sum_{n\le N}a_n\chi(n)\right|^2\ll\left(1+\frac{N}{q}\right)\sum_{n\le N}|a_n|^2$$

for characters $\chi\,(\text{mod }q)$

H.L. Montgomery. *Topics in multiplicative number theory*. Lecture Notes vol. 227, Springer, Berlin, 1971.

b) Let $S(a)=\left(\frac{1}{\varphi(q)}\right)\sum_{\chi(\text{mod }q)}\left|\sum_{n\le N}a_n\chi(n)\right|^2$ and suppose that $N>q$ and $a=u*v*w$, where $N=KLM$, u and v are arbitrary sequences supported on $[1, K]$ and $[1, L]$, respectively, and $W_m = 1$ for $1 \le m \le M$. Let S^* be the sum restricted to nonprincipal characters. Then

$$S^* \ll \|u\|\|v\|\|w\|(1 + q^{-3/4}(K+L)^{1/4}(KL)^{5/4} + q^{-1}(KL)^{7/4})q^{\varepsilon}$$

where $\|u\| = \sum_{n \le N, (n,q)=1} |u_n|^2$ etc.

J. Friedlander and H. Iwaniec. *A mean-value theorem for character sums*. Michigan Math. J. 39 (1992), 153–159.

§ XI.24 Large sieve-type inequalities of Selberg and Motohashi

1) a) For a character χ mod q let f be its conductor, and let χ be induced by χ^* mod f. For any character χ mod $f, r \in \mathbb{N}$ and for arbitrary complex numbers a_n, set $S_r(\chi) = \sum_{M<n\le M+N} a_n \chi(n) C_r(n)$, where

$$C_r(n) = \sum_{l=1,(l,r)=1}^{r} e(nl/r)$$

is Ramanujan's sum. Then we have

$$\sum_{\substack{rf \le Q\\(r,f)=1}} \frac{f}{\varphi(rf)} \sum_{\chi (\mathrm{mod} f)}^{*} |S_r(\chi)|^2 \le (N+Q^2) \sum_{M<n\le M+N} |a_n|^2$$

and

$$\sum_{\substack{rf \le Q\\(r,f)=1}} \left(\frac{N}{f} + \frac{3}{2} r Q\right)^{-1} \frac{1}{\varphi(rf)} \sum_{\chi (\mathrm{mod} f)}^{*} |S_r(\chi)|^2 \le \sum_{M<n\le M+N} |a_n|^2$$

A. Selberg. *Remarks on sieves*. Proc. Number Theory Conference (Univ. Colorado, Boulder, Colo. 1972), 205–216.

Corollary

$$\sum_{\substack{rf \le Q\\(r,f)=1}} \frac{\mu^2(r) f}{\varphi(rf)} \sum_{\chi (\mathrm{mod} f)}^{*} \left| \sum_{M<n\le M+N} a_n \chi(n) \Psi_r(n) \right|^2 \le (N+Q^2) \sum_{M<n\le M+N} |a_n|^2$$

where $\psi_r(n) := \mu((r,n))\,\varphi((r,n))$ and $\mu(r) \neq 0$

b) Let χ_j mod f_j, $f_j \le F (j = 1, 2, \ldots, J)$ be distinct primitive characters. Then

$$\sum_{r \le R} \sum_{\substack{j \le J\\(f_j,r)=1}} \frac{\mu^2(r) f_j}{\varphi(rf_j)} \left| \sum_{M<n\le M+N} a_n \chi_j(n) \Psi_r(n) \right|^2 \le$$

$$\le N + O(JFR^2 \log(FR)) \sum_{M<n\le M+N} |a_n|^2$$

Y. Motohashi. *On a density theorem of Linnik*. Proc. Japan Acad. 51 (1975), 815–817.

Note. These theorems have important applications in proving density theorems for Dirichlet's L-functions.

§ XI.25 A large sieve density estimate

a) $$\sum_{1\le q\le Q}\sum_{\chi}{}^{*}\left|\sum_{x\le p\le x+h}\chi(p)\log p\right| \ll h\exp(-a\log x/\log Q)$$

provided that $x/Q\le h\le x$ and $\exp(\sqrt{\log x})\le Q\le x^b$, where a and b are positive constants. Here the term with $q=1$ must be read as

$$\sum_{x\le p\le x+h}\log p-h$$

and if there is an exceptional zero $1-\delta$ of $L(s,\chi')$, with $\delta\log Q\le d$, then the corresponding term must be read as

$$\sum_{x\le p\le x+h}\chi'(p)\log p+h\,\xi^{-\delta}$$

for some $\xi\in[x,x+h]$. In the latter case, the bound on the right may be reduced, e.g. by a factor of $(\delta\log Q)^2$

P.X. Gallagher. *A large sieve density estimate near* $\sigma=1$. Invent. Math. 11 (1970), 329–339.

Remark. For related results, see

I. Allakov and È. Khamzaev. *Generalization of a theorem of Gallagher for the primes of an arithmetic progression.* (Russian.) Izv. Akad. Nauk. UzSSR Ser. Fiz.-Mat. Nauk 1987, 13–18, 93, no. 1.

§ XI.26 A theorem by Wolke

Let χ be a character modulo k and χ_0 the principal character mod k, $r\ge 1$ an integer and Q, N both not less than 2. Then

$$\sum_{k\le Q}\sum_{\chi\ne\chi_0}\left|\sum_{n\le N}\chi(n)\right|^{2r} \ll (\log Q)^c\cdot\left(Q^2N^r+Q\cdot N^{2r\left(1-\frac{1}{r+4}\right)}\right)$$

where c and the $\ll$-constant may depend on r.

D. Wolke. *Über eine Ungleichung von A.I. Vinogradov*. Arch. Math. (Basel) 23 (1972), 625–629.

§ XI.27 Character sums involving $\psi(X, \chi) = \sum_{n \le x} \Lambda(n)\chi(n)$

a) Let $T(Y, Q) = \sum_{q \le Q} \frac{q}{\varphi(q)} \cdot \sum_{\chi}^{*} \max_{X \le Y} |\Psi(X, \chi)|$, where χ is a primitive character mod q and $\psi(X, \chi) = \sum_{n \le X} \Lambda(n)\chi(n)$. Then

$$T(Y, Q) \ll L^4 \cdot (Y + Y^{5/6} \cdot Q + Y^{1/2} \cdot Q^2)$$

where $Q \ge 1, Y \ge 2$ and $L = \log YQ$

R.C. Vaughan. *An elementary method in prime number theory*. Acta Arith. 37 (1980), 111–115.

Remark. The proof is based on methods of:

R.C. Vaughan. *Mean value theorems in prime number theory*. J. London Math. Soc. (2) 10 (1975), 153–162.

b) $$\sum_{q \le Q} \frac{Dq}{\varphi(Dq)} \cdot \sum_{\chi (\text{mod } Dq)}^{*} \max_{X \le Y} |\psi(X, \chi)| \ll$$

$$\ll (Y + Y^{1/2} \cdot Q^2 \cdot D + Y^{5/6} \cdot Q \cdot D^{1/2}) (\log YQD)^4$$

for all integers $D \ge 1$ and reals $Q \ge 1, Y \ge 2$

P.D.T.A. Elliott. *Arithmetic functions and integer products*. Springer Verlag, 1985. (See pp. 116–119.)

Remark. The proof is again based on Vaughan's metod.

c) $$\sum_{\substack{q \le Q \\ q \text{ prime}}} \frac{q}{\varphi(q)} \cdot \sum_{\chi \ne \chi_0}^{*} \max_{X \le Y} |\psi(X, \chi)| \le 1.93 Q^2 Y (\log Y)^4$$

uniformly for $Y^{1/3} \le Q \le Y^{1/2}$, $Y \ge (202)^6$

(See P.D.T.A. Elliott, p. 410.)

§ XI.28 An estimate involving $\chi_1 * \chi_2$

Let χ_1 and χ_2 be characters modulo q_1 and q_2, respectively (q_1, q_2 positive integers). Let $f(n) = \sum_{d|n} \chi_1(d)\chi_2(n/d)$ Then

$$\sum_{n \le x} f(n) \log \frac{x}{n} = c_1(\chi_1, \chi_2) x \log x + c_2(\chi_1, \chi_2) x + c_3(\chi_1, \chi_2) \log x + c_4(\chi_1, \chi_2) + R(x)$$

where:

a) $R(x) = O(x^{-2/3})$

D. Redmond. *A generalization of a theorem of Ayoub and Chowla*. Proc. Amer. Math. Soc. 86 (1982), 574–580.

b) $R(x) = O(x^{-1/4})$

D. Redmond. *Corrections and additions to: "A generalization of a theorem of Ayoub and Chowla"*. Proc. Amer. Math Soc. 90 (1984) 345–346.

Note. The above paper contains also a generalization for $\sum \chi_1(d_1) \ldots \chi_k(d_k)$

§ XI.29 Number of primitive characters mod n, and the number of characters with modulus $\leq x$

1) a) Let χ be a character of the group M_n of reduced residue classes mod n. Suppose that there exists a character χ' of $M_{n'}$, with $n' < n$, such that χ and χ', considered as functions, coincide. Then χ is called an improper character of M_n. Let $\nu(n)$ denote the number of proper (i.e. not improper) characters of M_n. Then

$$\sum_{n \leq x} \nu(n) = \frac{1}{2}\rho x^2 + O(x\sqrt{x})$$

(ρ-constant)

H. Jager. *On the number of Dirichlet characters with modulus not exceeding x*. Indag. Math. 35 (1973), 452–455.

2) a) The number of primitive characters (mod n) is given by

$$J(n) = \sum_{d|n} \mu(d)\varphi(n/d)$$

b) Let $D = \{n : 0 \neq J(n)|n\}$. Then

$$D(x) = \frac{4}{\log 2 \log 3} \cdot \log^2 x + O(\log x)$$

where $D(x)$ denotes the counting function of D

c) Let $A(n) = \{q : J(q) = n\}$. Then

$$\sum_{n \leq x} A(n) \sim Cx$$

where $C = \frac{3}{2}\prod(1 + 1/(p-1)^2 + 1/p(p-2))$ (p prime)

W. Zhang. *On the average estimation of an arithmetical function*. (Chinese.) J. Shandong Univ., Nat. Sci. Ed. 24 (1989), 23–29.

§ XI.30 Continuous additive characters of a topological abelian group

Let $\chi_1, \ldots, \chi_p$ be continuous (or discontinuous) additive characters of a topological abelian group G

a) A necessary and sufficient condition that the set of solutions x of the inequalities
$$|\chi_i(x) - a_i| \leq \varepsilon (\text{mod } 1), i = 1, \ldots, p$$
where $a_1, \ldots, a_p \in \mathbb{R}$, for every $\varepsilon > 0$ be not empty, is that for arbitrary integers $n_1, \ldots, n_p$ with
$$n_1\chi_1(x) + \cdots + n_p\chi_p(x) \equiv 0 \,(\text{mod } 1)$$
for $x \in G$ also $n_1a_1 + \cdots + n_pa_p \equiv 0$ (mod 1)

b) Let G_1 be the discrete abelian group of the continuous additive characters χ of G. If $x_1, \ldots, x_p \in G$ and $a_1, \ldots, a_p \in \mathbb{R}$, then a necessary and sufficient condition that the inequalities
$$|\chi(x_i) - a_i| \leq \varepsilon \,(\text{mod } 1)$$
for $i = 1, \ldots, p$ have a solution $\chi \in G_1$, is that for arbitrary integers $n_1, \ldots, n_p$ with $\chi(n_1x_1 + \cdots + n_px_p) \equiv 0 \,(\text{mod } 1)$ for $\chi \in G_1$ also $n_1a_1 + \cdots + n_pa_p \equiv 0$ (mod 1)

E. Følner. *Generalization of the general diophantine approximation theorem of Kronecker.* Math. Scand. 68 (1991), 148–160.

§ XI.31 An estimate for perturbed Dirichlet characters

Let h be a totally multiplicative function which differs from the principal character χ_0 (mod k) only on the primes $p_1, \ldots, p_m (p_i \nmid k, i = 1, 2, \ldots, m)$. Then
$$\sum_{n \leq x} h(n) = \prod_p \left(1 - \frac{1}{p}\right) \cdot \left(1 - \frac{h(p)}{p}\right)^{-1} \cdot x + O(\log^m x)$$
for $x \to \infty$

I.V. Elistratov. *Estimates of the remainder terms in the theory of perturbed Dirichlet characters.* (Russian.) Mat. Zametki 23 (1978) 505–514.

§ XI.32 Estimates on Hecke characters

a) Let χ be a normalized Hecke character of an algebraic number field K. Then
$$\sum_{N(I) \leq x} \chi(I) = \alpha(\chi)x + o(x)$$

and

$$\sum_{N(p)\leq x} \chi(p) = \beta(x)\frac{x}{\log x} + o\left(\frac{x}{\log x}\right)$$

where I and p are ideals, and prime ideals of R_k, respectively.

E. Hecke. *Über die L-Funktionen und den Dirichletschen Primzahlsatz für einen beliebigen Zahlkörper.* Nachr. Ges. Wiss. Göttingen (1917), 90–95.

b) If K/Q is normal of degree n and χ is a character with conductor f, then the error term in a) can be improved to

$$O(Dx \log^2 x \exp(-cn(\log x)^{1/2} D^{-1}))$$

where $D = n^3(|d(K)|N(f))^{\varepsilon} \cdot c^{-n}$, provided that the exponents in the complex component of χ are bounded by $A, c = c(A, \varepsilon)$ and A, ε are given positive constants.

L.J. Goldstein. *A generalization of the Siegel-Walfisz theorem.* Trans. Amer. Math. Soc. 149 (1970), 417–429.

§ XI.33 Character sums in finite fields

1) a) Let p be a prime and E be the Galois field of order p^n (n positive integer). If χ is any nonprincipal character on E and B is any box in the n-dimensional space E over the Galois field F of order p, relative to a fixed basis then

$$\left|\sum_{t\in B} \chi(t)\right| < (p^{1/2} \cdot (\log p + 1))^n$$

H. Davenport and D.J. Lewis. *Character sums and primitive roots in finite fields.* Rend. Circ. Mat. Palermo (2) 12 (1963), 129–136.

b) If additionally, n is considered fixed, then

$$\sum_{y\in F} \chi(x+y) = O\left(p^{1-\frac{1}{2(n+1)}}\right)$$

H. Davenport. *On primitive roots in finite fields.* Quart. J. Math. (Oxford) 8 (1937), 308–312.

c) If χ is any nonprincipal character on E then for any fixed $\varepsilon > 0$ there exists a $\delta > 0$ for which

$$\sum_{f(x)\in V_m} \chi(f(x)) = O\left(p^{m(1-\delta)}\right)$$

where V_m is the set of polynomials of degree $< m$, provided

$$m > n \cdot \left(\frac{1}{4} + \varepsilon\right)$$

D.A. Burgess. *A note on character sums over finite fields.* J. Reine Angew. Math. 255 (1972), 80–82.

2) Let $f(x_1, \ldots, x_n)$ be a polynomial with coefficient in the finite field F_q and degree d and suppose that f defines a smooth affine hypersurface and that its terms of highest degree define a smooth projective hypersurface over F_q. Let χ be a non-trivial multiplicative character on F_q with order k and suppose k is relatively prime to d and q. Then

$$\left| \sum_{x_1,\ldots,x_n \in F_q} \chi(f(x_1, \ldots, x_n)) \right| \leq (d-1)^n \cdot q^{n/2}$$

G.I. Perel'muter. *Estimation of a multidimensional sum with multiplicative character.* (Russian.) Studies in number theory, Algebraic number theory, Interuniv Collect. Sci. Works, 9, Saratov 1987, 111–128.

3) Let F be a finite field; let B be a finite étale F-algebra, of dimension n over F. Let χ be any nontrivial complex valued multiplicative character of B^* (extended by zero to all of B), and x in B any regular element. Then

$$\left\| \sum_{t \in F} \chi(t - x) \right\| \leq (n-1)\sqrt{|F|}$$

N.M. Katz. *An estimate for character sums.* J. Amer. Math. Soc. 2 (1989), 197–200.

§ XI.34 On Kloosterman sums

Let F_q be the finite field of order q, and for a fixed additive character χ of F_q let

$$K(q, a) = \sum_{b \in F_q^*} \chi(b + ab^{-1})$$

a) $K(q, a) = 2q^{1/2} w(q, a)$
with $w(q, a) \in [-1, 1]$, and as $q \to \infty$ the $q - 1$ numbers $w(q, a)$, $a \in F_q^*$, have an asymptotic distribution given by the measure $(2/\pi)(1 - t^2)^{1/2}\,dt$ on $[-1, 1]$

N.M. Katz. *Gauss sums, Kloosterman sums and monodromy groups.* Princeton, 1988.

b) For $x \in [-1, 1]$ let $G(x) = (2/\pi) \int_1^x (1 - t^2)\,dt$,
$A([-1, x); q) = \text{card}\{a \in F_q^* : -1 \leq w(q, a) < x\}$,
$R_q(x) = A([-1, x); q)/q - G(x)$ and $D_q^* = \sup_{x \in [-1,1]} |R_q(x)|$. Then

$$D_q^* < 10q^{-1/4}$$

H. Niederreiter. *The distribution of values of Kloosterman sums.* Arch. Math. 56 (1991), 270–277.

§ XI.35 Dirichlet characters on additive sequences

Let χ be a nonprincipal character (mod p), $0 \le a < p$, and let $f_1(x)$, $f_2(y)$ be arbitrary complex valued functions with $|f_1(x)| \le f_1$, $|f_2(y)| \le y$. If $0 < \delta < 1/2$, $p \ge p_0(8)$, $X \ge p^{\delta}$, and $\sqrt{XY} \ge p^{1/2+\gamma}$, then

$$\left|\sum_{x \le X}\sum_{y \le Y} f_1(x) f_2(y) \chi(x + y + a)\right| \le XY f_1 f_2 p^{-0.05\,\delta^2}$$

A.A. Karatsuba. *The distributions of values of Dirichlet characters on additive sequences.* Sov. Math. Dokl. 44 (1992), 145–148.

Chapter XII

BINOMIAL COEFFICIENTS, CONSECUTIVE INTEGERS AND RELATED PROBLEMS

§ XII. 1 On $p^a \| \binom{n}{k}$

1) If $p^a \| \binom{k}{n}$, then

$$p^a \le n$$

P. Erdős. *Beweis eines Satzes von Tchebischeff.* Acta Sci. Math. (Szeged) 5 (1932), 194–198.

Note This property has been applied and rediscovered many times. See e.g.

P. Erdős. *A theorem of Sylvester and Schur.* J. London Math. Soc. 9 (1934), 282–288.

F. Hering. *Eine Beziehung zwischen Binomialkoeffizienten und Primzahlpotenzen.* Arch. Math. (Basel) 19 (1968), 411–412.

W. Stahl. *Bemerkung zu einer Arbeit von Hering.* Arch. Math. (Basel) 20 (1969), 580.

2) Let $B(a,m)$ denote the number of pairs (j, k) with $0 \le j + k < m$ (integers) such that $p^a \| \binom{j+k}{k}$ (p prime). Then

$$\lim_{m\to\infty} \frac{B(a, m)}{m(m+1)} = 0$$

D. Singmaster. *Notes on binomial coefficients.* III. Any integer divides almost all binomial coefficients. J. London Math. Soc. (2) 8 (1974), 555–560.

Note. According to F.T. Howard (MR 53 # 153), P. Castevens and F.T. Howard (unpublished) have extended the results to q-binomial coefficients.

Remarks. *) For geometrical interpretations see

M. Sved and J. Pitman. *Divisibility of binomials by prime powers.* A geometrical approach. Ars. Comb. 26A (1988), 197–298.

**) For other geometrical aspects of divisibility of binomial coefficients see

F. Marko. *Divisibility of binomial coefficients near a half-line and in convex sets.* Acta Math. Univ. Comenianae 50/51 (1987), 267–275.

3) For any $\varepsilon > 0$ and positive integer a, we have that every binomial coefficient $\binom{n}{m}$, with $\left|m - \frac{n}{2}\right| < n^{1-\varepsilon}$, is divisible by p^a, where $p > \frac{1}{2}n^{1/(a+1)}$ is a prime and n is sufficiently large.

J.W. Sander. *Prime power divisors of binomial coefficients.* J. Reine Angew. Math. 430 (1992), 1–20.

§ XII. 2 Number of binomial coefficients not divisible by an integer

1) a) Let $A(t,N)$ be the number of binomial coefficients $\binom{n}{k}$, $n \leq N$, not divisible by t. Then

$$\lim_{n\to\infty} \frac{A(t,N)}{N} = 0$$

for all $t \geq 2$

H. Harborth. *Über die Teilbarkeit im Paskal-Dreick.* Math.-Phys. Semesterber. 22 (1975), 13–21.

b) Let p be a prime, $\theta = \log \frac{p(p+1)}{2} / \log p$, $T(n) =$ the number of binomial coefficients $\binom{n}{m}$ not divisible by p and

$f(x) = \sum_{n \leq x} T(n)$ (x an integer). Then:

b_1) $f(x) \leq x^\theta$

$f(x) > x^\theta / p$

b_2) $$\sum_{n=0}^{\infty} T(n)x^n = \prod_{i=0}^{\infty} \frac{1 - (p+1)x^{p^{i+1}} + px^{p^{i(p+1)}}}{(1 - x^{p^i})^2}$$

$|x| < 1$

Corollary. $f(x) = o(x^2)$

A.H. Stein. *Binomial coefficients not divisible by a prime.* Number theory (New York, 1985/1988), 170–177, Lecture Notes in Math., 1383, Springer, Berlin-New York, 1989.

2) a) Let $F(n)$ be the number of odd numbers in the first n rows of Pascal's triangle. Put $\theta = \log 3 / \log 2$. Then

$$\limsup F(n)/n^\theta = 1$$

and

$$\liminf F(n)/n^\theta = 0.812556$$

H. Haborth. *Number of odd binomial coefficients.* Proc. Amer. Math. Soc. 62 (1976), 19–22.

b) Let $N(t)$ be the number of times the integer $t > 1$ occurs as a binomial coefficient. Then

$$N(t) = O(\log t / \log\log t)$$

H.L. Abbot, P. Erdős and D. Hanson. *On the number of times an integer occurs as a binomial coefficient.* Amer. Math. Monthly 81 (1974), 256–261.

3) Let p, n be positive integers and define polynomials in x as follows:

$$\binom{px}{n} = px(px-1)\dots(px-n+1)/n!$$

$$\binom{px}{0} = 1$$

There exist integers $a(n, p, k)$ such that

$$\binom{px}{n} = a(n, p, 0)\cdot\binom{x}{0} + a(n, p, 1)\cdot\binom{x}{1} + \dots + a(n, p, n)\cdot\binom{x}{n}$$

If p is prime and $p^{t(n,p,k)}\|a(n, p, k)$, then

$$t(n, p, k) \geq k - \frac{n-k}{p-1}$$

C.S. Weisman. *A divisibility property of binomial coefficients.* A collection of manuscripts related to the Fibonacci sequence, p. 57, Fib. Assoc. Santa Clara, Calif., 1980.

§ XII. 3 Number of distinct prime factors of binomial coefficients

1) a) For $2 < 2k \leq n$ (n, k positive integers) we have

$$\omega\left(\binom{n}{k}\right) > \frac{k\log 2}{\log 2k}$$

where $\omega(m)$ is the number of distinct prime factors of m

H. Scheid. *Die Anzahl der Primfaktoren in* $\binom{n}{k}$. Arch. Math. (Basel) 20 (1969), 581–582.

b) For all $\varepsilon > 0$ there exists $k_0(\varepsilon)$ such that for $k > k_0(\varepsilon)$ and $n \geq 2k$ we have

$$\omega\left(\binom{n}{k}\right) > (1-\varepsilon)\frac{k\log 4}{\log k}$$

For $k > k_0(\varepsilon)$ one has

$$\omega\left(\binom{2k}{k}\right) < (1+\varepsilon)\frac{k\log 4}{\log k}$$

(Thus the above estimate cannot be improved.)

P. Erdős. *Über die Anzahl der Primfaktoren von* $\binom{n}{k}$. Arch. Math. (Basel), 24 (1973), 53–56.

c) For all $\varepsilon > 0$ there exists $k_0(\varepsilon)$, such that for all

$$k > k_0(\varepsilon)$$

and

$$n > (2+\varepsilon)k$$

$$\omega\left(\binom{n}{k}\right) > \omega\left(\binom{2k}{k}\right)$$

(See P. Erdős.)

d) Let n_k be the least positive integer n with $\omega\left(\binom{n}{k}\right) \geq k$ and let N_k be the least for which $\omega\left(\binom{n}{k}\right) \geq k$ for every $n \geq N_k$. Then for all $\varepsilon > 0$ there exist $k_0(\varepsilon)$ and $k_1(\varepsilon)$ such that

$$n_k > (1-\varepsilon)k^2 \log k$$

for $k > k_0(\varepsilon)$; and

$$N_k < (\varepsilon + \mathrm{e})^k$$

for $k > k_1(\varepsilon)$

We also have

$\limsup_{k\to\infty} \log n_k / \log k \leq \mathrm{e}$; $\liminf_{k\to\infty} \log N_k / \log k \geq \mathrm{e}$, and $n_k > k^2$ for $k > 4939$

P. Erdős, H. Gupta and S.P. Khare. *On the number of distinct prime divisors of* $\binom{n}{k}$. Utilitas Math. 10 (1976), 51–60.

Remark. The first inequality on N_k, written in the form $\limsup_{k\to\infty} N_k^{1/k} \leq \mathrm{e}$ is due to Erdős and Szemerédi.

P. Erdős and E. Szemerédi. Problem 192, Mat. Lapok 25 (1974), 182–183.

e) $n_k > c\,k^2(\log k)^{4/3}(\log\log k)^{-4/3}(\log\log\log k)^{-1/3}$
for sufficiently large $k(c > 0$, constant), and

$$n_k < k^{\mathrm{e}+\varepsilon}$$

P. Erdős and A. Sárközy. *On the prime factors of* $\binom{n}{k}$ *and of consecutive integers*. Utilitas Math. 16 (1979), 197–215.

f) For $n \geq k! + k$ one has

$$\omega\left(\binom{n}{k}\right) \geq k$$

M. Mignotte. *Sur les coefficients du binôme.* Arch. Math. (Basel) 24 (1973), 162–163.

Remark. In the above result, the factorial can be replaced with $P(k) = \prod_{p^m \leq k} p$ (p prime, $m = 1, 2, \ldots$)

E.S. Selmer. *On the number of prime divisors of a binomial coefficient.* Math. Scand. 39 (1976), 271–281.

g) $\omega\left(\binom{n}{k}\right) \geq \omega(n)$
for $1 \leq k \leq n-1$

P.A.B. Pleasants. *The number of prime factors of binomial coefficients.* J. Number Theory 15 (1982), 203–225.

Note. This result was proved by the author about 1975 (unpublished).

h) There is a positive function $f(k)$ such that for all n,
$\omega\left(\binom{n}{s}\right) \geq \omega\left(\binom{n}{k}\right)$ whenever $k + f(k) < s \leq \frac{1}{2}n$
Moreover, $f(k) = O(k(\log k)^{-1/2})$. For all sufficiently large n, $f(k)$ can be selected so as to satisfy $f(k) = O(k^{c+\varepsilon})$ with
$c = 0.4801\ldots$ and $\varepsilon > 0$

i) There is a positive function $g(k)$ such that

$$\omega\left(\binom{n}{s}\right) \geq \omega\left(\binom{n}{k}\right) - g(k)$$

whenever $k \leq s \leq \frac{1}{2}n$

(See P.A.B. Pleasants.)

2) a) Let $g(k)$ be the least integer exceeding $k+1$ such that all prime factors of $\binom{g(k)}{k}$ are $> k$. Then

$$k^{1+c} < g(k) < \exp(k(1+o(1))$$

E.F. Ecklund, Jr., P. Erdős and J.L. Selfridge. *A new function associated with the prime factors of* $\binom{n}{k}$. Math. Comp. 28 (1974), 647–649.

Remark. For a method of computing $g(k)$ and a table of values of $g(k)$ for $k \leq 140$, see

R. Scheidler and H.C. Williams. *A method of tabulating the number-theoretic function* $g(k)$. Math. Comp. 59 (1992), 251–257.

b) Let $F(k)$ be the largest k with the property

$$\omega\left(\binom{F(k)}{k}\right) < k$$

Then $F(k) = o(A_k)$, where A_k denotes the l.c.m. of integers not less than k.

P. Erdős and E. Szemerédi.
(See P. Erdős. *On prime factors of binomial coefficients.* II. (Hungarian.) Mat. Lapok 30 (1978–1982), 307–316.)

§ XII. 4 Divisibility properties of $\binom{2n}{n}$

1) Let $g(n) = \sum \frac{1}{p}$, where $p < n$ is a prime with

$$p \nmid \binom{2n}{n}$$

a) $\frac{1}{x} \sum_{n=1}^{x} g(n) \to c$
(constant) as $x \to \infty$

b) $\frac{1}{x} \sum_{n=1}^{x} g(n)^2 \to c$

Corollary. $g(n) \to c$ for allmost all values of n

P. Erdős. *Quelques problèmes de théorie des nombres.* Monographie de L'Enseignement Mathématique No. 6 (1963), 81–135, p. 124.

Remark. It is not known if $g(n)$ is uniformly bounded.

2) a) For any two odd primes p and q there are infinitely many values of n for which $p \nmid \binom{2n}{n}$ and $q \nmid \binom{2n}{n}$

P. Erdős, R.L. Graham, I.Z. Ruzsa and E. Strauss. *On the prime factors of* $\binom{2n}{n}$. Math. Comp. 29 (1975), 83–92.

Note. If $p(n)$ is the smallest prime factor of $\binom{2n}{n}$, then $p(3160) = 13$ and $p(n) \leq 11$ for $3160 < n < 10^{110}$

b) $$\sum_{p \leq n,\, p \nmid \binom{2n}{n}} \frac{1}{p} = \sum_{k \geq 2} \frac{\log k}{2^k} + o(1)$$

for almost all n

See the reference from a).

c) $$\sum_{p \leq n,\, p \nmid \binom{2n}{n}} \frac{\log p}{p} = (1 - \log 2 + o(1)) \log n$$

for all n

J.W. Sander. *On primes not dividing binomial coefficients.* Math. Proc. Cambridge Phil. Soc. 113 (1993), 225–232.

3) Denote by $g(m)$ the smallest integer $n > m$ for which $\binom{2m}{m} \mid \binom{2n}{n}$ For all m we have $g(m) \geq 2m$, and for $m > m_0$,

$$m^{1+c} < g(m) < (2m)^{\log m/\log 2}$$

for a certain absolute constant $c > 0$

P. Erdős. *On some divisibility properties of* $\binom{2n}{n}$. Canad. Math. Bull. 7 (1964), 513–518.

Corollary. For $a > n/2$, $\binom{2a}{a} \nmid \binom{2n}{n}$

Note. Moser gave a simple proof that $\binom{2n}{n} = \binom{2a}{a} \cdot \binom{2b}{b}$ has no solutions.

L. Moser. *Insolvability of* $\binom{2n}{n} = \binom{2a}{a} \cdot \binom{2b}{b}$. Canad. Math. Bull. 6 (1963), 167–169.

4) Define $a_k = \omega\left(\binom{2k}{k}\right)\frac{\log 2k}{2k}$. Then:

a) $\lim_{k\to\infty} a_k = \log 2$

b) $a_k > \log 2$
for $k \geq 6$

$$a_k > \log 2 + \frac{\log 2k}{2k}$$

for $k \geq 8$

c) $a_k < \dfrac{(\log 2k)\log 2}{\log 2k - \log\log 2k} < \log 2 + \dfrac{\log\log 2k}{\log 2k}$

for $k \geq 200$

H.-J. Kanold. *Über einige Ergebnisse aus der kombinatorischen Zahlentheorie.* Abh. Braunsch. Wiss. Gesell. 36 (1984), 203–229.

5) Define $\mu_n = 2^{-2n} \cdot \binom{2n}{n}$ $(n = 0, 1, 2, \ldots)$. Then, if $(q, r) \not\equiv (p, s)$ (where (q, r) denotes an ordered pair), then

$$\mu_q \mu_r \neq \mu_p \mu_s$$

$(p, q, r, s = 0, 1, 2, \ldots)$

T. Bang and B. Fuglede. *No two quotients of normalized binomial mid–coefficients are equal.* J. Number Theory 35 (1990), 345–349.

§ XII. 5 Squarefree divisors of $\binom{2n}{n}$

a) Let $(s(n))^2$ denote the square part of $\binom{2n}{n}$ for $n = 1, 2, \ldots$ i.e. $(s(n))^2 | \binom{2n}{n}$ and $\binom{2n}{n} / (s(n))^2$ is squarefree.

If $\varepsilon > 0$, there exists $n_0(\varepsilon)$ such that

$$\exp((c - \varepsilon)\sqrt{n}) < (s(n))^2 < \exp((c + \varepsilon)\sqrt{n})$$

for $n > n_0(\varepsilon)$, where

$$c = \sqrt{2} \cdot \sum_{k=1}^{+\infty} \left(\frac{1}{\sqrt{2k-1}} - \frac{1}{\sqrt{2k}} \right) > 0$$

Corollary. $\binom{2n}{n}$ is not squarefree if $n > n_0$ (i.e. $s(n) > 1$ for $n > n_0$)

A. Sárközy. *On divisors of binomial coefficients.* I.J. Number Theory 20 (1985), 70–80.

Remark. Goetgheluck has shown that $\binom{2n}{n}$ is not squarefree if n is a power of 2, and for $4 < n \le 2^{42205184}$. It was conjectured by Erdős that $\binom{2n}{n}$ is not squarefree for all $n > 4$.

P. Goetgheluck. *On prime divisors of binomial coefficients.* Math. Comp. 51 (1988), 325–329.

b) There exists a computable constant $c > 0$ such that for $n > n_0$, each binomial coefficient $\binom{2n+d}{n}$ with $|d| < n^{1/2+c}$ is divisible by the square of some prime p between $\sqrt{n}$ and $\sqrt{2n + |d|}$

J.W. Sander. *Prime power divisors of* $\binom{2n}{n}$. J. Number Theory 39 (1991), 65–74.

c) For any positive integer a and sufficiently large n, there is a prime p such that

$$p^a \| \binom{2n}{n}$$

J.W. Sander. *Prime power divisors of binomial coefficients.* Reprise. J. Reine Angew. Math. 437 (1993), 217–220.

§ XII. 6 Divisibility properties of consecutive integers

1) Define $A_k(m) = \prod p^\alpha$, where $p^\alpha \| m$, $p \le k$ (p prime.)

a) $\min\limits_{1 \le i \le k} A_k(n+i) < ck$

for some constant $c > 0$

P. Erdős. *On consecutive integers*. Nieuw Arch. Wisk. 3 (1955), 124–128.

b) For every k and $\varepsilon > 0$,

$$A_k\left(\prod_{i=1}^{k}(n+i)\right) < n^{1+\varepsilon}$$

K. Mahler. *Ein Analogon zu einem Schneiderschen Satz*. Indag. Math. (1936), 633–640 and 729–737.

c) $\min\limits_{n} \max\limits_{1 \le i \le k} A_k(n+i) = k$

d) The normal order of $\sum\limits_{i=1}^{k} \dfrac{1}{A_k(n+i)}$ (as n varies) is

$$(1+o(1))e^{-\gamma}\frac{\pi^2}{6} \cdot \frac{k}{\log k}$$

where γ is Euler's constant.

e) $\lim\limits_{k \to \infty} A_k\left(\binom{n}{k}\right)^{1/k} = c$

holds for almost all n and large k

P. Erdős. *Problems and results on consecutive integers*. Publ. Math. Debrecen 23 (1976), 271–282.

f) Denote by $G(k)$ the largest integer for which there are $G(k)$ consecutive integers $n+i$, $1 \le i \le G(k)$ for which the integers $A_k(n+i)$, $1 \le i \le G(k)$ are all different.
Let $2 = p_1 < p_2 < \ldots < p_s \le k < p_{s+1} < p_{s+2}$ be the sequence of consecutive primes. Then

$$p_{s+2} - 2 \le G(k) \le (2+o(1))k$$

B. Gordon and P. Erdős. (See P. Erdős, (1976).)

2) Let $f(n,k,\lambda) = \prod p^\alpha$, where $p^\alpha \| \binom{n}{k}$, $p < n^\lambda$ (p prime)

Let $0 < \lambda \le \dfrac{1}{2}$.

Then for all $\varepsilon > 0$ and $\eta > 0$ there is a $k_0 = k_0(\varepsilon, \eta)$ such that, for $k > k_0$, the density of numbers n with the property

$$n^{k\lambda(1-\varepsilon)} < f(n, k, \lambda) < n^{k\lambda(1+\varepsilon)}$$

is greater than $1 - \eta$

P. Erdős. *On prime factors of binomial coefficients.* II. (Hungarian) Mat. Lapok 30 (1978–1982), 307–316.

3) A sequence S_k of k positive integers, all $\leq k$, has the consecutive integer property (CIP) if k consecutive natural numbers exist which give S_k after division by all prime powers $> k$ of their prime power factorizations. Then:

a) Only permutations of the first k positive integers are possible for

$$S_k$$

b) There exist only six types of CIP sequences.

c) There exist infinitely many values k with two CIP sequences.

P. Erdős, C.B. Lacampagne and J.L. Selfridge. *Prime factors of binomial coefficients and related problems.* Acta Arith. 49 (1988), 507–523.

§ XII. 7 The theorem of Sylvester and Schur

1) a) For every pair of integers $h, k; h \geq k \geq 1$, at least one of the integers

$$h+1, h+2, \ldots, h+k$$

is divisible by some prime $p > k$

J.J. Sylvester. *On arithmetical series.* Messenger of Math. 21 (1892), 1–19, 87–120 and Mathematical Papers 4 (1912), 687–731,

and

I. Schur. *Einige Sätze über Primzahlen mit Anwendungen auf Irreduzibilitätsfragen.* I. Sitz. Preuss. Akad. Wiss. 1929, 125–136.

Remarks. *) For an elementary proof of the theorem by Sylvester and Schur, see

P. Erdős. A theorem of Sylvester and Schur. J. London Math. Soc 9 (1934), 282–288.

**) The theorem may be reformulated as follows: If $n \geq 2k$ then $\binom{n}{k}$ contains a prime divisor greater than k.

b) Let p_k be the least prime $\geq 2k$. Then, if $n \geq p_k$, then $\binom{n}{k}$ has a prime divisor $\geq p_k$ with the exceptions $\binom{9}{2}$ and $\binom{10}{3}$

M. Faulkner. *On a theorem of Sylvester and Schur.* J. London Math. Soc 41 (1966), 107–110.

c) If $n \geq 2k$ then $\binom{n}{k}$ contains a prime divisor greater than $\frac{3}{2}k$ with the exception $\binom{9}{8}$.

<u>Corollary</u>. For all $k > 1, n \geq 2k$, $\binom{n}{k}$ has a prime divisor $\geq \frac{7}{5}k$

D. Hanson. *On a theorem of Sylvester and Schur.* Canad. Math. Bull. <u>16</u> (1973), 195–199.

2) a) If $n \geq 2k > 2$, then $\binom{n}{k}$ has a prime divisor $p \leq \frac{n}{2}$, except for $\binom{7}{3}$

E.F. Ecklund, Jr. *On prime divisors of the binomial coefficient.* Pacific J. Math. <u>29</u> (1969), 267–270.

b) There is an absolute constant c so that for $n \geq 2k$ and $k > k_0(c)$, $\binom{n}{k}$ has a prime factor $< n/k^c$

P. Erdős and J.L. Selfridge. *Some problems on the prime factors of consecutive integers.* II. Proc. Wash. State Univ. Conf. Number Theory, Pullman, Wash. 1971.

3) With the unique exception $a = 2, d = 7, k = 3$, at least one of $a, a + d, \ldots, a + (h - 1)d$ is divisible by a prime greater than k, provided that $a > 0, k > 2, d > 1$, (these conditions cannot be weakened.)

T.N. Shorey and R. Tijdeman. *On the greatest prime factor of an arithmetical progression.* A tribute to Paul Erdős, Cambridge Univ. Press, Cambridge 1990, pp. 385–389.

§ XII. 8 On the prime factorization of binomial coefficients

1) Let us consider the representation $\binom{n}{k} = u_n(k) \cdot v_n(k)$, where

$$u_n(k) = \prod_{p^\alpha \| \binom{n}{k}, p<k} p^\alpha$$

$$v_n(k) = \prod_{p^\alpha \| \binom{n}{k}, p\geq k} p^\alpha$$

(p prime.) Then:

a) $\max\limits_{1\leq k\leq n} v_n(k) = \exp\left(\frac{1}{2}n(1 + o(1))\right)$

b) There exists $k \leq (2 + o(1)) \log n$, such that

$$u_n(k) > 1$$

c) There exist infinitely many n such that for all $k \leq \left(\frac{1}{2} + o(1)\right) \log n$ one has

$$u_n(k) = 1$$

d) If $\hat{k}(n)$ denotes the value of k for which $u_n(k)$ attains its maximum, then

$$\hat{k}(n) = (1 + o(1)) \frac{\mathrm{e}}{\mathrm{e} + 1} n$$

P. Erdős and R.L. Graham. *On the prime factors of* $\binom{n}{k}$. Fib. Quart. 14 (1976), 348–352.

e) Let $n \geq 2k$. Then $u_n(k) > v_n(k)$ holds just in 12 cases, namely $\binom{8}{3}$, $\binom{9}{4}$, $\binom{10}{5}$, $\binom{12}{5}$, $\binom{21}{7}$, $\binom{21}{8}$, $\binom{30}{7}$, $\binom{33}{13}$, $\binom{33}{14}$, $\binom{36}{13}$, $\binom{36}{17}$, $\binom{56}{13}$

E.F. Ecklund, Jr., R.B. Eggleton, P. Erdős and J.L. Selfridge. *On the prime factorization of binomial coefficients.* J. Austral. Math. Soc. Ser. A 26 (1978), 257–269.

2) Let $\binom{n}{k} = U \cdot V$, where $P(U) \leq k$, $p(V) > k$, where $P(m)$ and $p(m)$ denote the greatest and the least prime factors of m, respectively. (In fact $U = U_n(k)$, $V = V_n(k)$). Then:

a) For all k and $\varepsilon > 0$ there exists an $n_0 = n_0(k, \varepsilon)$ such that for all $n > n_0$,

$$U < n^{1+\varepsilon}$$

K. Mahler. *Ein Analogon zu einem Schneiderschen Satz.* Indag. Math. (1936), 633–640.

Remark. The same is true for $u = u_n(k)$ of 1).

b) $V > U$

excepting a finite number of cases. For $k \neq 3, 5, 7$ the exceptions are the same as in 1) e).

(See E.F. Ecklund, Jr., R.B. Eggleton, P. Erdős and J.L. Selfridge.)

3) Let $\binom{n}{k} = U \cdot V \cdot W$, where $P(U) \leq k$, $p(V) > k$, but $P(V) \leq n - k$,

$$W = \prod_{n-k<p\leq n} p$$

a) For sufficiently large absolute constant C and $n > Ck$, $k \geq 4$ we have

$$V > \max(U, W)$$

P. Erdős. *Some unconventional problems in number theory.* Acta Math. Acad. Sci. Hungar. 33 (1979), 71–80.

b) Let max $(U, V, W) = M$. If $k \geq 4$, then excepting a finite number of pairs n, k we have

$$M \neq U$$

c) Let $\delta > 0$ be arbitrary. If $k > k_0(\delta)$ and $n < \left(\frac{14}{3} - \delta\right) k$, then

$$M = W$$

If $n > \left(\frac{14}{3} + \delta\right) k$, then

$$M = V$$

d) If $k > 10$, or $k \in \{6, 8\}$, then

$$M = V$$

e) For each positive integer r there exist infinitely many pairs of numbers n, k such that

$$U(n, k) > W(n, k) > n^r$$

where $U = U(n, k)$, $W = W(n, k)$

Remark. The proof of e) uses the following property:

$$\prod_{i=1}^{k} U(n + i, k) > \exp(c\, k^2)$$

for some absolute constant c

P. Erdős. *On prime factors of binomial coefficients.* (Hungarian). Mat. Lapok 28 (1977–1980), 287–296.

f) Let $m(n)$ be the greatest number such that

$$V(m(n), k) \leq V(n, k)$$

Then

$$\limsup_{n \to \infty} \frac{m(n)}{n} = +\infty$$

for all k

P. Erdős. Ibid. II. Mat. Lapok 30 (1978–1982), 307–316.

Remark. The same proof gives

$$m(n) > c \cdot n^{1+\frac{1}{k}}$$

for infinitely many n

On the other hand, one has

$$m(n) < n^{1+\varepsilon_k}$$

where $\varepsilon_k \to 0$ for $k \to \infty$

4) Let $n > k^2$ and write $\binom{n}{k} = U(n,k) \cdot V'(n,k) \cdot W'(n,k)$, where

$$V'(n,k) = \prod_{\substack{p^\alpha \| \binom{n}{k},\\ k<p<n/k}} p^\alpha$$

$$W'(n,k) = \prod_{\substack{p | \binom{n}{k},\\ p \geq n/k}} p$$

a) For almost all n we have

$$W'(n,k) = 1$$

b) Let n'_k be the least n with $W'(n,k) = 1$. Then

$$n'_k < k^{ck}$$

(See P. Erdős. II. Mat. Lapok 30 (1978–1982).)

§ XII. 9 Inequalities and estimates involving binomial coefficients

1) a) Let $g_{n,r} = \binom{n-r}{r}$, where $n \geq 0$, $0 \leq r \leq \left[\frac{n}{2}\right]$ are integers. Then

$$g_{n,r}^2 > g_{n,r-1} \cdot g_{n,r+1}$$

for $1 \leq r \leq \left[\frac{n}{2}\right]$

S.M. Tanny and M. Zucker. *On an unimodal sequence of binomial coefficients.* Discrete Math. 9 (1974), 79–89.

b) Let $n_1 \geq 3h, n_2 \geq 3h, h \geq 1$, where n_1, n_2, h are integers. Then for all integers $0 \leq x \leq h, 0 \leq y \leq h$ one has

$$\binom{n_1 - x}{n_1 - h} \cdot \binom{n_2 - y}{n_2 - h} \geq \binom{2h}{x + y}$$

I. Tomescu. *Problems in combinatorics and graph theory.* (Romanian), Bucureşti 1981 (See p. 15).

2) a) If n and $k < n$ are positive integers, then

$$\binom{n}{k} \leq \frac{n^n}{k^k(n-k)^{n-k}}$$

N. Åslund. *The fundamental theorems of information theory.* II. Nord. Mat. Tidskr. 9 (1961), 105.

b) Let k be positive integer and a a real number such that $a > k$. Then

$$\binom{a}{k} \leq \frac{a^a}{bk^k(a-k)^{a-k}}$$

with $b = (1 + 1/k)^k$

G. Kalajdžić. (See D.S. Mitrinović. *Analytic inequalities.* Springer-Verlag, 1970, p. 195.)

c) If n and r are natural numbers with $n \geq r > 2$, then

$$\frac{1}{r!} - \frac{1}{2n(r-2)!} < \binom{n}{r} \cdot n^{-r} < \frac{1}{r!}$$

J.K.L. Mac Donald. *Elementary rigorous treatment of the exponential limit.* Amer. Math. Monthly 47 (1940), 157–159.

d) If $n \geq 2$ is an integer and a is a positive number, then

$$\left(\frac{n+a}{a}\right)^a < \binom{r}{n} < \frac{1}{n!}\left(\frac{(n+1)(n+a)}{2a}\right)^n$$

with $r = \dfrac{n(n+1)}{2a} + n$

P. Bartoš and Š. Znám. *On symmetric and cyclic means of positive numbers.* Mat.-Fyz. Časopis Sloven. Akad. Vied. 16 (1966), 291–298.

e) Let n and k be natural numbers, and let

$$Q(n,k) = \sqrt{\frac{n}{2\pi k(n-k)}} \cdot \left(\frac{n}{k}\right)^k \cdot \left(\frac{n}{n-k}\right)^{n-k}$$

Then

$$Q(n,k) \cdot \exp\left(\frac{1}{12n + \frac{1}{4}} - \frac{1}{12k} - \frac{1}{12(n-k)}\right) < \binom{n}{k} <$$

$$< Q(n,k) \cdot \exp\left(\frac{1}{12n} - \frac{1}{12k + \frac{1}{4}} - \frac{1}{12(n-k) + \frac{1}{4}}\right)$$

P. Buchner. *Bemerkungen zur Stirlingschen Formel.* Elem. Math. 6 (1951), 8–11.

3) a) If m and n are positive integers, then

$$2^{n-1} \cdot \binom{m+n-1}{n-1} < \sum_{k=0}^{n-1} 2^k \cdot \binom{m+k}{k} < 2^n \cdot \binom{m+n-1}{n-1}$$

E. Makai. *The first main theorem of P. Turán.* Acta Math. Acad. Sci. Hung. 10 (1959), 405–411.

b) Let $r > 0$ and let n be a natural number. Denote

$$I_{r,n} = \sum_{k=0}^{n} \frac{1}{rk+1}\binom{n}{k}$$

Then

$$\frac{2^{n+1}-1}{n+1} < I_{r,n} < \frac{2^{n+1}-1}{r(n+1)}$$

for $0 < r < 1$ and $n \geq 1$;

$$\frac{2^n}{n} < I_{r,n} < \frac{2^{n+1}-1}{n+1}$$

for $1 < r \leq 2$ and $n \geq 3$;

$$\frac{2^{n+1}}{r(n+1)} < I_{r,n} < \frac{2^n}{n-1}$$

for $r \geq 2$ and $n \geq 2$.

D.S. Mitrinović. Problem 94. Mat. Vesnik 6 (21) (1969), 89–90.

Remarks. *) The solutions of the above inequalities were obtained by M.R. Tasković.

**) For the inequalities $\frac{2^n}{n} < I_{2,n} < \frac{2^n}{(n-1)}$ for $n \geq 3$, see

H.W. Smith and J. Barlaz. Problem 4378. Amer. Math. Monthly 58 (1951), 498–499.

c) If m, n are nonnegative integers, and $a \geq 0$, then

$$\sum_{k=0}^{s} \binom{m-k+a}{m-k} \cdot \binom{n-k+a}{n-k} \cdot \binom{k-a-2}{k} \geq 0$$

where $s = \min(m, n)$

G.G. Lorentz and K. Zeller. *Abschnittslimitierbarkeit und der Satz von Hardy-Bohr.* Arch. Math. (Basel) 15 (1964), 208–213.

d) Let $n \geq 13$. Then

$$\sum_{i=0}^{h-1} \binom{n}{i} > \binom{n}{h}$$

iff $h \geq \left[\frac{n}{3}\right] + 2$

E.L. Johnson, D. Newman and K. Wiston. *An inequality on binomial coefficients. Algorithmic aspects of combinatorics.* (Conf. Vancouver Island, B.C. 1976), and
Ann. Discrete Math. 2 (1978), 155–159.

Note. For a generalization of this result, see

L. Vannucci. *A generalization of an inequality on binomial coefficients.* Riv. Mat. Sci. Econom. Social. 2 (1979), 113–126.

4) Let a_i and b_i $(1 \leq i \leq m)$ be positive integers such that $b_i \leq a_i$. Put

$$A = \sum_{i=1}^{m} a_i \text{ and } B = \sum_{i=1}^{m} b_i$$

Then

$$0 < \prod_{i=1}^{m} \binom{a_i}{b_i} \leq \binom{A}{B}$$

E. Bareiss and F. Goldner. Problem 132. Elem. Math. 7 (1952), 117.

5) a) Let n, k be positive integers, $k < n$. If k and n tend to infinity such that $\lim_{n\to\infty} \frac{k - n/2}{\sqrt{n}} = \lambda$, then

$$\lim_{n\to\infty} \frac{\sqrt{n}}{2^n} \cdot \binom{n}{k} = \sqrt{\frac{2}{\pi}} \cdot \exp(-2\lambda^2)$$

G. Pólya and G. Szegő. *Problems and theorems in analysis*. Springer-Verlag, 1972 (Problem 58, Part II).

b) Let k and l be real numbers, $k > 1$. Then

$$\binom{nk+l}{n} \sim \frac{(k-1)^n}{\sqrt{2\pi n}} \cdot \left(\frac{k}{k-1}\right)^{nk+l+\frac{1}{2}}$$

as $n \to \infty$

(See G. Pólya and G. Szegő. Problem 206, Part II.)

6) a) Let $m > 0$ be a fixed real number, and n a positive integer. Then

$$\sum_{k=0}^{n} \binom{n}{k}^m \sim \frac{2^{mn}}{\sqrt{m}} \cdot \left(\frac{2}{\pi n}\right)^{\frac{m-1}{2}}$$

as $n \to \infty$

(See G. Pólya and G. Szegő. Problem 40, Part II.)

b) Let A_n and G_n denote the arithmetic and geometric means respectively, of the binomial coefficients

$$\binom{n}{0}, \binom{n}{1}, \ldots, \binom{n}{n}$$

Then

$$\lim_{n\to\infty} \sqrt[n]{A_n} = 2 \text{ and } \lim_{n\to\infty} \sqrt[n]{G_n} = \sqrt{e}$$

(See G. Pólya and G. Szegő. Problem 51, Part II.)

7) Let P_k denote the product of the first k primes. Then

$$\binom{k^2}{k} < P_k$$

for $k \geq 1794$. For $2 < k < 1794$, the reverse inequality is true.

H. Gupta and S.P. Khare. *On* $\binom{k^2}{k}$ *and the product of the first k primes*. Univ. Beograd Publ. Elektr. Fak. Ser. Mat. Fiz. No. 577–598 (1977), 25–29.

Corollary. For $k \geq 1794$, $\binom{n}{k}$ cannot have k or more distinct prime divisors for $n < k^2$.

8) Let (c_n) and (γ_n) be defined by

$$n! = \gamma_n \cdot \left(\frac{n}{e}\right)^n \cdot \sqrt{n}, \binom{2n}{n} = c_n \cdot \frac{2^{2n}}{\sqrt{n}}.$$

Then:

a) For $1 \leq k \leq \frac{n}{2}$ we have

$$\binom{n}{k}\sqrt{\frac{n}{2}} \cdot \frac{1}{2^n} \cdot \exp\left(\left(1 - \frac{1.5}{n}\right)\frac{(n-2k)^2}{2n}\right) \leq c_{[\frac{n+1}{2}]} < \frac{1}{\sqrt{\pi}}$$

b) For $n \geq 169$, $\frac{n}{2} - \sqrt{n}\log n \leq k \leq \frac{n}{2}$,

$$c_{[k/2]} \leq \binom{n}{k} \cdot \sqrt{\frac{n}{2}} \cdot \frac{1}{2^n} \cdot \exp\left(\left(\frac{\sqrt{n}}{\sqrt{n} - \log n}\right)\frac{(n-2k)^2}{2n}\right)$$

c) For $n \geq 164$, $1 \leq k \leq 0.4n$,

$$\binom{n}{k} \cdot \sqrt{\frac{n}{2}} \cdot \frac{1}{2^n} \cdot \exp\left(\frac{(n-2k)^2}{n}\right) < c_{[(n+1)/2]}$$

d) For all $k, n \geq 1, k < n$,

$$c_{[k/2]} \leq \frac{\gamma_n\sqrt{2}}{\gamma_k\gamma_{n-k}} \leq c_{[(n+1)/2]}$$

H.-J. Kanold. *Über einige Ergebnisse aus der kombinatorischen Zahlentheorie.* Abh. Braunschw. Wissensch. Gesell. 36 (1984), 203–229.

e) $$\frac{1}{\sqrt{\pi\left(n + \frac{1}{4} + \frac{1}{32n}\right)}} < \frac{c_n}{\sqrt{n}} < \frac{1}{\sqrt{\pi\left(n + \frac{1}{2}\right)}}$$

for all n

L. Panaitopol. *A refinement of Stirling's formula.* (Romanian). Gaz. Mat. 9/1985, pp. 329–331.

§ XII.10 On unimodal sequences of binomial coefficients

1) For all $h \geq 2$, integer,

$$\max_{(n_1,\ldots,n_k)} \sum_{i=1}^{k} \binom{n_i}{h} = \binom{n-k+1}{h}, \text{ and } \min_{(n_1,\ldots,n_k)} \sum_{i=1}^{k} \binom{n_i}{h} = (k-r) \cdot \binom{t}{h} + r \cdot \binom{t+1}{h}$$

where max and min are taken over all representations of n in the form $n = n_1 + \cdots + n_k$; $n_1, \ldots, n_k \geq 1, t = \left[\frac{n}{k}\right]$ and $r = n - kt$.

I. Tomescu. *Problems in combinatorics and graph theory.* (Romanian). Bucureşti, 1981, p. 15.

2) a) Let $g_{n,r} = \binom{n-r}{r}$ for $n \geq 0, 0 \leq r \leq \left[\frac{n}{2}\right]$, integers.

If r_n is the largest integer such that g_{n,r_n} is a maximum, then $r_n = \left\lceil \frac{n}{2} - \frac{3}{10} - \frac{1}{10} \cdot \sqrt{5n^2 + 10n + 9} \right\rceil$, where $\lceil x \rceil$ denotes the smallest integer greater than or equal to x.

S.M. Tanny and M. Zuker. *On an unimodal sequence of binomial coefficients.* Discrete Math. 9 (1974), 79–89.

b) Let $g(n, r, k) = \binom{n-kr}{r}$. If $r_{n,k}$ is the least integer at wich $g(n, r, k)$ attains its maximum (for fixed n and k), then the sequence $(r_{n,k}/n)_{n=1,2,\ldots}$ converges to the (unique) root of the polynomial $(1 - (k+1)x)^{k+1} - x \cdot (1 - kx)^k$, in the interval

$$(0, 1/(k+2))$$

S.M. Tanny and M. Zuker. *On an unimodal sequence of binomial coefficients.* II. J. Combinatorics Inform. Syst. Sci. 1 (1976), 81–91.

c) Let $n \geq 4$ be given and denote

$$A_n = \left\{ \binom{n}{0}, \binom{n-1}{1}, \ldots, \binom{n - \left[\frac{n+1}{2}\right]}{\left[\frac{n+1}{2}\right]} \right\}$$

If $\binom{n-k_n}{k_n}$ is the maximal element of A_n, then

$$k_n = \left[\frac{5n + 7 - \sqrt{5n^2 + 10n + 9}}{10} \right]$$

(Here $[x]$ denotes the integer part of x)

K.T. Atanassov. *One extremal problem.* Bull. Number Theory Rel. Topics 8 (1984), 6–12.

§ XII.11 A theorem of Pillai and Szekeres

a) In a sequence of $k \leq 16$ consecutive integers there exists a term which is relatively prime with all the others.

(S. Pillai and G. Szekeres. (See S. Pillai). *On m consecutive integers*. I, II, III. Proc. Indian Acad. Sci. Sect. A, 11 (1940), 6–12; 73–78 and 13 (1941), 530–533.)

b) The property from a) is not true for $k > 16$

A. Brauer and S. Pillai. (See A. Brauer. *On the property of k consecutive integers*. Bull. Amer. Math. Soc. 47 (1941), 328–331).

Remark. For $k = 17$ the smallest counterexample is $2184, 2185, \ldots, 2200$

§ XII.12 A sum on a function connected with consecutive integers

Let $t_k(n) = \min\{t \geq 1 : n \mid t(t+1)\ldots(t+k-1)\}$. Then

$$\frac{1}{x}\sum_{n\leq x} t_2(n) \ll x \log\log\log x / \log\log x$$

P. Erdős and R.R. Hall. *On some unconventional problems on the divisors of integers*. J. Austral. Math. Soc. Ser. A 25 (1978), 479–485.

§ XII.13 On consecutive integers. Theorems of Erdős-Rankin and Shorey

Let $f(k)$ be the least positive integer such that for all

$$n \geq k$$

$$P\left(\prod_{i=1}^{f(k)}(n+i)\right) > k$$

where $P(m)$ denotes the greatest prime factor of m. Then:

a) $f(k) \leq k$

J.J. Sylvester. *On arithmetical series*. Messenger of Math. 21 (1892), 1–19, 87–120,

and

I. Schur. *Einige Sätze über Primzahlen mit Anwendungen auf Ireduzibilitätsfragen*. I. Sitz. Preuss. Akad. Wiss. 1929, 125–136.

b) $f(k) < ck/\log k$
($c > 1$, constant.)

P. Erdős. *On consecutive integers*. Nieuw. Arch. Wisk. (3) 3 (1955), 124–128.

c) $f(k) \ll \dfrac{k}{\log k} \cdot \dfrac{\log\log\log k}{\log\log k}$

T.N. Shorey. *On gaps between numbers with a large prime factor*. II. Acta Arith. 25 (1974), 365–373.

d) $f(k) > c_1 \cdot \dfrac{\log k \cdot \log\log k \cdot \log\log\log\log k}{(\log\log\log k)^2}$

Note. This is an easy consequence of a theorem of Rankin on the difference of consecutive primes.

R.A. Rankin. *The difference between consecutive prime numbers*. J. London Math. Soc. 13 (1938), 242–247. See also P. Erdős.

§ XII.14 On prime factors on consecutive integers

1) Let n_k denote the smallest integer n for which $P(n+i) > k$ for all $i = 1, 2, \ldots, k$ (where $P(m)$ is the greatest prime factor of m). Then

a) $n_k < k^{\log k/\log\log k}$
for $k > k_0$

P. Erdős. *Problems and results on consecutive integers*. Publ. Math. Debrecen 23 (1976), 271–282.

b) $n_k > \dfrac{k^{5/2}}{16}$
for $k > k_1$

P. Erdős an A. Sárközy. *On the prime factors of $\binom{n}{k}$ and of consecutive integers*. Utilitas Math. 16 (1979), 197–215.

2) Let $f(n,k)$ be the number of integers $n+i, (i = 1, 2, \ldots, k)$ which have at least one prime factor greater than k.

a) Let $\alpha > 1, n > k^\alpha - k$. Then ($\alpha$ is an integer)

$$f(n,k) > k \cdot \left(1 + \frac{1}{\alpha}\right) - \frac{2k}{\log k}$$

b) For every $\alpha > 1$ there is an $\varepsilon_\alpha > 0$ so that for $k > k_0(\varepsilon)$ and $n > k^\alpha - k$,

$$f(n,k) > k \cdot \left(1 - \frac{1}{\alpha} + \varepsilon_\alpha\right)$$

(See P. Erdős.)

c) If $n > \exp(c(\log k)^2)$, then

$$f(n,k) \geq k - \pi(k)$$

K. Ramachandra, T.N. Shorey, and R. Tijdeman. *On Grimm's probelm relating to factorization of a block of consecutive integers*. J. Reine Angew. Math. 273 (1975), 109–124.

d) If $n > \exp k^{\varepsilon}$ then

$$f(n,k) > k \cdot \left(1 - \frac{ck\log\log k}{(\log k)^2}\right)$$

T.N. Shorey. *On gaps between numbers with large prime factor.* II. Acta Arith. 25 (1974), 365–373.

e) For $n \le k^{\alpha} - k(\alpha > 1)$ we have

$$f(n,k) < k(\alpha - 1) + \frac{k}{\log k}$$

(See P. Erdős.)

3) Let $g(n,k,r)$ denote the number of integers $n+i$, $1 \le i \le r$ with $P(n+i) \le k$. Then

$$\lim_{k\to\infty} \frac{1}{rk^{\alpha}} \sum_{n=1}^{k^{\alpha}} g(n,k,r) = c_{\alpha}$$

(Here c_{α} is a constant depending on α. E.g. for $1 \le \alpha \le 2$, $c_{\alpha} = 1 - \log\alpha$)

(See P. Erdős.)

Remark. Let $U(n,k)$ be the number of integers $m \le n$ with $P(m) \le k$. The proof of the above result uses the following estimate:

$$U(k^{\alpha}, k) = (c_{\alpha} + O(1))k^{\alpha}$$

$(k \to \infty)$

N.G. de Brujin. *On the number of positive integers $\le x$ an free of prime factors $> y$.* Indag. Math. 13 (1951), 50–60.

§ XII.15 The Grimm conjecture and analogues problems

1) Let $g(n)$ denote the largest number so that for each $m \in \{n+1, n+2, \ldots, n+g(n)\}$ there corresponds a prime factor q_m such that the q_m's are all different. Then:

a) $(\log n/\log\log n)^3 \ll g(n)$

K. Ramachandra, T.N. Shorey, and R. Tijdeman. *On Grimm's probelm relating to factorization of a block of consecutive integers.* J. Reine Angew. Math. 273 (1975), 109–124.

b) $g(n) \ll (n/\log n)^{1/2}$

P. Erdős and J.L. Selfridge. *Some problems on the prime factors of consecutive integers.* II. Proc. Wash. State Univ. (Conf. Number Theory at Pullman (1971), 13–21.)

c) $g(n) < n^{1/2-c}$

for some fixed $c > 0$ and all large n.

(See P. Erdős and J.L. Selfridge, where a result of Ramachandra is used: K. Ramachandra. *A note on numbers with a large prime factor*. J. London Math. Soc. (2) 1 (1969), 303–306.)

Remarks. 1) The famous Grimm conjecture (equivalently) asserts that $p + g(p) \geq p'$ when p, p' are consecutive primes.

C.A. Grimm. *A conjecture on consecutive composite numbers*. Amer. Math. Monthly 76 (1969), 1126–1128.

2) From b) it follows that if Grimm's conjecture is true, it must lie very deep. Indeed, Grimm's conjecture and b) imply $p' - p \ll (p/\log p)^{1/2}$. While this is undoubtedly true, it is generally recognized as probably hopeless at present.

d) There exists a positive constant c_1 such that

$$g(n) \geq \exp\left(c_1(\log n \cdot \log\log n)^{1/2}\right)$$

for infinitely many n

(See P. Erdős and J.L. Selfridge.)

e) There exists a positive constant c_2 such that

$$g(n) \leq \exp\left(c_2(\log n \cdot \log\log n)^{1/2}\right)$$

for infinitely many n

C. Pomerance. *Some number theoretic matching problems*. Queen's Papers in Pure and Applied Math. 54 (1980), 237–247.

2) Let $f(n)$ denote the largest integer such that for each composite $m \in \{n+1, n+2, \ldots, n+f(n)\}$ there is a divisor d_m of m with $1 < d_m < m$ such that the d_m's are all different.

a) For each $\varepsilon > 0$ we have

$$n^{1/2} \ll f(n) \ll n^{7/12+\varepsilon}$$

b) $\liminf_{n\to\infty} \dfrac{f(n)}{\sqrt{n}} \geq 4$

Equality holds if and only if there are infinitely many twin primes.

c) There exists a set A of integers of (asymptotic) density 1 such that

$$f(n) > 4\sqrt{2n}$$

for $n \in A$, n large.

P. Erdős and C. Pomerance. *An analogue of Grimm's problem of finding distinct prime factors of consecutive integers*. Utilitas Math. 24 (1983), 45–65.

§ XII.16 Great values of a function connected with consecutive integers

1) Let $f(n)$ be the least integer such that at least one of the numbers $n, n+1, \ldots, n+f(n)$ divides the product of the others. Then:

a) $f(k!) = k; f(n) > k$
for $n > k!$

b) $f(n) > \exp((\log n)^{1/2-\varepsilon})$
for an infinity of values of n

P. Erdős. *How many pairs of products of consecutive integers have the same prime factors?* Amer. Math. Monthly 87 (1980), 391–392.

See also:

R.K. Guy. *Unsolved problems in number theory*. Springer-Verlag 1981, p. 50.

2) Denote by $g(n)$ the largest integer k for which there is an m so that $n | \prod_{i=1}^{k}(m+i)$, but n does not divide the product of any $k-1$ of the integers $m+1, \ldots, m+k$. Then:

a) $\sum_{n \le x} f(n) = (x \log\log x)(1 + o(1))$

b) $$\max_{n \le x} f(n) = \frac{1}{2} e^{\gamma/2} \cdot \frac{\log x}{\sqrt{\log\log x}} + \frac{\gamma e^{\gamma}}{4} \cdot \frac{\log x}{\log\log x}(1 + O(1))$$

where γ is Euler's constant.

P. Erdős and J.-L. Nicolas. *Grandes valeurs d'une fonction liée au produit d'entiers consecutifs*. Ann. Fac. Sci. Toulouse 8 (1981), 173–199.

§ XII.17 A theorem of Erdős and Selfridge on the product of consecutive integers

1) a) The product of two or more consecutive positive integers is never a square, i.e. the equation

$$(n+1)\ldots(n+k) = x^{l} (*)$$

has no solutions in integers with $k \ge 2, n \ge 0, l = 2$.

V. Rigge. *Über die diophantisches Problem*. 9-th Congress Math. Scand. 1939, 155–160.

and

P. Erdős. *On the product of consecutive integers*. J. London Math. Soc. 14 (1939), 194–198.

b) For fixed l there are at most finitely many solutions to (*).

P. Erdős and V. Rigge. (See P. Erdős. *Notes on the product of consecutive integers*. I, II. J. London Math. Soc. 14 (1939), 194–198 and 245–249).

c) The product of two or more consecutive positive integers is never a power, i.e. equation (*) has no solution for $l \geq 2$

P. Erdős and J.L. Selfridge. *The product of consecutive integers is never a power*. Illinois J. Math. 19 (1975), 292–301.

Note. This was conjectured about 150 years ago, see Dickson's history.

L.E. Dickson. *History of the theory of numbers*. vol. 1, reprint, Chelsea, New York, 1952.

Remark. Erdős and Selfridge obtain in fact a stronger result, namely:

Let k, l, n be integers such that $k \geq 3, l \geq 2$ and $n + k \geq p^{(k)}$, where $p^{(k)}$ is the least prime satisfying $p^{(k)} \geq k$. Then there is a prime $p \geq k$ for which $\alpha_p \not\equiv 0 (\bmod l)$, where α_p is the power of p dividing $(n+1)\cdots(n+k)$.

2) For fixed t, $(n+d_1)\ldots(n+d_k) = x^l$, $1 = d_1 < \cdots < d_k \leq k+t$ has only a finite number of solutions.

(See P. Erdős and J.L. Selfridge (p. 300).)

3) Let $b, d, m, y, k > 2, l \geq 2$, with $(m, d) = 1$ and $P(b) \leq k$ be positive integers. Then there exists an effectively computable absolute constant c such that

$$m(m+d)\ldots(m+(k-1)d) = by^l$$

with $l \geq c$ implies that $k \leq K(m, n)$ an effectively computable number depending only on m and n

T.N. Shorey. *Some exponential diophantine equations*. II. Number theory and related topics, Pap. Ramanujan Colloq., Bombay/India 1988, Stud. Math., Tata Inst. Fundam. Res. 12 (1989), 217–229.

§ XII.18 Products terms in an arithmetical progression

1) Let $m(m+d)\cdots(m+(k-1)d) = by^l$, where b, d, k, l, m, y are positive integers satisfying $P(b) \leq k$, $(m, d) = 1$, $k > 2$, $l > 1$, $y > 1$ and $P(y) > k$. Further, it is assumed that l is prime. (Here $P(x)$ denotes the greatest prime factor of x)

If $l \in \{2, 3, 5\}$, then k is bounded by an effectively computable number depending on $\omega(d)$. If $l \geq 7$, then k is bounded by an effectively computable number only on l and $\omega(d_1)$, where d_1 is the maximal divisor of d such that all prime factors of d_1 are $\equiv 1 (\bmod l)$.

T.N. Shorey and R.T. Tijdeman. *Perfect powers in products of terms in an arithmetical progression*. Compositio Math. 75 (1990), 307–344.

Remark. The proofs use results of
J.-H. Evertse. Compositio Math. 47 (1982), 289–315, when $l = 3$ and $l = 5$.

2) Let a, d, k be positive integers and $\Delta = a(a+d)\cdots(a+(k-1)d)$. Then

$$\omega(\Delta) \geq \pi(k)$$

where $\omega(\Delta)$ is the number of distinct prime factors of Δ, and $\pi(k)$ is the number of primes $\leq k$. This inequality is best possible.

T.N. Shorey and R. Tijdeman. *On the number of prime factors of an arithmetic progression.* J. Sichuan Univ., Nat. Sci. Ed. 26, Spec. Issue (1989), 72–74.

Corollary. The greatest prime factor of Δ is $\geq k$

3) Suppose $l \geq 3$ is a prime, m and d are positive coprime integers with $k \geq 3$ and such that each of the numbers $m, m+d, \ldots, m+(k-1)d$ is a perfect l-th power. Further, let d_1 denote the largest divisor of d with the property that all its prime factors are $\equiv 1 (\bmod\, l)$ and let m_1 be similarly defined. Let $\varepsilon > 0$. Then there exists an effectively computable number $c = c(\varepsilon)$ such that $k \geq c$ implies $2^{\omega(d_1)} > (1-\varepsilon)k$.

T.N. Shorey and R. Tijdeman. *Perfect powers in arithmetical progression.* II. Compositio Math. 82 (1992), 107–117.

§ XII.19 On the sequence $n! + k$, $2 \leq k \leq n$

Let $n \geq 6$ and $2 \leq k \leq n$. Then $n! + k$ is divisible by two primes, one being larger than n.

M.R. Chowdhury. *Über die Zahlenfolge* $n! + k, 2 \leq k \leq n$. Elem. Math. 44 (1989), 129–130.

§ XII.20 Decomposition of $n!$ into prime factors

1) a) Let $n! = \prod_{i=1}^{k} p_i^{\alpha_i} (p_1 < p_2 < \cdots < p_k)$ be the prime factorization of $n!$. If $\alpha_i > \alpha_j$, then

$$p_i^{\alpha_i} > p_j^{\alpha_j}$$

P. Erdős. Problem 4226. Amer. Math. Monthly 53 (1946), 594. Solution by W.J. Harrington. Amer. Math. Monthly 55 (1948), 433–435.

b) Let $n! = \prod p_i^{\beta_i}$, $i = 1, 2, \ldots, n$. Let p^{β} be the smallest $p_i^{\beta_i}$ in the factorization and write $\max\{p^{\beta}\}$ over all such decompositions as $n^{\alpha(n)}$. Then

$$\lim_{n\to\infty} \alpha(n) = e^{\theta-1}$$

where $\theta = \sum_{k=2}^{\infty} \frac{1}{k} \log \frac{k}{k-1}$

K. Alladi and C. Grinstead. *On the decomposition of n! into prime powers.* J. Number theory 9 (1977), 452–458.

c) Let $k \geq 4$ be a positive integer. If $p \leq k/2$ is a prime number, and $p^{\alpha} \| k!$, then

$$p^{\alpha} > k$$

P. Erdős and J.-L. Nicolas. *Grandes valeurs d'une fonction liée au produit d'entiers consecutifs.* Ann. Fac. Sci. Toulouse 8 (1981), 173–199 (See p. 178.)

d) Let $n!$ be expressed as the product of n factors, with the least one l, as large as possible. Then:

*) For $n > n_0 = n_0(\varepsilon)$, we have

$$l > \frac{n}{e+\varepsilon}$$

P. Erdős, J.L. Selfridge and E.G. Straus (See P. Erdős. *Some problems in number theory.* Academic Press. London and New York, 1971, 405–414).

**) By changing the positions of powers of 2 only,

$$l \geq 3n/16$$

E.G. Straus (See P. Erdős (1971).)

2) For $1 \leq k \leq n$ positive integers, let

$$r(n,k) = \max\left\{t : t \quad \text{nonnegative integer and } (k+t)! | \frac{(n+k)!}{n!}\right\}$$

Further, let $R(n) = \max\{r(n,k) : 1 \leq k \leq n\}$. Then

$$R(3^s) \leq 2$$

for all s; and for all d there exists n such that

$$R(n) \geq d$$

H. Gupta and K. Singh. *The largest r for which* $\frac{(n+k)!}{n!(k+r)!}$ *is an integer.* Indian J. Pure Appl. Math. 10 (1979), 1249–1265.

3) Let $f(n)$ be the largest power of n which divides $n!$, further let

$$F(N) = \max_{2\leq n\leq N} f(n) \text{ and } B(N) = \frac{N}{F(N)}.$$

a) $B(N) \sim \dfrac{\log N}{\log\log N}$

Problem 4 at the Contest in Memoriam Schweitzer Miklós in year 1973.

b) If there is a constant A such that for $x > x_0(A)$ the interval $[x, x + x^A]$ contains a prime number, then

$$B(N) = \Gamma^{-1}(N) + O(\Gamma^{-1}(N)^{\max(1/2,A)})$$

where Γ^{-1} is the inverse of the Γ-function.

Corollary 1. $B(N) = \Gamma^{-1}(N) + O(\Gamma^{-1}(N)^{13/23})$

Corollary 2. If the density hypothesis for the zeta-function is true (i.e. if $N(\alpha, T)$ denotes the number of zeros of $\zeta(s)$ in the domain $\sigma > \alpha \geq 1/2, |t| \leq T$, then $N(\alpha, T) \leq c_1 \cdot T^{2(1-\alpha)}$ holds with constants c_1 and c_2) then

$$B(N) = \Gamma^{-1}(N) + O(\Gamma^{-1}(N)^{1/2+\varepsilon})$$

J. Pintz. *On the asymptotic behaviour of a number theoretical function.* Ann. Univ. Sci. Budap. Eötv. Nom. Sect. Math. 24 (1981), 51–56.

§ XII.21 Divisibility of products of factorials

1) a) Let a, b, n be positive integers. If $(a!\,b!)\,|\,n!$, then there exists a constant C such that

$$a + b < n + C\log n$$

P. Erdős. Problem 557. Elem. Math. 23 (1968). (Solution by P. Bundschuh, p. 111).

b) There is a constant C_1 such that with $a = [C_1 \log n]$

$$((n!)\,(a+n)!)\,|\,(2n)!$$

for almost all n

P. Erdős. *Solution of the second part of problem 557.* Elem. Math. 23 (1969), 112–113.

Corollary. There is a constant C' such that $a + b > n + C'\log n$ and $(a!\,b!)\,|\,n!$ for infinitely many n

2) Assume $(a_1!\,a_2!\cdots a_k!)\,|\,n$, $(a_i \geq 2, 1 \leq i \leq k)$. Then

$$\max \sum_{i=1}^{k} a_i < \frac{5}{2}n$$

where the maximum is taken with respect to all choices of the a_i's and k

P. Erdős. Problem H-208. Fib. Quart. 11 (1973), 73. Solution by O.P. Lossers, Fib. Quart. 12 (1974), 399–400.

3) For a subset $A \subseteq [1, n]$, let $m(A) = \prod_{a \in A} a!$. Then

$$\operatorname{card}\{m(A) : A \subseteq [1, n]\} = \exp\left((1 + O(1))\frac{n \log\log n}{\log n}\right)$$

P. Erdős and R.L. Graham. *On products of factorials*. Bull. Inst. Mat. Acad. Sinica 4 (1976), 337–355.

4) a) There are only finitely many integers n such that the equation $n! = a_1 a_2 \cdots a_k$ has solutions satisfying (*) $n < a_1 < a_2 < \cdots < a_k \leq 2n$

P. Erdős. *Some problems in number theory. Carleton coordinates*. Math. Dept., Carleton Univ. Ottawa, 1977.

b) (i) For $n > 239$ there are no solutions

(ii) If (*) is replaced by $n < a_1 \leq a_2 \leq \cdots \leq a_k \leq 2n$ then there exist solutions for all $n > 13$

(iii) If one removes the restriction $a_k \leq 2n$ and denotes by $f(n)$ the size of the smallest a_k that can occur, then there are positive constants c_1 and c_2 such that

$$2n + c_1 \cdot n/\log n < f(n) < 2n + c_2 \cdot n/\log n$$

P. Erdős, R.K. Guy and J.L. Selfridge. *Another property of* 239 *and some related problems*. Proc. Eleventh Manitoba Conf. on Numerical Math. and Computing (Winnipeg, Man. 1981), Congr. Numer. 34 (1982), 243–257.

§ XII.22 Powers and factorials

1) a) The relation $n! = x^4 + y^4$, $(x, y) = 1$ has only a finite number of solutions.

b) The relation $n! + m! = x^k$ has only a finite number of solutions in integers x, m, n, k.

P. Erdős and R. Obláth. *Über diophantische Gleichungen der Form* $n! = x^p \pm y^p$. Acta Szeged 8 (1937), 241–255.

2) a) For all positive integer $a \leq n!$ there exists a representation

$$a = d_1 + d_2 + \cdots + d_k,\ d_1 < \cdots < d_k,$$

where $d_i \mid n!$ and $1 \leq k \leq n (1 \leq i \leq k)$

P. Erdős. *On a diophantine equation*. (Hungarian). Mat. Lapok 1 (1950), 192–210.

b) Let $f(n) \to \infty$ $(n \to \infty)$ and $\beta \in (0, 3/2)$. Let $D(n, z)$ be the number of divisors, d, of $n!$ such that

$$z(1 - \exp(-(\log n)^{\beta})) < d < z(1 + \exp((\log n)^{\beta}))$$

Then there exists a constant C such that for all sufficiently large n and z satisfying

$$\sqrt{n!}\exp(-n/f(n)) \leq z \leq \sqrt{n!}\exp(n/f(n))$$

we have

$$D(n,z) \geq n^{-C} \cdot \exp(-(\log n)^{\beta}) d(n!)$$

G. Tenenbaum. *Sur une problème extrêmal en arithmétique*. Ann. Inst. Fourier (Grenoble) 37 (1987), 1–18.

3) For $k \geq 1$ (integer) define

$$F_k$$

by $F_k = \{n$: for some $A \subseteq [1,n]$ with $\max_{a \in A}\{a\}$ and

$$\text{card } A \leq k, m(A) = y^2$$

for some integer y}, where

$$m(A) = \prod_{a \in A} a!$$

Define $D_k = F_k \backslash F_{k-1}$ and $F_0 = \emptyset$. Then:

a) $D_k = \emptyset$
for $k > 6$

b) $D_2 = \{n^2 : n > 1\}$

c) $D_3(n) = O(n)$

d) $\lim_{n \to \infty} D_4(n)/D_3(n) = +\infty$

e) For almost all primes p,

$$13p \notin F_5$$

f) Let n^* be the least element of D_6. Then

$$n^* = 527 = 17 \cdot 31$$

P. Erdős and R.L. Graham. *On products of factorials*. Bull. Inst. Math. Acad. Sinica 4 (1976), 337–355.

4) For every positive integer r there is an $n_0 = n_0(r)$ so that none of the integers $\sum_{i=1}^{r} n_i!$, $n_0 < n_1 < \ldots < n_r$ are powerful, that is, each has a prime factor which divides $\sum_{i=1}^{r} n_i!$ exactly (to the first power).

B. Brindza and P. Erdős. *On some diophantine problems involving powers and factorials. New advances in transcendence theory*. (Durham, 1986), Cambridge Univ. Press, Cambridge-New York, 1988.

5) There exists an effectively computable absolute constant C such that all solutions of the equation

$$(p-1)! + a^{p-1} = P^k$$

(a, k positive integers, $p > 2$ prime) satisfy

$$\max\{p, a, k\} < C$$

(See B. Brindza and P. Erdős.)

§ XII.23 Distribution of divisors of $n!$

1) The probability that (in the sense of asymptotic density) the first digit of $n!$ is equal to $a (1 \le a \le 9)$ is equal to

$$\log((a+1)/a)$$

S. Kunoff. *N! has the first digit property*. Fib. Quart. 25 (1987), 365–367.

2) Let X_N be a random variable uniformly distributed over the set $\{\log d : d|N\}$, and let F_N be the distribution function of X_N normalized to have expectation 0 and variance 1. Let $N_j = j!$. Then the sequence $F_{j!}$ has a limit distribution F satisfying, for every $\varepsilon > 0$,

$$\sup_{x} F_{j!}(x) - F(x)| \ll_\varepsilon j^{-1/3+\varepsilon}$$

M.D. Vose. *The distribution of divisors of N!*. Acta Arith. 50 (1988), 203–209.

§ XII.24 Stirling's formula and power of factorials

1) a) $n! \sim \sqrt{2\pi n} \cdot n^n \cdot e^{-n}$

$(n \to \infty)$

J. Stirling (1764), See e.g. E.T. Whittaker and G.N. Watson. *A course of modern analysis 4-th ed.* London Cambridge Univ. Press, 1952.

b) $$\sqrt{2\pi n} \cdot \frac{n^n}{e^n} \cdot \exp\left(\frac{1}{12n + 1/4}\right) < n! < \sqrt{2\pi n} \cdot \frac{n^n}{e^n} \cdot \exp\frac{1}{12n}$$

for all $n \ge 2$. For $n \ge 3$, in the lower bond $12n + 1/4$ may be replaced with $12n + 0.6/n$

P. Buchner. *Bemerkungen zur Stirlingschen Formel*. Elem. Math. 6 (1951), 8–11.

Remark. The proof is based on a method of Cesàro, see

E. Cesàro. *Elementares Lehrbuch der algebraischen Analysis und der Infinitesimalrechnung*. Leipzig 1922, p. 154.

c)
$$\sqrt{2\pi n}\cdot n^n\cdot e^{-n}\cdot \exp\left(\frac{1}{12n}-\frac{1}{360\,n^3}\right) < n! <$$
$$< \sqrt{2\pi n}\cdot n^n\cdot e^{-n}\cdot \exp\left(\frac{1}{12n}-\frac{1}{(360+\gamma_n)\,n^3}\right)$$
where $\gamma_n = 30(7n(n+1)+1)/n^2\cdot(n+1)^2$

P.R. Beesack. *Improvement of Stirling formula by elementary methods*. Univ. Beograd Publ. Elektrotehn. Fak. Ser. Mat. Fiz. No. 274–301 (1969), 17–21.

Note. For other references concerning Stirling's formula, see

D.S. Mitrinović (in cooperation with P.M. Vasić). *Analytic Inequalities*. Springer-Verlag, 1970, (pp. 184–185.)

2) a) If $n > 1$ is a natural number, then
$$((n+1)!)^{1/(n+1)} > 2^{1/(n+1)}\cdot(n!)^{1/n}$$

A.K. Gupta. Problem 833. Math. Mag. 46 (1973), 107, Solutions by J. Vogel and M.T. Bird, Same journal, pp. 107–108.

b) Let $L_n = ((n+1)!)^{1/(n+1)} - n!^{1/n}$; $n = 1, 2, 3, \ldots$. Then
$$\left|L_n - \frac{1}{e}\right| < \frac{1}{\sqrt{n}}$$
for $n \geq 2$; and
$$\left|L_n - \frac{1}{e}\right| < \varepsilon$$
for all $n \geq N(\varepsilon)$, where $N(\varepsilon) = 1 + \left[8 + \frac{1}{2}\varepsilon^2 + \left|8 - \frac{1}{2}\varepsilon^2\right|\right]$

A. Lupaş. *On problem 579.* (1901) in Gazeta Matematică. (Romanian.) Gaz. Mat. Ser. B. 81 (1976), 281–286.

c) $L_n > L_{n+1}$
for $n \geq 7$

J. Sándor. *Sur la fonction Gamma*. Publ. C.R.M.P. Neuchâtel. Série I, 21, pp. 4–7 (1989).

Remark. A simple computation shows that $L_n > L_{n+1}$ holds also for $n \in \{1, 2, 3, 4, 5, 6\}$

§ XII.25 The Wallis sequence and related inequalities on gamma function

1) Define the sequence $(W_n)_{n=1,2,3,\ldots}$, given by
$$W_n = \left(\frac{(2n)!!}{(2n-1)!!}\right)^2\cdot\frac{1}{2n+1},\quad n\geq 1$$

(The Wallis sequence.) Then:

a) (W_n)
is strictly increasing and $\lim_{n\to\infty} W_n = \frac{\pi}{2}$

J. Wallis. *Arithmetica infinitorum*. Oxford, 1656.

See also

K. Knopp. *Theorie und Anwendungen der unendlichen Reihen. #.* Aufl. Springer, Berlin 1931 (p. 397.)

b) $\frac{\pi(n+1/4)}{2n+1} < W_n < \frac{\pi}{2}$

D.K. Kazarinoff. *On Wallis' formula*. Edinburgh Math. Notes 40 (1956), 19–21.

c) Let n be a positive integer and let $c \geq -1$ be independent of n. Then

$$W_n > \frac{\pi(n+c)}{2n+1}$$

for $c \leq 1/4$; and

$$W_n < \frac{\pi(n+c)}{2n+1}$$

for $c \geq \frac{n+1}{4n+3}$

J.T. Chu. *A modified Wallis product and some applications*. Amer. Math. Monthly 69 (1962), 402–404.

d) $\frac{(4n+1)\pi}{4(2n+1)} < W_n < \frac{(2n+1)\pi}{4n+3}$

J. Gurland. *An inequality satisfied by the gamma function*. Skand. Aktuarietidskr. 39 (1956), 171–172.

Remark. Since $\frac{\Gamma(n+1)}{\Gamma(n+\frac{1}{2})} = \frac{1}{\sqrt{\pi}} \cdot \frac{(2n)!!}{(2n-1)!!}$, the Wallis sequence may be studied by using the theory of gamma function, specially the function $\frac{\Gamma(x+1)}{\Gamma(x+s)}$, $0 < s < 1$, see the section with Euler's gamma function.

e) The asymptotic expansion of W_n is given by

$$W_n = \frac{\pi}{2} \exp\left(\frac{a_1}{n} + \frac{a_2}{n^2} + \dots \frac{a_k}{n^k} + \dots\right)$$

where $a_k = 1/k \cdot 2^k$, for k even, and

$$a_k = (2(2^{k+1} - 1)B_{k+1} - (k+1)/k(k+1)2^k$$

for k odd. Here B_k are the Bernoulli numbers, i.e.

$$B_1 = 0, B_2 = 1/6, B_3 = 0, B_4 = -1/30, \dots$$

(See e.g. L. Tóth and A. Vernescu. *The asymptotic expansion of the Wallis sequence*. (Romanian.) Gaz. Mat. Perf. Met. Method. (1989), 26–29.)

2) a) If b and c are real numbers such that $c > 0$ and $c - 2b > 0$, then

$$\frac{\Gamma(c-2b)\Gamma(c)}{(\Gamma(c-b))^2} \geq \frac{b^2+c}{c}$$

with equality holding if $b = 0$ or $b = -1$

(See J. Gurland.)

b) Let b, c real numbers such that $c > 2, c - 2b > 0, b \neq 0, b \neq -1$. Then

$$\frac{\Gamma(c-2b)\Gamma(c)}{(\Gamma(c-b))^2} > 1 + \frac{b^2(c-2)}{(c-b-1)^2}$$

D. Gonkale. *On an inequality for gamma functions*. Skand. Aktuarietidskr. (1962), 213–215.

c) If b and c are real numbers such that $c > 0$ and $c - 2b > 0$, then

$$\frac{\Gamma(c-2b)\Gamma(c)}{(\Gamma(c-b))^2} \geq \sum_{k=0}^{n-1} \frac{1}{k!} \cdot \frac{((b)_k)^2}{(c_k)}$$

where $(x)_y$ denotes $\dfrac{\Gamma(x+y)}{\Gamma(x)}$

H. Ruben. *Variance bounds and orthogonal expansions in Hilbert space with an application to inequalities for gamma functions and* π. J. Reine Angew. Math. 225 (1967), 147–153.

Remark. For other references see also

D.S. Mitrinović. *Analytic inequalities*. Springer-Verlag, 1970.

3) $$\frac{3^{3^2} \cdot 5^{5^2} \ldots (2n+1)^{(2n+1)^2}}{2^{2^2} \cdot 4^{4^2} \ldots (2n)^{(2n)^2}} \sim \left(\exp\left(\frac{3}{2} - \frac{7\zeta(3)}{4\pi^2}\right)\right) \cdot (2n)^{2n^2+3n+1} \cdot e^{3n/2}$$

as $n \to \infty$

U. Balakrishnan. *A series for* $\zeta(\beta)$. Proc. Edinburgh Math. Soc. (2) 31 (1988), 205–210.

§ XII.26 A special sequence of Cesáro

For n a positive integer, denote $(n^n)! = 1^1 \cdot 2^2 \cdots n^n$. Then there exists $\theta_n \in (0, 1)$ such that

$$(n^n)! = c \cdot n^{\frac{n(n+1)}{2} + \frac{1}{12}} \cdot \exp\left(-\frac{n^2}{4} + \frac{1}{720\, n^2} - \frac{\theta_n}{5040\, n^4}\right)$$

where

$$c = 2^{-5/36} \cdot \pi^{1/6} \cdot \exp\left(\frac{2}{3}\int_0^{1/2} \log\Gamma(t)\,dt\right)$$

A. Lupaş and L. Lupaş. *On certain special functions*. (Romanian.) Sem. Itin. Ec. funcţ. Aprox. convex. 1980, Timişoara, 55–68.

Remark. Cesàro obtained that

$$(n^n)! \sim n^{\frac{n(n+1)}{2}} \cdot \exp\left(-\frac{n^2}{4}\right)$$

$(n \to \infty)$

E. Cesàro. Problem. Nouv. Ann. Math. (3) 17 (1888), 112. Solution by G. Pólya. Nouv. Ann. Math. (4) 11 (1911), 373–381.

See also

G. Pólya and G. Szegő. *Problems and theorems in analysis*. Springer-Verlag, 1972 (Problem 15, Part II.)

§ XII.27 Inequalities on powers and factorials related to the gamma function

1) a) $(k!)^m \cdot m^{km} \geq (m!)^k \cdot k^{km}$
for $k \geq m$ positive integers

b) $$\left(\frac{k^{k-1}}{(k-1)!}\right)^{m(m-1)} \leq \left(\frac{m^{m-1}}{(m-1)!}\right)^{k(k-1)}$$

for $k \leq m$

J. Sándor. *On some diophantine equations involving the factorial of a number*. Seminar Arghiriade No. 21., 1989, Timişoara (Romania.)

2) a) $((k+1)!)^{1/(k+2)} - ((m+1)!)^{1/(m+2)} \geq (k!)^{1/(k+1)} - (m!)^{1/(m+1)}$
for $k \geq m$ positive integers.

b) $$((k+2)!)^{1/(k+2)} - ((m+2)!)^{1/(m+2)} \leq \leq ((k+1)!)^{1/(k+1)} - ((m+1)!)^{1/(m+1)}$$
for $k \geq m$

(See J. Sándor.)

Remark. The proof is based on the facts that the function $x \to \Gamma(x)^{1/x}$ is strictly convexe for $x > 0$, while the function $x \to \Gamma(x+1)^{1/x}$ is strictly concave for $x > 7$ (Γ is the Euler's gamma function).

J. Sándor. *Sur la fonction gamma*. Publ. C.R.M.P. Neuchâtel Série I, 21 pp. 4–7 (1989.)

§ XII.28 Arithmetical products involving the gamma function

1) Let $S(n) = \prod_{1 \leq k \leq n, (k,n)=1} \Gamma(k/n)$, where Γ is Euler's gamma function. Then:

a) $\log S(n) \sim \varphi(n) \log \sqrt{2\pi}$

$(n \to \infty)$

b) $$\sum_{n \le x} \log S(n) = \frac{3 \log 2\pi}{2\pi^2} \cdot x^2 + O(x \log x)$$

J. Sándor and L. Tóth. *A remark on the gamma function*. Elem. Math. 44 (1989), 73–76.

Note. a) is a simple consquence of the following result:
If $f : [0,1] \to \mathbb{R}$ is integrable, then

$$\lim_{n\to\infty} \frac{1}{\varphi(n)} \sum_{\substack{1\le k\le n \\ (k,n)=1}} f(k/n) = \int_0^1 f(x)\mathrm{d}x$$

G. Pólya and G. Szegő. *Problems and theorems in analysis*. Springer-Verlag, Berlin, Heidelberg 1972, (Problem 188, Part II.)

2) Denote $(\Gamma(n))! = \Gamma(1)\Gamma(2)\cdots\Gamma(n)$. Then

$$(\Gamma(n))! = \frac{(2\pi)^{n/2}}{C} \cdot n^{n^2/2-1/12} \cdot \exp\left(-\frac{3n^2}{4} + \frac{1}{12} - \frac{1}{240n^2} + \frac{z}{1008n^4}\right)$$

where $z \in (0,1)$ and $C = 2^{-5/36} \cdot \pi^{1/6} \cdot \exp\left(\frac{2}{3} \cdot \int_0^{1/2} \log \Gamma(t)\, \mathrm{d}t\right)$

A. Lupaş and L. Lupaş. *On certain special functions*. (Romanian.) Sem. Itin. Ec. funcţ. Aprox. convex. 1980, Timişoara, 55–68.

§ XII.29 Monotonicity and convexity results of certain expressions of gamma function

1) a) Let $f(x) = \Gamma(x+1)^{1/x}$ and $g(x) = f(x+1)/f(x)$. Then $g(x)$ is strictly decreasing for $x > 1$.

b) $f(x)/x$
is strictly decreasing for $x > 1$, and $f^2(x)/x$ is strictly increasing for $x \ge 6$

Remark. For x a positive integer, the first result was discovered by Minc and Sathre.

H. Minc and L. Sathre. *Some inequalities involving* $(r!)^{1/r}$. Proc. Edinburgh Math. Soc. (2) 14 (1965/66), 41–46.

See also

D.S. Mitrinović (in cooperation with P.M. Vasić). *Analytic inequalities*. Springer-Verlag, New-York, Heidelberg, Berlin, 1970.

c) $x \cdot g(x)$
is strictly increasing for $x > 1$

J. Sándor. *Sur la fonction gamma*. Publ. C.R.M.P. Neuchâtel, Série I, 21, pp. 4–7 (1989.)

2) a) $f(x)$
is strictly concave for $x > 7$ and $f_1(x) = \Gamma(x)^{1/x}$ is strictly convexe for $x > 0$

b) $g(x)$
is strictly convexe for $x \geq 6$

c) $x \cdot g(x)$
is strictly concave for $x \geq 5$

d) $f(x)$ and $f_1(x)$
are logarithmically concaves for $x \geq 6$

e) $\Gamma(x)^{f(x+1)/2f(x)}$ and $\Gamma(x+1)^{f(x+1)/2f(x)}$
are logarithmically concaves for $x \geq 6$.

(See J. Sándor.)

3) Let $H(x) = \psi(x+1) + \gamma$, where $\psi(x) = \dfrac{\Gamma'(x)}{\Gamma(x)}(x > 0)$ is Euler's digamma function. Then

$$\log\left(x + \frac{1}{2}\right) + \gamma < H(x) < \log(x + a(x)) + \gamma$$

where $a(x) = (8x+9)/(16x+16)$, and γ is Euler's constant.

J. Sándor. *Remark on a function which generalizes the harmonic series.* C.R. l'Acad. Bulg. Sci. 41 (1988), 19–21.

4) a) $\left(\Gamma\left(1 + \frac{1}{x}\right)\right)^x$
is strictly decreasing for $x > 0$

b) $x \cdot \left(\Gamma\left(1 + \frac{1}{x}\right)\right)^x$
is strictly increasing for $x > 0$

c) $x^{1-\gamma} \cdot \left(\Gamma\left(1 + \frac{1}{x}\right)\right)^x$
is strictly decreasing for $0 < x < 1$
(Here γ denotes Euler's constant.)

D. Kershaw and A. Laforgia. *Monotonicity results for the gamma function.* Atti della Acad. Sci. Torino 119 (1985), 127–134.

5) a) $n^{1-s} < \dfrac{\Gamma(n+1)}{\Gamma(n+s)} < \exp((1-s)\psi(n+1))$

where $0 < s < 1,\ n = 1, 2, 3, \ldots$ and $\psi = \dfrac{\Gamma'}{\Gamma}$

W. Gautschi. *Some elementary inequalities relating to the gamma and incomplete gamma function.* J. Math. Phys. 38 (1959), 77–81.

$$\text{b)}\quad \left(x + \frac{1}{4}\right)^{1/2} < \frac{\Gamma(x+1)}{\Gamma(x+1/2)} < \left(x + \frac{1}{\pi}\right)^{1/2}$$

for $x > 0$

G.N. Watson. *A note on gamma function.* Proc. Edinburgh Math. Soc. (2) 11 (1958–59); Edinburgh Math. Notes No. 42 (1959), 7–9.

$$\text{c)}\quad \frac{\Gamma(n+1)}{\Gamma(n+s)} < \frac{4(n+s)}{4n+(s+1)^2} \cdot (n+1)^{1-s}$$

for $0 < s < 1, n = 1, 2, \ldots$

T. Erber. *The gamma function inequalities of Gurland and Gautschi.* Skand. Aktuarietidskr. 1960 (1961), 27–28.

$$\text{d)}\quad \begin{aligned} &\frac{(x+1)^x}{(x+s)^{x+s-1}} \cdot \exp(-(1-s)) < \frac{\Gamma(x+1)}{\Gamma(x+s)} < \\ &\quad < \frac{(x+1)^{x+1/2}}{(x+s)^{x+s-1/2}} \cdot \exp(-(1-s)) \end{aligned}$$

for $0 < s < 1, x > 0$

J.D. Kečkić and P.M. Vasić. *Some inequalities for the gamma function.* Publ. Inst. Math. (Beograd) 11 (1971), 107–114.

$$\text{e)}\quad \exp\left((1-s)\,\psi(x+s^{1/2})\right) < \frac{\Gamma(x+1)}{\Gamma(x+s)} < \exp\left((1-s)\,\psi\left(x + \frac{s+1}{2}\right)\right)$$

and $\left(x + \dfrac{s}{2}\right)^{1-s} < \dfrac{\Gamma(x+1)}{\Gamma(x+s)} < \left(x - \dfrac{1}{2} + \left(s + \dfrac{1}{4}\right)^{1/2}\right)^{1-s}$, for all $x > 0$,

$0 < s < 1$.

D. Kershaw. *Some extensions of W. Gautschi's inequality for the gamma function.* Math. Comp. 41 (1983), 607–611.

Remarks. *) Independently, Lorch has proved the following weaker inequalities

$$(k+c)^{s-1} < \frac{\Gamma(x+s)}{\Gamma(x+1)} < \left(k + \frac{s}{2}\right)^{s-1} \quad \text{for } 0 < s < 1,\ k = 0, 1, 2, \ldots$$

L. Lorch. *Inequalities for ultraspherical polynomials and the gamma function.* J. Approx. Theory 40 (1984), 115–120.

**) For $s = \dfrac{1}{2}$ we have

$$\left(x+\frac{1}{4}+\frac{1}{32k+32}\right)^{1/2} < \frac{\Gamma(k+1)}{\Gamma\left(x+\frac{1}{2}\right)} < \left(\frac{\left(m+\frac{1}{2}\right)^2}{k+\frac{3}{4}+\frac{1}{32k+48}}\right)^{1/2}$$

$k = 1, 2, 3, \ldots$

A.V. Boyd. *Note on a paper by Uppuluri.* Pacific J. Math. 22 (1967), 9–10.

***) A more general inequality states that

$$\sqrt{x}\exp\sum_{k=1}^{2m}\frac{(1-2^{-2k})\cdot B_{2k}}{k(2k-1)x^{2k-1}} < \frac{\Gamma(x+1)}{\Gamma\left(x+\frac{1}{2}\right)} <$$

$$< \sqrt{x}\exp\sum_{k=1}^{2l-1}\frac{(1-2^{-2k})\cdot B_{2k}}{k(2k-1)x^{2k-1}}$$

for $m, l = 1, 2, 3, \ldots, x > 0$ (Here B_{2k} are Bernoulli numbers.)

D.V. Slavić. *In inequalities for* $\Gamma(x+1)/\Gamma(x+1/2)$. Univ. Beograd Publ. Elektrotehn. Fak. Ser. Mat. Fiz. No. 498–541 (1975), 17–20.

f) $$\left(x+\frac{2}{3}s\right)^{s-1} < \frac{\Gamma(x+s)}{\Gamma(x+1)}$$

for $0 < s < 1, x \geq 1$; and

$$\left(x+\frac{-11+\sqrt{121+24s}}{2}\right)^{s-1} < \frac{\Gamma(x+s)}{\Gamma(x+1)}$$

for $0 < s < 1, x \geq 5$

g) $$\frac{\Gamma(x+s)}{\Gamma(x+1)} < \left(x+\frac{s}{2}+\frac{1}{8}\right)^{s-1}$$

for $1 < s < 2, x \geq 0$; and

$$\frac{\Gamma(x+s)}{\Gamma(x+1)} < \left(x+\frac{s}{2}+\frac{1}{10}\right)^{s-1}$$

for $1 < s < 2, x \geq 1$

A. Laforgia. *Further inequalities for the gamma function.* Math. Comp. 42 (1984), 597–600.

h) Let $a = \dfrac{s(1-s)}{2}, b = a\dfrac{(s-2)(3s-1)}{12}$. Then

$$\frac{\Gamma(x+s)}{\Gamma(x+1)} < x^{s-3}(x^2-ax-b)$$

if $2 < s < 3$; and

$$x > x_0(s), \frac{\Gamma(x+s)}{\Gamma(x+1)} > x^{s-3}(x^2-ax-b)$$

if $0 < s < 2, s > 3; x > x_1(s)$ where $x_0(s)$ and $x_1(s)$ are constants depending on s

Remark. For $0 < s < 0.77$, the first inequality from h) is valid for all $x \geq 1$. If $0.77 \leq s \leq 0.9$, it is valid for $x \geq 3$; if $0.9 < \lambda < 1$, it is valid for $x \geq 1/3(1-s)$

A. Laforgia and S. Sismondi. *Monotonicity results and inequalities for the gamma and error functions.* J. Comp. Appl. Math. 23 (1988), 25–33.

i) Let b as in h) and put $c = a\dfrac{(s+1)(3s-2)}{12}$. Then

$$\frac{\Gamma(x+s)}{\Gamma(x+1)} > \frac{x^{s+1}}{x^2+ax-c}$$

$(s > 0)$ is valid in the following situations: $0 < s < 2/3, x > 0$; $2/3 \leq s \leq 0.88$, $x \geq 1$; $0.88 < s \leq 0.95$, $x \geq 3$; $0.95 < s < 1$, $x \geq 1/8(1-s)$;

(See A. Laforgia and S. Sismondi.)

6) Let m, n, r, s be positive real numbers and $k = 1, 2, 3, \ldots$. Then:

a) $(\Gamma(n+k+1)/k!)^{1/n} \leq (\Gamma(r+k+1)/k!)^{1/r}$
if $0 < n \leq r$

b) $\Gamma(n+k+1)^{r-m} \leq \Gamma(m+k+1)^{r-n} \cdot \Gamma(r+k+1)^{n-m}$
if $0 \leq m \leq n \leq r$

c) $$\left(\frac{\Gamma(r+k+1)}{\Gamma(m+k+1)}\right)^{1/(r-m)} \leq \left(\frac{\Gamma(s+k+1)}{\Gamma(n+k+1)}\right)^{1/(s-n)}$$

if $m \leq n, r \leq s$

D.S. Mitrinović and J.E. Pečarić. *Note on the Gauss-Winckler inequality.* Anzeiger Österr. Akad. Wiss. Math.-Naturwiss. K1. (1986), Nr. 6, 89–92.

7) a) If $p > 0, q > 0, p > r, q > s$, then for $r > 0$ and $s > 0$, we have

$$B(p,q) \leq \left(\frac{r}{r+s}\right)^r \cdot \left(\frac{s}{r+s}\right)^s \cdot B(p-r, q-s)$$

where B is the beta function. For $r < 0$ and $s < 0$, the reverse inequality holds.

P. Kesava Menon. *Some inequalities involving the Γ and ζ functions.* Math. Student 11 (1943), 10–12.

b) For all $x > 1, y > 1$, one has

$$\frac{(x-1)^{x-1}(y-1)^{y-1}}{(x+y-1)(x+y-2)^{x+y-2}} \le \frac{\Gamma(x)\Gamma(y)}{\Gamma(x+y)} \le$$

$$\le \frac{(x+1-x/(x+y))^{x+1-x/(x+y)} \cdot (y+1-y/(x+y))^{y+1-y/(x+y)}}{(x+y)^{x+y}}$$

A.U. Afuwape and C.O. Imoru. *Bounds for the beta function.* Boll. U.M.I. (5) 17–A (1980), 330–334.

Remark. For inequalities related to the gamma and beta functions, see also
D.S. Mitrinović. (In coop. with P.M. Vasić). *Analytic Inequalities.* Springer-Verlag, 1970, (pp. 285–289.)

§ XII.30 Left factorial function

The left factorial function $!(z)$ is defined as follows:

$$!(z) = \int_0^{+\infty} \mathrm{e}^{-t} \cdot (t^z - 1)\, \mathrm{d}t/(t-1)$$

Re $z > 0$. We have:

a) $!(z+1) - !(z) = \Gamma(z+1)$

b) $\lim_{x\to\infty} !(x)/\Gamma(x+1) = 0$

c) $!(x) \sim \left(\frac{2\pi}{x}\right)^{1/2} \cdot \left(\frac{x}{\mathrm{e}}\right)^x$

for $x \to \infty$

Đ. Kurepa. *Left factorial function in complex domain.* Math. Balcanica 3 (1973), 297–307.

Chapter XIII

ESTIMATES INVOLVING FINITE GROUPS AND SEMI-SIMPLE RINGS

§ XIII. 1 Maximal order of an element in the symmetric group

Let $o(\sigma)$ be the order of the element σ in the symmetric group S_n and denote $M(n) = \max_{\sigma \in S_n} o(\sigma)$. Then:

a) Let $\varepsilon > 0$. Then there exist $n_0(\varepsilon)$ such that for $n \geq n_0(\varepsilon)$,

$$(1-\varepsilon)\sqrt{n \log n} \leq \log M(n) \leq (1+\varepsilon)\sqrt{n \log n}$$

b) $\lim_{n\to\infty} \dfrac{\log M(n)}{\sqrt{n \log n}} = 1$

E. Landau. *Handbuch der Lehre von der Verteilung der Primzahlen.* I. pp. 222–229, Teubner, Leipzig, 1909.

c) For almost all n we have

$$\log M(n) \geq \sqrt{n \log n}$$

O. Herrmann. *Aufgabe 546.* Elemente der Math. 23 (1968), 41–42.

d) Let $\varepsilon > 0$. Then for $n \geq n_0(\varepsilon)$,

$$\log M(n) = (n(\log n + \log\log n + \delta(n)))^{1/2}$$

where

$$-1 + (\log\log n - 2 - \varepsilon)/\log n < \delta(n) < 1/4$$

M. Szalay. *On a theorem of Landau.* (Hungarian.) Mat. Lapok 22 (1971), 317–321.

e) $\log M(n) \leq 1.05314 \cdot (n \log n)^{1/2}$
and the maximum of $(\log M(n))/(n \log n)^{1/2}$ is attained for

$$n = 1319766$$

J.-P. Massias. *Majoration explicite de l'ordre maximum d'un élément du groupe symétrique.* Ann. Fac. Sci. Toulouse Math. (5) 6 (1984) 269–281.

f) $\log M(n) = \sqrt{\mathrm{li}^{-1}(n)} + O(n \exp(-a\sqrt{\log n}))$

($a > 0$, constant.)

J.-P. Massias, J.-L. Nicolas and G. Robin. *Évaluation asymptotique de l'ordre maximum d'un élément du groupe symétrique*. Acta Arith. 50 (1988), 221–242.

See also

J.-P. Massias, J.-L. Nicolas and G. Robin. *Effective bounds for the maximal order of an element in the symmetric group*. Math. Comp. 53 (1989), no. 188, 665–678.

Remark. In these papers appear also results on $P(M(n))$, where $P(k)$ denotes the greatest prime factor of k.

g) $\log M(n) \leq \sqrt{n \log n} \cdot \left(1 + \frac{\log\log n - 0.975}{2\log n}\right)$
for $n \geq 3$; and

$$\log M(n) \geq \sqrt{n \log n} \cdot \left(1 + \frac{\log\log n - 1.2}{2\log n}\right)$$

for $n \geq 93898$
(See J.-P. Massias, J.-L. Nicolas and G. Robin (1989).)

§ XIII. 2 A sum on the order of elements of S_n

1) Let S_n be the symmetric group on n letters, and let $N(\sigma)$ denote the group-theoretic order of σ, for $\sigma \in S_n$.
Let $\mu_n = \sum_{\sigma \in S_n} N(\sigma)/n!$. Then

$$\log \mu_n < 7.7(n/\log n)^{1/2}$$

E. Schmutz. *Proof of a conjecture of Erdős and Turán*. J. Number Theory 31 (1989), 260–271.

Remarks. (i) Erdős and Turán promised a proof of the estimation

$$\log \mu_n = O((n/\log n)^{1/2})$$

but this proof never materialized.

P. Erdős and P. Turán. *On some problems of a statistical group-theory*. IV. Acta Math. Acad. Sci. Hungar. 19 (1968), 413–435.

(ii) Schmutz raises the following problem: Prove or disprove that

$$\log \mu_n \sim 2\pi \cdot 6^{-1/2} \cdot (n/\log n)^{1/2}$$

2) $\log \mu_n \sim c(n/\log n)^{1/2}$

where $c = 2\left(2\int_0^\infty \log\log(e/(1-e^{-t}))dt\right)^{1/2}$

W.M.Y. Goh and E. Schmutz. *The expected order of a random permutation*. Bull. London Math. Soc. 23 (1991), 34–42.

§ XIII. 3 Statistical problems in S_n

Let S_n be the symmetric group on n letters and let $\mathrm{o}(x)$ denote the group theoretic order of x for $x \in S_n$

a) Let $k(n, y)$ be the number of $x \in S_n$ satisfying

$$\log \mathrm{o}(x) \le \frac{1}{2}\log^2 n + y \log^{3/2} n/\sqrt{3}$$

Then

$$\lim_{n\to\infty} \frac{k(n, y)}{n!} = \frac{1}{\sqrt{2\pi}} \cdot \int_{-\infty}^{y} \exp\left(-\frac{1}{2}t^2\right) dt$$

P. Erdős and P. Turán. *On some problems of a statistical group theory*. III. Acta Math. Acad. Sci. Hung. 18 (1967), 309–320.

b) For $x \in S_n$

$$\log \mathrm{o}(x) = \sum_{k=1}^{n} \varepsilon_k(x)(\log n - \log k) + A_n(x)$$

where the ε_k are independent random variables on S_n such that $\varepsilon_k = 1$ with probability $1/k$ and 0 with probability $1 - 1/k$;
and $|A_n(x)| < 3 \log n \cdot (\log\log n)^4$ for all except $o(n!)$ values of x (Here $o(n!)$ has the usual asymptotic meaning and should not be confused with the order of a group element!)

J.D. Bovey. *An approximate probability distribution for the order of elements of the symmetric group*. Bull. London Math. Soc. 12 (1980), 41–46.

c) $\log \mathrm{o}(x) < \sum_{c|x} \log \mathrm{o}(c) < \log \mathrm{o}(x) + 3 \log n \cdot (\log\log n)^4$
for all but $o(n!)$ elements $x \in S_n$

P. Erdős and P. Turán. *On some problems of a statistical group theory*. I. Z. Wahrscheinlichkeitstheorie und verw. Gebiete 4 (1965), 175–186.

d) For $x \in S_n$

$$\sum_{c|x} \log \mathrm{o}(c) = \sum_{k=1}^{n} \varepsilon_k(x)(\log n - \log k - \gamma) + B_n(x)$$

where

$$E(B_n^2) = O(\log n)$$

Here $E(f)$ is the expectation of f, i.e. $E(f) = \frac{1}{n!}\sum_{x\in S_n} f(x)$, and $\varepsilon_k : S_n \to \{0, 1\}$ is defined by

$$\varepsilon_k(x) = \begin{cases} 1 & \text{if } a_k(x) = k, \text{ with } a_n = x(n) \text{ for } x \in S_n \\ 0 & \text{otherwise} \end{cases}$$

(See J.D. Bovey.)

§ XIII. 4 Probability of generating the symmetric group

1) a) For any fixed real number t let $h(n, t)$ be the number of permutations $\sigma \in S_n$ satisfying

$$g(\sigma) \leq \log n + t\sqrt{\log n}$$

where $g(\sigma)$ denotes the number of cycles in the canonical decomposition of σ. Then

$$\lim_{n\to\infty} \frac{1}{n!} h(n, t) = \frac{1}{\sqrt{2\pi}} \cdot \int_{-\infty}^{t} e^{-\lambda^2/2} d\lambda$$

Corollary. If $k(n)$ tends to infinity arbitrarily slowly monotonically then for all but $o(n!)$ permutations σ the inequality $|g(\sigma) - \log n| \leq k(n)\sqrt{\log n}$ holds.

V.I. Gončarov. *On the field of combinatory analysis*. Izv. Akad. Nauk. SSSR. Ser. Mat. 8 (1944), 3–48 (See also Translations of the Amer. Math. Soc. Ser. 2, 19 (1962), 1–46.)

b) If $k_1(n)$ and $k_2(n)$ tend arbitrarily slowly monotonically to $+\infty$, then the canonical decomposition of all but $o(n!)$ permutations σ have the double property that no two cycles of length $> k_1(n)$ are equally long and at most $k_2(n)$ cycles can have the same length $\leq k_1(n)$.

c) Let $d(\sigma)$ be the number of the different cycle length of σ and denote by $H_n(t)$ the number of σ's satisfying

$$|d(\sigma) - \log n| < k(n)\sqrt{\log n}$$

where $k(n)$ tends to $+\infty$ arbitrarily slowly, and suppose that σ satisfies the two requirements of b). Then

$$\lim_{n\to\infty} \frac{1}{n!} H_n(t) = \frac{1}{\sqrt{2\pi}} \cdot \int_{-\infty}^{t} e^{-\lambda^2/2} d\lambda$$

(t real)

P. Erdős and P. Turán. *On some problems of a statistical group-theory*. I. Z. Wahrscheinlichkeitstheorie verw. Geb. 4 (1965), 175–186.

2) Let $Q \subset \{1, 2, \ldots, n\}$ with sup $Q \leq n/\log n$. Define $f(n, Q, k)$ to be the proportion of permutations $\sigma \in S_n$ containing exactly k q-cycles ($q \in Q$). Then

$$f(n, Q, k) = \frac{\lambda^k}{k!}e^{-\lambda} + O(n^{5-\log\log n})$$

where $\lambda = \sum_{q \in Q} 1/q$

J.D. Bovey. *The probability that some power of a permutation has small degree.* Bull. London Math. Soc. 12 (1980), 47–51.

Remark. This result is a generalization of a Lemma from
J.D. Bovey and A. Williamson. *The probability of generating the symmetric group.* Bull. London Math. Soc. 10 (1978), 91–96.

3) Let $\sigma \in S_n$ and suppose $\sigma = c_1 \cdots c_w$ is a decomposition of σ into disjoint cycles. Define $B(\sigma) = \prod_{j=1}^{w} \mathrm{o}(c_j)$, where $\mathrm{o}(c_j)$ is the order of c_j.
Let N be a positive integer. Then

$$P\{\sigma \in S_n : N | B(\sigma)\} \leq \frac{C(\Omega(N))(\log n)^{\Omega(N)}}{N}$$

where $\Omega(N)$ is the total number of prime factors of N and C is a function of $\Omega(N)$ only. (P denotes probability.)
(See J.D. Bovey.)

§ XIII. 5 Primitive subgroups of S_n

1) a) Let b be a primitive permutation group of minimal degree m and of degree n. Then

$$n < \frac{1}{4}(4+m)\left(4 + m\log\frac{1}{2}m\right)$$

C. Jordan. *Sur la limite du degré des groups primitifs qui contiennent une substitution donée.* J. Reine Angew. Math. 79 (1875), 248–253.

Note. For the notion of primitive degree (or class) see e.g.
H. Wielandt. *Finite permutation groups.* Academic Press, New York, 1964.

b) The proportion of pairs (x, y) which generate a primitive subgroup of S_n is $1 + \frac{1}{n} + O\left(\frac{1}{n^2}\right)$

c) The proportion of pairs (x, y) which fail to generate a primitive subgroup of S_n is $\frac{1}{n} + O\left(\frac{1}{n^2}\right)$

J.D. Dixon. *The probability of generating the symmetric group.* Math. Z. 110 (1969), 199–205.

2) Given $\varepsilon > 0$ and $0 < \alpha < 1$. Then

$$n^{\alpha} \ll P\{\sigma \in S_n : \text{minimal degree of } < \sigma \gg n^{\alpha}\} \ll n^{\varepsilon-\alpha}$$

as $n \to \infty$

J.D. Bovey. *The probability that some power of a permutation has small degree.* Bull. London Math. Soc. 12 (1980), 47–51.

Corollary. Given $\varepsilon > 0$, the proportion of ordered pairs $(x, y)(x, y \in S_n)$ which generate either A_n or S_n is greater than $1 - n^{-1+\varepsilon}$ for all sufficiently large n.

Remark. This improves an estimate of Bovey and Williamson.

J.D. Bovey and A. Williamson. *The probability of generating the symmetric group.* Bull. London Math. Soc. 10 (1978), 91–96.

§ XIII. 6 Number of solutions of $x^k = 1$ in symmetric groups

Let $T_n(k)$ denote the number of solutions of $x^k = e$ in the symmetric group of order n (e is the unit element.) Then:

1) a) $T_n(k) = \exp\big((1 + o(1))(1 - k^{-1})n \log n\big)$

for any fixed $k \geq 2$, as $n \to \infty$.

A.I. Pavlov. *On the number of solutions of the equation $x^k = a$ in the symmetric group.* (Russian.) Mat. Sb. 112 (1980), 380–395.

Notes. (i) Chowla, Herstein and Scott posed the problem of the asymptotic behaviour of $T_n(k)$, for fixed integer $k \geq 2$, as $n \to \infty$. They found the generating function $\exp\left(\sum_{d|k} z^d/d\right)$ for $T_n(k)/n!$.

S. Chowla, I.N. Herstein and W.R. Scott. *The solutions of $x^d = 1$ in symmetric groups.* Norske Vid. Selsk. 25 (1952), 29–31;

(ii) The problem of the asymptotic behaviour of $T_n(k)$ was solved for $k = 2$ by Chowla, Herstein and Moore and for $k = p$ (fixed prime) by Moser and Wyman.

S. Chowla, I.N. Herstein and K. Moore. *On the recursions connected with symmetric groups.* I. Canad. J. Math. 3 (1951), 328–334;

L. Moser and M. Wyman. *On the solutions of $x^d = 1$ in symmetric groups.* Canad. J. Math. 7 (1955), 159–168;

L. Moser and M. Wyman. *Asymptotic expansions.* Canad. J. Math. 8 (1956), 225–233.

b) $T_n(k) \sim \left(\frac{n}{e}\right)^n \cdot n^{-n/k} \cdot k^{-1/2} \cdot \exp\left(\sum_{d|k} \frac{n^{d/k}}{d} - \frac{1+(-1)^k}{4k}\right)$

for any fixed $k \geq 2$, as $n \to \infty$

L.M. Volynets. *On the number of solutions of the equation $x^s = e$ in the symmetric group.* (Russian.) Mat. Zametki 40 (1986), 155–160 (English translation: Math. Notes 40 (1986), 586–589);

and

H.S. Wilf. *The asymptotic of $e^{P(z)}$ and the number of elements of each order in S_n.* Bull. Amer. Math. Soc. 15 (1986), 228–232.

2) a) $T_n(n) = \left(1 + (1+(-1)^n) \cdot \frac{1}{n} + O(n^{\varepsilon-2})\right) \cdot (n-1)! \sim (n-1)!$

R. Warlimont. *Über die Anzahl der Lösungen $x^n = 1$ in der symmetrischen Gruppe S_n.* Arch. Math. 30 (1978), 591–594.

b) Let $\varepsilon > 0$ be fixed, $0 < \varepsilon < 1/100$. For $1 \leq k \leq n^{\frac{1}{4}-\varepsilon}$ and $n \to \infty$, we have

$$T_n(k) = (1 + o(1) + f(k,n)) \cdot \frac{n!}{n^{n/k}} \cdot \sqrt{2\pi nk} \exp\left(\sum_{d|k} \frac{n^{d/k}}{d} - \frac{1+(-1)^k}{4k}\right)$$

where

$$f(n,k) = 2\mathrm{Re}\left(\sum_{j=1}^{[\frac{k-1}{2}]} \exp\left(-i\frac{2jn\pi}{k} + \sum_{d|k} \frac{n^{d/k}}{d}\left(\exp\left(i\frac{2jd\pi}{k}\right) - 1\right)\right)\right) +$$

$$+ (1+(-1)^k) \cdot \frac{(-1)^n}{2} \cdot \exp\left(-2 \sum_{d|k, 2\nmid d} n^{d/k}/d\right)$$

P. Erdős and M. Szalay. *On some problems of the statistical theory of partitions.* Colloq. Math. Soc. János Bolyai, 51, Number Theory, Budapest, 1987, pp. 93–110.

§ XIII. 7 On the dimensions of representations of S_n

Let S_n be the symmetric group of n letters. Let $\Gamma_1, \ldots, \Gamma_{p(n)}$ by the pairwise nonequivalent irreducible representations of S_n, where $p(n)$ is the number of unrestricted partitions of n. Then for almost all representations

$$\dim \Gamma_j \geq \exp\left(\frac{1}{2} n \log n - n \log\log n - n\right)$$

M. Szalay. *A note on the dimensions of representations of S_n.* Proc. Colloq. János Bolyai, Debrecen, pp. 383–388, Amsterdam (1976.)

§ XIII. 8 Conjugacy classes of the alternating group of degree n

1) a) The number of conjugacy classes of the alternating group of degree n is asymptotically equal to $\frac{1}{2}p(n)$, where $p(n)$ denotes the number of unrestricted partitions.

J. Dénes, P. Erdős and P. Turán. *On some statistical properties of the alternating group of degree n.* L'Enseign Math. 15 (1969), 89–99.

Remarks. (i) More precisely, for the number $g(n)$ of these conjugacy classes, one has

$$g(n) = \frac{1}{2}p(n) + O\left(\frac{1}{\sqrt{n}}\right)\exp\left(\frac{\pi}{\sqrt{3}}\sqrt{n}\right) \sim \frac{1}{8n\sqrt{3}}\exp\left(\frac{2\pi}{\sqrt{6}}\sqrt{n}\right)$$

(ii) We have the inequality $g(n) - \frac{1}{2}p(n) > \exp(B\sqrt{n})$, with an explicit numerical B

(See J. Dénes, P. Erdős and P. Turán.)

b) Almost all conjugacy classes of the alternating group of degree n contain a pair of generators. See:

L.B. Beasley, J.L. Brenner, P. Erdős, M. Szalay and A.G. Williamson. *Generation of alternating groups by pairs of conjugates*. Periodica Math. Hung. 18 (1987), 259–269.

Note. The result from b) is due to Beasley, Brenner and Williamson.

c) For almost all conjugacy classes of A_n (i.e. with the exception of $o(g(n)) = o(p(n))$ classes at most) the elements can be commuted exactly with

$$\exp\left((1+o(1))\frac{\sqrt{6}}{4\pi}\cdot\sqrt{n}\log^2 n\right)$$

elements of A_n

(See J. Dénes, P. Erdős and P. Turán.)

2) Let $F(n,x)$ be the number of $\sigma \in A_n$ with

$$\log \mathrm{o}(\sigma) \le \frac{1}{2}\log^2 n + \frac{x}{\sqrt{3}}\log^{3/2} n$$

Then

$$\lim_{n\to\infty}\frac{F(n,x)}{\frac{1}{2}n!} = \frac{1}{\sqrt{2\pi}}\int_{-\infty}^{x} e^{-\lambda^2/2}d\lambda$$

(See J. Dénes, P. Erdős and P. Turán.)

§ XIII. 9 An estimate for the order of rational matrices

Let A be an $n \times n$ matrix, with rational elements, satisfying $A^m = I$ (unit-matrix.) Then

$$m \le e^{\gamma} \cdot (\log(n+1)) \cdot (1 + 1/\log^2(n+1)) \cdot n^{\pi(n+1)}$$

where $\pi(x)$ denotes the number of primes $\le x$

R. Putz. *An estimate for the order of rational matrices*. Canad. Math. Bull. 10 (1967), 459–461.

§ XIII.10 On kth power coset representatives mod p

Let p be an odd prime, k a positive integer, and $d = (k, p-1) > 1$. Let $C(p)$ be the multiplicative group consisting of the residue classes mod p that are relative prime to p, and $C_k(p)$ the multiplicative subgroup consisting of the kth power residues, and $H_0 (= C_k(p))$, $H_1, \ldots, H_{d-1}$, the cosets of $C_k(p)$ in $C(p)$. Let $g_m(p, k)$ be the smallest positive representative of H_m. (It can be assumed that
$1 = g_0(p, k) < g_1(p, k) < \cdots < g_{d-1}(p, k)$.) Then

$$g_m(p, k) < 2(dm/(d-m))^{1/2} \cdot p^{1/2}$$

and if -1 is a kth power residue mod p, then

$$g_m(p, k) < (dm/(d-m))^{1/2} \cdot p^{1/2}$$

R. Metsänkylä. *On k-th power coset representatives* mod p. Ann. Acad. Sci. Fenn. Sert. A I, No. 557 (1973), 6 pp.

§ XIII.11 Arithmetical properties of permutations of integers

1) Let $M(n)$ denote the number of permutations $a_1 a_2 \ldots a_n$ of $\{1, 2, \ldots, n\}$ containing no monotone 3-term arithmetic progression. Then

$$M(n) \ge 2^{2n-1};\ M(2n-1) \le (n!)^2 \text{ and } M(2n) \le (n+1)(n!)^2$$

for $n = 1, 2, 3, \ldots$

J.A. Davis, R.C. Entringer, R.L. Graham and G.J. Simmons. *On permutations containing no long arithmetic progressions*. Acta Arith. 34 (1977/78), 81–90.

2) Let $f(n)$ denote the largest integer k such that every permutation $a_1 a_2 \ldots a_n$ of $\{1, 2, \ldots, n\}$ satisfies $[a_i, a_{i+1}] \ge k$ for some i. Then

$$f(n) = (1 + o(1)) \frac{n^2}{4 \log n}$$

(as $n \to \infty$)

P. Erdős, R. Freud and N. Hegyváry. *Arithmetical properties of permutations of integers*. Acta Math. Hungar. 41 (1983), 168–176.

3) If $a_1, a_2, \ldots$ is a permutation of the positive integers, let

$$T = \{m \in \mathbb{N} : m = a_i + a_{i+1} \text{ for some } i\}.$$

Let $a_1, \ldots, a_n$ be a permutation of $1, \ldots, n$ and let $f(n)$ be the maximum over all permutations of $1, \ldots, n$ of card$T$ divided by $\max\limits_{1 \le i \le n} (a_i + a_{i+1})$. Also, let $ng(n)$ denote the maximal number of different values $\le n$ of $a_i + a_{i+1}$, where $a_1, \ldots, a_m$ is a permutation of $1, \ldots, m$, with the maximum taken over all n and all permutations. Then

$$f(n) = \frac{2}{3} + O\left(\frac{1}{n}\right)$$

and

$$g(n) = \frac{2}{3} + O\left(\frac{1}{n}\right)$$

R. Freud. *On sums of subsequent terms of permutations*. Acta Math. Hungar. 41 (1983), 177–185.

§ XIII.12 Number of non-isomorphic abelian groups of order n

Let $a(n)$ denote the number of non-isomorphic abelian groups of order n. Then:

1) a) $\limsup\limits_{n \to \infty} \log a(n) = \dfrac{\log 5}{4} \cdot \dfrac{\log n}{\log \log n}$

E. Krätzel. *Die maximale Ordnung der Anzahl der wesentlich verschiedenen Abelschen Gruppen gegebener Ordnung Quart*. J. Math. (Oxford) 21 (2) (1970), 273–275.

b) For all $\varepsilon > 0$ we have

$$a(n) < (\log n)^{\frac{1}{4} \log 5 + \varepsilon}$$

for almost all n

E. Krätzel. *Zahlentheorie*. Berlin 1981 (See p. 216.)

Remarks. (i) The first important bound on $a(n)$ was obtained by Kendall and Rankin, who proved, that there exists a positive constant c with

$$a(n) = O(n^{c/\log \log n})$$

D.G. Kendall and R.A. Rankin. *On the number of abelian groups of a given order*. Quart. J. Math. (Oxford) 18 (72) (1947), 197–208.

(ii) In 1958, Drozdova and Freiman found a more precise result:

$$\log a(n) \le c \cdot \frac{\log n}{\log \log n}\left(1 + O\left(\frac{\log \log \log n}{\log \log n}\right)\right)$$

The value $c = \dfrac{\log 5}{4}$ was given by Krätzel (1970.)

A.A. Drozdova and G.A. Freiman. (See A.G. Postnikov. *Introduction to analytic number theory*. (Russian.) Moscow 1971.

c) Let $A = A(n)$ be the smallest integer such that

$$\theta(A) \geq \frac{1}{4} \log n$$

(where $\theta(x) = \sum_{p \leq x} \log p$ is Chebysheff's function.) Then

$$\log a(n) \leq \log 5 \cdot \pi(A) + O((\log n)^{\theta})$$

where

$$\theta = \frac{2 \log 11}{5 \log 5} < 0.994$$

(Here $\pi(x) = \sum_{p \leq x} 1$); and there are infinitely many integers n for which one has

$$\log a(n) = \log 5 \cdot \pi(A)$$

Corollary. $\log a(n) \leq \log 5 \cdot \mathrm{li}\left(\frac{1}{4} \log n\right) + O(\log n \cdot \exp(-c_1 (\log \log n)^{1/2}))$ and there are infinitely many values of n such that

$$\log a(n) \geq \log 5 \cdot \mathrm{li}\left(\frac{1}{4} \log n\right) + O(\log n \cdot \exp(-c_2 (\log \log n)^{1/2}))$$

W. Schwarz and E. Wirsing. *The maximal number of non-isomorphic abelian groups of order n*. Arch. Math. 24 (1973), 59–62.

Remark. Heppner proved that this corollary is also correct for a general class of prime—independent multiplicative functions.

E. Heppner. *Die maximale Ordnung primzahl-unabhängiger multiplikativer Funktionen*. Arch. Math. 24 (1973), 63–66.

2) $A(x) = \sum_{n<x} a(n) = c_1 x + R(x)$
where:

a) $c_1 = \zeta(2)\,\zeta(3)\,\zeta(4)\ldots = 2.29485$ and $R(x) = O(\sqrt{x})$

P. Erdős and G. Szekeres. *Über die Anzahl der Abelschen Gruppen gegebener Ordnung und über ein verwandtes zahlentheoretisches Problem*. Acta Sci. Math. Szeged 7 (1935), 95–102.

b) $A(x) = c_1 x + c_2 x^{1/2} + O(x^{1/3} \log^2 x)$

where $c_n = \prod_{\substack{k=1 \\ k \neq n}}^{\infty} \zeta\left(\frac{k}{n}\right)$, $(n = 1, 2, \ldots)$

(See D.G. Kendall and R.A. Rankin 1) b).)

c) $A(x) = c_1 x + c_2 x^{1/2} + c_3 x^{1/3} + O((x \log^3 x)^{3/10})$
where $c_n (n = 1, 2, 3)$ are given as in b).

H.-E. Richert. *Zur Anzahl Abelscher Gruppen gegebener Ordnung*. I. Math. Z. 56 (1952), 21–32; and Ibid. II. Math. Z. 58 (1953), 71–84.

d) $A(x) = c_1 x + c_2 x^{1/2} + c_3 x^{1/3} + O((x^{20} \cdot \log^{63} x)^{1/69})$

W. Schwarz. *Über die Anzahl Abelscher Gruppen gegebener Ordnung*. I. Math. Z. 92 (1966), 314–320; and Ibid. II. J. Reine Angew. Math. 228 (1967), 133–138.

e) The same estimate as in d) holds, with remainder term

$$O(x^{7/27} \cdot \log^2 x)$$

P.G. Schmidt. *Zur Anzahl Abelscher Gruppen gegebener Ordnung*. I. J. Reine Angew. Math. 229 (1968), 34–42, and Ibid. II. Acta Arith. 13 (1968), 405–417.

f) The same holds with $O(x^{105/407} \cdot \log^2 x)$

B.R. Srinivasan. *On the number of abelian groups of a given order*. Acta Arith. 23 (1973), 195–205.

g) The same holds, with the error term

$$O(x^{97/381} \cdot \log^{35} x)$$

G. Kolesnik. *On the number of abelian groups of a given order*. J. Reine Angew. Math. 329 (1981), 164–175.

Remark. Recently Liu obtained the stronger term $O(x^{40/159+\varepsilon})$, and even $O(x^{50/159+\varepsilon})$.

H.-Q. Liu. *On the number of abelian groups of a given order.* Acta Arith. 59 (1991), 261–277.
H.-Q. Liu. bid. Acta Arith. 64 (1993), 285–296.

h) Assuming the Riemann Hypothesis, we have

$$A(x) = \sum_{k=1}^{6} c_k x^{1/k} + \Omega(x^{1/6-\varepsilon})$$

for each $\varepsilon > 0$

(See W. Schwarz. II.)

i) Let $\Delta(x)$ denote the error term in the asymptotic formula for $\sum_{n \le x} a(n)$.

Then

$$\int_1^x \Delta^2(t)\mathrm{d}t \ll x^{39/29} \cdot \log^2 x$$

A. Ivić. *The number of finite nonisomorphic abelian groups in mean square.* Hardy-Ramanujan J. 9 (1986), 17–23.

Remark. Assuming the Lindelöf hypothesis, Ivić states without proof that

$$\int_1^x \Delta^2(t)\mathrm{d}t \ll x^{4/3} \cdot \log^3 x$$

j) $\displaystyle\int_1^x \Delta^2(t)\mathrm{d}t \ll x^{4/3}(\log x)^{89}$

for $x \ge 2$

D.R. Heath-Brown. *The number of Abelian groups of order at most x.* Journées arithmétiques, Exp. Congr., Luminy/Fr. 1989, Astérisque 198–200, 153–163 (1991.)

Remark. Since Ivić proved that $\int_1^x \Delta^2(t)dt = \Omega(x^{4/3}\log x)$. (See A. Ivić.) The above result is essentially the best possible.

3) Denote $A(x;q,k) = \sum_{n\le x, n\equiv k(\mathrm{mod}\ q)} a(n)$. Then there exist constants b_1, b_2, b_3 with

$$A(x;q,k) = b_1 x + b_2 x^{1/2} + b_3 x^{1/3} + \Delta(x,q)$$

where:

a) $\Delta(x,q) = O(x^{1/2+\varepsilon}\cdot q^{1/2})$

I.I. Pjateckij-Šapiro. *An asymptotic formula for the number of abelian groups with order less than n.* (Russian.) Mat. Sbornik (N.S.) 26 (68) (1950), 479–486.

b) $\Delta(x,q) = O((x^3 q^{16}\log^9 x)^{1/10})$

(See H.-E. Richert. II.)

c) $\Delta(x,q) = O((x^{34}q^{165})^{1/123}\log q)$

J. Duttlinger. *Über die Anzahl Abelscher Gruppen gegebener Ordnung.* J. Reine Angew. Math. 273 (1974), 61–76.

4) Let $A_k(x) = \sum_{n\le x, a(n)=k} 1$. Then:

a) $A_k(x) = d_k x + O(\sqrt{x}\log x)$

where $d_k = \lim_{x\to\infty}\frac{1}{x}\sum_{n\le x, a(n)=k} 1$

A. Ivić. *The distribution of value of the enumerating function of non-isomorphic abelian groups of finite order.* Arch. Math. 30 (1978), 374–379.

Remark. The existence of d_k was first shown by Kendall and Rankin. They proved also $d_0 = 0$, $d_1 = 6/\pi^2$, and

$$d_k = \sum_{a(n_2)=k}\frac{1}{n_2}\cdot\sum_{p|n_2}\left(1+\frac{1}{p}\right)^{-1} \quad (k\ge 2),$$ where n_2 runs over all squarefull numbers. Moreover, $\sum_{k=0}^{\infty} d_k = 1$,

$$\sum_{k=0}^{\infty} k d_k = c_1 = \lim_{x\to\infty}\frac{1}{x}\sum_{n\le x} a(n)$$

(See D.G. Kendall and R.A. Rankin.)

b) Let $a > 0$ and $\delta(x) = (\log x)^{3/5}\cdot(\log\log x)^{-1/5}$. Then, for $k \equiv 1(\mathrm{mod}\ 2)$,

$$A_k(x) = d_k x + O(\sqrt{x}\exp(-a\delta(x)))$$

If $s \geq 1, k = 2^s \cdot k', k' \equiv 1 \pmod 2$, then

$$A_k(x) = d_k x + O\left(\sqrt{x} \cdot \frac{(\log\log x)^{s-1}}{\log x}\right)$$

E. Krätzel. *Die Werterveteilung der nicht-isomorphen Abelschen Gruppen endlicher Ordnung und ein verwandtes Zahlentheoretisches Problem.* Publ. Inst. Math. (Belgrade) 31 (45) (1982), 93–101.

c) Let k be odd. Asumming the Riemann Hypothesis, one has

$$A_k(x) = d_k x + O(x^{1/3} \log x)$$

d) Let $k \equiv \pm 1 \pmod 6$. Asumming the Riemann Hypothesis,

$$A_k(x) = d_k x + O(x^{7/22+\varepsilon})$$

E. Krätzel. *The distribution of values of $a(n)$.* Arch. Math. 57 (1991), 47–52.

§ XIII.13 Abelian groups of a given order

Let $a(n)$ denote the number of nonisomorphic Abelian groups of order n. Then:

a) $d_k = \lim\limits_{x\to\infty} \dfrac{1}{x} \sum\limits_{n\leq x, a(n)=k} 1$
exists and $0 < d_k < \infty$

D.G. Kendall and R.A. Rankin. *On the number of Abelian groups of a given order.* Quart. J. Math. Oxford Ser. (2) 18 (1947), 197–208.

b) There exist absolute constants $c_1, c_2, c_3, (> 0)$ such that

$$d_k \leq c_1 \exp(-c_2 \log k \cdot \log\log k)$$

for all $k \gg 3$
For an infinity of k's we have

$$d_k > \exp(-c_3 \log k \cdot \log\log k)$$

A. Ivić. *On the number of abelian groups of a given order and on certain related multiplicative functions.* J. Number Theory 16 (1983), 119–137.

§ XIII.14 Number of non-isomorphic abelian groups in short intervals

Let $A_k(x,h) = A_k(x+h) - A_k(x)$, where $A_k(x) = \sum\limits_{n\leq x, a(n)=k} 1$

a) If $h \geq x^\theta \cdot \log^\alpha x$, where $\theta = \dfrac{581}{1744}$, $\alpha = 1$, then
$(^*) A_k(x,h) \sim d_k \cdot h$

$$(\text{Here } d_k = \lim_{x\to\infty} \frac{1}{x} \sum_{\substack{a(n)=k,\\ n\le x}} 1)$$

A. Ivić. *On the number of finite non-isomorphic abelian groups in short intervals*. Math. Nachr. 101 (1981), 257–271.

b) (*) holds true also in the following cases:

$$\theta = \frac{581}{1744} = 0.3331\ldots,\ \alpha = 1+\varepsilon(\varepsilon > 0), \text{ for } k \equiv 0 \pmod 6;$$

$$\theta = \frac{1740}{5229} = 0.332\ldots,\ \alpha = \varepsilon, \text{ for } k \equiv \pm 2 \pmod 6;$$

$$\theta = \frac{105}{407} = 0.257\ldots,\ \alpha = 2+\varepsilon, \text{ for } k \equiv 3 \pmod 6;$$

$$\theta = \frac{577}{2436} = 0.236\ldots,\ \alpha = \varepsilon, \text{ for } k \equiv \pm 1 \pmod 6,\ k \equiv 0 \pmod 5;$$

$$\theta = \frac{109556}{494419} = 0.221\ldots,\ \alpha = \varepsilon, \text{ for } k \equiv \pm 1 \pmod 6,\ k \not\equiv 0 \pmod 5.$$

E. Krätzel. *Die Werteverteilung der Anzahl der nicht-isomorphen Abelschen Gruppen endlicher Ordnung in kurzen Intervallen*. Math. Nachr. 98 (1980), 135–144.

c) Suppose that $x^\gamma < h = o(x)$, $(\gamma > 0)$. Then for every $\varepsilon > 0$,

$$A_k(x,h) = (d_k + o(1)) \cdot h + O(x^{11/42+\varepsilon})$$

In case $k \equiv \pm 1 \pmod 6$, we have

$$A_k(x,h) = (d_k + o(1)) \cdot h + O(x^{2/9+\varepsilon})$$

E. Krätzel. *Lattice points*. Berlin 1988 (See p. 300 and p. 303.)

§ XIII.15 Number of representations of n as a product of k-full numbers

Let $a_k(n)$ denote the number of representations of n as a product of k-full number ($a_1(n) = a(n)$ is the number of nonisomorphic abelian groups of order n), and put

$$A_{k,m}(x) = \sum_{n\le x, a_k(n)=m} 1,\ A_{1,m}(x) = A_m(x)$$

Let $\zeta_k(s) = \sum_{n_k=1}^{\infty} 1/n_k$, where n_k denotes a k-full number, and $\gamma_{\nu,k} = \operatorname{Re}_{s=1/\nu} \zeta_k(s)$, $\nu = k, k+1, \ldots, 2k-1$. Then:

a) $A_{k,m}(x) = \sum_{\nu=k}^{k+q} \gamma_{\nu,k} \cdot P_{\nu,k,m} \cdot x^{1/\nu} + O(x^{1/(k+q)-\varepsilon})$

for $k \geq 1, m \geq 1, q = \begin{cases} k-1 & \text{for } 1 \leq k \leq 4 \\ \left[\sqrt{\frac{8k}{3}}\right] & \text{for } k > 4 \end{cases}$ and

$$P_{\nu,k,m} = F_k\left(\frac{1}{\nu}\right) \cdot B_{k,m}\left(\frac{1}{\nu}\right), \text{ with } F_k(s) = \prod_p \left(1 - \frac{p^{-2ks}}{1 - p^{-s} + p^{-ks}}\right)$$

(for Re $s > 1/2\,k$), and

$$B_{k,m}(s) = \sum_{\substack{n_{2k}=1 \\ a_k(n_{2k})=m}} 1/n_{2k}^s \cdot \prod_{p|n_{2k}} \left(1 + p^{-ks} \cdot \frac{1 - p^{-ks}}{1 - p^{-s}}\right)^{-1}$$

for $m > 1$ and $B_{k,1}(s) = 1$

b) For m odd and $1 \leq k \leq 4$ we have for all $a > 0$,

$$A_{k,m}(x) = \sum_{\nu=k}^{2k-1} \gamma_{\nu,k} \cdot P_{\nu,k,m} \cdot x^{1/\nu} + O(x^{1/2k} \cdot e^{-a\delta(x)})$$

where $\delta(x) = (\log x)^{3/5} \cdot (\log\log x)^{-1/5}$

c) For $m = 2^d \cdot m'(m' \equiv 1 (\text{mod } 2)), d \geq 1$ and $1 \leq k \leq 4$ one has

$$A_{k,m}(x) = \sum_{\nu=k}^{2k-1} \gamma_{\nu,k} \cdot P_{\nu,k,m} \cdot x^{1/\nu} + O\left(x^{1/2k} \cdot \frac{(\log\log x)^{d-1}}{\log x}\right)$$

E. Krätzel. *Die Werteverteilung der nicht-isomorphen Abelschen Gruppen endlicher Ordnung und ein verwandtes Zahlentheoretisches Problem.* Publ. Inst. Math. (Belgrade) (N.S.) 31 (45) (1982), 93–101.

§ XIII.16 Number of distinct values taken by $a(n)$ and related problems

1) Let $a(n)$ be the number of nonisomorphic Abelian groups of order n and let $b(n)$ denote the number of solutions in squarefull s of the equation

$$n = a(s)$$

for a fixed n. Then:

a) $\sum_{n \leq x} b(n) = \exp((B + o(1)) \log^{2/3} x)$

where $B = (3/2)(6\zeta(3)/\pi^2)^{1/3}$

A. Ivić. *On the number of abelian groups of a given order and on certain related multiplicative functions.* J. Number Theory 16 (1983), 119–137.

Corollary. Let $D(x)$ be the number of $n \leq x$ such that $n = a(m)$ for some m. Then

$$D(x) = \exp((B + o(1)) \log^{2/3} x)$$

(See A. Ivić.)

b) $$\log \sum_{n \le x} b(n) = B \log^{2/3} x + B^* \log^{1/3} x \cdot \log\log x + + O(\log^{1/3} x \cdot \sqrt{\log\log x})$$

J. Herzog and W. Schwarz. *Über eine spezielle Partitionenfunktion, die mit der Anzahl der Abelschen Gruppen der Ordnung n zusammenhängt.* Analysis 5 (1985), 153–161.

2) Let $C(x)$ denote the number of distinct values taken by $a(n)$ for $n \le x$ and $D(x)$ the number of $n \le x$ such that $n = a(m)$ for some integer m (See Corollary 1) a).) Then:

a) $$C(x) \le \exp\left(\left(\frac{2\pi}{\sqrt{3}} + \varepsilon\right)(\log x / \log\log x)^{1/2}\right)$$
for $x \ge x_0(\varepsilon)$

(See A. Ivić.)

b) $$D(x) \ge \frac{1}{3} C(x) \log\log x$$
for $x \ge x_0$

c) For any fixed $A > 0$ and $x \ge x_0(A)$,

$$C(x) > (\log x)^A$$

P. Erdős and A. Ivić. *The distribution of values of a certain class of arithmetic functions at consecutive integers.* Colloq. Math. Soc. János Bolyai, 51. Number Theory, Budapest, 1987, pp. 45–91.

Remark. Assuming a certain conjecture on prime factors of the partition function, Erdős and Ivić obtain the asymptotic order of magnitude of $C(x)$ and $D(x)$, namely

$$C(x) = \exp((\log x)^{1/2+o(1)})$$

$$D(x) = \exp((\log x)^{2/3+o(1)})$$

§ XIII.17 Number of $n \le x$ with $a(n) = a(n+1)$. The functions $a(n)$ at consecutive integers

Let $a(n)$ denote the number of nonisomorphic Abelian groups with n elements. Then:

1) a) $$\sum_{\substack{n \le x \\ a(n)=a(n+1)}} = Ax + O(x^{3/4} \cdot \log^4 x)$$

where $A > 0$ is a constant, which may be explicitly evaluated.

b) $\sum_{n \le x} a^r(n)a^s(n+1) = C_{r,s} \cdot x + O(x^{3/4+\varepsilon})$
where $C_{r,s} > 0$ and r, s are fixed real numbers.

c) $\sum_{n \le x} (a(n) - a(n+1))^{2m} = C_m \cdot x + O(x^{3/4+\varepsilon})$
with $C_m > 0$ and m a fixed positive integer.

See P. Erdős and A. Ivić. *The distribution of values of a certain class of arithmetic functions at consecutive integers.* Colloq. Math. Soc. János Bolyai, 51. Number Theory, Budapest, 1987, pp. 45–91.

2) a) There exist at least $x^{1/2}$ numbers n from $[x, 2x]$ such that

$$a(n+1) = a(n+2) = \ldots = a(n+k)$$

where

$$k = \left[\frac{\log x \cdot \log\log\log x}{40(\log\log x)^2} \right]$$

b) There exist at least $x^{1/2}$ numbers n from $[x, 2x]$ such that for a suitable $C > 0$ the values $a(n+1), a(n+2), \ldots, a(n+t)$ are all distinct for

$$t = \left[C \cdot \left(\frac{\log x}{\log\log x} \right)^{1/2} \right]$$

(See P. Erdős and A. Ivić.)

§ XIII.18 Sums involving $(\Omega(n+1) - \omega(n+1)) \cdot a(n), d(n+1)\, a(n), \omega(n+1)\, a(n)$

1) a) $\sum_{n \le x} a(n)(\Omega(n+1) - \omega(n+1)) = Cx + O(x^{3/4+\varepsilon})$
$(C > 0)$

b) $$\sum_{\substack{n \le x \\ a(n) = \Omega(n+1) - \omega(n+1)}} 1 = Dx + O(x^{3/4} \log^4 x)$$
$(D > 0)$ where $\omega(m)$ and $\Omega(m)$ denote the number of distinct, and total, number of prime factors of m, respectively.

P. Erdős and A. Ivić. *The distribution of values of a certain class of arithmetic functions at consecutive integers.* Colloq. Math. Soc. János Bolyai, 51. Number Theory, Budapest, 1987, pp. 45–91.

2) a) $\sum_{n \le x} a(n)\, d(n+1) = C_1 x \log x + C_2 x + O(x^{8/9+\varepsilon})$

b) $\sum_{n\le x} a(n)\omega(n+1) = D_1 x \log\log x + D_2 x + O\left(\frac{x}{\log x}\right)$
where $C_1, D_1 > 0$ (Here $d(m)$ denotes the number of all divisors of m)
(See P. Erdős and A. Ivić.)

Remark. For other results, see the Chapters with $d(n)$ and $\omega(n)$

§ XIII.19 On sums involving $\frac{1}{a(n)}$ and $\frac{1}{\log a(n)}$

1) $\sum_{n\le x} \frac{1}{a(n)} = Ax + O(x^{1/2}\log^{-1/2} x)$
where $A > 0$ is a constant.

J.-M. de Koninck and A. Ivić. *Topics in arithmetical functions*. Notas de Matemática (72) North. Holland 1980 (See p. 16.)

Remarks. (i) In fact,

$$A = \prod_p \left(1 - \sum_{j=2}^{\infty}\left(\frac{1}{P(j-1)} - \frac{1}{P(j)}\right) p^{-js}\right)$$

where $P(n)$ is the number of unrestricted partitions of n. (p runs over primes.)

(ii) For a more precise result, see

W.G. Nowak. *On the average number of finite Abelian groups of a given order.* Ann. Sci. Math. Qué. 15 (1991), 193–202.

2) $\sum_{n\le x}' \frac{1}{\log a(n)} = x \cdot \int_{-\infty}^{0} (C(t) - 6/\pi^2)dt + O(x^{1/2}\log^{1/2} x)$
where

$$C(t) = \prod_p \left(1 + \sum_{k=2}^{\infty}(P^t(k) - P^t(k-1))p^{-k}\right)$$

(with $P(n)$ as the number of partitions of n), and $\sum'$ denotes summation over those values of n for which $a(n) > 1$
(See J.-M. de Koninck and A. Ivić, p. 81 and p. 88.)

§ XIII.20 The iterates of $a(n)$

1) Let $a(n)$ denote the number of non-isomorphic Abelian groups of order n. Then, if $a^{(r)}(n) = a(a^{(r-1)}(n))$, $a^{(1)}(n) = a(n)$, $r = 2, 3, \ldots$

a) $a^{(2)}(n) \ll \exp(B(\log n)^{7/8}/(\log\log n)^{19/16})$
with a positive constant B

b) $\log a^{(r)}n \ll (\log n)^{c_r}$
with $c_1 = 1$, $c_2 = 7/8$ and $c_r \le \frac{1}{2}c_{r-1} + \frac{3}{8}c_{r-2}$ for $r \ge 3$

P. Erdős and A. Ivić. *On the iterates of the enumerating function of finite Abelian groups.* Bull. Cl. Sci. Math. Nat., Sci. Math. 17 (1989), 13–22.

Note. The authors establish also an asymptotic representation for the mean value of $K(n) = \min\{r : a^{(r)}(n) = 1\}$

2) $\sum_{n\le x} a^{(2)}(n) = \sum_{k=1}^{\infty} a(k) \cdot d_k x + O(x^{1/2}\log^4 x)$
where $d_k = \lim_{x\to\infty} \frac{1}{x} \sum_{\substack{n\le x\\ a(n)=k}} 1$

A. Ivić. *On the number of Abelian groups of a given order and on certain related multiplicative functions.* J. Number Theory 16 (1983), 119–137.

3) $\sum_{n\le x} a(n + a(n)) = Cx + O(x^{11/12+\varepsilon})$
where $C > 0$ is effectively computable constant.

A. Ivić. *An asymptotic formula involving the enumerating function of finite Abelian groups.* Publ. Elektr. Fak. Univ. Beograd. Ser. Mat. 3 (1992), 61–66.

§ XIII.21 Statistical theorems on the embedding of abelian groups into symmetrical ones

Let A be the set of non-isomorphic Abelian groups, and let $A_n = \{G \in A : |G| \le n\}$;

$$m(G) = \min\{m \in \mathbb{N} : G \text{ is embeddable into } S_m\}$$

(Here S_m is the symmetric group of order m);

$$m(k) = \max\{m(G) : G \in A, |G| = k\}$$

Further, let $F(n,m) = \text{card}\{k \le n : m(k) > m\}$ and
$G(n,m) = \text{card}\{G \in A_n : m(G) > m\}$

a) $G(n,m) = O(|A_n|)$

if $\lim_{n\to\infty} \frac{\log m}{\log n} = 1$; and

$G(n,m) \ge c(\delta)|A_n|$

if $\dfrac{\log m}{\log n} \leq 1 - \delta$ holds with some $\delta > 0$

P. Erdős and P. Turán. *On some problems of a statistical group-theory*. IV. Acta Math. Acad. Sci. Hungar. 19 (1968), 413–435.

b) $F(n, m) = \theta(u)n + O\left(\log(1+u)\dfrac{n}{\log m}\right)$

and

$G(n, m) = \theta(u)|A_n| + O\left(\log(1+u)\dfrac{|A_n|}{\log m}\right)$

where $n, m \geq 2$, $u = \dfrac{\log n}{\log m}$, the constants in the O-symbols are absolute.

Here $\theta(u) = 1 - \rho(u)$, where $\rho(u)$ is uniquely defined, for $u > 0$ by

$$\rho(u) = \begin{cases} 1 & \text{if } 0 < u \leq 1, \\ 1 - \displaystyle\int_0^{u-1} \frac{\rho(t)}{t+1} dt & \text{if } 1 < u \end{cases}$$

A. Balog. *Statical theorems about the embedding of Abelian groups into symmetrical ones.* Acta Math. Acad. Sci. Hungar. 39 (1–3) (1982), 117–124.

Remark. $\theta(u)$ is a continuous increasing function with the limit 1 when u tends to infinity and $\theta(u) = \log u$, if $1 \leq u \leq 2$

$\theta(u) = 1 - \exp(-u \log u - u \log\log u + O(u))$, if $u > 2$. See:

N.G. Bruijn. *The asymptotic behaviour of a function occurring in the theory of primes.* J. Indian Math. Soc. (N.S.) 15 (1951), 25–32.

c) $\dfrac{1}{n}\displaystyle\sum_{k \leq n} m(k) = c_1 \frac{n}{\log n} + O\left(\frac{n}{\log^2 n}\right)$

and

$$\frac{1}{A_n} \sum_{G \in A_n} m(G) = c_2 \frac{n}{\log n} + O\left(\frac{n}{\log^2 n}\right)$$

where $n \geq 2$, $c_1 = \pi^2/12$, $c_2 = 1/2\zeta(3)\zeta(5)\ldots$.

(See A. Balog.)

§ XIII.22 Probabilistic results in group theory

Let $(G, +)$ be an Abelian group of order n. Let $a_1, \ldots, a_k$ be elements of G and denote by $V_k(g)$ the number of representations of $g \in G$ in the form

$$g = \varepsilon_1 a_1 + \cdots + \varepsilon_k a_k, \ \varepsilon_i \in \{0, 1\}, i = 1, 2, \ldots, k.$$

a) If $k \geq \left(2\log n + 2\log\dfrac{1}{\varepsilon} + \log\dfrac{1}{\delta}\right)/\log 2$, where $\varepsilon > 0$ and $\delta > 0$ are arbitrary small positive numbers then

$$P\left(\max_{g\in G}\left|V_k(g)-\frac{2^k}{n}\right|\le\varepsilon\cdot\frac{2^k}{n}\right)>1-\delta$$

(Here $P(\dots)$ denotes the probability of the random variable in the bracket.)

Corollary. There exists in every Abelian group of order n for each $k\ge\left(2\log n+2\log\frac{1}{\varepsilon}+\log\frac{1}{\delta}\right)/\log 2$, k elements $a_1,\dots,a_k$ such that each element $g\in G$ can be represented in the form

$$g=\varepsilon_1a_1+\cdots+\varepsilon_ka_k$$

$(\varepsilon_i\in\{0,1\})\frac{2^k}{n}\cdot(1+\varepsilon_g)$ times, where $|\varepsilon_g|\le\varepsilon$.

P. Erdős and A. Rényi. *Probabilistic methods in group theory.* J. Analyse Math. 14 (1965), 127–138.

Remark. In the special case, when G is the additive group of residues mod n we have:
For any $\delta>0$, if $k\ge\left(\log n+2\log\frac{1}{\delta}+\log\frac{\log n}{\log 2}\right)/\log 2+5$, then

$$P\left(\min_{g\in G}V_k(b)>0\right)>1-\delta$$

(See P. Erdős and A. Rényi.)

b) Let $\eta>0$ be fixed. For all but $o(n^k)$ choices of elements $a_1,\dots,a_k$, the inequalities

$$(1-\eta)\frac{2^k}{n}<V_k(g)<(1+\eta)\frac{2^k}{n}$$

hold for every $g\in G$ provided

$$k\ge\frac{\log n}{\log 2}\cdot\left(1+O\left(\frac{\log\log\log n}{\log n}\right)\right)$$

P. Erdős and R. Hall. *Probabilistic methods in group theory.* II. Houston J. Math. 2 (1976), 173–180.

Note. See also
P. Erdős and R. Hall. *Some new results in probabilistic group theory.* Comment. Math. Helv. 53 (1978), 448–457.

§ XIII.23 Finite abelian group cohesion

1) Let G be a finite Abelian group. For $A,B\subset G$, let $m(x,A,B)=\text{card}\{(a,b):a+b=x,a\in A,b\in B\}$. For $E\subset G$ let E' denote its complement.

a) $\min_{E \subset G} \max_{x \in G} | m(x, E, E) + m(x, E', E') - 2m(x, E, E') | \geq p^{1/2}$
where $p = \text{card}\, G$

b) Let $\lambda > \frac{1}{2}$. Let G be a finite group with no elements of order 2. Then

$$\min_{E \subset G} \max_{x \in G} | m(x, E, E) + m(x, E', E') - 2m(x, E, E') | \leq K \cdot p^{\lambda}$$

where K depends only on λ

P. Erdős and B. Smith. *Finite abelian group cohesion.* Israel J. Math. 39 (1981), 177–185.

Note. The proof of b) uses, among other results, the following inequality (obtained by Erdős and Smith): Let card $G = n + 1$ and let there be no elements of order 2 in G. Let $\alpha = (n-1)/n$. Then

$$\text{Exp}\left(\sum_{\substack{x \in G \\ x \neq 0}} \left(m(x) - \frac{\alpha n}{4}\right)^{2k}\right) \leq K \cdot n^{k+1}$$

where $m(x) = m(x, E, F)$, with $G \backslash \{0\} = E \cup F$,
card E = card F =. $n/2$
Here Exp denotes expectation and K depends only on k.

2) Let G be a finite group. Denote by $r(G)$ the least cardinality of a subset A satisfying $A^2 = G$. Then

$$r(G) \leq (4/\sqrt{3})|G|^{1/2}$$

G. Kozma and A. Leo. *Bases and decomposition numbers of finite groups.* Arch. Math. (Basel) (to appear.)

§ XIII.24 Number of non-isomorphic groups of order n

Let $G(n)$ denote the number of non-isomorphic groups of order n.

1) If $F_k(x) = \text{card}\,\{n \leq x : G(n) = k\}$, then:

a) $F_1(x) \sim xe^{-\gamma} / \log\log\log x$
where γ is Euler's constant.

P. Erdős. *Some asymptotic formulas in number theory.* J. Indian Math. Soc. (N.S.) 12 (1948), 75–78.

b) $F_{2^k}(x) \sim x \cdot c(k)/(\log\log\log x)^{k+1}$
where $c(k)$ is a constant depending on $k \geq 1$

P. Erdős, M.R. Murty. and V.K. Murty. *On the enumeration of finite groups.* J. Number Theory 25 (1987), 360–378.

c) $F_2(x) = \dfrac{xe^{-\gamma}}{(\log\log\log x)^2} + O\left(\dfrac{x \cdot (\log\log\log\log x)^2}{(\log\log\log x)^3}\right)$

M.G. Lu. *The asymptotic formula for* $F_2(x)$. Sci. Sinica Ser. A 30 (1987), 262–278.

d) $F_{2^k}(x) = \dfrac{x \cdot e^{-\gamma}}{k!(\log_3 x)^{k+1}} + O\left(\dfrac{x \cdot (\log_4 x)^{k+1}}{(\log_3 x)^{k+2}}\right)$

where $k \geq 2$ and $\log_r$ denotes the r-fold iterated logarithm.

H.-Q. Liu. *An asymptotic formula for* $F_{2^a}(x)$. (Chinese.) Acta Math. Sinica 30 (1987), 695–705.

Remark. Results for F_k, when $k - 2$ is prime and k does not lie in a certain set S, have been obtained by Spiro.

C.A. Spiro. *The probability that the number of groups of squarefree order is two more than a fixed prime.* Proc. London Math. Soc. (3) 60 (1990), 444–470.

2) a) $\sum_{n \leq x} \mu^2(n) \log G(n) = (c + o(1))x \log\log x$

where c is a positive constant.

M.R. Murty and V.K. Murty. *On groups of squarefree order.* Math. Ann. 267 (1984), 299–309.

b) $x^{1.68} \leq \sum_{n \leq x} \mu^2(n) G(n) \leq x^2 / \exp\left((1 + o(1)) \dfrac{\log x \cdot \log\log\log x}{\log\log x}\right)$

as $x \to \infty$

C. Pomerance. *On the average number of groups of squarefree order.* Proc. Amer. Math. Soc. 99 (1987), 223–231.

Note. Pomerance conjectured that the upper estimate is asymptotically correct.

c) If n is squarefree, then

$$G(n) = \Omega(n^{1-\varepsilon})$$

for each $\varepsilon > 0$; and

$$\log G(n) \sim (\log\log n) \cdot \sum_{p|n} (\log p)/(p-1)$$

as $n \to \infty$

(See P. Erdős, M.R. Murty and V.K. Murty.)

Remark. Previously, Murty and Murty had shown that

$$G(n) \leq \varphi(n)$$

M.R. Murty and V.K. Murty. J. Number Theory 18 (1984), 178–191.

d) $G(n) = O(\varphi(n)/(\log n)^{A \log\log\log n})$

($A > 0$, constant) when n is squarefree; and there is a constant $B > 0$, such that

$G(n) > \varphi(n)/(\log n)^{B \log\log\log n}$ for infinitely many squarefree n.

M.R. Murty and S. Srinivasan. *On the number of groups of squarefree order.* Canad. Math. 30 (1987), 412–420.

3) a) $G(n) \leq n^{\frac{\lambda(\lambda-1)}{2}}$
for all n, where $\lambda = \Omega(n)$ denotes the number of prime factors of n, counting multiplicities.

P.M. Neumann. Quart. J. Math. Oxford Ser. (2) 20 (1969), 395–401.

b) Let $\nu_p(n) = \max\{a : p^a | n\}$ and
$\nu = \nu(n) = \max\{\nu_p(n) : p \text{ prime}\}$. Then

$$G(n) \leq n^{\nu^2+\nu+2}$$

A. McIver and P.M. Neumann. *Enumerating finite groups.* Quart. J. Math. Oxford Ser. (2) 38 (1987), no. 152, 473–488.

4) Let $Q_k(x) = \text{card}\{n \leq x : G(n) = k, n \text{ squarefree}\}$ and let $\Psi(n, p)$ denote the number of primes q, dividing the natural number n for which $q \equiv 1(\text{mod} p)$, where p is a prime.
Put $S = \{G(n) : n \text{ odd, squarefree}, \psi(n, p) \leq 1 \text{ for all } p|n\}$. Then, if $k \in S$, there exists a computable constant $c(k)$ with

$$F_k(x) \geq Q_k(x) \gg x(\log\log x)^{-c(k)}$$

C.A. Spiro-Silverman. *When the group-counting function assumes a prescribed integer value at squarefree integers frequently, but not extremely frequently.* Acta Arith. 61 (1992), 1–12.

§ XIII.25 Density of finite simple group orders

Denote by $S(x)$ the number of integers $n < x$ for which there is a non-cyclic simple group of order n. Then

a) $S(x) = o(x)$

L. Dornhoff. *Simple groups are scarce.* Proc. Amer. Math. Soc. 19 (1968), 692–696.

b) $S(x) < cx \cdot \left(\dfrac{\log\log\log x}{\log\log x}\right)^{1/2}$
where $c > 0$ is an absolute constant.

L. Dornhoff and E.E. Spitznagel, Jr. *Density of finite simple group orders.* Math. Z. 106 (1968), 175–177.

c) $S(x) < x \exp\left(-\left(\dfrac{1}{2} + o(1)\right)(\log x \log\log x)^{1/2}\right)$

P. Erdős. *Remarks on some problems in number theory.* Math. Balcanica 4 (1974), 197–202.

Remarks. (i) Erdős asserts he can prove

$$S(x) \leq x \exp(-(1 + o(1))(\log x \cdot \log\log x)^{1/2})$$

(ii) The classical result of Feit and Thompson states that if there is a non-cyclic simple group of order n, then n must be even.

§ XIII.26 Large cyclic subgroups of finite groups

If $0 < \delta < 1$ is given, then almost all integers $n \leq x$ have the property that every group of order n contains a characteristic cyclic subgroup of squarefree order

$$n^{1-1/(\log n)^{1-\delta}}$$

Corollary. Given $\varepsilon > 0$, almost all integers $n \leq x$ (as $x \to \infty$) have the property that each group of order n has more than

$$n^{1-\varepsilon}$$

conjugacy classes.

E.A. Bertram. *On large cyclic subgroups of finite groups.* Proc. Amer. Math. Soc. 56 (1976), 63–66.

§ XIII.27 Counting solvable, cyclic, nilpotent groups orders

1) Let $H(x)$ be the number of integers $n \leq x$ with the property that every group of order n is solvable. Then

$$H(x) = c_1 x + O(x \exp(-c_2(\log\log x \cdot \log\log\log x)^{1/2}))$$

with c_1, c_2 positive constants.

E. Heppner. *Über die Anzahl der Naturlichen Zahlen n kleiner oder gleich x, für die jede Gruppe der Ordnung n auflösbar ist.* Arch. Math. (Basel) 32 (1979), 548–550.

Remark. For related results, see also

M.E. Mays. *Counting abelian, nilpotent, solvable and supersolvable group orders.* Arch. Math. 31 (1978/79), 536–538.

2) Let C be the set of positive integers n such that every group of order n is cyclic. Let A be the set of numbers n such that every group of order n is Abelian. Then:

a) $$\sum_{\substack{n \leq x \\ n \in C}} 1 \sim e^{-\gamma} \cdot x / \log_3 x$$

where $\log_k$ denotes the k-fold iterated natural logarithm.

P. Erdős. *Some asymptotic formulas in number theory.* J. Indian Math. Soc. (N.S.) 12 (1948), 75–78.

b) $$x \cdot (\log_2 x)^{-1} \cdot (\log_3 x)^{-2} \ll \sum_{\substack{n \leq x \\ n \in A \setminus C}} 1 \ll x (\log_2 x)^{-1} \cdot (\log_3 x)^{\varepsilon - 1/2}$$

R. Warlimont. *On the set of natural numbers which only yield orders of abelian group.* J. Number Theory 20 (1985), 354–362.

c) $$\sum_{\substack{n\le x\\ n\in \bar{A}\setminus C}} 1 \ll x\cdot(\log_2 x)^{-1}\cdot(\log_3 x)^{-2}$$

S. Srinivasan. *On orders solely of abelian groups.* Glasgow Math. J. 29 (1987), 105–108.

d) $$\sum_{\substack{n\le x\\ n\in \bar{A}\setminus C}} 1 \sim x\cdot e^{-\gamma}\cdot(\log_2 x)^{-1}\cdot(\log_3 x)^{-2}$$

as $z \to \infty$

M.J. Narlikar and S. Srinivasan. *On orders solely of abelian groups.* II. Bull. London Math. Soc. 20 (1988), no. 3, 211–216.

Remark. This result has been obtained also by Erdős and Mays, but with a certain constant c in place of $e^{-\gamma}$.

P. Erdős and M.E. Mays. *On nilpotent but not abelian groups and abelian but not cyclic groups.* J. Number Theory 28 (1988), 363–368.

3) Let N be the set of positive integers n such that every group of order n is nilpotent. Then

$$\sum_{\substack{n\le x\\ n\in \bar{N}\setminus A}} 1 \sim cx/(\log\log x)^2\cdot(\log\log\log x)^2$$

as $x \to \infty$.

(See P. Erdős and M.E. Mays.)

4) The number of $n \le x$ such that every group of order n is metacyclic is $\sim \dfrac{6}{\pi^2}x$

S. Srinivasan. *On orders solely of Abelian groups.* III. J. Number Theory 39 (1991), 175–180.

§ XIII.28 On C-groups

1) A finite group is called a C-group if all its Sylow subgroups are cyclic. Let $C(n)$ be the number of non-isomorphic C-groups of order n. Then

a) $$\sum_{n\le x} \log C(n) = (a+o(1))x\log\log x$$

where $a > 0$ is a constant.

M.R. Murty and V.K. Murty. *On groups of squarefree order.* Math. Ann. 267 (1984), 299–309.

b) There is a constant $A > 0$ such that

$$C(n) = O(\varphi(n)/(\log n)^{A\log\log\log n})$$

where φ is Euler's function.

M.R. Murty and S. Srinivasan. *On the number of squarefree order.* Canad. Math. Bull. 30 (1987), 412–420.

c) $C(n) = \Omega(\varphi(n)/(\log n)^{B \log\log\log n})$

where $B > 0$ is a constant.

(See M.R. Murty and S. Srinivasan.)

2) Let $f_A(n)$ be the number of (non-isomorphic) groups of order n all of whose Sylow subgroups are Abelian. Then

$$f_A(n) \leq n^{\Omega(n)+1}$$

where $\Omega(n)$ denotes the number of prime factors of n, counting multiplicities.

A. McIver and P.M. Neumann. *Enumerating finite groups.* Quart. J. Math. Oxford Ser. (2) 38 (1987), no. 152, 473–488.

3) Almost all odd numbers n have

$$(1 + o(1)) \prod_{p|n} \left(1 - \frac{1}{p-1}\right) \log\log n$$

prime divisors such that the corresponding Sylow subgroup is a direct factor in each group of order n.

P. Erdős and P.P. Pálfy. *On the order of directly indecomposable groups.* (Hungarian.) Mat. Lapok 33 (1982/86), no. 4, 289–298.

4) Let $r(x)$ be the number of integers $n \leq x$ such that every group of order n has a normal Sylow subgroup. Then

$$\lim_{x\to\infty} \frac{r(x)}{x} = 1$$

L. Dornhoff. *Simple groups are scarce.* Proc. Amer. Math. Soc. 19 (1968), 692–696.

§ XIII.29 The order of directly indecomposable groups. Direct factors of a finite abelian groups

1) If $n = n_1 n_2$ is a factorization such that all groups of order n decompose into the direct product of subgroups of order n_1 and n_2, then for almost all n one of the direct factors is always a cyclic group.

P. Erdős and P.P. Pálfy. *On the order of directly indecomposable groups.* (Hungarian.) Mat. Lapok 33 (1982/86), no. 4, 289–298.

2) For an Abelian group G let $\tau(G)$ denote the number of direct factors of G and let $T(x)$ be the sum of $\tau(G)$ over all groups of order $\leq x$. Then

a) $T(x) = ax \log x + bx + O(x^{1/2} \log^2 x)$

where a, b are constants.

E. Cohen. On the average number of direct factors of a finite abelian group. Acta Arith. 6 (1960), 159–173.

b) $T(x) = ax \log x + bx + cx^{1/2} \log x + dx^{1/2} + O(x^{5/12} \log^4 x)$

E. Krätzel. *On the average number of direct factors of a finite abelian group.* Acta Arith. 51 (1988), 369–379.

3) Let $t(G)$ be the number of direct decompositions of the Abelian groups G, into relatively prime direct factors (two factors H_1 and H_2 of G are called relatively prime if the identity is the only common direct factor of H_1 and H_2). Then

$$\sum{}' t(G) = ax \log x + bx + O(x^{1/2} \log x)$$

where a, b are constants, and the dash indicates that the summation is extended over all Abelian groups of order $\leq x$.

(See E. Cohen.)

Remark. For more general results, see

C.N. Yeung. *An asymptotic formula for the numbers of k-free groups of order $\leq x$.* J. Natur. Sci. and Math. 11 (1971), 243–256.

4) Let $\Delta_1(x)$ denote the error term in the asymptotic formula for $T(x)$ defined as in 2). Then

a) $\displaystyle\int_1^A \Delta_1(x)dx \ll_\varepsilon A^{7/6+\varepsilon}$

b) $\displaystyle\int_1^A \Delta_1^2(x)dx = \Omega(A^{3/2} \log^4 A)$

A. Ivić. *On the error term for the counting functions of finite Abelian groups.* Monatsh. Math. 114 (1992), 115–124.

c) $\Delta_1(x) \ll x^{3/8} \cdot \log^{7/2} x$

P.G. Schmidt. *Zur Anzahl unitärer Faktoren abelscher Gruppen.* Acta Arith. 64 (1993), 237–248.

§ XIII.30 On a family of almost cyclic finite groups

1) For a fixed integer $c > 1$ let F_c be the family of finite groups G having the property that, for each d dividing the order of G, $x^d = e$ has less than cd solutions in G (e is the unit element.)

a) If N is the number of cyclic subgroup of order p^α (p-prime) in a group $G \in F(c)$, then

$$N \leq c - 1 + \frac{c-2}{p-1}$$

If equality holds, then the intersection of these N subgroups has order $p^{\alpha-1}$.

M. Hausman and H.N. Shapiro. *On a family of almost cyclic finite groups.* Comm. Pure Appl. Math. 33 (1980), 635–649.

b) For a finite group G let $[G : C]$ denote the index in G of a cyclic subgroup C, and define

$$K(c) = \sup_{G \in F(c)} \min_{C \subset G} [G : C]$$

For every fixed integer $c \geq 2$, $K(c)$ is finite.

c) Let C^* denote an invariant cyclic subgroup of G, and define

$$K^*(c) = \sup_{G \in F(c)} \min_{C^* \subset G} [G : C^*]$$

Then for each integer $c \geq 2$, $K^*(c)$ is finite.

(See M. Hausman and H.N. Shapiro.)

§ XIII.31 Asymptotic results for elements of a semigroup

a) Let G be a multiplicative free commutative semigroup with a countable basis $a_1, a_2, \ldots$ of generators and let N be such a homomorphism of G into a multiplicative semigroup $\overline{G}$ of positive reals that for each real x there are at most finitely many elements $g \in G$ with the property $N(g) \leq x$. Moreover, let $\pi_G(x) = \sum_{\substack{N(a) \leq x \\ a \in P}} 1$; $\omega_G(x) = \sum_{\substack{N(g) \leq x \\ g \in G}} 1$, where $P = \{a_1, a_2, \ldots\}$ and assume

$$\pi_G(x) = \frac{ax^b}{b \log x} + O\left(\frac{x^b}{\log^{1+\varepsilon} x}\right) \text{ with } a, b, \varepsilon > 0. \text{ Then}$$

$$\omega_G(x) = C_G \cdot x^b \log^{a-1} x + O\left(\frac{x^b \log^{a-1} x}{(\log \log x)^{\varepsilon_1}}\right)$$

where $\varepsilon_1 = \min(1, \varepsilon)$, and C_G is a constant depending on G.

B.M. Bredihin. (See A.G. Postnikov. *Introduction to analytic number theory.* (Russian.) Moscow, 1971.)

b) With the above notations, suppose that

$$\omega_G(x) = C \cdot x^\theta + O(x^{\theta_1})$$

$(C > 0, 0 \leq \theta_1 < \theta)$ Then

$$\pi_G(x) = \left(\frac{1}{\theta} + o(1)\right) \cdot \frac{x^\theta}{\log x}$$

(See B.M. Bredihin.)

Remark. For applications of the above theorems in the theory of k-full numbers and quadratic residues, see

A. Ivić. *An asymptotic formula for elements of a semigroup of integers.* Mat. Vesnik 10 (25) (1973), 255–257;
A. Ivić. *On an arithmetical semigroup connected with quadratic residues.* C.R. Acad. Bulg. Sci. 29 (1976), 1257–1259.

§ XIII.32 Number of non-isomorphic semi-simple finite rings of order n

Let $S(n)$ denote the number of non-isomorphic semisimple finite rings of order n. Then:

1) a) $\sum_{n\le x} S(n) = \alpha_1 x + \alpha_2 x^{1/2} + O(x^{1/3}\log^2 x)$

where α_1, α_2 are (known) positive constants.

J. Knopfmacher. *Arithmetical properties of finite rings and algebras, and analytic number theory.* J. Reine Angew. Math. 252 (1972), 16–43.

b) $\sum_{n\le x} S(n) = \alpha_1 x + \alpha_2 x^{1/2} + \alpha_3 x^{1/3} + O(x^{7/27}\log^2 x)$

J. Duttlinger. *Eine Bemerkung zu einer asymptotischen Formel von Herrn Knopfmacher.* J. Reine Angew. Math. 266 (1974), 104–106.

2) Let $d_k = \sum_{n=1}^{\infty} \frac{g_k(n)}{n}$, where $g_k(n) = \sum_{\substack{t\mid n \\ S(t)=k}} \mu\left(\frac{n}{t}\right)$ (Here μ denotes the Möbius function.) Then

(*) $\sum_{\substack{x<n\le x+h \\ S(n)=k}} 1 = (d_k + o(1))\cdot h$

where $h = x^{\theta}\log^{\alpha} x$, with the following numerical values:

$$\theta = \frac{581}{1744} = 0.3331\ldots, \alpha = 1+\varepsilon(\varepsilon>0) \text{ for } k \equiv 0 \pmod 6;$$

$$\theta = \frac{1740}{5229} = 0.332\ldots, \alpha = \varepsilon \text{ for } k \equiv \pm 2 \pmod 6;$$

$$\theta = \frac{105}{407} = 0.257\ldots, \alpha = 2+\varepsilon \text{ for } k \equiv 3 \pmod 6;$$

$$\theta = \frac{109556}{494419} = 0.221\ldots, \alpha = \varepsilon \text{ for } k \equiv \pm 1 \pmod 6.$$

E. Krätzel. *Die Werteverteilung der Anzahl der nicht-isomorphen Abelschen Gruppen endlicher Ordnung in kurzen Intervallen.* Math. Nachr. 98 (1980), 135–144.

Remarks. (i) For the method of proof, see also

A. Ivić. *On the number of finite non-isomorphic abelian groups in short intervals.* Math. Nachr. 101 (1981), 257–271.

(ii) (*) holds true for $h \ge x^{1/(3-\alpha)}\cdot\log^2 x$, uniformly in $k \ge 0$ (fixed integer), where α satisfies

$$\sum_{n\leq x} d(n) = x\log x + (2\gamma - 1)x + O(x^{\alpha}\log^2 x)$$

with $d(n)$ denoting the divisor function.

J.-M. de Koninck and A. Ivić. *Topics in Arithmetical functions (Asymptotic formulae for sums of reciprocals of arithmetical functions and related results.)* Notas de Matemática (72), North-Holland 1980, (See p. 186.)

§ XIII.33 On a problem of Rohrbach for finite groups

a) Every finite group of order n has a basis of order two such that its cardinality is less than $2\sqrt{n\log n} + 2$.

b) For every $h \geq 3$ and $\delta > 0$, there exists an integer $M = M(h, \delta)$ such that every finite group of order $n \geq M$ has a basis of order h with cardinality $< (h + \delta)\sqrt[h]{n\log n}$.

M.B. Nathanson. *On a problem of Rohrbach for finite groups.* J. Number Theory 41 (1992), 69–76.

§ XIII.34 On cocyclity of finite groups

Let G be a finite group and let $q_0 < q_1 < \ldots < q_h$ be the prime divisors of $|G|$.
Let $\gamma_G = \max_{d\,||G|} \{\gamma : x^d = 1$ has γd solutions in $G\}$ be the multiplicity of G and $l_G = \min_{x\in G} [G :< x >]$ the cocyclity of G. Then

$$l_G < \frac{1 - 1/q_0 + \varepsilon(\gamma_G)}{h!\log q_0 \cdots \log q_h}\gamma_G \log^h \gamma_G$$

where $\varepsilon(\gamma_G) \to 0$ as $\gamma_0 \to \infty$.

P. Cellini. *High order elements and number of roots of unity in a finite group.* Ann. Mat. Pura Appl. IV. Ser. 162 (1992), 105–114.

Chapter XIV

PARTITIONS

§ XIV. 1 Unrestricted partitions of an integer

Let $P(n)$ be the number of unrestricted partitions of a positive integer n. (Define $P(0) = 1$.)

1) a) $$P(n) = \frac{e^{K\lambda_n}}{4\sqrt{3}\cdot\lambda_n^2} + O\left(\frac{e^{K\lambda_n}}{\lambda_n^3}\right)$$

where $K = \pi\sqrt{\dfrac{2}{3}}$ and $\lambda_n = \sqrt{n - \dfrac{1}{24}}$, $n \geq 1$.

G.H. Hardy and S. Ramanujan. *Asymptotic formulae in combinatory analysis.* Proc. London Math. Soc. (2) 17 (1918), 75–115.

Remark. The above formula may be written more simply as

$$P(n) = \frac{e^{K\sqrt{n}}}{4\sqrt{3}n}\cdot\left(1 + O\left(\frac{1}{\sqrt{n}}\right)\right)$$

b) $$P(n) = \frac{1}{2\pi\sqrt{2}}\frac{d}{dn}\left(\frac{e^{K\lambda_n}}{\lambda_n}\right) + O(e^{Hn^{1/2}}),\ H < K$$

J.V. Uspensky. Bull. de l'Acad. des Sciences de l'URSS, (6) 14 (1920), 199–218.

c) $$P(n) = \frac{1}{\pi\sqrt{2}}\sum_{q=1}^{\infty} A_q(n)\cdot q^{\frac{1}{2}}\cdot\frac{\mathrm{d}}{\mathrm{d}n}\frac{\sinh\left(\dfrac{K}{q}\cdot\lambda_n\right)}{\lambda_n}$$

where K, λ_n are defined as above, and

$A_q(n) = \sum_p \rho_{p,q}\cdot e^{-2\pi i mp/q}$, with $0 \leq \dfrac{p}{q} < 1, (p,q) = 1, m = \lambda_n^2$,

$\rho_{p,q} = \omega_{p,q}\cdot e^{\frac{\pi i}{4} - \frac{\pi i a}{12q}}$ and $(\omega_{p,q})^{24} = 1$.

Here a is the least positive integer such that $-ap - bq = 1$ with some integer b.

H. Rademacher. *Convergent series for the partition function.* Proc. Nat. Acad. Sci. (USA) 23 (1937), 78–84;
H. Rademacher. *On the partition function.* Proc. London Math. Soc. (2) 43 (1937), 241–254.

Remarks: (i) It is immediate that $|A_q(n)| \leq q$. This was improved by Lehmer to

$$|A_q(n)| < 2q^{5/6}.$$

D.H. Lehmer. *On the series for the partition function.* Trans. Amer. Math. Soc. 43 (1938), 271–295.

(ii) Rademacher, in his Bombay lectures, has described the work of A. Selberg on $A_q(n)$. It follows from the work of Selberg that

$$A_q(n) = O\left(q^{\frac{1}{2}+\varepsilon}\right), \varepsilon > 0.$$

H. Rademacher. *Lectures on analytic number theory.* Tata Institute of Fundamental Research, Bombay (1955).

2) a) $P(n) < e^{K\sqrt{n}}$ for all $n \geq 1$, where $K = \pi\sqrt{\frac{2}{3}}$.

P. Erdős. *On an elementary proof of some asymptotic formulas in the theory of partitions.* Ann. Math. 43 (1942), 437–450.

Remark. Erdős obtained an elementary argument to prove the asymptotic formula for $P(n)$ with a positive constant $K > 0$.

Newman obtained $K = \pi\sqrt{\frac{2}{3}}$. Ingham used a Tauberian theorem and Petersson developed a function-theoretic method.

D.J. Newman. *The evaluation of the constant in the formula for the number of partitions of n.* Amer. J. Math. 73 (1951), 599–601;
D.J. Newman. *A simplified proof of the partition formula.* Michigan Math. J. 9 (1962), 283–287;
A.E. Ingham. *A Tauberian theorem for partitions.* Ann. Math. (2) 42 (1941), 1075–1090;
H. Petersson. *Über Partitionen-probleme in Verbindung mit Potenzresten nach einem Primzahlmodul.* Math. Z. 66 (1956), 241–268.

b) $P(n) \leq 5^{n/4}$
for $n \geq 1$

E. Krätzel. *Die maximale Ordnung der Anzahl der wesentlich verschiedenen Abelschen Gruppen n-ter Ordnung.* Quart. J. Math. (2) Oxford Ser. 21 (1970), 273–275.

Remark. Krätzel used this inequality in obtaining the maximal order of magnitude for the number of non-isomorphic Abelian groups of order n.

c) $P(n) < \dfrac{\pi \cdot e^{K\sqrt{n}}}{\sqrt{6(n-1)}}$
for $n > 1$

J.H. van Lint. Combinatorial Theory Seminar (Eindhoven Univ. of Technology), Lect. Notes Math. 382, Springer-Verlag 1974. (See Chapter 4.)

3) a) $P(n) + P(n-2) \geq 2P(n-1)$
for $n \geq 2$

b) $P(n) \geq P(n-k) + P(k)$
for $n > k \geq 1$

c) $P(n) + P(n-4) + P(n-5) > P(n-1) + P(n-2) + P(n-6)$
for $n \geq 6$.

H.S. Manzur. *On the function $c(n) = q_k^e(n) - q_k^0(n)$ and some inequalities in partition theory.* Panjab. Univ. J. Math. (Lahore) 3 (1970), 49–57.

d) Let $\Delta P(n) = P(n) - P(n-1)$, $\Delta P(0) = P(0) = 1$ and define recursively $\Delta^k P(n)$ by $\Delta(\Delta^{k-1} P(n))$. Then there is an h_0 such that, if $h \geq h_0$, there is an integer $n_0(h)$ such that

$$(-1)^n \Delta^h P(n) > 0$$

for $0 \leq n \leq n_0(h)$; and

$$\Delta^k P(n) \geq 0$$

for $n \geq n_0(h)$.

A.M. Odlyzko. *Differences of the partition functions.* Acta Arith. 49 (1988), 237–254.

e) For $n_0(h)$ of d) one has

$$n_0(h) \sim \frac{6}{\pi^2} \cdot k^2 \log^2 k$$

as $h \to \infty$.
(See A.M. Odlyzko.)

Remark. For a recursion formula for $\Delta^k P(n)$ and related asymptotic results, see also

Ch. Knessl and J.B. Keller. *Asymptotic behavior of high-order differences of the partition function.* Commun. Pure Appl. Math. 44 (1991). No. 8/9, 1033–1045;

and

G. Almkvist. *On the differences of the partition function.* Acta Arith. 61 (1992), 173–181.

§ XIV. 2 Partitions of n into exactly k positive parts

Let $P(n;k)$ be the number of partitions of n into exactly k positive integer parts, and let $P(n;r,k)$ be the number of partitions of n into r parts with maximal summand k.

1) a) $\binom{n-1}{k-1} \leq k!P(n;k) \leq \binom{n+\frac{k(k-1)}{2}-1}{k-1}$ for $k \geq 1$

H. Gupta. *On asymptotic formula in partitions.* Proc. Indian Acad. Sci. 16 (1942), 101–102.

Corollary. $P(n;k) \sim \frac{1}{k!} \cdot \binom{n-1}{k-1}$ for $k = o(n^{1/3})$

P. Erdős and J. Lehner. *The distribution of the number of summands in the partitions of a positive integer.* Duke Math. J. 8 (1941), 335–345.

b) For all $k \geq 4$ and $n \geq n_k$ we have

$$\binom{n+k-1+a_{k-1}}{k-1} \leq k!\ P(n+k;k) \leq \binom{n+k-1+a_k}{k-1}$$

where $a_k = \frac{k(k-1)}{4}$

c) For any $\varepsilon > 0$ and all $k \geq 4, n \geq n(\varepsilon, k)$,

$$\binom{n+k-1+a_k-\varepsilon}{k-1} \leq k!\ P(n+k;k) \leq \binom{n+k-1+a_k}{k-1}$$

H. Gupta. *An inequality in partitions.* J. Univ. Bombay 11 (1942), Part III, pp. 16–18.

Corollary. $k!P(n;k) \sim \binom{n-1+a_k}{k-1}$ as $n \to \infty$, where $a_k = \frac{k(k-1)}{4}$.

d) $P(n;k) = \frac{n^{k-1}}{k!(k-1)!}\left(1 + \frac{k(k-1)(k-3)}{4n} + O(1/n^2)\right)$

for fixed $k \geq 3$ and n large.

Remark. Earlier, K. Iseki proved that

$$P(n;k) \sim \frac{n^{k-1}}{k!(k-1)!}$$

as $n \to \infty$

K. Iseki. *Ein Theorem der Zahlentheorie.* Tôhoku Math. J. 48 (1941), 60–63.

e) $P(n;k) \leq \frac{1}{k!(k-1)!}\left(n + \frac{k(k-3)}{4}\right)^{k-1} + 1$

G.J. Rieger. *Über Partitionen.* Math. Ann. 138 (1959), 356–362.

2) Let $\lambda = ck/\sqrt{n}$, $\mu = cr/\sqrt{n}$.

a) If $(\varepsilon + 1/4)\log n < \lambda$, $\mu < \log n$ for some $\varepsilon > 0$, then

$$P(n;r,k) \sim P(n)\exp(-(\lambda+\mu) - (\sqrt{n}/c)(e^{-\lambda} + e^{-\mu}))$$

where $P(n)$ is the number of unrestricted partitions of n.

b) If $(\varepsilon + 1/4)\log n < \lambda < \log n$ then

$$P(n;k) \sim P(n)\exp(-\lambda - (\sqrt{n}/c)e^{-\lambda})$$

G. Szekeres. *Asymptotic distribution of partitions by numbers and size of parts.* Number theory, vol. I, Elementary and analytic, Proc. Conf., Budapest/Hung. 1987, Colloq. Math. Soc. János Bolyai, 51 (1990), 527–538.

3) a) Let m run over each of the $kP(n;k)$ parts occuring in the $P(n;k)$ partitions of n into exactly k parts, and define $S_r(n;k) = \sum_m m^r$, $A_r(n;k) = S_r(n,k)/kP(n;k)$. Then, for a large n and a fixed k,

$$A_r(n;k) = \binom{r+k-1}{r}^{-1} n^r + O(n^{r-1})$$

H. Gupta. *Certain averages connected with partitions.* Research Bull. Panjab Univ. No. 124 (1957), 427–430.

b) $$\frac{1}{P(n)}\sum_{k=1}^{n} kP(n;k) = \left(CN^{1/2} + \frac{1}{2}C^2\right)\cdot(\log(CN^{1/2}) + \gamma) + $$
$$+ \frac{1+C^2}{4} + O(N^{-1/2}\cdot\log N)$$

where $N = n - \frac{1}{24}$, $C = \sqrt{6}/\pi$ and γ is Euler's constant.

S.M. Luthra. *On the average number of summands in partitions of n.* Proc. Nat. Inst. Sci. India Part. A, 23 (1957), 483–498.

c) $$\frac{1}{P(n)}\sum_{k=1}^{n} kP(n;k) = \sqrt{\frac{3n}{2\pi}}\cdot\left(\log n + 2\gamma - \log\frac{\pi}{6}\right) + O(\log^3 n)$$

I. Kessler and M. Livingston. *The expected number of parts in a partition of n.* Monatsh. Math. 81 (1976), 203–212.

§ XIV. 3 Partitions of n into at most k summands

Let $P(n;k)$ denote the number of partitions of n into k parts and $q(n;k)$ the number of partitions of n into k distinct parts. Further, let $P^*(n;k)$ be the number of partitions of n into at most k summands.

1) a) $$P(n;k) \sim \frac{1}{k!}\binom{n-1}{k-1}$$

for $k = o(n^{1/3})$

b) $P^*(n;k)/P(n) \sim \exp\left(-\frac{1}{C} \cdot e^{-CX}\right)$

where $C = \pi/\sqrt{6}$ and X is given by relation

$$Ck = \sqrt{n}\log\sqrt{n} + CX\sqrt{n}$$

P. Erdős and J. Lehner. *The distribution of the number of summands in the partitions of a positive integer.* Duke
Math. J. 8 (1941), 335–345.

c) $P(n,k)/P(n) \sim t \cdot n^{-1/2} \cdot \exp(-t/C)$

where $t = \lim_{n,k\to\infty} \sqrt{n} \cdot \exp(-Ckn^{-1/2})$, $C = \pi/\sqrt{6}$, and $P(n)$ is the number of unrestricted partitions of n.

H. Gupta. *An asymptotic formula in partitions.* J. Indian Math. Soc. (N.S.) 10 (1946), 73–76.

d) $P^*(n;k) \sim P(n;k)$

for $k = o(n^{1/2})$

G. Szekeres. *An asymptotic formula in the theory of partitions.* Quart. J. Math. Oxford Ser. (2) 2 (1951), 85–108.

2) a) For fixed n, a maximum of $P(n;k)$ occurs at

$$k = \frac{\sqrt{n}}{C} \cdot L + \frac{1}{C^2}\left(\frac{3}{2} + \frac{3}{2}L - \frac{1}{4}L^2\right) - \frac{1}{2} + O\left(\frac{\log^4 n}{\sqrt{n}}\right)$$

where $L = \log(\sqrt{n}/C)$, and $C = \pi/\sqrt{6}$.

G. Szekeres. *Some asymptotic formulae in the theory of partitions.* II. Quart J. Math. Oxford Ser. (2) 4 (1953), 96–111.

See also

G. Szekeres. *An asymptotic formula in the theory of partitions.* Quart. J. Math. Oxford Ser. (2) 2 (1951), 85–108.

Remark. This sharpens an earlier result of Erdős and Lehner.

See also

P. Erdős. *On some asymptotic formulas in the theory of partitions.* Bull. Amer. Math. Soc. 52 (1946), 185–188.

b) $q(n,k)$ has a maximum (for fixed n) at

$$k = \frac{2\sqrt{3n} \cdot \log 2}{\pi} + \beta + O(n^{-1/2})$$

where β is a fixed constant.

(See G. Szekeres. (1953).)

Remark. According to Gupta, in 1954, Haselgrove and Temperley developed a method which enabled them to prove that $P(n;k)$ attains its greatest value for at most two consecutive values of k when n is large and fixed.

H. Gupta. *Partitions—a survey.* Journal of Research Nat. Bureau of Standards. Mat. Sci. 74B (1970), 1–29.

3) In almost all (i.e. with the exception of $o(q(n))$ partitions at most) unequal partitions of n the maximal summand is

$$O(\sqrt{n}\log n)$$

P. Erdős and M. Szalay. *On some problems of J. Dénes and P. Turán.* in: Studies in Pure Mathematics (To the memory of Paul Turán.) Akadémiai Kiadó, Budapest, 1983, pp. 187–212.

4) Let $z = z_{p,q} = \exp(-2(D - i\pi p/q))$, where $D = d/d\zeta, \zeta = 2j - 1/12$. Then

$$P^*(n,k) = \frac{4\sqrt{3}}{\pi}\sum_{q=1}^{\infty}\sum_{p}{}' q^{3/2}\omega_{p,q}(\exp -2np\pi i/q)\cdot$$

$$\cdot\prod_{\nu=k+1}^{n-1}(1-z^{\nu})D^2\cosh\left(\frac{\pi}{q}\sqrt{\frac{\zeta}{3}}\right)$$

where $\sum'$ indicates summation restricted to p with $(p,q)=1$, and $\omega_{p,q}$ are 24-th roots of 1 arising in the transformation formulae of Dedekind's $\eta(\tau)$.

G. Almkvist and G.E. Andrews. *A Hardy-Ramanujan formula for restricted partitions.* J. Number Theory 38 (1991), 135–144.

§ XIV. 4 Unequal partitions of n containing each a_j as a summand

1) For arbitrary positive integers n, k and $a_1, a_2, \ldots, a_k$ with $1 \le a_1 < a_2 < \ldots a_k \le n$, let $q(n;\overline{a_1},\ldots,\overline{a_k})$ denote the number of unequal partitions of n not having summands from $\{a_1,\ldots,a_k\}$.

Let $0 < \varepsilon < 1/100$ be fixed. For $1 \le k \le n^{\frac{1}{6}-\varepsilon}$, we have

$$q(n;\overline{a_1},\overline{a_2},\ldots,\overline{a_k}) = (1+o(1))\cdot\frac{q(n)}{\prod_{j=1}^{k}\left(1+\exp\left(-\frac{\pi a_j}{2\sqrt{3n}}\right)\right)}$$

where $q(n)$ denotes the number of unrestricted unequal partitions of n.

P. Erdős and M. Szalay. *On some problems of the statistical theory of partitions.* Colloq. Math. Soc. János Bolyai, 51, Number theory, Budapest, 1987, pp. 93–110.

Corollary. Let $q(n;a_1,\ldots,a_k)$ denote the number of unequal partitions of n, containing each a_j as a summand.

Let $0 < \varepsilon < 1/100$ be fixed. For $1 \le k \le n^{1/6-\varepsilon}$ and $a_1 + \cdots + a_k \le n^{\frac{3}{4}-\varepsilon}$, we have

$$q(n; a_1, \ldots, a_k) = (1 + o(1)) \cdot \frac{q(n)}{\prod_{j=1}^{k} \left(1 + \exp\left(\frac{\pi a_j}{2\sqrt{3n}}\right)\right)}$$

Remark. The corollary follows easily, since

$$q(n; a_1, \ldots, a_k) = q(n - a_1 - \cdots - a_k; \bar{a}_1, \ldots, \bar{a}_k)$$

§ XIV. 5 Partitions of n into members of a finite set

1) Let $P(n, A)$ denote the number of partitions of n into members of the set $A = \{a_0, a_1, a_2, \ldots, a_m\}$ (a_i distinct positive integers); $a_0 = 1$.

$$\binom{n+m}{m} \leq P(n, A) \prod_{j=1}^{m} a_j \leq \binom{n + \sum_{j=1}^{m} a_j}{m}$$

H. Gupta. *Partitions in general.* Research Bull. Panjab Univ. No. 67, 31–38 (1955).

2) If A is any non-empty set of positive integers, then $P(n, A)$ is a non-decreasing function of n for large n, if and only if A either:

(i) contains the element 1 or

(ii) A contains more than one element and, if we remove any single element from A, the remaining elements have greatest common divisor 1.

P.T. Bateman and P. Erdős. *Monotonicity of partition functions.* Mathematika 3 (1956), 1–14.

§ XIV. 6 Partitions of n without a given subsum

For $A = \{a_1, \ldots, a_k\} \subset N^*$, let $r(n, A)$ be the number of partitions of n with no parts belonging to A. If each part is allowed to occur at most once, then we use the notation $\rho(n, A)$.

1) There exists $\lambda_2 > 0$ such that if $A = \{a_1, \ldots, a_k\} \subset N^*$ satisfies $s = a_1 + \cdots + a_k \leq \lambda_2 n$ then, as $n \to \infty$

$$\exp\left(O\left(\frac{s}{\sqrt{n}}\right)\right) \leq \frac{r(n, A)}{\left(\prod_{i=1}^{k} a_i\right) P(n) \left(\frac{\pi}{\sqrt{6n}}\right)^k} \leq 1 + O\left(\frac{1}{\sqrt{n}}\right)$$

P. Erdős, J.-L. Nicolas and A. Sárközy. *On the number of partitions of n without given subsum* (I). Discrete Math. 75 (1989), 155–166.

2) For $0 < \varepsilon < 1/100$, fixed, and for k satisfying $1 \le k \le n^{1/6-\varepsilon}$

$$\rho(n, A) = (1 + o(1)) \cdot \frac{q(n)}{\prod_{1 \le j \le k} (1 + \exp(-\pi a_j / 2\sqrt{3n}))}$$

P. Erdős and M. Szalay. *On some problems of the statistical theory of partitions.* Colloq. Math. Soc. János Bolyai, Number Theory, Budapest 51 (1987), 93–110.

§ XIV. 7 Partitions of n which no part is repeated more than t times

1) Let $W(n)$ denote the number of partitions of n into positive integers each of which occurs only an odd number of times. Then

$$W(n) \sim \frac{\alpha}{2\pi n} \cdot \exp(2\alpha\sqrt{n})$$

where $\alpha^2 = \frac{\pi^2}{12} + \int_0^1 \log(1 + y - y^2)\,dy/y$.

F.C. Auluck, K.S. Singwi and B.K. Agarwala. *On a new type of partition.* Proc. Nat. Inst. Sci. India 16 (1950), 147–156.

2) Let $p(n, t)$ be the number of partitions of n in which no part is repeated more than t times. Then

$$p(n, t) = 2\sqrt{3}t^{1/4}(t + 1)^{-3/4}(24n + t)^{-3/4}\cdot$$
$$\cdot \left(\exp\left(\frac{\pi}{6}\sqrt{t(24n + t)/(t + 1)}\right)\right) \cdot (1 + O(n^{-1/2}))$$

P. Hagis, Jr. *Partitions with a restriction on the multiplicity of the summands.* Trans. Amer. Math. Soc. 155 (1971), 375–384.

§ XIV. 8 Partitions of n whose parts are $\ge m$

Let $r(n, m)$ denote the number of partitions of n whose parts are $\ge m$.

a) For $m = O(n^{3/8} \cdot (\log n)^{1/4})$,

$$\log r(n, m) = \pi\sqrt{\frac{2}{3}} \cdot \sqrt{n} - \frac{1}{2}m \log n + m \log m -$$
$$- m\left(1 + \log\frac{\sqrt{6}}{\pi}\right) + O(n^{1/4} \cdot \sqrt{\log n})$$

J. Herzog. *Gleichmässige asymptotische Formeln für parameterabhängige Partitionenfunktionen*, Thesis, Univ. J. W. Göthe, Frankfurt am Main, 1987.

b) We have uniformly for $1 \leq m \leq n^{1/4}$,

$$r(n,m) = P(n)\left(\frac{\pi}{\sqrt{6n}}\right)^{m-1} \cdot (m-1)!(1 + O(m^2/\sqrt{n}))$$

J. Dixmier and J.-L. Nicolas. *Partitions without small parts.* Colloq. Math. Soc. János Bolyai, Number Theory, Budapest 51 (1987), pp. 9–33.

c) For $0 < \varepsilon < 1/3$ and $m \leq n^{1/3-\varepsilon}$,

$$r(n,m) \sim P(n)(m-1)! \cdot \left(\frac{C}{2\sqrt{n}}\right)^{m-1} \cdot \exp\left(-\left(\frac{C}{8} + \frac{1}{2C}\right) \cdot \frac{m^2}{\sqrt{n}}\right)$$

where $C = \pi \cdot \sqrt{\frac{2}{3}}$

J. Dixmier and J.-L. Nicolas. *Partitions sans petits sommants.* A tribute to Paul Erdős (Edited by A. Baker, B. Bollobás, A. Hajnal). Cambridge Univ. Press, 1990.

d) There exists $\alpha > 0$ such that as $n \to \infty$, we have uniformly for $1 \leq m \leq \alpha\sqrt{n}$

$$\exp\left(O\left(\frac{m^2}{\sqrt{n}}\right)\right) \leq \frac{r(n,m)}{P(n)\left(\frac{\pi}{\sqrt{6n}}\right)^{m-1} \cdot (m-1)!} \leq 1 + O\left(\frac{1}{\sqrt{n}}\right)$$

P. Erdős, J.-L. Nicolas and A. Sárközy. *On the number of partitions of n without a given subsum* (I), Discrete Math. 75 (1989), 155–166.

e) $$\log r(n,m) = \sqrt{\frac{2}{3}} \cdot \pi\sqrt{n} - m\frac{\sqrt{n}}{m} + 1 + \log\frac{\sqrt{6}}{\pi} + \\ + O(m^2 \cdot n^{-1/2} + n^{1/4} \cdot \sqrt{\log n})$$

uniformly for $m = o(\sqrt{n})$

J. Herzog. *Weak asymptotic formulas for partitions free of small summands.* Acta Arith. 54 (1990), 257–271.

f) For a real number $x > 0$ put $r(n,x) = r(n, \lceil x \rceil)$, where $\lceil x \rceil$ denotes the least integer $\geq x$.
Let $m = \lambda\sqrt{n}$, where $\lambda > 0$ is fixed. Then there exists a function g such that for $n \to \infty$, one has

$$\log r(n,m) \sim g(\lambda) \cdot \sqrt{n}$$

(See J. Dixmier and J.-L. Nicolas (1990).)

Remark. The function g is analytical for $\lambda > 0$, verifies a second order differential equation and admits an asymptotical development near the origin.

§ XIV. 9 Partitions of n into unequal parts $\geq m$

Let $\rho(n, m)$ be the number of partitions of n into unequal parts $\geq m$.

a) For $m \leq n^{1/5}$ one has

$$\rho(n, m) \sim q(n)/2^m$$

where $q(n)$ is the number of unrestricted, unequal partitions of n.

P. Erdős and M. Szalay. *On the statistical theory of partitions.* Colloq. Math. Soc. János Bolyai, Topics in classical number theory, Budapest 34 (1981), 397–450.

b) For all $n \geq 1$ and $1 \leq m \leq n$,

$$\frac{q(n)}{2^{m-1}} \leq \rho(n, m) \leq \frac{1}{2^{m-1}} \cdot q\left(n + \frac{m(m-1)}{2}\right)$$

and

$$\rho(n, m) \leq \frac{1}{2^{m-2}} \cdot q\left(n + \left[\frac{m(m-1)}{4}\right]\right)$$

P. Erdős, J.-L. Nicolas and M. Szalay. *Partitions into parts which are unequal and large.* Number Theory, Ulm 87, (Edited by H.P. Schickewei and E. Wirsing), Springer Lecture Note 1380 (1987), 19–30.

c) For $m = o\left(\frac{n}{\log n}\right)^{1/3}$

$$\rho(n, m) = (1 + o(1)) \cdot \frac{1}{2^{m-1}} \cdot q\left(n + \left[\frac{m(m-1)}{4}\right]\right)$$

d) For $0 < \varepsilon < 1/100$, fixed, and $m \leq n^{3/8-\varepsilon}$,

$$\rho(n, m) = (1 + o(1)) \cdot \frac{q(n)}{\prod_{j=1}^{m-1}\left(1 + \exp\left(-\frac{\pi j}{2\sqrt{3n}}\right)\right)}$$

(See P. Erdős, J.-L. Nicolas and M. Szalay.)

e) For $m = o(\sqrt{n})$,

$$\log \rho(n, m) = \pi\sqrt{\frac{n}{3}} - m \log 2 + \frac{\pi}{8\sqrt{3}} \cdot \frac{m^2}{\sqrt{n}} - \frac{\pi^2}{288} \cdot \frac{m^3}{n} + $$
$$+ O(m^4 n^{-3/2} + n^{1/4}\sqrt{\log n})$$

J. Herzog. *On partitions into distinct parts $\geq Y$*, preprint.

f) Let $\sigma = \sigma(n, m)$ be the unique real number such that $n = \sum_{j=m}^{n} j/(1 + e^{\sigma j})$. Put $B^2 = \sum_{j=m}^{n} j^2 e^{\sigma j}/(1 + e^{\sigma j})^2$.

Then, if $1 \le m \le n(2\log n)^{-4}$, we have

$$\rho(n,m) \sim \frac{1}{\sqrt{2\pi B^2}} e^{\sigma n} \cdot \prod_{j=m}^{n} (1+e^{-\sigma j})$$

G. Freiman and J. Pitman. *Partitions into distinct large parts*, preprint.

§ XIV.10 On the subsums of a partition

Let $R(n,a)$ be the number of partitions of n such that $n = n_1 + \cdots + n_t$, whose subsums $n_{i_1} + \cdots + n_{i_j}$ are all different from a.

a) If a is fixed, $n \to \infty$, then

$$R(n,a) \sim P(n) \left(\frac{\pi}{\sqrt{6n}}\right)^{\psi(a)} \cdot u(a)$$

where $\Psi(a) = [a/2] + 1$ and the constant $u(a)$ depends only on a.

J. Dixmier. *Sur les sous-sommes d'une partition.* Mémoire de la Soc. Math. France no. 35, supplément au Bull. S.M.F. 116, 1988.

Remarks: (i) One has the following inequalities:

$$\left(\left[\frac{a}{3}\right] - 1\right)! a^{a/6+3} \le u(a) \le 2^{a/2} \cdot a!/\left(\frac{a}{2} - 1\right)!, \text{ for } a \text{ even;}$$

$$\left(\left[\frac{a}{3}\right] - 1\right)! a^{a/6+3} \le u(a) \le 2^{a/2} \cdot a!/\left(\frac{a-1}{2}\right)!, \text{ for } a \text{ odd;}$$

(See J. Dixmier (1988).)

(ii) When a is odd and $a \to \infty$, $u(a)$ has the asymptotical development
$u(a) = (1 \cdot 3 \cdot 5 \cdot \ldots \cdot a)\left(\alpha_0 + \frac{\alpha_1}{a} + \frac{\alpha_2}{a^2} + \frac{\alpha_3}{a^3} + \cdots\right)$, where $\alpha_i \ge 0$ are integers, e.g. $\alpha_0 = 1$, $\alpha_1 = 2$, $\alpha_2 = 24$.
If a is even, and $a \ge a_0$, one has

$$u(a) \le \frac{a^{a/2}}{e^{a/2}} \cdot e^{-0.006a}$$

J. Dixmier. *Sur les sous-sommes d'une partition* III. Bull. Sci. Math. 113 (1989), 125–149.

b) There exists $\lambda_0 > 0$, such that we have, uniformly for $1 \le a \le \lambda_0\sqrt{n}$ as $n \to \infty$

$$\log \frac{R(n,a)}{P(n)} \le \Psi(a) \log \frac{\pi a}{\sqrt{6n}} + O(1/\sqrt{n})$$

and

$$\log \frac{R(n,a)}{P(n)} \le \psi(a) \log \frac{\pi a}{\sqrt{6n}} - \gamma_a \cdot a + O(a^2/\sqrt{n})$$

where $\gamma_a = 1/2$ if a is odd,

and $\gamma_a = \frac{1}{2} + \log 3 - \frac{7}{6}\log 2 + c\frac{\log a}{a} = 0.79\ldots + c\frac{\log a}{a}$ (c constant), if a is even.

P. Erdős, J.-L. Nicolas and A. Sárközy. *On the number of partitions of n without a given subsum* (I). Discrete Math. 75 (1989), 155–166.

c) For $n > n_0$, and $10^{18} \cdot \sqrt{n} \le a \le n^{5/7}$,

$$R(n,a) \le P\left(\left[\frac{n}{2}\right]\right) \exp(4 \cdot 10^5 \cdot a^{-1/3} \cdot n^{2/3} \cdot \log(a^{1/3} \cdot n^{-1/6}))$$

and for $n > n_0$ and $n^{5/7} < a \le n/2$, one has

$$R(n,a) \le P\left(\left[\frac{n}{2}\right]\right) \exp(n^{1/2-1/30})$$

P. Erdős, J.-L. Nicolas and A. Sárközy. Ibid. (II). Analytic Number Theory, Proc. Conf. in Honor of P.T. Bateman, Progress in Math., Birkhäuser, 85 (1990), pp. 205–234.

Remark. Let $s(a)$ denote the least positive integer which does not divide a. For n sufficiently large, $s(a) \ge 40000$ and

$\frac{7}{100} \cdot n^{1/2} s(a)^{3/2} \le a \le \frac{1}{40} n/s(a)$, one has

$$R(n,a) \le \exp(301 n^{1/2} \cdot s(a)^{-1/2} \cdot \log s(a))$$

(See the above paper.)

d) Let λ_3 be fixed, $0 < \lambda_3 \le 1/2$ and $r \ge 1$ a fixed integer. If $n \to \infty$, $a = a_n$ such that $\lambda_3 n \le a \le n/2$, and $s(a) = r + 1$, then

(i) If $a \ge n/(r+1)$, then

$$\log R(n,a) \sim \log P(a);$$

(ii) If $a \le n/(r+1)$ and $(r+1) \nmid (n-a)$, then

$$R(n,a) \sim \log P\left(\frac{n}{r+1}\right);$$

(iii) If $\frac{n}{r+2} \le a \le \frac{n}{r+1}$ and $(r+1) | (n-a)$, then

$$\log R(n,a) \sim \log P(a);$$

(iv) If $a \le \frac{n}{r+2}$ and $(r+1) | (n-a)$, then

$$\log R(n,a) \sim \log P\left(\frac{n-a}{r+1}\right);$$

J. Dixmier. *Partitions avec sous-sommes interdites*, preprint IHES, 1989.

e) (i) If a is odd, and $a \sim \sqrt{n}$, then
$\log R(n,a) \geq 2.0138\sqrt{n}$

(ii) If a is odd, then there exists a function f such that if $a \sim \lambda\sqrt{n}$, we have $\log R(n,a) \geq f(\lambda)\sqrt{n}$

J. Dixmier and J.-L. Nicolas. *Partitions sans petits sommants, In: A tribute to Paul Erdős*, (Edited by A. Baker, B. Bollobás and A. Hajnal), Cambridge Univ. Press, 1990.

f) For $\varepsilon > 0$ there exists $\delta > 1$ such that, for large n one has

$$\varepsilon\sqrt{n} \leq a \leq n - \varepsilon\sqrt{n} \Rightarrow R(n,a) \leq P(a)^{\delta}$$

(See J. Dixmier, Preprint IHES.)

§ XIV.11 On other subsums of a partition

Let $Q(n,a)$ be a number of partitions of n such that $n = n_1 + \cdots + n_t$, whose subsums $n_{i_1} + \cdots + n_{i_j}$ are all different from a, and each part is allowed to occur at most once.

a) There exists $\lambda_1 > 0$, such that, for $1 \leq a \leq \lambda_1\sqrt{n}$,

$$\log\frac{Q(n,a)}{q(n)} \geq -\frac{a}{6}\log\frac{16}{3} + O(1 + a^2/\sqrt{n})$$

P. Erdős, J.-L. Nicolas and A. Sárközy. *On the number of partitions of n without a given subsum* (I). Discrete Math. 75 (1989), 155–166.

b) For $a \leq \frac{3}{5}\sqrt{n}$, and n sufficiently large,

$$\log\frac{Q(n,a)}{q(n)} \leq -a\log\frac{2}{\sqrt{3}} + \frac{\pi a^2}{8\sqrt{3n}}$$

P. Erdős, J.-L. Nicolas and A. Sárközy. Ibid. (II). Analytic Number Theory, Proc. Conf. in Honor of P.T. Bateman, Birkhäuser 85 (1990), 205–234.

c) (i) For $n > n_0$ and $10^{18} \cdot \sqrt{n} \leq a \leq n^{5/7}$,

$$Q(n,a) \leq q\left(\left[\frac{n}{2}\right]\right)\exp(4 \cdot 10^5 \cdot a^{-1/3} \cdot n^{2/3} \cdot \log(a^{1/3} \cdot n^{-1/6}));$$

(ii) For $n > n_0$ and $n^{5/7} < a \leq n/2$,

$$Q(n,a) \le q\left(\left[\frac{n}{2}\right]\right)\exp\left(n^{1/2-1/30}\right).$$

(See P. Erdős, J.-L. Nicolas and A. Sárközy. (II).)

Corollary. Let $s(a)$ denote the least positive integer which does not divide a. For $n \ge n_0$, $s(a) \ge 40000$ and $\frac{7}{100}n^{1/2}s(a)^{3/2} \le a \le \frac{1}{40}n/s(a)$, one has

$$Q(n,a) \le \exp(201 n^{1/2} s(a)^{-1/2} \log s(a))$$

§ XIV.12 Partitions of j-partite numbers into k summands

1) Let $P(m,n)$ denote the number of partitions of the bipartite number (m,n). If $m = o(n^{1/4})$, then

$$\sqrt{3}\cdot 4n(m!)P(m,n) \sim (6n/\pi^2)^{m/2}\cdot \exp(\pi(2n/3)^{1/2})$$

V.S. Nanda. *Bipartite partitions.* Proc. Cambridge Philos. Soc. 53 (1957), 273–277.

Remark. For an asymptotic formula for $P(m,n)$ when m is fixed and $n \to \infty$, and m, n have the same order of magnitude, see

F.C. Auluck. *On partitions of bipartite numbers.* Proc. Cambridge Philos. Soc. 49 (1953), 72–83.

2) Let $P(N_j;k)$ denote the number of partitions of the j-partite number $N = (n_1,\ldots,n_j)$ into exactly k non-degenerate summands (i.e. with non-zero components).

a) There exist numbers a_k and b_k (depending only on k) such that

$$\prod_{i=1}^{j}\binom{n_i - b_k}{k-1} \le k!P(N_j;k) \le \prod_{i=1}^{j}\binom{n_i + a_k}{k-1}$$

Corollary. For sufficiently large $n_1,\ldots,n_j$ and any fixed k,

$$k!P(N_j;k) \sim \prod_{i=1}^{j}\binom{n_i - 1}{k-1}$$

H. Gupta. *Partition of j-partite numbers into k summands.* J. London Math. Soc. 33 (1958), 403–405.

Remark. For $j = 1$ the above Corollary was proved by Erdős and Lehner.

P. Erdős and J. Lehner. *The distribution of summands in the partitions of a positive integer.* Duke Math. J. 8 (1941), 335–345.

b) Let $\exp_t(x)$ stand for the sum of the first t terms in the expansion of e^x, in ascending powers of x. Then

$$P(N_j;k) \geq \prod_{i=1}^{j} \binom{n_i - 1}{k - 1} \cdot \exp_{[\frac{k+1}{2}]}\left(\frac{k^{j+2}}{2^{2j+1} \cdot e^2 \cdot n_1 \cdots n_j}\right), \text{ and}$$

$$P(N_j;k) < \prod_{i=1}^{j} \binom{n_i - 1}{k - 1} \cdot \exp_k\left(\frac{k^{j+2}}{2n_1 \cdots n_j}\right)$$

Corollary. For some $0 < \mu < 1$,

$$P(N_j;k)/\prod_{i=1}^{j} \binom{n_i - 1}{k - 1} = \exp\left(\frac{\mu \cdot k^{j+2}}{2n_1 \cdots n_j}\right)$$

H. Gupta. *An inequality for $P(N_j;k)$*. Research Bull. (N.S.) Panjab Univ. 13 (1962), 173–178.

3) Let $P_r(n) = P_r(n_1, \ldots, n_j), r \in \{1, 2, 3, 4\}$, denote the number of partitions of the integral vector $(n_1, \ldots, n_j) = n, n_1 \geq \ldots \geq n_j \geq 0$ such that $P_1(n)$ is the total number of partitions of n;
$P_2(n)$ is the number of partitions in which no part has a zero component;
$P_3(n)$ is the number of partitions into different parts;
$P_4(n)$ is the number of partitions into different parts in which no part has a zero component.
Put $R = \frac{1}{2}(n_2 + \cdots + n_j)$, $S = \prod_{s=2}^{j} n_s!$,
$T = (n_1 \ldots n_{j-1})^{n_j - 1} \cdot ((n_j - 1)!)^{1-j} \cdot (n_j!)^{-1}$
Then, if $n_s = o(n_1^{1/4})$, $2 \leq s \leq j$,

a) $$P_1(n) \sim \left(\frac{6n_1}{\pi^2}\right)^R \cdot (4 \cdot 3^{1/2} \cdot n_1 \cdot S)^{-1} \cdot \exp\left(\pi\left(\frac{2n_1}{3}\right)^{1/2}\right)$$

b) $$P_3(n) \sim \left(\frac{12n_1}{\pi^2}\right)^R \cdot (4 \cdot 3^{1/4} \cdot n_1^{3/4} \cdot S)^{-1} \cdot \exp\left(\pi\left(\frac{n_1}{3}\right)^{1/2}\right), \text{ as } n_1 \to \infty$$

c) If $n_j = o(n_s^{1/3})$, $1 \leq s \leq j - 1$, then

$$P_2(n) \sim P_4(n) \sim T$$

as $n_s \to \infty$ for $1 \leq s \leq j - 1$

M.M. Robertson. *Asymptotic formulae for the number of partitions of a multi-partite number*. Proc. Edinburgh Math. Soc. (2) 12 (1960/61), 31–40.

§ XIV.13 On a result of Turán

1) Let Λ be an arbitrary partition of n with distinct parts, let $\overline{\Lambda}$ stand for the set of all summands in Λ, and let $|\overline{\Lambda}|$ be the number of summands in $\overline{\Lambda}$. Let $q(n)$ be the number of unequal partitions of n. The phrase "for almost all k-tuples $\Lambda_1, \ldots, \Lambda_k$ of unequal partitions" will mean "with the exception of at most $O(q(n)^k)$ k-tuples, k fixed, $n \to \infty$". Then, for fixed $k \geq 2$ and $n \to \infty$, for almost all k-tuples $\Lambda_1, \ldots, \Lambda_k$ of Λ the cardinality of the set $\overline{\Lambda}_1 \cap \ldots \cap \overline{\Lambda}_k$ is

$$\geq (1 - O(1))(\sqrt{3}/\pi k^{2k-1})\sqrt{n}$$

P. Turán. *On a property of partitions.* J. Number Theory 6 (1974), 405–411.

2) Let $P(n)$ be the number of unrestricted partitions of n.
Let $k = o(n^{1/2})$.
Then for all but $o\left(\binom{P(n)}{k}\right)$ choices of k distinct partitions of n they all contain a common summand.

R.R. Hall and K. Wild. *On a problem of Turán concerning partitions.* J. London Math. Soc. (2) 13 (1976), 472–474.

§ XIV.14 Statistical theory of partitions

1) Let Π: $n = \lambda_1 + \cdots + \lambda_m$, $\lambda_j \geq 1$, be a (unrestricted) partition of n. Denote by $S_1(\Pi, \lambda)$ the number of λ_j's satisfying $\lambda_j \geq \lambda$ and by $S_2(\Pi, \lambda)$ the number of λ_j's satisfying $\lambda_j \leq \lambda$.

a) If $11 \log n \leq \lambda \leq \frac{\sqrt{6}}{2\pi} \cdot \sqrt{n} \log n - \sqrt{n} \log\log n$, then for almost all Π (i.e. with the exception of $o(q(n))$ Π's at most) one has

$$\left| S_1(\Pi, \lambda) - \frac{\sqrt{6}}{\pi}\sqrt{n} \log M \right| < c \cdot \sqrt{(n \log n)/\lambda}$$

where $M = 1 - \exp(-\pi \cdot \lambda \cdot (6n)^{-1/2})$, and c is a computable constant.

b) If $A(n)$ is a function tending to ∞ (as $n \to \infty$) and $A(n) \leq \frac{1}{5} \log n$; $1000A(n) \leq \lambda \leq 13 \log n$, then with the exception of at most $8P(n) \exp(-A(n))$ partitions one has

$$\frac{\sqrt{6n}}{\pi}(\log\lambda - \log A(n)) \le S_2(\Pi, \lambda) \le \frac{\sqrt{6n}}{\pi}(\log\lambda + A(n))$$

M. Szalay and P. Turán. *On some problems of the statistical theory of partitions with application to characters of the symmetric group* II. Acta Math. Hungar. 29 (1977), 381–392.

2) For almost all Π's, if $\lambda_j > 1$ and $(\lambda_j, \lambda_i) = 1$ for each $i \ne j$, then λ_j is a prime.

P. Erdős and M. Szalay. *On some problems of the statistical theory of partitions.* Colloq. Math. Soc. János Bolyai 51. Number Theory, Budapest, 1987, pp. 93–110.

3) Let $\Pi^* : n = \alpha_1 + \cdots + \alpha_m, \alpha_1 > \alpha_2 > \ldots > \alpha_n (\ge 1)$, be a generic (unrestricted) unequal partition of n. Let $M(\Pi^*)$ denote the maximal number of consecutive summands in Π^*.
If $f(n) \to \infty$ (arbitrarily slowly), then for almost all Π^*'s we have

$$M(\Pi^*) = \frac{1}{2\log 2}\log n - \frac{1}{\log 2}\log\log n + O(f(n))$$

(See P. Erdős and M. Szalay.)

Remark. For the number $q(n)$ of unequal unrestricted partitions it is known that

$$q(n) \sim \frac{1}{4n^{3/4} \cdot 3^{1/4}} \cdot \exp\left(\frac{\pi}{\sqrt{3}} \cdot \sqrt{n}\right)$$

G.H. Hardy and S. Ramanujan. *Asymptotic formulae in combinatory analysis.* Proc. London Math. Soc. (2) 17 (1918), 75–115.

§ XIV.15 Partitions of n into distinct parts all $\equiv a_i \pmod m$

If gcd $(a_1, \ldots, a_s, M) = 1$, let $A(a_1, \ldots, a_s, M, m)$ be the number of partitions of n into distinct parts all $\equiv a_i \pmod m$
Then

$$A(a_1, \ldots, a_s, M, m) = 2^{((s-3)/2 + (\sum a_i)/4)} \cdot 3^{-1/4} \cdot m^{-3/4} \cdot \exp(\pi\sqrt{5M/3m} + O(M^{-1/2+\varepsilon}))$$

for any $\varepsilon > 0$

L.B. Richmond. *On a conjecture of Andrews.* Utilitas Math. 2 (1972), 3–8.

§ XIV.16 Partitions with congruences conditions

1) For $q \ge 2, 0 \le r < q$, let $E_{r,q}(N)$ be the number of positive integers $n \le N$ with $P(n) \equiv r \pmod q$, where $P(n)$ is the number of unrestricted partitions

of n. Then for each q, there exist at least two distinct values of r such that, whenever N is sufficiently large,

$$E_{r,q}(N) > \log\log N/q \log 2$$

L. Mirsky. *The distribution of values of the partition function in residue classes.* J. Math. Anal. Appl. 93 (1983), 593–598.

2) Let $A = \bigcup_{i=1}^{q}\{a(i) + \nu M : \nu = 0, 1, 2, \ldots\}$, where q, M and $a(i)$ are positive integers such that $a(1) < a(2) < \ldots < a(q) \leq M$. Let $p(n, k, A)$ be the number of partitions of n into k parts, and $p^*(n, k)$ the number of partitions of n into k unequal parts; where all the parts belong to A.
For all positive integer n define r by $r \equiv n(\text{mod} M)$, $1 \leq r \leq n$ and write

$$P(x, k, A) = \left(\sum_{i=1}^{q} x^{a(i)}\right)^k = \sum_{\nu=1}^{kM} C_\nu \cdot x^\nu,\ S_r(P(x, k, A)) = \sum_{\nu=0}^{k-1} C_{\nu M+r}$$

and d, δ for the greatest common divisors $(a(1), \ldots, a(q), M)$, $(a(2) - a(1), \ldots, a(q) - a(1), M)$ and $\delta = M$ for $q = 1$.

a) For a given k, r for which $S_r(P(x, k, A)) \neq 0$, we have

$$p(Mn + r, k, A) \sim p^*(Mn + r, k, A) \sim n^{k-1} \cdot S_r(P(x, k, A))/k!(k-1)!$$

as $n \to \infty$.

b) If $\delta | r$, $n \to \infty$ and $k \to \infty$ through multiples of δ/d subject to the condition $k = o(n^{1/4})$, then

$$p(Mn + r, k, A) \sim p^*(Mn + r, k, A) \sim n^{k-1} \cdot q^k \cdot \delta/M(k!(k-1)!$$

M.M. Robertson. *Partitions with congruence conditions.* Proc. Amer. Math. Soc. 57 (1976), 45–49.

§ XIV.17 Partitions of n whose parts are relatively prime, or prime to n, etc.

1) a) Denote by $P_M(n)$ the number of partitions of n into summands which are positive and prime to a square-free number M. Then

$$P_M(n) \sim \frac{1}{2}\tau(M) \cdot \left(\frac{\varphi(M)}{6M}\right)^{1/4} \cdot n^{-3/4} \cdot \exp\left(2\pi\left(\frac{\varphi(M)}{6M}\right)^{1/2} \cdot n^{1/2}\right)$$

where

$$\tau(M) = \begin{cases} M^{-1/2}, & \text{for } M \text{ prime} \\ 1, & \text{otherwise} \end{cases}$$

S. Iseki. *Partitions in certain arithmetic progressions.* Amer. J. Math. 83 (1961), 243–264.

b) Let $R(n)$ be the number of partitions of n into summands relatively prime to n. Then

$$\log R(n) = \pi\sqrt{\frac{2\varphi(n)}{3}} + O(n^{(1+\varepsilon)\log 2/\log\log n})$$

L.B. Richmond. *A problem of Erdős concerning partitions.* J. Number Theory 9 (1977), 525–534.

Remark. This improves the Erdős result:

$$\log R(n) \sim \pi\sqrt{2\varphi(n)/3}$$

P. Erdős. *On some asymptotic formulas in the theory of partitions.* Bull. Amer. Math. Soc. 52 (1946), 185–188.

2) Let $f(n)$ be the maximum least common multiple of $a_1, \ldots, a_k$ for all partitions $n = a_1 + \cdots + a_k$ in positive integers a_i. If $g(x) = \log f(x)$, then

$$g(x) \sim (x \log x)^{1/2}$$

E. Landau. *Handbuch der Lehre von Primzahlen*, Leipzig, 1909, vol. I, pp. 222–229.

3) Let $g(n)$ be the number of partitions of n into parts that are pairwise relatively prime (i.e. the number of solutions of $n = \sum \lambda_i$, $\lambda_{i+1} \leq \lambda_1$, $(\lambda_i, \lambda_j) = 1$ for all $i \neq j$).
Then

$$\log g(n) \sim \frac{2\pi}{\sqrt{6}} \cdot \sqrt{n/\log n}$$

E. Schmutz. *Partitions whose parts are pairwise relatively prime.* Discrete math. 81 (1990), 87–89.

§ XIV.18 Partitions of n whose parts $a_i (i = \overline{1,k})$ satisfy $a_1|a_2|\ldots|a_k$

1) Let $g(n)$ be the number of partitions $n = a_1 + \cdots + a_k$ into distinct positive integers a_j with $a_1|a_2|\cdots|a_k$ (where "|" means "divides") and let $g_1(n)$ be the number of partitions of this type with $a_1 = 1$ (Clearly, $g(n) = g_1(n) + g_1(n+1)$).

a) For $n \geq 6$, we have $g(n) \geq \log_2 n$, where $\log_2$ denotes logarithm in base 2.

b) For $n \geq 27$, we have $g_1(n) \geq \frac{1}{2}\log_2 n$, except when $n-1$ is prime, in which case $g_1(n) = 1$.

P. Erdős and J.H. Loxton. *Some problems in partitio numerorum.* J. Austral. Math. Soc. (Series A) 27 (1979), 319–331.

c) Let $G_1(x) = \sum_{1 \le n \le x} g_1(n)$. If ρ is the unique positive root of the equation $\zeta(x) = 2$, where $\zeta(s)$ is the Riemann zeta function, then

$$G_1(x) \sim cx^{\rho}$$

$(x \to \infty)$, for some positive constant c.

d) $G_1(x) = cx^{\rho} + \Omega_{\varepsilon}(x^{\rho-\varepsilon})$ as $x \to \infty$, for every $\varepsilon > 0$

(See P. Erdős and J.H. Loxton.)

2) Let $h(n)$ be the number of partitions $n = a_1 + \cdots + a_k$ into positive integers a_j with $a_1|a_2|\cdots|a_k$, repetitions being allowed.

Put $H(x) = \sum_{0 \le n \le x} h(n)$. Then

$$\log H(x) = \frac{1}{2\log 2}\left(\log \frac{x}{\log x}\right)^2 + \left(\frac{1}{2} + \frac{1}{\log 2} + \frac{\log\log 2}{\log 2}\right)\log x -$$
$$-\left(1 + \frac{\log\log 2}{\log 2}\right)\log\log x + V\left(\frac{\log x - \log\log x}{\log 2}\right) + o(1)$$

as $x \to \infty$, where $V(t)$ is a certain periodic function with period 1.

(See P. Erdős and J.H. Loxton.)

Remark. In order to obtain the above result, Erdős and Loxton first prove that $b(n) \le h(n) \le cb(n)$, where $b(n)$ is the number of partitions of n as sums of powers of 2 (c positive constant.)

They then use a Tauberian theorem of Ingham.

A.E. Ingham. *A Tauberian theorem for partitions*. Ann. Math. 42 (1941), 1075–1090.

3) a) Let $k(n)$ be the number of partitions of n into distinct positive integers such that each part divides the largest one. Then

$$\log k(n) \sim (\log 2) \cdot D(n)$$

where $D(n) = \max_{1 \le m \le n} d(m)$, $d(n)$ being the divisor function.

Corollary. $\log\log k(n) \sim \log 2 \cdot \log n / \log\log n$ as $n \to \infty$

b) Let $s(n)$ be the number of partitions of n into positive integers in which each part divides the largest part and repetitions are allowed. Then

$$\log s(n) \sim \frac{1}{2} D(n) \log n$$

as $n \to \infty$.

Corollary. $\log\log s(n) \sim \log 2 \cdot \log n / \log\log n$

(See P. Erdős and J.H. Loxton.)

§ XIV.19 Partitions of n as sums of powers of 2

1) a) Let $b(n)$ be the number of partitions of n as sums of powers of 2. (The function $b(n)$ satisfies the recurrence $b(2n+1) = b(2n)$, $b(2n) = b(2n-1) + b(n), n \geq 1$). Put

$$B(x) = \sum_{0 \leq n \leq x} b(n) = b(2[x])$$

Then

$$\log B(x) = \frac{1}{2\log 2} \cdot \left(\log \frac{x}{\log x}\right)^2 + \frac{1}{2} + \left(\frac{1}{\log 2} + \frac{\log\log 2}{\log 2}\right) \log x -$$

$$- \left(1 + \frac{\log\log 2}{\log 2}\right) \log\log x + U\left(\frac{\log x - \log\log x}{\log 2}\right) + o(1)$$

(as $x \to \infty$), where $U(t)$ is a certain periodic function with period 1.

N.G. de Bruijn. *On Mahler's partition problem.* Proc. Kon. Nederl. Akad. Wet. (A) 51 (1948), 659–669.

b) Let $c(n)$ be the number of partitions of $2n$ into powers of 2. Then $c(0) = 1, c(n) = c(n+1) + c\left[\frac{n}{2}\right]$, and

$$\log c(n) \sim \frac{\log^2 n}{\log 4}$$

$(n \to \infty)$.

D.E. Knuth. *An almost linear recurrence.* Fib. Quart. 4 (1966), 117–128 (Correction: 4 (1966), 354).

Remarks. (i) Clearly, b) is a simple consequence of a), but it can be obtained by elementary arguments.

(ii) For the function $b(n)$, see also

C.-E. Fröberg. *Accurate estimation of the number of binary partitions.* ≪BIT≫, 17 (1977), 386–391.

2) Let $b_r(n)$ be the number of partitions of n as sums of powers of r. Then

$$f_r(n) \ll b_r(n) \ll f_r(n)$$

where $f_r(n) = \sum_{k=0}^{\infty} x^k/(r^{k(k+1)/2}k!)$

J. Bohman. *A note on the number partitions other than binary.* ≪BIT≫, 17 (1977), 479–480.

Remark. For $r = 2$ this contains a result by C.-E. Fröberg.

§ XIV.20 Partitions of n into powers of r (≥ 2)

Let $P_r(n)$ be the number of partitions of n into powers of $r \geq 2$ (integer). Then

$$\log P_r(rn) = (1/2\log r)(\log(n/\log n))^2 +$$
$$+\left(\frac{1}{2} + \frac{1}{\log r} + \log\log r/\log r\right)\log n -$$
$$-(1 + \log\log r/\log r)\log\log n + O(1)$$

K. Mahler. *On a special functional equation.* J. London Math. Soc. 15 (1940), 115–123.

Remark. For extensions of Mahler's result, see

N.G. de Bruijn. *On Mahler's partition problem.* Nederl. Akad. Wetensch. Proc. 51, 659–669 = Indag. Math. 10 (1948), 210–220;

W.B. Pennington. *On Mahler's partition problem.* Ann. Math. (2) 57 (1953), 531–546.

See also

B. Richmond. *Mahler's partition problem.* Ars Combinatorica 2 (1976), 169–189.

§ XIV.21 On a problem of Frobenius

1) Let a_i be positive integers, $(a_1, \dots, a_n) = 1$ and denote by A_N the number of solutions in nonnegative integers of

$$a_1x_1 + \cdots + a_nx_n = N$$

Then

a) $\lim\limits_{N\to\infty} A_N/N^{n-1} = \dfrac{1}{a_1 \cdots a_n(n-1)!}$

E. Laguerre. *Oeuvres*, I, Paris: Gauthier-Villars 1898 pp. 218–220.

b) Suppose that $(a_i, a_j) = 1$ for $i \neq j$. Then

$$A_N = P(N) + Q_N$$

where $P(x)$ is a polynomial of degree $n-1$ with relational coefficients and the sequence (Q_N) is periodical with respect to $a_1 \cdots a_n$ (i.e. $Q_{N+a_1\cdots a_n} = Q_N$.)

E. Netto. *Lehrbuch der Combinatorik.* 2nd edition. Leipzig and Berlin, B.G. Teubner 1927, pp. 319–320.

2) Given n integers $0 < a_1 < \cdots < a_n$ with $(a_1, \dots, a_n) = 1$, let $G(a_1, \dots, a_n)$ denote the greatest integer N for which $N = \sum_{i=1}^{n} a_ix_i$ has no solutions in nonnegative integers x_i.

a) $G(a_1, \dots, a_n) \le \sum_{i=1}^{n-1} a_{i+1}d_i/d_{i+1}$

where $d_i = (a_1, \dots, a_i)$.

A. Brauer. *On a problem of partitions.* Amer. J. Math. 64 (1942), 299–312;
A. Brauer and J.E. Shockley. *On a problem of Frobenius.* J. Reine Angew. Math. 211 (1962), 215–220.
A. Brauer and B.M. Seelbinder. *On a problem of partitions.* II. Amer. J. Math. 76 (1954), 343–346.

b) $G(a_1, \ldots, a_n) \leq 2a_{n-1} \cdot [a_n/n] - a_n$

P. Erdős and R.L. Graham. *On a linear diophantine problem of Frobenius.* Acta Arith. 21 (1972), 399–408.

c) Define (after Erdős and Graham)

$$g(n, t) = \max G(a_1, \ldots, a_n)$$

where the maximum is taken over all $0 < a_1 < \cdots < a_n \leq t$ with $(a_1, \ldots, a_n) = 1$. Then

$$g(n, t) < 2t^2/n$$

and

$$g(n, t) \geq G(x, \ldots, x^*) \geq t^2/(n-1) - 5t$$

for $n \geq 2$, where $x^* = (n-1)[t/(n-1)] - 1$, and $x = [t/(n-1)]$ (See P. Erdős and R.L. Graham.)

See also

R.K. Guy. *Unsolved problems in number theory*, Springer-Verlag, 1981 (See pp. 63–64).

d) $g(3, t) = [(t-2)^2/2] - 1$

M. Lewin. *On a linear diophantine problem.* Bull. London Math. Soc. 5 (1973), 75–78.

Remarks. (i) Sylvester showed that $G(a_1, a_2) = (a_1 - 1)(a_2 - 1) - 1$ and that the number of nonrepresentable numbers is $(a_1 - 1)(a_2 - 1)/2$. Roberts and Bateman found the value of $G(a_1, \ldots, a_n)$ if the a_i are in the arithmetic progression.

J.J. Sylvester. Math. Quest. Educ. Times 41 (1884), 21;
J.B. Roberts. *Note on linear forms.* Proc. Amer. Math. Soc. 7 (1956), 465–469;
P.T. Bateman. *Remark on a recent note on linear forms.* Amer. Math. Monthly 65 (1958), 517–518.

(ii) For the exact value of $g(n, t)$ if $n - 1 | t$, or $t - 1$ or $t - 2$, see

J. Dixmier. *Proof of a conjecture by Erdős and Graham concerning the problem of Frobenius.* J. Number Theory 34 (1990), 198–209.

§ XIV.22 An Abel-Tauber problem for partitions

Let $0 < \lambda_1 < \lambda_2 < \cdots$ be a given sequence of real numbers and put $n(u) = \sum_{\lambda_k \leq u} 1$. Let $P(u)$ be the number of solutions of $n_1\lambda_1 + n_2\lambda_2 + \cdots \leq u$ in integers $n_i \geq 0$. Suppose $\lim_{x\to\infty} \limsup_{t\to\infty} \frac{n(tx)}{n(t)} < \infty$ and $\lim_{x\to\infty} n(x) = \infty$. Then

$$\log P(x) = \int_0^x s(t)\mathrm{d}t + O(xs(x))$$

$(x \to \infty)$, where $s(t) = \inf \left\{ u : \frac{1}{u} n \left(\frac{1}{u} \right) \leq t \right\}, t > 0$

J.L. Geluk. *An Abel-Tauber theorem for partitions*. II. J. Number Theory 33 (1989), 170–181.

§ XIV.23 On partitions of the positive integers with no x, y, z belonging to distinct classes satisfying $x + y = z$

Let $P^s = A_1 \cup \cdots \cup A_s$ be a partition of the positive integers into s non-empty classes and P_n^s a partition as above of the integers $1, 2, \ldots, n$. A partition as above is admissible if the equation $x + y = z$ has no solution with x, y, z belonging to three distinct classes.

a) There is no admissible P_n^s with $\min |A_i| > 2^{1-s} \cdot n$

b) If P_n^s is admissible, then $s \leq \log_2 n + 1$.

J. Schönheim. *On partitions of the positive integers with no x, y, z belonging to distinct classes satisfying $x + y = z$*. Number theory, Proc. 1st Conf. Can. Num. Theory Assoc., Banff/Alberta (Canada) 1988, 515–528.

Remark. The question whether for any given P_n^3 the equation $x + y = z$ has a solution with x, y, z belonging to the same class has been answered in the affirmative by Schur.

I. Schur. *Über die Kongruenz $x^m + y^m = z^m$ (mod p)*. Jahresbericht der D.M.V. 25 (1917).

§ XIV.24 On certain partitions of n into $r \geq 2$ distinct pairs

Let $a_1^{(j)} + a_2^{(j)} + \cdots + a_r^{(j)} = n, (1 \leq j \leq k)$; be k partitions of n into $r \geq 2$ distinct parts. Assume that the rk summands a are all distinct. Then the largest value of k for any given n is

$$k_n(r) = \left[\frac{2n - r}{r^2} \right]$$

H. Gupta. *On a partition-problem of Erdős*. Indian J. Pure Appl. Math. 12 (11) (1981), 1293–1298.

§ XIV.25 Additively independent partitions

Two partitions τ_1 and τ_2 of the same natural number are said to be additively independent if their sub-sums (excepting 0 and n) are distinct. Let $G(n)$ (resp. $H(n)$) denote the number of pairs of partitions (resp. partitions without repetitions) which are additively independent.

a) For all k, there exist $\alpha_1, \alpha_2, \ldots, \alpha_k$ such that

$$G_n = 2P(n)\left(1 + \frac{\alpha_1}{\sqrt{n}} + \frac{\alpha_2}{n} + \cdots + \frac{\alpha_k}{n^{k/2}} + O\left(\frac{1}{n^{(k+1)/2}}\right)\right)$$

(E.g. $\alpha_1 = 1.28\ldots$, $a_2 = 12.98\ldots$, $\alpha_3 = 91.46\ldots$, $\alpha_4 = 495.53\ldots$)

Remark. The result a) was conjectured by J. Dénes in 1967.

b) $H(n) = cq(n)\left(1 + O\left(\frac{\log^2 n}{\sqrt{n}}\right)\right)$

where c is a constant satifying $13.83 \le c \le 14.29$

P. Erdős, J.-L. Nicolas and A. Sárközy. *On the number of pairs of partitions of n without common subsums* Colloq. Math. 63 (1992), 61–83.

See also

J.-L. Nicolas. *Distribution des sous-sommes d'une partition.* Publ. Math. Orsay, 92–01 (1992), 85–103.

§ XIV.26 A problem in "factorisatio numerorum" of Kalmár

1) Let $1 < a_1 \le a_2 \le \cdots$ be a sequence of integers. Denote by $f(n)$ the number of representations of n as the product of the a's, where two representations are considered equal only if they contain the same factors in the same order. Let

$$F(x) = \sum_{n \le x} f(n)$$

a) If $a_k = k + 1$, then

$$F(x) = -\frac{x^\rho}{\rho\zeta'(\rho)} + O(x^\rho \cdot e^{-\alpha \log\log x \cdot \log\log\log x})$$

for $\alpha < \frac{1}{2(\rho - 1)\log 2}$, where ρ is defined as the (unique) positive root of $\zeta(s) = 2$ (with ζ the Riemann zeta-function) and $\sum_{n=1}^{\infty} f(n)/n^s = 1/(2 - \zeta(s))$.

L. Kalmár. *Über die mittlere Anzahl der Produktdarstellungen der Zahlen* (Erste Mitteilung). Acta Litt. ac Scient. Szeged 5 (1931), 95–107.

b) Let $p_1 < p_2 < \cdots$ be a sequence of primes and $a_1 < a_2 < \cdots$ the sequence of integers composed of these primes. Then

$$F(x) = cx^{\rho} \cdot (1 + o(1))$$

where $\sum_i 1/a_i^{\rho} = 1, \rho > 0.$

E. Hille. *A problem in "factorisatio numerorum"*. Acta Arith. 2 (1936), 136–144.

Remark. When $a_k = k + 1$, Hille shows

$$F(x) = -\frac{x^{\rho}}{\rho\zeta'(\rho)} + \Omega_{\varepsilon}(x^{\rho-\varepsilon})$$

for all $\varepsilon > 0$, with ρ the positive root of $\zeta(s) = 2$.

c) Assume that $\sum 1/a_i^{1+\varepsilon}$ converges for every $\varepsilon > 0$ and that the a's are not all powers of a_1. Then

$$F(x) = cx^{\rho}(1 + o(1))$$

where $\sum_i 1/a_i^{\rho} = 1, \rho > 0.$

P. Erdős. *On some asymptotic formulas in the theory of the "factorisatio numerorum"*. Ann. Math. 42 (1941), 989–993.

Remark. Kalmár and Hille use methods of analytic number theory, and the Tauberian theorem of Wiener and Ikehara, respectively. Erdős's proof is elementary.

d) Let $a_k = k + 1$. Then

$$F(x) = -\frac{x^{\rho}}{\rho\zeta'(\rho)} + O\left(x^{\rho} \cdot e^{-\alpha(\log\log x)^{\nu}}\right)$$

where $\nu^{-1} = 3/4 + \varepsilon$ for any $\varepsilon > 0$ and $0 < \alpha < 1/2$.

S. Ikehara. *On Kalmár's problem in "Factorisatio numerorum"*. II. Proc. Phys.-Math. Soc. Japan (3) 23 (1941), 767–774.

2) Let $n = p_1 \cdots p_m$, where the p's are distinct primes. If $g(m)$ denotes the number of ordered nontrivial factorizations of n, then

$$\limsup_{m\to\infty} \left| \frac{2g(m)}{m!} - \frac{1}{m+1} \right|^{1/m} = 1/(\log^2 2 + 4\pi^2)^{1/2};$$

and

$$\frac{2g(m)}{m!} = 1/(\log 2)^{m+1} + O(1/(2\pi)^m)$$

R.D. James. *The factors of a square-free integer.* Canad. Math. Bull. 11 (1968), 733–735.

3) a) Let $h(n)$ be the number of representations of n (positive integer) as a product of integral factors larger than 1, a change in the order of factors not counting as a distinct representation. Then

$$\sum_{n\le x, n\equiv l(\mathrm{mod}k)} h(n) = B_k x(\log x)^{-3/4} \cdot \exp(2(\varphi(k)\cdot k^{-1}\cdot \log x)^{1/2}) \cdot (1 + O(\log^{-1/2} x))$$

with $B_k = (16\pi^2 k\varphi^3(k))^{-1/4} \cdot e^g$, where

$$g = \lim_{s\to 1}\left(\sum_{m=1}^{\infty} \frac{L(ms)-1}{m} - \frac{\varphi(k)}{k(s-1)}\right)$$

and the L-function corresponds to the principal character mod k.

G.J. Rieger. *Über die Anzahl der Produktzerlegungen ganzer Zahlen.* Math. Z. 76 (1961), 226–234.

b) A number n is called highly factorable, if $h(m) < h(n)$ for all $m < n$. Then

$$h(n) = n \cdot (L(n))^{-1+o(1)}$$

for highly factorable n,
where $L(n) = \exp(\log n \cdot \log\log\log n / \log\log n)$.

E.R. Canfield, P. Erdős and C. Pomerance. *On a problem of Oppenheim concerning "factorisatio numerorum".* J. Number Theory 17 (1983), 1–28.

c) $h(n) \le n(\log n)^{-\alpha}$
for any fixed $\alpha > 0$ and n sufficiently large;

and $h(n) \ne O(n^{\beta})$

for $\beta < 1$

F.W. Dodd and L.E. Mattics. *Estimating the number of multiplicative partitions.* Rocky Mountain J. Math. 17 (1987), 797–813.

d) $h(n) \le n/4 + 1$
for all n.

W. Chen. *Upper bound of the number of multiplicative partitions* (Chinese). Acta Math. Sinica 32 (1989), 604–609.

e) If the smallest prime factor is > 3, then

$$h(n) < n/\log n$$

H.Z. Cao. *On a conjecture of multiplicative partitions.* Rend. Mat. Appl. 7. Ser. 11, No. 4 (1991), 729–735.

Remark. The conjecture that $h(n) \le n/\log n$ for $n \ne 144$ is still open.

4) Let $F_y(n)$ be the number of factorizations of a positive integer n into factors $d, 2 \le d \le y$, the order of factors not being counted.
Let $K(x, y) = \sum_{n\le x} F_y(n)$
Let $r = (\log x)^{1/2}/\log y$.

a) If $4(\log\log x)^{-1} \le r \le (\log x)^{1/2}/(12\log\log x)$, then there exists a constant c such that

$$\log K(x, y) \ge \log x + f(r)(\log x)^{1/2} - c(1 + r^2)\log\log x$$

where $f(r) = \sup_{h>r}(-h\log r - h\log h + h + h\log\rho(1/2 - r/h))$, where

$$\rho(a) = \inf_t \int_{-1/2-a}^{1/2-a} e^{tx}dx \text{ for } -1/2 \le a \le 1/2$$

b) If $r \le c_1(\log x)^{1/4}$, then

$$\log K(x, y) \le \log x + f(r)(\log x)^{1/2} + O((\log\log x)^2 + r^2\log\log x)$$

D. Hensley. *The number of factorizations of numbers less than x into factors less than y.* Trans. Amer. Math. Soc. 275 (1983). no. 2, 477–496.

Remark. The function $f(r)$ is concave and decreasing, with $f(0) = 2$, $f'(0) = 0$.

5) For positive integers m and n, define $f_2(m, n)$ to be the number of different ways to write $(m, n) = (a_1, b_1)\cdots(a_k, b_k)$, where the multiplication is done coordinate-wise, all a_i, b_i are positive integers, (1, 1) is not used as a factor of $(m, n) \ne (1, 1)$, and two such factorizations are considered the same if they differ only in the order of the factors. Then

$$f_2(m, n) < (mn)^{1.516}/\log(mn)$$

for $(m, n) \ne (1, 1)$.

B.M. Landman and R.N. Greenwell. *Multiplicative partitions of bipartite numbers.* Fib. Quart. 29 (1991), 264–267.

§ XIV.27 Cyclotomic partitions

Let $c(n)$ denote the number of solutions of $\sum_q k_q\varphi(q) = n$ in nonnegative integers k_q (where φ is Euler's function). Also, let $c_d(n)$ be the number of solutions for distinct k_q. Then

a) $c(n) = A(\log n)^{-1/2}\cdot n^{-1}\cdot\exp(B\sqrt{n})(1 + O(\log\log n/\log n))$
where $B = 2\zeta(2)\sqrt{\zeta(3)/\zeta(6)}$.

b) $c_d(n) = A_d n^{-3/4}\exp(B_d\sqrt{n})(1 + O(1/\log n))$
where $B_d = B/\sqrt{2}$.

D.W. Boyd and H.L. Montgomery. *Cyclotomic partitions.* Number theory, Proc. 1st Conf. Can. Num. Theory Assoc., Banff/Alberta (Canada) 1988, 7–25.

Remark. $c(n)$ denotes also the number of cyclotomic polynomials of degree n.

§ XIV.28 Multiplicative properties of the partition function

Let $P(n)$ be the number of unrestricted partitions of n. Then

$$\omega\left(\prod_{m=1}^{N} P(m)\right) > (1-\varepsilon)\frac{\log N}{\log 2}$$

if $N > N_0(\varepsilon)$

A. Schinzel and E. Wirsing. *Multiplicative properties of the partition function.* Proc. Indian Acad. Sci. Math. Sci. 97 (1987), 297–303.

Remark. The first important result of this type is due to Schinzel, who proved that

$$\omega\left(\prod_{m=1}^{N} P(m)\right) \to \infty$$

$(N \to \infty)$

This was conjectured by Erdős and Ivić. The proof by Schinzel appears in

P. Erdős and A. Ivić. *The distribution of values of a certain class of arithmetic functions at consecutive integers.* Colloq. Math. Soc. János Bolyai 51. Number Theory, Budapest, 1987, pp. 45–91.

§ XIV.29 Partitions into primes

For the "Partition into Primes", see the chapter with Additive and diophantine problems involving primes.

§ XIV.30 Partitions of N into terms of $1, 2, \ldots, n$, repeating a term at most p times

Let R denote the number of partitions of N into terms of $1, 2, \ldots, n$, repeating a term at most p times (i.e. the number of solutions of the equation $1 \cdot x_1 + 2 \cdot x_2 + \cdots + n \cdot x_n = N, 0 \le x_i \le p$ integers). Suppose $n \ge 2$, $p \ge 1$ and N are natural numbers.

Let $A_n = n(n+1)p/4$, $D_n^2 = n(n+1)(2n+1)p(p+2)/72$. Then

$$R = \frac{(p+1)^n}{D_n\sqrt{2\pi}}\left(\exp\left(-\frac{(N-A_n)^2}{2D_n^2}\right) + \frac{\theta}{n}\right)$$

where $|\theta| \le K$, with an universal constant K

I. Joó. *On the number of partitions of the number N into terms of* $1, 2, \dots, n$ *repeating a term at most p times*. Ann. Univ. Sci. Budapest, Eötvös Sect. Math. 28 (1985), 217–227.

§ XIV.31 Partition which assumes all integral values

a) Let $H(n)$ be the number of partitions of n into parts repeated exactly 1, 3, 4, 6, 7, 9 or 10 times with the parts repeated exactly 1, 4, 6 or 9 times being even in number minus the number of partitions of n into parts repeated exactly 1, 3, 4, 6, 7, 9 or 10 times with the parts repeated exactly 1, 4, 6 or 9 times being odd in number.
Then the set of n for which $H(n) \neq 0$ has density 0, and $H(n)$ takes on every integer value infinitely often.

M. Tamba. *On a partition function which assumes all integral values*. J. Number Theory 41 (1992), 77–86.

b) Let $T(n)$ denote the number of partitions of n into parts which are repeated exactly 1, 3, 4, 6, 7, 9 or 10 times with the parts repeated exactly 3, 4, 6 or 7 times being even in number minus the number of them with the parts repeated exactly 3, 4, 6, or 7 times being odd in number.
Then $T(n)$ is non-negative and assumes all non-negative integral values infinitely often.

M. Tamba. *Note on a partition function*. J. Number Theory 41 (1992), 280–282.

§ XIV.32 Partitions free of small summands

Let $0 < \lambda_1 < \lambda_2 < \cdots$ be an unbounded sequence of real numbers. For any real number l representable as a linear combination of the numbers λ_ν with non-negative integer coefficients, and for $y > 0$, let $p_y(l)$ denote the number of partitions of l into parts from the sequence $\{\lambda_\nu\}$ which are free of parts less than y.
Put $P_y(u) = \sum_{l \leq u} p_y(l)$. Then

$$\log P_y(u) = \frac{u}{y} \log \frac{y^2}{u} - \frac{u}{y} \log\log \frac{y^2}{u} + \frac{u}{y} + O_\varepsilon\left(\frac{u \log\log u}{y \log u}\right)$$

uniformly for $u^{1/2+\varepsilon} \leq y \leq u^{1-\varepsilon}$ as $u \to \infty$.

J. Herzog. *Weak asymptotic formulas for partitions free of small summands*. II. Acta Math. Hung. 62 (1993), 173–188.

Chapter XV

CONGRUENCES, RESIDUES AND PRIMITIVE ROOTS

§ XV. 1 Addition of residue classes mod p

1) Let $k = k(m)$ be the largest number of distinct residue classes, modulo m, so that no subset has sum zero.

a) If $m \geq 5$, then $k \geq [(-1 + \sqrt{8m + 9})/2]$

P. Erdős and H. Heibronn. *On the addition of residue classes* mod p. Acta Arith. 9 (1964), 149–159.

b) If $m = 2(l^2 + l + 1)$, then $k \geq 2l + 1 = \sqrt{2m - 3}$.

J.L. Selfridge. See R.K. Guy, *Unsolved problems in number theory,* 1981, (p. 73).

2) a) If $a_1, a_2, \ldots, a_k$, where $k \geq 3 \cdot (6p)^{1/2}$, are distinct residues $(\bmod p)$, where p is a prime, then every residue $(\bmod p)$ can be written in the form $\sum_{i=1}^{k} \varepsilon_i a_i$, $\varepsilon_i \in \{0, 1\}$.

(See P. Erdős and H. Heilbronn.)

b) The same holds for $k > 2 \cdot \sqrt{p}$, and this is the best possible.

J.E. Olsen. *An addition theorem, modulo p.* J. Combin. Theory 5 (1968), 45–52.

3) Let p be a prime number; u, v, S, T integers with $1 \leq u, v \leq p - 1, 1 \leq T \leq p$; furthermore, $a_1, a_2, \ldots, a_u, b_1, b_2, \ldots, b_v$ are integers with

$$a_i \not\equiv a_j (\bmod p) \text{ for } 1 \leq i < j \leq u,$$

$$b_i \not\equiv b_i (\bmod p) \text{ for } 1 \leq i < j \leq v.$$

Let $f(n)$ denote the number of solutions of

$$a_x b_y \equiv n (\bmod p),\ 1 \leq x \leq u, 1 \leq x \leq v.$$

Then

$$\left| \sum_{n=s+1}^{s+T} f(n) - \frac{u\,v\,T}{p} \right| < 2(p\,u\,v)^{1/2} \log p$$

A. Sárközy. *On the distribution of residues of products of integers*. Acta Math. Hung. 49 (3–4) (1987), 397–401.

4) Let $B(n)$ be the smallest integer so that there is a residue a_p for every prime p with $2 \le p \le B(n)$, and every positive integer $x \le n$ satisfies at least one of the congruences

$$x \equiv a_p (\text{mod}\, p)$$

a) $B(n) > c\sqrt{n}$ ($c > 0$ constant).

H. Iwaniec. See P. Erdős. *Some unconventional problems in number theory*. Acta Math. Acad. Sci. Hung. 33 (1–2) (1979), 71–80.

b) $B(n) < c\,n(\log\log\log n)^2 / \log n \cdot \log\log n \cdot \log\log\log\log n$

(See P. Erdős (1979).)

5) Denote by ε_n the smallest number so that there is a residue b_p for every prime p with $n^{\varepsilon_n} < p \le n$, and every positive integer $x \le n$ satisfies at least one of the congruences $x \equiv b_p (\text{mod}\, p)$. Then $\varepsilon_n > c \log\log\log n / \log\log n$.

(See P. Erdős.)

§ XV. 2 Residues of n^n

For an odd prime p, let $r(p)$ be the number of distinct residues (mod p) of n^n, where $1 \le n \le p-1$. Then $[\sqrt{(p-1)/2}] \le r(p)$ and $r(p) \le p-4$ for p sufficiently large. If $p \equiv 3 (\text{mod}\, 8)$,

$$r(p) \le p - 6$$

R. Crocker. *On residues of n^n*. Amer. Math. Monthly 76 (1969), 1028–1029.

§ XV. 3 Distribution of quadratic nonresidues

1) a) If $p \equiv 1 (\text{mod}\, 8)$ is a given prime, there must exist a prime $q < 2\sqrt{p} + 1$ such that q is a quadratic nonresidue of p.

C.F. Gauss. *Disquisitiones Arithmeticae*, Göttingen: Königlichen Gesellschaft der Wissenchaften, 1863 (original: 1801) (See Article 129).

Remark. Euler was the first to define residues and nonresidues and systematically investigate their properties. Gauss' introduction of the congruence notation considerably clarified the theory.

L. Euler. *Disquisitio accuratior circa residua ex divisione quadratorum altiorumque potestatum per numeros primos relicta*, Opera Omnia, I, 3, pp. 513–543 (original: 1783).

b) Let q be the smallest odd prime which is a quadratic nonresidue of p and r the smallest odd prime which is a quadratic residue. Then

(i) If $p \equiv 1 \pmod 8$, then $q < \sqrt{p}$;

(ii) If $p \equiv 5 \pmod 8$, then $q < \sqrt{2p}$;

(iii) If $7 < p \equiv -1 (\bmod 8)$, then $q < \sqrt{2p} - 1$;

(iv) If $7 < p \equiv -1 \pmod 8$, then $r \leq 2\sqrt{p} - 1$;

(v) If $p \equiv 3 \pmod 8$, then $r \leq \sqrt{(p+16)/3} - 2$,
except $p = 3, 11, 19, 43, 67, 163$ and perhaps one other value of p.

T. Nagell. *Sur les restes et les non-restes quadratiques suivant un module premier.* Ark. Math. 1 (1950), 185–193.

c) If $p \neq 3, 5, 7, 11, 13, 23, 59, 109, 131$ then

$$q < \sqrt{p}$$

L. Rédei. *Die Existenz eines ungeraden quadratischen Nichtrestes* mod$p$ *im Intervall* 1, $\sqrt{p}$. Acta Szeged 15 (1953), 12–19.

d) Let $p \not\equiv 1 (\bmod 8)$ be an odd prime. Then the smallest positive quadratic nonresidue q of p satisfies

$$q < (2p)^{2/5} + 3(2p)^{1/5} + 1$$

A. Brauer. *Über die kleinsten quadratischen Nichtreste*. Math. Z. 33 (1931), 161–176.

e) $q < p^{2/5} + 12p^{1/5} + 33$
where q is defined as in b).

R.H. Hudson and K.S. Williams. *On the least quadratic nonresidue of a prime* $p \equiv 3 \pmod 4$. J. Reine Angew. Math. 318 (1980), 106–109.

Remark. The proof is elementary.

2) Let $n(p)$ denote the smallest positive quadratic non-residue modulo p (prime). Then

a) $n(p) < p^{1/2\sqrt{e}} \cdot \log^2 p$
(where $e = 2.71828\ldots$, is the base of the natural logarithm.)

I.M. Vinogradov. *On the distribution of residues and non-residues of powers*. J. Physico-Mathematical Soc. of Perm 1 (1918), 94–96.

b) $n(p) = O\left(p^{\frac{1}{2}\beta} \cdot (\log p)^{\beta}\right)$
where $\beta = 1/\sqrt{e}$.

H. Davenport and P. Erdős. *The distribution of quadratic and higher residues*. Publ. Math. 2 (1952), 252–265.

c) Assuming the Extended Riemann Hypothesis, we have

$$n(p) = O(p^{\varepsilon})$$

for all $\varepsilon > 0$.

Ju.V. Linuik. *A remark on the least quadratic non-residue*. C. R. I' Acad. Sci. URSS, 36 (1942), 119–120.

d) On the same Hypothesis,

$$n(p) = O(\log^2 p)$$

N.C. Ankeny. *The least quadratic non-residue*. Ann. Math. 55 (1952), 65–72.

e) $n(p) = O(p^{\propto})$, for any fixed $\propto > \frac{1}{4}e^{-1/2}$.

D.A. Burgess. *The distribution of quadratic residues and non-residues*. Mathematica 4 (1957), 106–112.

§ XV. 4 Distribution of quadratic residues

1) a) Given s (positive integer) and $p > p_0(s)$, an odd prime, the number series $1, 2, \ldots, p-1$ always contains a sequence of not less than s consecutive quadratic residues or nonresidues.

A. Brauer. See A.O. Gelfond and Yu.V. Linnik. *Elementary method in the analytic theory of numbers*. Pergamon Press, Oxford, 1966. (See pp. 183–184);

and

A. Brauer. *Über Sequenzen von Potenzresten*. Akad. Wiss. Berlin, Sitz. (1928), 9–16.

b) If $2^s \le \sqrt{p}/\log^2 p$, $p > p_0$, then there always exists a sequence $\varepsilon_1, \ldots, \varepsilon_s$ of signs $+1$ or -1, corresponding to residues or nonresidues mod p among the numbers $1, 2, \ldots, p-1$.

H. Davenport. *On the distribution of quadratic residues* (mod p). J. London Math. Soc. 8 (1933), 46–52.

c) For each odd prime p there exists a sequence of $\frac{p-1}{2}$ consecutive relatively prime residue classes mod p among which the number of quadratic residues exceeds the number of quadratic nonresidues by at least $\left[\frac{\sqrt{p}+1}{2}\right]$.

K. Burde. *Zur Verteilung quadratischer Reste*. Math. Z. 105 (1968), 150–152.

d) The maximum number of consecutive quadratic residues or non-residues $(\bmod p)$ is $O(p^{1/4+\delta})$ for large p. ($\delta > 0$ fixed.)

D.A. Burgess. *The distribution of quadratic residues and non-residues.* Mathematika 4 (1957), 106–112.

e) If l_1 is a quadratic residue, l_2 a non-residue $\bmod q$, let $\varepsilon(n, q, l_1, l_2)$ take the value $+1$ or -1 according to $n \equiv l_1$ or $l_2 \pmod q$.

(i) If $q < 25$, then

(*) $$\lim_{x-\infty} \sum_p \varepsilon(p, q, l_1, l_2) \log p \cdot p^{-\alpha} \cdot \exp(-(\log^2 p)/x) = -\infty$$

for all $x \leq \alpha < 1/2$.

(ii) For general q, (*) holds, if all zeros $\rho = \beta + i\gamma$ of all $L(s, \chi \bmod q)$, q fix, satisfy the inequality $\beta^2 - \gamma^2 < 1/4$.

H.-J. Bentz and J. Pintz. *Quadratic residues and the distribution of prime numbers.* Monatsh. Math. 90 (1980), 91–100.

Remark. (i) confirms Shank's conjecture for $q < 25$ in a certain sense, that there are "more" primes in the non-quadratic residue classes $\bmod q$ than in the quadratic ones.

D. Shanks. *Quadratic residues and the distribution of primes.* Math. Tables and other Aids to Computation 13 (1959), 272–284.

2) If p is a prime, let $n_1, n_2, \ldots, n_{(p-1)/2}$ (respectively $r_1, r_2, \ldots, r_{(p-1)/2}$) denote distinct quadratic nonresidues (quadratic residues) of p and let $r_0 \equiv 0 \pmod p$. (For convenience, we call r_0 a residue.)

a) Among the integers $n_1 + a, n_2 + a, \ldots, n_{2k} + a$ there are k residues and k nonresidues if a is a residue; $k + 1$ residues (including 0) and $k - 1$ nonresidues if a is a nonresidue. (Here residue = quadratic residue, etc.).

b) Among the integers $r_0 + a, r_1 + a, \ldots, r_{2k} + a$ there are $k + 1$ residues (including 0) and k nonresidues if a is a residue; k residues and $k + 1$ nonresidues if a is a nonresidue.

O. Perron. *Bemerkung über die Verteilung der quadratischen Reste.* Math. Z. 56 (1952), 122–130.

3) If p is a sufficiently large prime and if $H \gg p^{11/24} \cdot (\log p)^{3/2}$, then the sequence $N + 1, N + 2, \ldots, N + H$ includes a pair of consecutive quadratic residues and a pair of consecutive non-residues $(\bmod p)$.

D.A. Burgess. *On Dirichlet characters of polynomials.* Proc. London Math. Soc. (3) 13 (1963), 537–548.

§ XV. 5 Sequences of consecutive quadratic nonresidues

a) Let $a(p)$ be the least positive integer a for which $\left(\frac{a}{p}\right) = \left(\frac{a+1}{p}\right) = -1$ (if there is no such a put $a(p) = 0$). Then

(i) $\frac{1}{\pi(x)} \sum_{p \le x} a(p) = A + O((\log\log x)^{-1/2})$
(A constant)

P.D.T.A. Elliott. *On the mean value of $f(p)$*. Proc. London Math. Soc. (3) 21 (1970), 28–96.

(ii) $a(p) \ll p^{\frac{1}{4}(1-\frac{1}{2}\cdot e^{-10})+\varepsilon}$
($\varepsilon > 0$, fixed)

P.D.T.A. Elliott. *On the least pair of consecutive quadratic non-residues* (mod p). Proc. 1972 Number Theory Conf. (Colorado), pp. 75–79.

Remark. The same result is valid with -9 in place of -10.

Z. Zheng. *On the least quadratic nonresidue* (mod p). Chin. Sci. Bull. 38 (1993), 621–627.

(iii) $a(p) \le p^{1/(4\sqrt{e})+\varepsilon}$
for $p \ge p_0(\varepsilon)$.

A. Hildebrand. *On the least pair of consecutive quadratic non-residues*. Michigan Math. J. 34 (1987), 57–62.

Note. The following elementary bound of Hudson is used (among other estimates)

$$a(p) \le (q_1 - 1)q_2$$

where q_1 and q_2 denote the smallest and second smallest primes which are quadratic non-residues mod p.

R. Hudson. *The least pair of consecutive character non-residues*. J. Reine Angew. Math. 281 (1976), 219–220.

b) Let l_n denote the maximum number of consecutive quadratic nonresidues for a prime p. Then

(i) $l_n < \sqrt{p}$
if $p \equiv 1 \pmod{24}$

(ii) $l_n < \sqrt{p} + \frac{3}{4}\sqrt{2} \cdot \sqrt[4]{p} + 2$
for any prime p

R.H. Hudson. *On sequences of consecutive quadratic nonresidues*. J. Number Theory 3 (1971), 178–181.

(iii) If $p > 2^{322}$, $p \equiv 13 \pmod{24}$ then

$$l_n < \sqrt{p}$$

(iv) If $p \equiv 5 \pmod{24}$, then

$$l_n < \left(\frac{12}{13}p\right)^{1/2} + 2^{3/2}\left(\frac{12}{13}p\right)^{1/4} + 2$$

(v) If $p \equiv 17 \pmod{24}$, then

$$l_n < \left(\frac{4}{5}p\right)^{1/2} + 2\left(\frac{36}{5}p\right)^{1/4} + 3$$

R.H. Hudson. *On a conjecture of Issai Schur.* J. Reine Angew. Math. 289 (1977), 215–220.

c) Let $b(p)$ denote the smallest positive integer a such that $\left(\frac{a}{p}\right) = \left(\frac{a+1}{p}\right) = \left(\frac{a+2}{p}\right)$. Then

$$b(p) < 17.5p^{1/4}\log p + 4$$

for all primes $p > 7$.

R.H. Hudson. *A bound for the first occurence of three consecutive integers with equal quadratic character.* Duke Math. J. 40 (1973), 33–39.

§ XV. 6 On residue difference sets

A residue difference set modulo p is defined to be any set $\{a_1, \ldots, a_k\} \subset \mathbb{I}^k$, with $1 \leq a_i < p$ such that all $a_i (1 \leq i \leq k)$ and all $a_i - a_j (1 \leq i, j \leq k, i \neq j)$ are quadratic residues mod p. Let p be a prime, $p \equiv 1 \pmod 4$. Denote by m_p the maximal cardinality of a residue difference set mod p. Then

a) $\frac{\log p}{2} < m_p < p^{1/2}\log p$
for all p

b) $m_p < (1+\varepsilon)\frac{p^{1/2}\log p}{4\log 2}$
for all $p > C(\varepsilon)$ and all $\varepsilon > 0$, where $C(\varepsilon) > 0$ is a constant.

D.A. Buell and K.S. Williams. *Maximal residue difference sets modulo p.* Proc. Amer. Math. Soc. 69 (1978), 205–209.

§ XV. 7 Sets which contain a quadratic residue mod p for almost all p

Let P be a set of primes with a certain property. We say that this property holds for almost all primes p if the density of P is unity, i.e.

$$\lim_{x\to\infty} \text{card}\,\{p : p \in P, p \leq x\}/\text{card}\,\{p : p \text{ prime}, p \leq x\} = 1$$

Let S be a finite set of non-zero integers. Then, for S to contain a quadratic residue $\bmod p$ for almost all primes p, it is necessary and sufficient that there is an odd-size subset T of S such that the product of the elements of T is a square.

M.A. Filaseta and D.R. Richman. *Sets which contain a quadratic residue modulo p for almost all p.* Math. J. Okayama Univ. 31 (1989), 1–8.

§ XV. 8 Least prime quadratic residue

1) Let $n'(p)$ denote the least prime quadratic residue. Then

$$n'(p) = O(p^{\alpha})$$

for fixed $\alpha > \frac{1}{4}e^{-1/2}$

A.I. Vinogradov and Yu.V. Linnik. *Hypoelliptical curves and the least prime quadratic residue.* Akad. Nauk USSR Doklady, 168 (1966), 259–261.

Remark. For improvments under certain assumptions, see

P.D.T.A. Elliott. *A note on recent paper of Yu.V. Linnik and A.I. Vinogradov.* Acta Arith. 13 (1967/1968), 103–105.

2) Let $p \equiv 1 \pmod{k}$ be a prime and denote by $r_k(\text{p})$ the least prime that is k-th power residue $\bmod p$; otherwise let $r_k(p) = 0$. Then

$$r_k(p) = O(p^{(k-1)/4+\varepsilon})$$

for all $\varepsilon > 0$.

P.D.T.A. Elliott. *The least prime k-th power residue.* J. London Math. Soc. (2) 3 (1971), 205–210.

§ XV. 9 Quadratic residues of squarefree integers

Let $0 < \varepsilon \leq \frac{1}{2}$, $\delta = \varepsilon^2/32$. Then for $x > p^{\frac{1}{4}+2\delta+\varepsilon}$, the number of quadratic residues in the sequence of squarfree integers not exceeding x equals

$$\frac{3}{\pi^2}x + O(xp^{-\delta})$$

O.V. Popov. *On quadratic residues and nonresidues in a sequence of square-free numbers.* (Russian.) Vestnik Moskov. Univ. Ser. I. Mat. Mekh. 1989, no. 5, 81–83.

§ XV.10 Least k-th power nonresidue

1) Let $n_k(p)$ be the least positive k-th power nonresidue modulo p (prime).
   a) If $k > 1$ is a divisor of $p-1$, then

$$n_k(p) < p^{1/2a} \cdot \log^2 p$$

   where $a = e^{(k-1)/k}$, and p is sufficiently large.

   b) If, in addition, $k > m^m$, $m > 8$, then

$$n_k(p) < p^{1/m}$$

   for all sufficiently large p.

I.M. Vinogradov. *On the bound of the least non-residue of n-th powers.* Trans. Amer. Math. Soc. 29 (1927), 218–226.

   c) If $k > 1$ and $(k, p-1) > 1$, then

$$n_k(p) < (p/3)^{1/2} + 2$$

   for $p \neq 23$ or 71.

R.H. Hudson. *On the least k-th power non-residue.* Ark. Math. 12 (1974), 217–220.

2) a) $$\sum_{p \le x} n_2(p) = (1 + o(1)) \sum_{k=1}^{\infty} \frac{p_k}{2^k} \cdot \frac{x}{\log x}$$

P. Erdős. *Remarks on number theory.* I. (Hungarian.) Mat. Lapok 12 (1967), 10–17.

   b) For all k, α with $\alpha < 4e^{1-1/k}$, one has

$$\sum_{p \le x} (n_k(p))^\alpha \sim C_{k,\alpha} \cdot \frac{x}{\log x}$$

   where $C_{k,\alpha}$ is a constant. If k is an odd prime, then

$$C_{k,\alpha} = \sum_{v=1}^{\infty} k^{-v} \cdot q_v^\alpha$$

   where q_v runs over all primes.

P.D.T.A. Elliott. *A problem of Erdős concerning power residue sums.* Acta Arith. 13 (1967/68), 131–149. Corrigendum: 14 (1967/68), 437.

3) For $k \ge 2$, integer, let $p \equiv 1 (\bmod q)$ be a prime for which -1 is a k-th power non-residue. Let a be a real number such that $2/5 < a \le 4/9$.

Then there are at least $\left[\frac{1}{4}(p^{5a/2-1}-2)\right]$ k-th power non-residues in the interval $[1, 2(p^a + p^{a/2} + 1)]$.

R.H. Hudson. *On the distribution of k-th power non-residues in the interval* $[1, p^a], 2/5 < a \le 4/9$. J. Reine Angew. Math. 260 (1973), 178–180.

4) Let χ be a non-principal character mod p of order k, where p is an odd prime $p \equiv 1 (\operatorname{mod} k)$, $k \ge 2$. Let $\alpha_j(k, p)$ be the least positive integer α such that $\chi(\alpha) \ne 1, \chi(\alpha+1) \ne 1, \ldots, \chi(\alpha+j-1) \ne 1$. Then, for all $\varepsilon > 0$,

$$\alpha_{k-1}(k, p) = O_{\varepsilon,k}\left(p^{\frac{1}{4}+\varepsilon}\right)$$

for all primes $p > p_0(\varepsilon, k)$.

R.H. Hudson. *A bound for the first k − 1 consecutive k-th power non-residues* (mod p). Acta Arith. 28 (1975/76), 341–343.

§ XV.11 Quadratic residues in arithmetic progressions

1) If n, k, and a are given positive integers, and p is a large prime with $p \equiv 1 (\operatorname{mod} n)$, then we define $r = r(n, k(a), p)$ as the least positive integer such that $r, r+a, r+2a, \ldots, r+(k-1)a$ are all n-th power residues mod p. Define $\Lambda(n, k(a)) = \limsup_{p\to\infty} r(n, k(a), p)$. Then

a) $\Lambda(2, 2(a)) < \infty$
for every finite a

b) $\Lambda(2, 2(a)) \le \left(\frac{a-1}{2}\right)^2$
for all $a \ge 7$

c) $\Lambda(2, 3(a)) = \infty$
whenever $a \equiv \pm 1 (\operatorname{mod} 3)$ or $\equiv \pm 2 (\operatorname{mod} 5)$.

S. Sahib. *Bounds of quadratic residues in arithmetic progressions.* J. Number Theory 2 (1970), 162–167.

Remarks: (i) The existence of k members of n-th power residues in arithmetic progression, with a given common difference a for every large prime p, follows from the Brauer theorem.

A. Brauer. *Über Sequenzen von Potenzresten.* Akad. Wiss. Berlin Sitz. (1928), 9–16.

(ii) The notation $\Lambda(n, k) = \Lambda(n, k(1))$ has been introduced by D.H Lehmer and by E. Lehmer, who showed that $\Lambda(2, 3) = \infty$ and

$\Lambda(k, 4) = \infty$ for $k \leq 1048909$ and conjectured that $\Lambda(k, 4) = \infty$ for all $k \geq 2$. This was obtained by Graham.

D.H. Lehmer and E. Lehmer. *On run of residues*. Proc. Amer. Math. Soc. 13 (1962), 102–106;

R.L. Graham. *On quadruples of consecutive k-th power residues*. Proc. Amer. Math. Soc. 15 (1964), 196–197.

2) If $r, r + a, \ldots, r + (k - 1)a$ belong to any one the n cosets determined with the help of the subgroup of n-th power residues mod p in the multiplicative group of Galois field of p elements $0, 1, 2, \ldots, (p - 1)$, then in analogy with $\Lambda(n, k(a))$ we define $\Lambda^*(n, k(a))$. Then

$$\Lambda^*(2g, 3(a)) = \infty$$

where $a \equiv \pm 1 \pmod 3$ and

$$\Lambda(2g, 3(a)) = \infty$$

where $a \equiv \pm 1 \pmod 3$ or $\pm 2 \pmod 5$, where g is any positive integer.

(See S. Sahib.)

Remark. The symbol $\Lambda^*(n, k) = \Lambda^*(n, k(1))$ appears in

J.H. Jordan. *Pairs of consecutive power residues or nonresidues*. Canad. J. Math. 16 (1964), 310–314;

see also

J.H. Jordan. *The distribution of cubic and quintic non-residues*. Pacific J. Math. 16 (1966), 77–85;

and

J.H. Jordan. *The distribution of k-th power residues and nonresidues*. Proc. Amer. Math. Soc. 19 (1968), 678–680.

3) a) $\Lambda(k, l) = \infty$ for $l \geq 4$ and all $k \geq 2$; and for $l = 3$ and all even values of k.

R.L. Graham. *On quadruples of consecutive k-th power residues*. Proc. Amer. Math. Soc. 15 (1964), 196–197.

See also

D.H. Lehmer and E. Lehmer. *On run of residues*. Proc. Amer. Math. Soc. 13 (1962), 102–106.

b) $\Lambda(k, 2) < \infty$ for all positive integers k.

A. Hildebrand. *On consecutive k-th power residues*, II. Michigan Math. J. 38 (1991), 241–253.

Remarks: (i) This was conjunctured by P. Chowla and by S. Chowla.

P. Chowla and S. Chowla. *On k-th power residues*. J. Number Theory 10 (1978), 351–353.

(ii) In fact, Hildebrand proves a more general theorem, namely, if $F_k = \{f : \mathbb{N} \to \mathbb{C} : f^k \equiv 1, f(nm) \equiv f(n)f(m)(n, m \in \mathbb{N})\}$ and k is a positive integer, then there exists a constant $c_0(k)$ such that for any function $f \in F_k$ there is a positive integer $n \leq c_0(k)$ with $f(n) = f(n + 1) = 1$

§ XV.12 Bounds on n-th power residues (mod p)

Let $p \equiv 1 \pmod k$ be a prime (k-positive integer) and denote by $n(p, k)$ the least integer n such that x^k and $(-x)^k (x = 1, 2, \ldots, n)$ yield all the non-zero k-th power residues $\pmod p$ (possible with repetitions). Then

$$\frac{1}{2}(p-1)/k \leq n(p, k) < \left(\frac{1}{2} - \frac{1}{2k}\right) \cdot p$$

S. Chowla and H. London. *Bounds on the n-th power residues* (mod p). Canad. Math. Bull. 12 (1969), 679–680.

§ XV.13 Positive d-th power residues $\leq x$, with $d|(p-1)$, which are prime to A

1) Let $N_d(x)$ denote the number of d-th power residues $\leq x$, a positive integer $< p$, with $d|(p-1)$ (p an odd prime.) Then

$$N_d(x) = \frac{x}{d} + O(\sqrt{p} \log p)$$

where the O is uniform in x, d, and p.

I.M. Vinogradov. *On the bound of the least non-residue of n-th powers*. Trans. Amer. Math. Soc. 29 (1927), 218–226.

2) For p an odd prime, d dividing $(p-1)$, and A a positive integer, let $N_d(x, A) =$ the number of positive d-th power residues mod p that are $\leq x$, and that are prime to A. Then

$$N_d(x, a) = \frac{\varphi(A)}{A} \cdot \frac{x}{d} + O(2^{\omega(A)} \cdot \sqrt{p} \log p)$$

where the O is uniform in x, d, A and p.

H.N. Shapiro. *Introduction to the theory of numbers*. John Wiley and Sons, 1983 (See p. 301).

§ XV.14 Distribution of r-th powers in a finite field

Denote by E the Galois field of order p^n, and by F the Galois field of order p.

a) If d divides $p^n - 1$ then there are at least

$$(1 - d^{-1}) p^{[n/2]+1} \cdot (1 + O(1/p))$$

d-th power non-residues of E which are polynomials of degree not exceeding $\frac{n}{2}$

C. Whyburn. *The distribution of r-th powers in a finite field.* J. für Mathematik 245 (1970), 183–187.

b) The number of elements of E of degree $< m$ in a given coset of the group of d-th power residues is

$$d^{-1}(p^m - 1) + O(P^{n/2}(\log p + 1)^n)$$

H. Davenport and D.J. Lewis. *Character sums and primitive roots in finite fields.* Rend. Circ. Math. Palermo (2) 12 (1963), 129–136.

c) There exist a $\delta > 0$ such that the number of d-th power residues (where d divides $p^n - 1$) of degree $< m$ is

$$(1 - d^{-1})(p^m - 1) + O(p^{m(1-\delta)})$$

D.A. Burgess. *A note on character sums over finite fields.* J. Reine Angew. Math. 255 (1972), 80–82.

§ XV.15 Pólya-Vinogradov inequality for quadratic characters

Let p be an odd prime, and N, H (> 0) integers. Then

a) $\left|\sum_{n=N}^{N+H} \left(\frac{n}{p}\right)\right| < p^{1/2} \log p$

G. Pólya. *Über die Verteilung der quadratischen Reste und Nichtreste.* Göttingen Nachrichten (1918), 21–29;

and

I.M. Vinogradov. *Sur la distribution des résidues et des non-résidues des puissances.* J. Physico-Math. Soc. Univ. Perm No. 1 (1918) 94–96.

Remark. For estimates on general Dirichlet characters, see the Chapter with Character sums.

b) Let δ and ε be any fixed positive numbers. Then, for all sufficiently large p and any N, we have

$$\left|\sum_{n=N}^{N+H} \left(\frac{n}{p}\right)\right| < \varepsilon H$$

provided $H > p^{1/4+\delta}$.

D.A. Burgess. *The distribution of quadratic residues and non-residues.* Mathematika 4 (1957), 106–112.

§ XV.16 Distribution questions concerning the Legendre symbol

1) Let p be an odd prime. We write $f(p)$ for the least positive integer z such that

$$\sum_{n\le z}\left(\frac{n}{p}\right)=0$$

a) $f(p) > \exp\left((\log p)^{\frac{1}{24}-\varepsilon}\right)$
for infinitely many p.

D. Wolke. *Eine Bemerkung über das Legendre-Symbol.* Monatsh. Math. 77 (1973), 267–275.

b) $f(p) > \exp\left((\log p)^{\frac{1}{2}-\varepsilon}\right)$ $(\varepsilon > 0)$
for infinitely many primes p.

R.C. Baker and G. Harman. *Unbalanced quadratic residues and non-residues.* Math. Proc. Camb. Phil. Soc. 98 (1985), 9–17.

2) Let $S_p(x) = \sum_{n\le px}\left(\frac{n}{p}\right)$ (p prime), and put
$\alpha(p) = \text{card}\,\{x \in [0,1] : S_p(x) > 0\}$. Then

a) For all $\delta > 0$ there exist infinitely many primes p such that

$$\alpha(p) < \frac{1}{3} + \delta$$

b) $\alpha(p) > \dfrac{1}{50}$
for all p.

H.L. Montgomery. *Distribution questions concerning a character sum.* Topics in number theory (Proc. Colloq. Debrecen, 1974), pp. 195–203. Colloq. Math. Soc. János Bolyai, vol. 13, North-Holland, 1976.

§ XV.17 A sum on $\left(\frac{n}{p}\right)\cdot n^k$

Let $S(k,p) = \sum_{n=1}^{p-1}\left(\frac{n}{p}\right)\cdot n^k$, where $\left(\frac{n}{p}\right)$ is the Legendre symbol. For $k > 2$, positive integer and $A > 1$, there exist $C, D > 0$ such that for all large x, there are at least $\exp(\log x - (\log\log x)^A)$ prime $p \equiv 3 (\text{mod}\, 4)$, $p \le x$ with $S(k,p) > C \cdot p^{k+1/2}$ and at least that many such that

$$S(k,P) < -D\cdot p^{k+1/2}$$

R.J. Cook. *A note on character sums.* J.Number Theory 11 (1979), 505–515.

Remark. Fine has proved earlier that, for $k > 2$ there exist infinitely many primes $p \equiv 3\,(\text{mod}\,4)$ with $S(k,p) > 0$; and infinitely many with $S(k,p) < 0$. Ayoub, Chowla and Walum have proved that $S(k,p) < 0$ for $k = 1, 2$ and for $k \ge p-2$.

R. Ayoub, S. Chowla and H. Walum. *On sums involving quadratic residues.* J. London Math. Soc. 42 (1967), 152–154;

N.J. Fine. *On a question of Ayoub, Chowla and Walum concerning character sums.* Illinois J. Math. 14 (1970), 88–90.

§ XV.18 An exponential polynomial formed with the Legendre symbol

Let $S_p(\alpha) = \sum_{n=1}^{p-1} \left(\frac{n}{p}\right) \cdot e(n\,\alpha)$, $\alpha \in \mathbb{R}$ (p prime.)

Then for all sufficiently large p one has

$$\max |S_p(\alpha)| \frac{2}{\pi} \sqrt{p} \log\log p$$

H.L. Montgomery. *An exponential polynomial formed with the Legendre symbol.* Acta Arith. 37 (1980), 375–380.

§ XV.19 A mean value of a quadratic character sum

Let p denote a prime, and $\left(\frac{n}{p}\right)$ be Legendre's symbol. Then

$$\sum_{2<p\leq p} \max_N \left| \sum_{n=1}^{N} \left(\frac{n}{p}\right) \right|^{2k} \ll_k \pi(P) \cdot P^{2k}$$

where $k > 0$ is a fixed real number.

H.L. Montgomery and R.C. Vaughan. *Mean values of character sums.* Canad. J. Math. 31 (1979), 476–487.

§ XV.20 Two sums involving Legendre's symbol with primes

1) Let q be an odd prime, k an integer not divisible by q. Then

$$\left| \sum_{p\leq x} \left(\frac{p+k}{q}\right) \right| \ll x^{1+\varepsilon} \cdot (q^{-1} + q\, x^{-1/3})^{1/2}$$

($\varepsilon > 0$)

I.M. Vinogradov. *On the distribution of quadratic rests and non-rests of the form $p+k$ to a prime modulus.* (Russian.) Rec. Math. Moscou, (2) 3 (1938), 311–319.

2) Let p, q be odd primes and $\left(\frac{p}{q}\right)$ the Legendre symbol. Then

$$\sum_{p\le x}\sum_{q\le x}\left(\frac{p}{q}\right) \ll x^{7/4}(\log x)^{-5/4}$$

H. Heilbronn. *On the averages of soms arithmetical functions of two variables.* Mathematatika 5 (1958), 1–7.

§ XV.21 Least primitive roots mod p. Least primitive roots mod p^2. Number of solutions of congruence $x^{n-1} \equiv 1 (\mathrm{mod}\, n)$ for n composite

1) Let $g(p)$ denote the least positive primitive root mod p (prime). Then

a) $g(p) \le 2^m \dfrac{p-1}{\varphi(p-1)} \cdot p^{1/2}$
where $m = \omega(p-1)$.

I.M. Vinogradov. *On the least primitive root of a prime.* Dokl. Akad. Nauk S.S.S.R. (1930), 7–11.

b) $g(p) < 2^{m+1} \cdot p^{1/2}$

L.K. Hua. *On the least primitive root of a prime.* Bull. Amer. Math. Soc. 48 (1942), 726–730.

c) $g(p) < p^{1/2}(\log p)^{17}$
for p sufficiently large.

P. Erdős. *Least primitive root of a prime.* Bull. Amer. Math. Soc. 55 (1945), 131–132.

d) $g(p) > \log\log p$
for infinitely many p.

S. Pillai. *On the smallest primitive root of a prime.* J. Indian. Math. Soc. 8 (1944), 14–17.

e) $g(p) = O(m^c \cdot p^{1/2})$
(c a constant.)

P. Erdős and H.N. Shapiro. *On the least primitive root of a prime.* Pacific J. Math. 7 (1957), 861–865.

f) $g(p) = O(p^{1/4+\varepsilon})$
for all $\varepsilon > 0$.

Y. Wang. *A note on the least primitive root of a prime.* Science Record, China (N.S.) 3 (1959), 174–179;

and

D.A. Burgess. *On character sums and primitive roots.* Proc. London Math. Soc. 12 (1962), 179–192.

g) Assuming the Riemann Hypothesis, one has

$$g(p) = O(m^6(\log p)^2)$$

Y. Wang. *On the least primitive root of a prime.* Acta Math. Sinica 10 (1961), 1–14.

h) $g(p) = \Omega(\log p)$

P. Turán. *30 years of mathematics in the Soviet Union.* III. Results of number-theory in the Soviet Union. Mat. Lapok 1 (1950), 243–266.

2) a) Let $h(p)$ denote the least primitive root modulo p^2. Then

$$\frac{1}{\pi(x)} \sum_{p \leq x} h(p) \ll (\log x)^3 (\log\log x)^6$$

(p prime)

D.A. Burgess. *The average of the least primitive root modulo p^2*. Acta Arith. 18 (1971), 263–271.

b) $h(p) < p^{1/4+\varepsilon}$, for all sufficiently large p; and a) holds with $(\log x)^2 (\log\log x)^6$ in place of $(\log x)^3 (\log\log x)^6$.

S.D. Cohen, R.W.K. Odoni and W.W. Stothers. *On the least primitive root modulo p^2*. Bull. London Math. Soc. 6 (1974), 42–46.

Remark. The weaker result $h(p) = O(p^{1/2+\varepsilon})$ was obtained by Burgess.

D.A. Burgess. *On character sums and primitive roots.* Proc. London Math. Soc. (3) 12 (1962), 179–192;
D.A. Burgess. *On character sums and L-series.* II. Proc. London. Math. Soc. (3) 13 (1963), 524–536.

3) $$\frac{1}{\pi(x)} \sum_{p \leq x} g(p) \ll (\log x)^2 (\log\log x)^4$$

(p prime.)

D.A. Burgess and P.D.T.A. Elliott. *The average of the least primitive root.* Mathematika 15 (1968), 39–50.

4) Let $o_m(a) = \min\{h \in \mathbb{N} : a^h \equiv 1 \pmod{m}\}$; $g(m) = \min\{a : 1 \leq a \leq m, (a,m) = 1$ and $o_m(a) \geq o_m(b)$ for all b with $1 \leq b \leq m, (b,m) = 1\}$; $h(m) = \min\{a : 1 \leq a \leq m, (a,m) = 1, x^2 \equiv a \pmod{m}$ is unsolvable $\}$. Then

a) $$\sum_{m \leq x} g(m) \ll x^{1+\varepsilon}$$

(all $\varepsilon > 0$)

b) $$\sum_{m \leq x, m \text{ odd}} h(m) \sim x$$

($x \to \infty$)

R. Warlimont. *Über die kleinste natürliche Zahl maximaler Ordnung* mod m. Monatsh. Math. 85 (1978), 253–258.

5) An integer t is said to possess weak order mod m ($m \geq 1$, fixed integer), if there exists a natural number n such that $t^{n+1} \equiv t \pmod{m}$. Let $k(m)$ be the number of incongruent elements which possess weak order (mod m). Let $x \geq 1$ and $K(x) = \sum_{m \leq x} k(m)$. Then

$$K(x) = \alpha x^2 + R(x)$$

where:

a) $R(x) = O(x \log^3 x)$
and $\alpha = \frac{1}{2}\zeta(2) \sum_{n=1}^{\infty} C(n)/n^2$
with $C(n) = \prod_{p|n} p/(p+1)$

V.S. Joshi. *Number theory.* (Mysore, 1981), 93–100. Lecture Notes in Math., 938, Springer, Berlin, 1982.

b) $R(x) = O(x \log^2 x)$

Z.H. Yang. *A note on order-free integers* (mod m). J. China Univ. Sci. Tech. 16 (1986), 116–118.

6) Let n be a composite number. Denote by $F(n)$ the number of solutions of the congruence $x^{n-1} \equiv 1 (\bmod n)$. (E.g. $F(15) = 4$.) Put $A(x) = \frac{1}{x} \sum_{\substack{n \le x \\ n \text{ composite}}} F(n)$.

Then

a) $A(x) > x^{15/23}, x \ge x_0$;

b) $A(x) \le x \exp(-(1 + O(1)) \log x \cdot \log\log\log x / \log\log x)$ as $x \to \infty$.

P. Erdős and C. Pomerance. *On the number of false witnesses for a composite number.* Number theory (New York, 1984–1985), 97–100. Lecture Notes in Math., 1240, Springer, Berlin-New York, 1987.

7) For d an odd natural number, let $l(d)$ denote the exponent to which 2 belongs modulo d i.e $2^{l(d)} \equiv 1 (\bmod d)$ and $2^k \not\equiv 1 \pmod d$ for all $1 \le k < l(d)$. Put $E(n) = \sum_{l(d)=n} 1/d$. Then

a) $\sum_{n \le x} E(n) \le (e^{\gamma} + o(1)) \log x$
where γ is Euler's constant.

P. Erdős. *On some problems of Bellman and a theorem of Romanoff.* J. Chinese Math. Soc. (N.S.) 1 (1951), 409–421.

b) $\sum_{n \le x} E(n) \le (e^{\gamma} - c_1 + o(1)) \log x$
for $c_1 > 0$ a positive constant.

C. Pomerance. *On primitive divisors of Mersenne numbers.* Acta Arith. 46 (1986), 355–367.

c) For infinitely many n, one has
$nE(n) \ge \exp(1 + o(1)\sqrt{\log\log n}),\ nE(n) \le c_2 (\log n)^{-17/24}$

d) There is a set of natural numbers S of logarithmic density 1 such that

$$\lim_{n \in S, n \to \infty} nE(n) = 0$$

(See C. Pomerance.)

Remark. For prime values of l at primes, Pomerance proves that

$$\operatorname{card}\{p \leq x : l(p) \text{ is prime}\} = O\left(\frac{x \cdot \log\log\log x}{\log x \cdot \log\log x}\right)$$

§ XV.22 Distribution of primitive roots of a prime

1) Let p be a prime, let $g_1, g_2, \ldots, g_\lambda$, where $\lambda = \varphi(p-1)$, denote the primitive roots of p. If a is an integer $(p \nmid a)$, then $\overline{a}$ will denote a solution of the congruence $a\,\overline{a} \equiv 1 (\mathrm{mod}\, p)$.

a) If $p = 4k + 1$ such that $\varphi(p-1)/(p-1) > \dfrac{1}{4}$ and b is a quadratic residue modulo p, then there is at least one primitive root of p among the integers

(*) $$g_1 + b, g_2 + b, \ldots, g_\lambda + b, g_1 + \overline{b}, g_2 + \overline{b}, \ldots, g_\lambda + \overline{b}$$

b) If $p = 4k + 3 > 3$ such that $\varphi(p-1)/(p-1) > \dfrac{1}{3}$ and b is an integer $(p \nmid b)$, then there is at least one primitive root of p among the integers (*).

E. Vegh. *A note on the distribution of the primitive roots of a prime.* J. Number Theory 3 (1971), 13–18.

c) Let p be an odd prime, b an integer $(p \nmid b)$ and $g_1, g_2, \ldots, g_\lambda$ be the primitive roots of p. Let $N(p, b)$ be the number of integers $g_i + b$ $(i = 1, 2, \ldots, \lambda)$ that are primitive roots of p. Then

$$\left| N(p,b) - (p-2)\left(\frac{\varphi(p-1)}{p-1}\right)^2 \right| < p^{1/2} \cdot 4^{\omega(p-1)} \left(\frac{\varphi(p-1)}{p-1}\right)^2$$

M. Szalay. *On the distribution of the primitive roots of a prime.* J. Number Theory 7 (1975), 184–188.

2) Let p be an odd prime and let $M(p, n)$ denote the number of sequences $x, x+1, \ldots, x+n-1$ from $1, 2, \ldots, p-1$ that consist of primitive roots mod p. Then

$$\left| M(p,n) - p\left(\frac{\varphi(p-1)}{p-1}\right)^n \right| < n + n \cdot p^{1-1/2n} \cdot d^n(p-1)\left(\frac{\varphi(p-1)}{p-1}\right)^n$$

M. Szalay. *On the distribution of primitive roots* (mod p). (Hungarian.) Math. Lapok 21 (1970), 357–362.

§ XV.23 Artin's conjecture on primitive roots

Let $N_a(x)$ denote the number of primes $p \leq x$ for which a is a primitive root mod p, where a is a given non-zero integer other than $1, -1$, or a perfect square.

1) a) If the Riemann Hypothesis holds for Dedekind zeta-function of each field $Q(\sqrt[q]{a})$ (where q runs over primes), then

$$N_a(x) = A(a)\frac{x}{\log x} + O\left(\frac{x \log\log x}{\log^2 x}\right)$$

where $A(a)$ is a certain constant depending on a.

C. Hooley. *On Artin's conjecture.* J. Reine Angew. Math. 226 (1967), 209–220.

Remarks: (i) The formula $N_a(x) \sim A(a)\frac{x}{\log x}(x \to \infty)$ (without any hypothesis) was conjectured by Artin in 1927. So, by (the unproved) Artin's conjecture there exist infinitely many primes p for which a is a primitive root mod p. See

E. Artin. *Collected Papers*, MA: Addison-Wesley 1965, pages VIII–X.

(ii) Hooley's theorem implies that if Artin's conjecture is false, then the generalized Riemann hypothesis is false.

b) For any distinct primes q, r, s at least one element in the set

$$\{qs^2, q^3r^2, q^2r, r^3s^2, r^2s, q^2s^3, qr^3, q^3rs^2, rs^3, q^2r^3s, q^3s, qr^2s^3, qrs\}$$

is a primitive root (mod p) for infinitely many primes p.

R. Gupta and M.R. Murty. *A remark on Artin's conjuncture.* Invent. Math. 78 (1984), 127–130.

c) Let E be the set of integers, which are not perfect squares, for which Artin's conjecture is false. Let $E(x)$ be the counting function of the set E. Then

$$E(x) = O(\log^6 x)$$

M.R. Murty and S. Srinivasan. *Some remarks on Artin's conjecture.* Canad. Math. Bull. 30 (1987), 80–85.

d) With the notations of c),

$$E(x) = O(\log^2 x)$$

D.R. Heath-Brown. *Artin's conjecture for primitive roots.* Quart. J. Math. Oxford (2) 37 (1986), 27–38.

Remark. Artin's conjecture provides us with a rich interplay of algebraic and analytic number theory. See

M.R. Murty. *Artin's conjecture for primitive roots.* Math. Intelligencer 10 (1988), 59–67.

2) a) $N_a(x) = c \cdot \operatorname{li} x + O(x/\log^D x)$ for each $D \geq 1$, for all integers $a \leq A \leq x$ with at most $C_1 \cdot A^{9/10} \cdot (5 \log x + 1)^{\log x/\log A + D + 2}$ exceptions. Here $c = \prod_p (1 - 1/p(p-1))$, and c_1 depends only on D, and $N_a(x)$ denotes the number of primes $p \leq x$ such that a is a primitive root mod p.

M. Golfeld. *Artin's conjecture on the average.* Mathematika 15 (1968), 223–226.

b) $N_a(x) - c\,\pi(x) = O(x \log^{-D} x)$ ($D > 1$) for all a with $1 \leq a \leq y$ and $\log^8 x \leq y \leq x^2$, with at most $O(x^{1/k} \cdot k^2 (\log x)^{2D+1/k^{-2}} \cdot T^{1-1/k})$ exceptions; where $k = [\log x^2 \cdot \log^{-1} y]$ and $T \max\limits_{1 \leq n \leq x^2} \sum\limits_{d|n} 1$.

R. Warlimont. *On Artin's conjecture.* J. London Math. Soc. (2) 5 (1972), 91–94.

§ XV.24 Number of primitive roots $\leq x$ which are $\equiv l (\operatorname{mod} k)$

a) Let prim(x) = the number of positive primitive roots modulo a fixed prime p that are $\leq x$. Then

$$\operatorname{prim}(x) = \frac{\varphi(p-1)}{(p-1)}(x + O(2^{\omega(p-1)} \cdot \sqrt{p} \log p))$$

where the O term is uniform in x and p.

b) Let prim(x, k, l) = the number of positive primitive roots modulo an odd prime p that are $\leq x$ and $\equiv l (\operatorname{mod} k)$. For $k > 0, l$, and p given, such that $(p, k) = 1$, we have

$$\operatorname{prim}(x, k, l) = \frac{\varphi(p-1)}{(p-1)} \left(\frac{x}{k} + O(2^{\omega(p-1)} \sqrt{p} \log p)\right)$$

H.N. Shapiro. *Introduction to the theory of numbers.* John Wiley and Sons, Inc., 1983, pp. 304–305.

§ XV.25 Number of squarefull (squarefree) primitive roots $\leq x$

a) For p an odd prime, the number of positive primitive roots $\leq x$ which are squarefull equals

$$\frac{\varphi(p-1)}{(p-1)} \left(c\sqrt{x} + O\left(\sqrt[3]{x}\, p^{1/6} (\log p)^{1/3} 2^{\omega(p-1)}\right)\right)$$

where $c = 2 \left(\sum\limits_{\substack{q \text{ squarefree} \\ \left(\frac{q}{p}\right) = -1}} 1/q^{3/2} \right) \left(1 - \frac{1}{p}\right)$

b) The number of positive squarefree primitive roots modulo p that are $\leq x$ equals

$$\frac{\varphi(p-1)}{(p-1)}(c_1 x + O(2^{\omega(p-1)} \cdot p^{1/4} \cdot (\log p)^{1/2} \cdot x^{1/2}))$$

where $c_1 = \prod_p (1 - 1/p^2)$

H.N. Shapiro. *Introduction to the theory of numbers.* John Wiley and Sons, Inc., 1983 (pp. 305 and 307–308).

§ XV.26 Number of integers in $[M+1, M+N]$ which are not primitive roots (mod p) for any $p \leq N^{1/2}$

Let $E(M, N)$ denote the number of integers in the interval $[M+1, M+N]$ which are not primitive roots (mod p) for any odd prime $p \leq N^{1/2}$. Then

$$E(M, N) \ll N^{1/2} \cdot (\log N)^{\alpha}$$

where $\alpha = 0.37\ldots$

R.C. Vaughan. *Some applications of Montgomery's sieve.* J. Number Theory 5 (1973), 64–79.

Remark. For $\alpha = 1$ this result is due to Gallagher.

P.X. Gallagher. *The large sieve.* Mathematika 14 (1967), 14–20.

§ XV.27 Least prime primitive roots

1) Let $G(p)$ denote the least prime primitive root modulo a prime p. Assuming that the Generalized Riemann Hypothesis is true, for any monotone increasing function f satisfying $\lim_{x \to \infty} f(x) = \infty$, there exists A such that $f(x) \ll (\log x)^A$; $f(x) \ll f(x/\log x)$, the estimate

$$G(p) \ll f(p)$$

holds for all primes $p < x$ with at most $O(\pi(x)/\log f(x))$ exceptions.

L. Murata. *On the magnitude of the least prime primitive root.* J. Number Theory 37 (1991), 47–66.

Remark. The method is a substantial extension of Hooley's proof (assuming GRH) on Artin's conjecture.

C. Hooley. *On Artin's conjecture.* J. Reine Angew. Math. 226 (1967), 209–220.

2) Assuming the Riemann Hypothesis for the Dedekind zeta-function of appropriate number fields, one has:

a) $\sum_{p \leq x} G(p) \ll (\log x) \cdot (\log\log x)^{1+\varepsilon}$

b) $\sum_{p \le x} G(p)^{\delta} \sim C_{\delta} \cdot \pi(x)$

for $0 < \delta < \frac{1}{2}$, where c_{δ} is a constant depending on δ.

L. Murata. *On the magnitude of the least primitive root.* Journées arithmétiques, Exp. Congr., Luminy / Fr. 1989, Astérisque 198–200, 253–257 (1991).

§ XV.28 Fibonacci primitive roots

Let $F(x) = \text{card}\{p \le x :$ there exists a primitive root $g(\text{mod}\, p)$ with $g^2 \equiv g + 1(\text{mod}\, p), p \text{ prime}\}$. Then

$$\frac{F(x)}{\pi(x)} = \frac{27}{38}A + O\left(\frac{\log\log x}{\log x}\right)$$

where $A = \prod_p (1 - 1/p(p-1))$, if we assume the truth of the Riemann Hypothesis for all fields $Q(\sqrt[n]{\rho}, \zeta)$ with $\rho = \frac{1+\sqrt{5}}{2}$ and primitive $2n$-th root of unity ζ.

J.W. Sander. *On Fibonacci primitive roots.* Fib. Quart. 28 (1990), 79–80.

Remark. The proof relies on the work

G. Göttsch. *Über die mittlere Periodenlänge der Fibonacci-Folgen modulo p.* (Dissertation), Hanover, 1982.

§ XV.29 Distribution of primitive roots in finite fields

Let p be a prime and θ a root of an irreducible polynomial of degree n over the finite field $GF(p)$, so that θ is an element of $GF(p^n)$. Denote by $N(P)$, $1 \le P < p$ the number of primitive roots of the field $GF(p^n)$ among the elements $\theta + t$, $t = 0, 1, \ldots, P-1$. For any $\varepsilon > 0$ and $P > p$ one has

$$N(P) = P\varphi(p^n - 1)/p^n + O\left(P^{1/2+\varepsilon}\right)$$

G.I. Perel'muter and I.E. Shparlinskij. *The distribution of primitive roots in finite fields.* Russ. Math. Surv. 45 (1990), 223–224; translation from Usp. Mat. Nauk 45 (1990), No. 1 (271), 185–186.

§ XV.30 Number of solutions to $f(x) \equiv 0(\text{mod}\, m)$ counted $\text{mod}\, m$

1) Let $f(x)$ be a polynomial with integer coefficients and let $N(f, m)$ be the number of solutions to the congruence $f(x) \equiv 0 \pmod m$ ($m > 1$, integer), counted modulo m, and including multiplicities. We define $c = c(f)$ as the smallest positive integer that is representable as $A(x)f(x) + B(x)f'(x) = c$,

where $A(x)$ and $B(x)$ are polynomials with integer coefficients. If $f(x)$ is a polynomial of degree $n \geq 2$ with no multiple roots, then

a) $N(f, m) \leq n^{\omega(m)} \cdot (c(f))^2$
where $\omega\ (m)$ denotes the number of distinct prime divisors of m.

O. Ore and T. Nagell (independently, in 1921). See T. Nagell. *Introduction to Number Theory*, New York: Wiley, 1951.

Remark. The following more general result also holds: Let $f_i(x), i = 1, 2, \ldots, s$, be polynomials with integer coefficients of degree n_i, $i = 1, 2, \ldots, s$, respectively, such that each $f_i(x)$ has no multiple roots. Then

$$N(f_1(x)\ldots f_s(x), m) \leq \left(\prod_{i=1}^{s} c^2(f_i)\right) \sum_{d_1\ldots d_s=m} n_1^{\omega(d_1)} \ldots n_s^{\omega(d_s)}$$

b) $N(f, m) \leq m^{\omega(m)} \cdot |c(f)|^{1/n}$

M.N. Huxley. *A note on polynomial congruences.* Recent progress in analytic number theory, vol. I (Durham, 1979), pp. 193–196, Academic Press, London 1981.

2) Let $f(x) = \sum_{k=0}^{n} a_k x^k$ be a polynomial with integral coefficients and with $(a_0, \ldots, a_n, m) = 1$ where $m > 1$ is a positive integer. Let $N(n, m) = \max N(f, m)$. Then

a) $n(n, m) \leq m^{1-1/n} \cdot d^{n-1}(m)$

E. Kamke. *Zur Arithmetik der Polynome.* Math. Z. 19 (1924), 247–264.

b) If $c_n = \sup m^{1/n-1} \cdot N(n, m)$, then

$$c_n = n/e + O(\log^2 n)$$

for $n \geq 2$

S.V. Konjagin. *Letter to the editors: "The number of solutions of congruences of the n-th degree with one unknown".* (Russian.) Mat. Sb. (N.S.) 110 (152) (1972), no. 1, 158.

c) $N(f, m) \leq m^{1-1/n} \cdot \exp\left(c\, n((\log)^{1/n} / \log\log m + 1)\right)$
where c is a constant.

S.B. Stečkin. *An estimate of a complete rational trigonometrical sum.* (Russian.) Trudy Mat. Inst. Steklov. 143 (1977), 188–207, 211.

3) Let $f(x)$ as in 2) and let $N_n(f, P, m)$ denote the number of solutions x in integers $0 \leq x \leq P - 1$ of the congruence $f(x) = 0 \pmod m$ $(m > 1)$. Then, for $\varepsilon > 0$, one has

$$N_n(f, P, m) \ll p^{\varepsilon}\left(P^{1-1/n-\theta_n} + P\, m^{-1/n}\right)$$

where $\theta_n = (n-1)/n(n^3 - n^2 + 1)$ and the constant implied in the symbol $\ll$ depends only on n and ε.

I.E. Shparlinskij. *On polynomial congruences.* (Russian.) Acta Arith. 58 (1991), 153–156.

§ XV.31 Estimates on Legendre symbols of polynomials

1) a) Let $\Phi(x)$ be a fixed polynomial with integer coefficients that is not a constant multiple of a perfect square. If all the roots of $\Phi(x)$ are rational, then, for each sufficiently large prime p there is a positive integer satisfying $x = O(p^{1/4} \cdot \log p)$ for which $\left(\frac{\Phi(x)}{p}\right) = -1$

D.A. Burgess. *On Dirichlet characters of polynomials.* Proc. London Math. Soc. (3) 13 (1963), 537–548.

Remark. It is a consequence of the work of Weil on the roots of the ζ-functions associated with function fields over finite constant fields that, (without the assumption that the roots of $\Phi(x)$ are rational) for any choice of the odd prime p and the positive integer N,

$$\sum_{n=0}^{N} \left(\frac{\Phi(x)}{p}\right) = O(p^{1/2} \cdot \log p)$$

Thus, for each sufficiently large prime p, if N is greater than a certain multiple of $\sqrt{p}\log p$, then there is a positive integer $x < N$ for which $\left(\frac{\Phi(x)}{p}\right) = -1$.

A. Weil. *Sur les courbes algébriques et les variétés qui s'en déduisent.* Actualités Math. Sci., No. 1041 (Paris, 1945), Deuxiéme Partie, IV.

b) Let $g_2(m, p)$ be the smallest positive integer x such that $\left(\frac{x^2 - m}{p}\right) = -1$, ($m$ integer, $p \nmid m$, p odd primc.) Then

$$g_2(m, p) \ll p^{1/2\sqrt{e}+\varepsilon}$$

J.B. Friedlander. *On characters and polynomials.* Acta Arith. 25 (1973/74), 31–37.

Remark. For similar results, see

D.A. Burgess. *On the quadratic character of a polynomial.* J. London Math. Soc. 42 (1967), 73–80.

2) a) Let $f(x) = \sum_{i=0}^{n} a_i x^i \in \mathbb{I}[x]$, where $(a_n, p) = 1$ with $n \geq 3$ an odd integer and p a prime. Put $|S| = \left|\sum_{x \leq p} \left(\frac{f(x)}{p}\right)\right|$. Then if $p \geq \frac{n^2+9}{2}$, then

$$|S| \leq (n-1) \cdot (p - (n-3)(n-4)/4)^{1/2}$$

N.M. Korobov. *An estimate of the sum of the Legendre symbols.* (Russian.) Dokl. Akad. Nauk SSSR 196 (1971), 764–767.

b) If n is even, and for the odd prime p one has $p \geq \dfrac{n^2 - 2n}{2}$, then

$$|S| \leq (n-2) \cdot (p + 1 - n(n-4)/4)^{1/2} + 1$$

provided that $(a_n, p) = 1$ and $f(x) \not\equiv a_n g^2(x) (\mathrm{mod}\, p)$ for any polynomial g.

D.A. Mit'kin. *Estimate of a sum of Legendre symbols of polynomial of even degree.* Math. Notes 14 (1973), 597–602; translation from Mat. Zametki 14 (1973), 73–81.

3) If q is a prime, and $f(x) = (x-a)(x-b)$ with a, b integers, $a \not\equiv b (\mathrm{mod}\, q)$, put $S_N = \sum_{p \leq N} \left(\dfrac{f(p)}{q}\right)$ (p prime). If $N > q^{0.75+\lambda}$, where $\lambda \in (0, 1/2)$, then

$$S_N \ll \pi(N) q^{-0.01\lambda^3}$$

A.A. Karacuba. *Distribution of the value of the Legendre symbol in polynomials with prime numbers.* (Russian.) Dokl. Akad. Nauk SSSR 238 (1978), 524–526.

§ XV.32 Number of solutions to $f(x) \equiv a(\mathrm{mod}\, p^b)$ (p prime)

1) If $f(x)$ is a polynomial of degree n modulo p, with integral coefficients where p is a prime, then the number of roots of $f(x) \mod p$ (i.e. the number of incongruent solutions of $f(x) \equiv 0 \ (\mathrm{mod}\, p)$), counted with their multiplicities, is at most n.

J.L. Lagrange. *Nouvelle Méthode pour résoudre les Problèmes Indéterminés en Nombres Entiers.* Oeuvres de Lagrange, II, Paris: Gauthiers-Villars, 1848 (original: 1770).

2) Let $f(x)$ be a polynomial of degree n and nonzero discriminant divisible by exactly p^δ. Let $N(f, p^\alpha)$ be the number of solutions of $f(x) \equiv O(\mathrm{mod}\, p^\alpha)$ Then

a) $N = N(f, p^\alpha) \leq n\, p^\delta$

b) If $\alpha > \delta$, then

$$N \leq n\, p^{\delta/2}$$

N being independent of α in this case.

Gy. Sándor. *Über die Anzahl der Lösungen einer Kongruenz.* Acta Math. 87 (1952), 13–16.

3) Let f be a polynomial of degree n with integral coefficients and p be a prime such that $f - f(0)$ is primitive modulo p. Let t be the integer ≥ 0 such that $p^{-t} f'$ is primitive modulo p. If $p^{-t} f'$ has $r \geq 1$ distinct roots modulo p with

multiplicities $m_1, \ldots, m_r$ put $m = m_1 + \cdots + m_r$ and $M = \max\{m_1, \ldots, m_r\}$. If the polynomial $p^{-t} f'$ has no roots modulo p, let $m = 1$ and let M be any number ≥ 1. For any positive integers a and b, let $N_a(f, p^b)$ be the number of solutions x modulo p^b of the congruence $f(x) \equiv a (\bmod p^b)$.

a) $N_a(f, p^b) < (2+\sqrt{2}) \cdot C$
where $C = n^3 \cdot p^{b(1-1/n)}$

L.K. Hua. *Additive theory of prime numbers*, English translation, Amer. Math. Soc., Providence, RI; 1965.

b) $N_a(f, p^b) < (2+\sqrt{2}) \cdot D$
where $D = m\, n\, p^{t/(m+1)+b\left(1-\frac{1}{m+1}\right)}$, for $b \geq 2$.

H.H.J. Chalk. *Quelques remarques sur les congruences polynômes modulo p^α*. C. R. Acad. Sci. Paris Sér. I. Math. 307 (1988), 513–515.

§ XV.33 Number of residue classes $k (\bmod r)$ with $f(k) \equiv 0 (\bmod r)$

1) Let $g(x)$ be a polynomial with integral coefficients which is irreducible over the rational number field. For each positive integer r let $\rho(r)$ denote the number of residue classes $k (\bmod r)$ for which $g(k) \equiv 0 (\bmod r)$.

a) $\sum_{p \leq x} \rho(p) = \frac{x}{\log x} + O(x/\log^2 x)$
(p prime)

b) $\sum_{p \leq x} \frac{\rho(p)}{p} = \log\log x + C + O(1/\log x)$
(C constant)

c) $\prod_{p \leq x} (1 + \rho(p)/p) = O(\log x)$

d) $\sum_{n \leq x} \rho(n) = A\,x + O\left(x^{1-\alpha}\right)$
($A > 0$, $\alpha > 0$ constants)

E. Landau. *Neuer Beweis der Primzahlsatzes und Beweis der Primidealsatzes.* Math. Annalen 56 (1903), 645–670;

E. Landau. *Einführung in die elementare und analytische Theorie der algebraischen Zahlen und der Ideale.* (Leipzig: Teubner 1927).

2) Let $g(x)$ be an irreducible primitive polynomial of degree $n > 1$ and discriminant D. Let $\rho(k)$ be defined as in 1). If q denotes a prime such that $\rho(q) = n$ and $q \nmid D$, then there exists a positive constant C_2 such that

$$\prod_{q \le x}\left(1 + \frac{1}{q}\right) > C_2(\log x)^{1/n!}$$

C. Hooley. *The distribution of the roots of polynomial congruences*. Mathematika 11 (1964), 39–49.

3) Let g be a polynomial as in 2), and consider the numbers of the form ν/k, where $g(\nu) \equiv 0 \pmod{k}$, $0 < \nu \le k$. Arrange these numbers as a sequence $s_1, s_2, \ldots, s_m, \ldots$. This sequence is uniformly distributed in the interval (0,1). (See C. Hooley.)

§ XV.34 Zeros of polynomials over finite fields

1) Let $F_i(x_1, \ldots, x_n)\,(i = 1, \ldots, k)$ be polynomials in $x_1, \ldots, x_n$ with integer coefficients, such that F_i is of degree $d_i \ge 1$, and $F_i(0, \ldots, 0)$. Then if $n > \sum_{k=1}^{d} d_i$, for any given prime p the system of congruences

$$F_i(x_i, \ldots, x_n) \equiv 0 \pmod{p},\ i = 1, \ldots, k$$

has a nontrivial solution for $x_1, \ldots, x_n$.

C. Chevalley. *Démonstration d'une hypothèse de M. Artin*, Hamburg: Universität, Mathematisches Seminar, Abhandlungen 11 (1936), 73–75.

Remark. For certain divisibility properties (and improvement of Chevalley's theorem) of the number of solutions of the above congruences, see

E. Warning. *Bemerkung zur vorstehenden Arbeit von Herrn Chevalley*, Hamburg: Universität, Mathematisches Seminar, Abhandlungen 11 (1936), 76–83;

and

J. Ax. *Zeros of polynomials over finite fields*. Amer. J. Math. 86 (1964), 255–261.

2) Let $F(x_1, \ldots, x_n)$ be an absolutely irreducible polynomial with integer coefficients (i.e. which can be decomposed in no extension of the field of rationals into nontrivial factors). Then the number $N(F, p)$ of solutions of the congruence

$$F(x_1, \ldots, x_n) \equiv 0 \pmod{p}\ (p \text{ prime})$$

satisfies

$$|N(F, p) - p^{n-1}| < c(F) \cdot p^{n-3/2}$$

where $c(F)$ is a constant depending only on F (and not on p).

S. Lang and A. Weil. *Number of points of varieties in finite fields*. Amer. J. Math. 76 (1954), 819–827;

and

L.B. Nisnevič. *On the number of points of an algebraic variety over a finite prime field*. (Russian.) Dokl. Akad. Nauk SSSR 99 (1954), 17–20.

Remarks: (i) The first important results for $n = 2$ were obtained by Weil.
A. Weil. *Sur les courbes algébriques et les variétés qui s'en déduisent.* Act. Sci. Ind. 1041, Paris, Hermann, 1948.

(ii) For particular polynomials much better estimates are known e.g. the number N of solutions of $a_1x_1^{r_1} + \cdots + a_nx_n^{r_n} = 0 \pmod p)\ (p \nmid a_i$ for all $i = 1, 2, \ldots, n)$ satisfies the inequality

$$|N - p^{n-1}| \le c \cdot (p-1) \cdot p^{n/2-1}$$

where $c = (d_1 - 1) \cdots (d_n - s)$ with $d_i = (r_i, p-1)$.
(See e.g. Z.I. Borevič and I.R. Šafarevič. *Number theory* (Russian), Moskva, 1964.)

3) a) Let T be the number of solutions, in integers, of the system of congruences

$$x_1 + \cdots + x_k \equiv \mu_1 (\operatorname{mod} p), \ldots, x_1^k + \cdots + x_k^k \equiv \mu_k (\operatorname{mod} p)$$

where $0 \le x_j \le M\,p^k - 1$ for $j = 1, 2, \ldots, k$; $M \ge 1$ integer; $\mu_1, \ldots, \mu_k$ given integers; p prime, $p > k$, and $x_j \not\equiv x_{j'} (\operatorname{mod} p)$ for $j \ne j'$. Then

$$T \le k!M^k \cdot p^{\frac{1}{2}k(k-1)}$$

A.A. Karacuba and N.M. Korobov. Doklady Akad. Nauk SSSR, 149 (1963), 245–248.

Remarks: (i) Karacuba and Korobov are influenced, to a certain extent, by earlier papers by Vinogradov, Linnik and Hua.

(ii) Estimations of the above type are used in obtaining various forms of the Vinogradov mean-value theorem. (See the Chapter with Exponential sums.)

b) Let $a_{jr_j} \ge 3$, $n_j \ge 3$, $s \ge 3$ be integers, $(a_{jr_j}, p) = 1$ and $3 \le n_1 < \cdots < n_k < p$ (p prime), where $j = 1, 2, \ldots, k$; $r_j = 1, 2, \ldots, \alpha_j$. Let $t = \alpha_1 + \cdots + \alpha_k$, $l_j = \min(n_j, s)$ and $N = t\left(\sum_{j=1}^{k} a_j/l_j\right)^{-1}$. Let $T_q(N)$ denote the number of solutions of the congruence $\sum_{j=1}^{k}\sum_{r_j=1}^{\alpha_j} a_{jr_j} x_{jr_j}^{n_j} \equiv y (\operatorname{mod} p^s)$, where $M_j \le x_j \le M_j + Q_j - 1$ with $M_1, \ldots, M_t$, $Q_1, \ldots, Q_t$ integers satisfying $0 \le M_j < M_j + Q_j \le p^s (j = 1, \ldots, t)$; and where $m \le y \le m + q - 1$ with m, q integers such that $0 \le m \le m + q \le p^s$. Then, for $t \ge N, \varepsilon > 0, 1 \le q \le p^s$, and $p^{s(1/2+1/l_j+\varepsilon)} \le Q_{jr_j} \le p^s, r_j = 1, \ldots, \alpha_j, j = 1, \ldots, k$, we have

$$T_q(N) = qQ_1 \cdots Q_t p^{-s} \cdot \left(1 + O\left(\gamma(q)\,\eta(t) \max\left(p^{-t/2+1}, p^{-l_1(t/N-1)}\right)\right)\right)$$

where $\gamma(q) = \begin{cases} 1 & \text{if } q = 1 \\ q^{-1} \cdot \log q & \text{if } q \geq 2 \end{cases}$ and $\eta(t) = \begin{cases} s & \text{if } t = N \\ 1 & \text{if } t > N \end{cases}$. The constant in the symbol O depends only on $n_1, \ldots, n_k, \alpha_1, \ldots, \alpha_k$ and ε.

B.G. Kocarev. *On the problem of an asymptotic formula for the number of solutions of a congruence of Waring type.* Soviet Math. Dokl. 11 (1970), No. 3, 758–762.

§ XV.35 Congruences on homogenous linear forms

1) Let $L_1, \ldots, L_n$ be n homogeneous liniar forms in x, y, z with integer coefficients. Let $q_1, \ldots, q_n$ be any positive integers and r_1, r_2, r_3 any positive number such that $r_1 r_2 r_3 \geq q_1 \cdots q_n$. Then the congruences

$$L_s \equiv 0 \pmod{q_s},\ s = 1, 2, \ldots, n$$

have a non-trivial solution, such that $0 \leq x < r_1, 0 \leq y < r_2, 0 \leq z < r_3$.

L.J. Mordell. *On the equation* $ax^2 + by^2 - cz^2 = 0$. Monatsh. Math. 55 (1951), 323–327.

Remark. Similar results hold for homogeneous liniar forms in many variables.

2) Let D be the area of the hyperplane $\sum_{i=1}^{r} a_i x_i = 0$ wich lies inside the cube $|x_i| \leq m^{1/r}$, where m and the a_i are integers. Let $A_r(m)$ be the number of solutions of $\sum_{i=1}^{r} a_i x_i = 0 \pmod{m}$ lying inside the cube. Then

$$A_r(m) = D/\left(\sum_{i=1}^{m} a_i^2\right)^{1/2} + O(m^{1-2/r})$$

L. Fjellstedt. *Einige Sätze über liniare Kongruenzen.* Ark. Math. 3 (1956), 271–274.

3) Let $q_s(p)$ be the minimum of $\prod_{0 \leq i \leq s} \max(1, |x_i|)$ taken over all nontrivial solutions of the congruence $x_0 + a_1 x_1 + \cdots + a_s x_s \equiv 0 \pmod{p}$ with $-\frac{1}{2}p < x_i \leq \frac{1}{2}p$, and let $Q_s(p)$ be the minimum of $\prod_{0 \leq i \leq s} |k_i|$ taken over all nontrivial solutions of the congruences $a_i k_0 \equiv k_i \pmod{p}$, $1 \leq i \leq s$, in the same range. If p is a power of a prime, then

$$q_s(p) \geq Q_s(p)^s/(2s+3)^{s+1} \cdot p^{s^2-1}$$

N.M. Korobov. *An estimate of A.O. Gelfond.* (Russian.) Vestnik Moskow Univ. Ser. I. Mat. Mekh. 1983, no. 3, 3–7.

§ XV.36 Waring's problem (mod p)

1) Let p be a prime, k a positive integer, $d = (k, p-1)$. Let $\theta(k, p)$ be thc least positive integer s such that the congruence $x_1^k + \cdots + x_s^k = 0 \pmod p$ has a nontrivial solution. Denote $\theta(k) = \max\{\theta(k, p) : p > 1 + 2k\}$.

a) $\theta(k) = O(k^{1-c+\varepsilon})$
$\varepsilon > 0$, where $c = (103 - 3\sqrt{641})/220$.

I. Chowla. *On Waring's problem* (mod p). Proc. Indian Nat. Acad. Sci. A 13 (1943), 195–220.

b) $\theta(k) = O(k^{2/3+\varepsilon})$

M. Dodson. *On Waring's problem in* GF [p]. Acta Arith. 19 (1971), 147–173.

c) $\theta(k) = O(k^{1/2+\varepsilon})$

A. Tietäväinen. *Proof of a conjecture of S. Chowla.* J. Number Theory 7 (1975), 353–356.

Remark. This was conjectured by S. Chowla.

S. Chowla. *Proceedings of the 1963 Number Theory Conference.* Univ. of Colorado. Boulder, Colorado, 1963.

2) Let $\gamma^*(k, p)$ be the least s such that every congruence $a_1x_1^k + \cdots + a_sx_s^k \equiv 0 \pmod p$, where $a_1, \ldots, a_s \in \mathbb{I}$, has a nontrivial solution.

a) If $d < p - 1$ then, for k sufficiently large,

$$\gamma^*(k, p) < k^{2/3+\varepsilon}$$

(See M. Dodson.)

b) $\gamma^*(k, p) < k^{1/2+\varepsilon}$

(See A. Tietäväinen.)

§ XV.37 Estimate of Mordell on congruences

a) The congruence

$$(*) \qquad f(x) = a_1x_1^{l_1} + a_2x_2^{l_2} + \cdots + a_nx_n^{l_n} \equiv 0 \pmod p$$

(p prime), $a_1a_2 \ldots a_n \not\equiv \pmod p$ has always a solution with $x_1 \ldots x_n \not\equiv 0 \pmod p$ if $(p-1)^{n-1} \cdot p^{-n/2} > l_1 \ldots l_n$

b) The congruence

$$a_1 x_1^{l_1} + \cdots + a_n x_n^{l_n} + a \equiv 0 \pmod p$$

(p prime) with $a\,a_1\,a_2 \ldots a_n \not\equiv 0 \pmod p$ has always a solution with $x_1 \ldots x_n \not\equiv 0 (\mathrm{mod}\, p)$ if $(p-1)^{n-1/2} \cdot p^{-n/2} > l_1 \ldots l_n$.

L.J. Mordell. *Diophantine equations.* Academic Press, 1969 (pp. 39–40).

Remark. Let $N(a) = N(a_1, \ldots, a_n)$ be the number of solutions of (*) with $x_1 \ldots x_n \not\equiv 0 \pmod p$. Then, Mordell proves, in fact, that

$$\sum_{(a)} |pN(a) - (p-1)^n|^2 \leq p^n \cdot (p-1)^{2+n} \cdot l_1 \ldots l_n$$

§ XV.38 Distribution of solutions of congruences

Let $f(X_1, \ldots, X_n) \in \mathbb{I}[X_1, \ldots, X_n]$ be a homogeneous polynomial of degree $d \geq 2$ in the $n \geq 2$ indeterminates $X_1, \ldots, X_n$ and let p be a prime satisfying $p \geq (20\,d)^n$. Let $N_p(f)$ denote the number of solutions $x \in \mathbb{I}^n$ of the congruence

$$(*) \qquad f(x) \equiv 0 \pmod p$$

in the cube $R^n(p) = \{\underline{x} = (x_1, \ldots, x_n) \in \mathbb{R}^n : 0 \leq x_i < p, i = 1, 2, \ldots, n\}$. If $N_p(f) \geq \frac{1}{2} p^{n-1}$ and f is absolutley irreducible (mod p), then every subcube

$$S(i_1, \ldots, i_n) = \{x \in \mathbb{R}^n(p) : i_j \mu \leq x_j \leq (i_j + 1)\mu, j = 1, 2, \ldots, n\}$$

$i_1, \ldots, i_n = 0, 1, 2, \ldots, \lambda - 1$, where $\lambda = \left[\dfrac{p^{1/n}}{10\,d}\right]$ and $\mu = \dfrac{P}{\lambda}$, contains a solution $\underline{x} \in \mathbb{I}^n$ of (*).

K.S. Williams. *A distribution property of the solutions of a congruence modulo a large prime.* J. Number Theory 3 (1971), 19–32.

Remark. The proof is based on a metod of Tietäväinen. For results on the distribution of the solutions of (*) in the fundamental cube $R^n(p)$, see also Vinogradov, Mordell, Chalk and Williams, Smith, Tietäväinen and Williams.

I.M. Vinogradov. Elements of number theory. Chap. 5, p. 103, Dover, New York, 1954;

L.J. Mordell. *On the number of solutions in incomplete residue systems for congruences.* Czechoslovak Math. J. 14 (1964), 235–242;

J.H.H. Chalk and K.S. Williams. *The distribution of solutions of congruences.* Mathematika 12 (1965), 176–192;

R.A. Smith. *The circle problem in an arithmetic progression.* Canad. Math. Bull. 11 (1968), 175–184;

A. Tietävänien. On the solvability of equations in incomplete finite fields. Ann. Univ. Turku. Ser. A1, 102 (1967), 3–13;

K.S. Williams. Small solutions of the congruence $ax^2 + by^2 \equiv c (\mathrm{mod}\, k)$. Canad. Math. Bull. 21 (1969), 311–320.

§ XV.39 On a set of congruences related to character sums

Let d, r and h be positive integers. Let X denote an indeterminate. For each $\underline{m} \in \mathbb{I}^{2r}$, with each component satisfying $0 < m_i \leq h$, write

$$f_1(X) = \prod_{i=1}^{r}(X + m_i),\ f_2(X) = \prod_{i=r+1}^{2r}(X + m_i),\ F(X) = f_1'(X)f_2(X) - f_2'(X)f_1(X)$$

a) For each prime $p > r$ and each h satisfying $0 < h \leq p$ we have

$$\text{card}\,\{\underline{m} : 0 < m_i \leq h, \text{ there is } x \text{ such that } p \nmid f_1(x)f_2(x),$$
$$0 \equiv F(x) \equiv F'(x) \equiv \ldots \equiv F^{(d)}(x) \not\equiv F^{(d+1)}(x) (\text{mod}\, p)\} \ll h^{2r-d}$$

D.A. Burgess. *On a set of congruences related to character sums.* J. London Math. Soc. (2) 37 (1988), 385–394.

b) Let r and d be positive integers. Let p be a prime with $p > r$. Let h be an integer satisfying $0 < h \leq p/(2r)$. Then

$$\text{card}\,\{\underline{m} : 0 < m_i \leq h, m_1 + \cdots + m_r = m_{r+1} + \cdots + m_{2r}$$

there is x such that

$$p \nmid f_1(x)f_2(x), 0 \equiv F(x) \equiv F'(x) \equiv \ldots \equiv$$
$$\equiv F^{(d)}(x) \not\equiv F^{(d+1)}(x) (\text{mod}\, p)\} \ll h^{2r-1-d}$$

D.A. Burgess. *On a set of congruences related to character sums.* II. Bull. London Math. Soc. 22 (1990), 333–338.

§ XV.40 Small zeros of quadratic congruences mod p

Let $Q(x_1, \ldots, x_n)$ be an integral quadratic form. Let $\|x\| = \max |x_i|$. Then:

a) The congruence $Q = 0\,(\text{mod}\, q)$ has nonzero solution with

$$\|x\| \leq q^{1/2+1/2n}$$

for n odd;

$$\|x\| \leq q^{1/2+1/2(n-1)}$$

for n even.

A. Schinzel, H.P. Schlickewei and W.M. Schmidt. *Small solutions of quadratic congruences and small fractional parts of quadratic forms.* Acta Arith. 37 (1980), 241–248.

b) For $n \geq 4$, and q prime,

$$\|x\| \ll q^{1/2} \log q$$

guarantees the existence of nonzero solutions.

D.R. Heath-Brown. *Small solutions of quadratic congruences.* Glasgow Math. J. 27 (1985), 87–93.

Remark. If $B_n(q)$ is a bound for the smallest nonzero solution, then

$$B_3(p) \geq p^{2/3} + O(p^{1/3})$$

(See D.R. Heath-Brown.)

c) The factor $\log q$ can be removed in b).

T. Cochrane. *Small zeros of quadratic forms modulo p*. III. J. Number theory 37 (1991), 92–99.

Remark. For a generalization of Cochrane's result for arbitrary finite fields see

Y. Wang. *On small zeros of quadratic forms over finite fields*. II. Acta Math. Sinica, New Ser. 9 (1993), 382–389.

See also

Y. Wang. *Small solutions of congruences*. J. Number Theory 45 (1993), 261–280.

d) $\|x\| \ll q^{1/2+\varepsilon}$

suffices whenever $n \geq 4$ and q has at most two prime factors.

D.R. Heath-Brown. Mathematika (to appear.)

Remark. For similar, but weaker results, see

T. Cochrane. *Small zeros of quadratic congruences modulo p q*. Mathematika 37 (1990), 261–272.

§ XV.41 Congruence-preserving arithmetical functions

An integer-valued function $f(n)$ is said to be congruence-preserving if, for all natural numbers a, b, and m, the condition $a \equiv b \pmod m$ implies that $f(a) \equiv f(b) \pmod m$. If $f(n)$ is congruence-preserving and not a polynomial then:

a) for infinitely many n and every $\varepsilon > 0$,

$$|f(n)| > (c-1)^{n(1-\varepsilon)}$$

$(c > 1)$

b) $f(n)/n^k \to \infty$

as $n \to \infty$, k fixed.

I. Ruzsa. *On congruence-preserving functions*. (Hungarian.) Mat. Lapok 22 (1971), 125–134.

Remark. For similar results, see also

R.R. Hall. *On pseudo-polynomials*. Mathematika 18 (1971), 71–77.

§ XV.42 On a congruence of Mirimanoff type

Assuming the generalized Riemann Hypothesis, there exists a positive constant c such that for any sufficiently large prime p there is a prime $q < c\,(\log p)^2$ satisfying

$$q^{p-1} \not\equiv 1 \pmod{p^2}$$

S.G. Hahn. *On Mirimanoff type congruences*. J. Number Theory 41 (1992), 167–171.

Chapter XVI

ADDITIVE AND MULTIPLICATIVE FUNCTIONS

§ XVI. 1 Erdős' theorem on additive functions with difference tending to zero, generalizations, extensions and related results

Let $f(n)$ be a real valued additive function. Denote $\Delta f(n) = f(n+1) - f(n)$, $\Delta^k f(n) = \Delta(\Delta^{k-1} f(n)), k \geq 2, \Delta^1 f(n) = \Delta f(n)$.

1) a) If $\Delta f(n) \geq 0$, for all n, or $\Delta f(n) \to 0 (n \to \infty)$, then $f(n)$ is a constant multiple of $\log n$

P. Erdős. *On the distribution function of additive functions*. Ann. Math. 47 (1946), 1–20.

b) If liminf $\Delta^k f(n) \geq 0$ with some $k \in \{1, 2, \ldots\}$, then $f(n)$ is a constant multiple of $\log n$

I. Kátai. *A remark on additive arithmetical functions*. Annales Univ. Sci. Budapest, Sectio Math. 10 (1967), 81–83.

c) If $\Delta f(n) \geq -K$ with some constant K, then f can be written as $f(n) = c \log n + u(n)$, where $u(n)$ is bounded and c is a suitable constant.

E. Wirsing. *A characterization of* $\log n$ *as an additive arithmetic function*, Symposia Mathematica, IV (1970), Instituto Nazionale de Alta Mathematica; pp. 45–57, Academic Press, London.

Remark. This was conjectured by Erdős.

d) If f and g are additive functions and $g(n+1) - f(n) \to 0$ as $n \to \infty$, then

$$f(n) = g(n) = c \log n$$

e) If f, g are additive and $g(n+1) - f(n)$ is bounded, then there exist bounded additive functions u, v and a suitable c such that

$$f(n) = c \log n + u(n)$$
$$g(n) = c \log n + v(n)$$

I. Kátai. *On additive functions*. Publ. Math. Debrecen 25 (1978), 251–257.

I. Kátai. *Characterization of* $\log n$. Studies in Pure Mathematics (to the memory of Paul Turán), Budapest, 1984, 415–421.

f) If $f(n+1) - f(n) \to 0 (n \to \infty)$, through a set of density one, then

$$f(n) = c \log n$$

A. Hildebrand. *An Erdős-Winter theorem for difference of additive functions*. Trans. Amer. Math. Soc. 310 (1988), 257–276.

2) a) If (f being additive)

$$\frac{1}{x} \sum_{n \le x} |\Delta f(n)| \to 0$$

then

$$f(n) = c \log n$$

I. Kátai. *On a problem of P. Erdős*. J. Number Theory 2 (1970), 1–6.

b) Assume that there exists a constant $\gamma > 1$ and a sequence $x_1 < x_2 < \cdots$ such that

$$\frac{1}{x_i} \sum_{x_i < n \le \gamma x_i} |\Delta f(n)| \to 0$$

$(i \to \infty)$ Then

$$f(n) = c \log n$$

E. Wirsing. *Characterization of the logarithm as an additive function*. Proc. 1969 Summer Inst. Number Theory, prepared by Amer. Math. Soc. (1971), 375–381.

c) If f, g are additive and

$$\frac{1}{x} \sum_{n \le x} |g(n+1) - f(n)| \to 0$$

$(x \to \infty)$, then

$$f(n) = g(n) = c \log n$$

(See I. Kátai (1984).)

d) For a sequence s_n and a polynomial $P(x) = \sum_{j=0}^{r} c_j x^j$, define

$$P(E)s_n = \sum_{0}^{r} c_j s_{n+j}$$

If $P(x) \in \mathbb{R}[x]$ and f additive satisfying

$$\frac{1}{x} \sum_{n \le x} |P(E) f(n)| \to 0$$

$(x \to \infty)$, then

$$f(n) = c \log n + u(n)$$

where $P(E)u(n) = 0 (n = 1, 2, \ldots)$
If $P(1) \neq 0$, then

$$c = 0$$

Furthermore u is of finite support (i.e. it vanishes on the set of prime powers except possibly on the powers of finitely many primes.)

P.D.T.A. Elliott. *On sums of an additive arithmetic function with shifted arguments*, J. London Math. Soc. (2) 22 (1980), 25–38

and

I. Kátai. *Characterization of* log *n*. Studies in Pure Mathematics (to the memory of P. Turán), Budapest, 1984, 415–421.

Remark. For similar results on additive functions with values in a compact group, see:

Z. Daróczy and I. Kátai. *On additive functions taking values from a compact group.* Acta Sci. Math. 53 (1989), 59–65.

See also

I. Kátai. *Characterization of arithmetical functions, problems and results.* Théorie des nombres (Québec, PQ, 1987), 544–555, de Gruyter, Berlin-New York, 1989.

3) a) Let $f : \mathbb{N} \to \mathbb{R}$ be additive, and let h be a positive and nondecreasing function such that

$$h(x^2) \le 2^{\alpha} h(x)$$

where $\alpha = 6/5$, and $\Delta f(n) \le h(n)$, $f(2) \ge 0$ for all x and n. Then there is an effective constant C such that

$$\left| \frac{f(m)}{\log m} - \frac{f(n)}{\log n} \right| \le C \left(\frac{h(m)}{\log m} + \frac{h(n)}{\log n} \right)$$

for all m, n with $2 \le m \le n \le e^m$

E. Wirsing. *Additive and completely additive functions with restricted growth.* Recent progress in analytic number theory, vol. 2 (Durham, 1979), pp. 231–280, London 1981.

b) If $|f(n+1) - f(n)| \le \log^a n$ for $n \le x^{3/2}$, $a \ge 3$, then

$$|f(n)| \le C(a) \log^a n$$

for $n \le x$

(See E. Wirsing.)

c) Put $u^+(x) = \max_{n \le x}(f(n+1) - f(n))$, $u^-(x) = \max_{n \le x}(f(n) - f(n+1))$

With a suitable absolute constant C we have

$$u^-(x) = C(u^+(2x^2) + |f(2)|)$$

I.Z. Ruzsa. *Additive functions with bounded difference.* Periodica Math. Hung. 10 (1979), 67–70.

d) There exists an absolute constant C such that the inequality

$$\inf_{\lambda \in \mathbb{R}} \max_{n \le x} |f(n) - \lambda \log n| \le C \max_{n \le x^C} |f(n+1) - f(n)|$$

holds for all $x \ge 1$

P.D.T.A. Elliott. *Arithmetic Functions and Integer Products.*
Springer, New York 1985 (Theorem 14.1).

Corollary. For any additive function $f : \mathbb{N} \to \mathbb{R}$ and any $\alpha \ge 1$ the estimate $f(n+1) - f(n) = O((\log n)^{\alpha})$ implies $f(n) = O((\log n)^{\alpha})$

Remark. The proof of Elliott's theorem is extremely complicated and takes up a large part of the above mentioned book.

4) Let $f : \mathbb{N} \to \mathbb{C}$ be additive.

a) If there exists $L \in \mathbb{C}$ such that $\lim_{n\to\infty}(f(2n+1) - f(n)) = L$, then

$$f(n) = L \log n / \log 2$$

J.-L. Mauclaire. *Sur la régularité des fonctions additives.* Enseignement Math. (2) 18 (1972), 167–174.

b) If there is an $M \in \mathbb{R}$ with $|f(2n+1) - f(n)| \le M$ for all $n = 1, 2, \ldots$ then

$$f(n) = C \log n + g(n)$$

where C is a constant and g is a bounded additive function.

J.-L. Mauclaire. Ibid. C. R. Acad. Sci. Paris Sér. A–B 276 (1973), A431–A433; and Sémin. Delange-Pisot-Poitou (1973/74), Fasc. 1, Exp. No. 23, 4 pp., Paris, 1975.

c) Let f be an additive function, for which the limit $\lim_{n\to\infty}(f(an+b) - f(cn+d))$ exists, where a, b, c, d are positive integers satisfying $ad - bc \neq 0$. Then

$$f(n) = \lambda \log n$$

for some λ and $(n, (ad - bc)ac) = 1$

P.D.T.A. Elliott. *Arithmetic functions and integer products.* Springer, New York 1985.

Remark. This solves a conjecture of Kátai.

§ XVI. 2 Completely additive functions with restricted growth

Let $f(n)$ be a completely additive real valued function.

a) If $\Delta f(n) = o(\log n)$, then

$$f(n) = c \log n$$

(c-constant.)

E. Wirsing. *Additive and completely additive functions with restricted growth.* Recent Progress in Analytic Number Theory, vol. 2, London (1981), 231–280.

b) If, f, g are completely additive and $g(n+1) - f(n) = o(\log n)$, then

$$f(n) = g(n) = c \log n$$

I. Kátai. *Characterization of* $\log n$. Studies in Pure Mathematics (to the memory of P. Turán), Budapest, 1984, 415–421, Akadémiai Kiadó.

c) If $f(n)$ is monotonic in infinitely many intervals of type $[K, K + (2+\varepsilon)\sqrt{K}]$, then

$$f(n) = c \log n$$

A. Iványi and I. Kátai. *On monotonic additive functions satisfying linear recursion in short intervals.* Ann. Univ. Sci. Budapest, Eötvös Sect. Math. 31 (1988), 135–149.

§ XVI. 3 Turán-Kubilius inequality

1) Let f be a complex-valued additive arithmetical function. Then:

a) $\sum_{n \le x} |f(n) - A(x)|^2 \le c \cdot x \cdot B^2(x)$

where $A(x) = \frac{1}{x} \sum_{n \le x} f(n)$, $B(x) = \left(\sum_{p^k \le x} |f(p^k)|^2 \cdot p^{-k} \right)^{1/2}$, with $x \ge 1$

and c an absolute constant (Turán-Kubilius inequality.)

P. Turán. *Über einige Verallgemeinerungen eines Satzes von Hardy und Ramanujan.* J. London Math. Soc. 11 (1936), 125–133;

J. Kubilius. *Probabilistic methods in the theory of numbers.* (Russian), Vilnius, 1959, English edition: AMS, Providence 1964.

Remark. The same inequality holds when $A(x)$ is replaced by $A_1(x) = \sum_{p^k \le x} \left(1 - \frac{1}{p}\right) f(p^k) p^{-k}$ or $A_2(x) = \sum_{p \le x} f(p)/p$ (where p denotes a prime). See

P.D.T.A. Elliott. *Probabilistic number theory* I. Springer Verlag, New York, 1979.

b) $\frac{1}{x} \sum_{n \le x} |f(n) - D(x)|^\beta \le c_1 (B(x))^\beta + c_2 \sum_{p^k \le x} p^{-k} |f(p^k)|^\beta$

for all $\beta \ge 0$, where $D(x) = \sum_{p^k \le x} p^{-k} \cdot f(p^k)$, $B(x)$ as defined as in a), and c_1, c_2 are constants depending at most upon β. If $0 \le \beta \le 2$, the second term on the right-hand side can be omitted.

P.D.T.A. Elliott. *High-power analogues of the Turán-Kubilius inequality, and an application to number theory.* Canad. J. Math. 32 (1980), 893–907.

c) Put $B(x) \equiv B(f, x)$ in a). There exist positive constants c_1 and c_2 such that

$$c_2 \min_{\lambda \in \mathbb{C}} \{B^2(f - \lambda \log, x) + |\lambda|^2\} \le \frac{1}{x} \sum_{n \le x} |f(n) - A(x)|^2 \le$$
$$\le c_1 \min_{\lambda \in \mathbb{C}} \{B^2(f - \lambda \log, x) + |\lambda|^2\}$$

for all $x \ge 1$

I.Z. Ruzsa. *On the variance of an additive function.* In: Studies in Pure Mathematics (to the memory of P. Turán), Editor: P. Erdős, Akadémiai Kiadó, Budapest 1983.

d) Let $f_s(m) = \sum_{p^k \le s, p^k \| m} f(p^k)$ and use the notations of c).

If $\sum_{p^k \le n} B^2(p^k)/p^k = O(B^2(n))$ or $\sum_{p^k \le n} |f(p^k)|p^k = O(B(n))$, then

$$\sum_{m \le n} \max_{s \le n} |f_s(m) - D(s)|^2 \le cnB^2(n)$$

J.E. Collisohn. *An analogue of Kolmogorov's inequalities for a class of additive arithmetic functions.* Pacific J. Math. 97 (1981), 319–325.

e) With the notations of a), if F is nonnegative, increasing function which satisfies $F(2x) \le cF(x)$(c-constant), then we have

$$\frac{1}{x}\sum_{n \le x} F(|f(n) - A|) \le c'\left(F(B(x)) + \sum_{p^k \le x} p^{-k} F(|f(p^k)|)\right)$$

where c' depends on c and A is any of $A(x), A_1(x), A_2(x)$. (See Remark.)

I.Z. Ruzsa. *Generalized moments of additive functions.* J. Number Theory 18 (1984), 27–33.

Remark. For $F(x) = x^\beta$ this gives the result b).

2) Let $\sup_{x \ge 1} C_x$ be the optimal constant in 1) a) (where $C_x = \sup_f \sum_{n \le x} |f(n) - A(x)|/xB^2(x)$ with $B(x) \ne 0$). Then:

a) with $C = \sup_{x \ge 1} C_x$ one has

$$1.47 \le C \le 2.08$$

J. Kubilius. *On an inequality for additive arithmetic functions.* Acta Arith. 27 (1975), 371–383.

b) $C \le 2$

P.D.T.A. Elliott. *The Turán-Kubilius inequality.* Proc. Amer. Math. Soc. 65 (1977), 8–10.

Remark. According to Elliott, Kubilius can prove $c < 1.764$

c) $\lim_{x \to \infty} C_x = 1.5$

H.L. Montgomery. See J. Kubilius. *Estimating the second central moment for arbitrary additive functions.* Litov. Mat. Sb. 23 (1983), (2), 110–117.

Remark. This was a conjecture of Kubilius. Numerical calculations carried out by him show that $C_x < 1.5$ for small values of x, and thus suggests the conjecture that $C \le 1.5$

d) Let $\tau_x = \sup_f \frac{1}{xB^2(x)} \sum_{n \le 2} |f(n) - A_1(x)|$, where f is additive. Then

$$\tau(x) = \frac{3}{2} + O(\log^{-1/2} x)$$

J. Kubilius. Litov. Mat. Sb. 23 (1983), 122–133 and ibid. 110–117 (No. 2).

e) $\frac{3}{2} - a/\log x \le \tau_x \le \frac{3}{2} - b/\log x$
for sufficiently large x, with positive absolute constants a and b $(b \le 0.644\ldots)$

J. Lee. *The second central moment of additive functions.* Proc. Amer. Math. Soc. 114 (1992), 887–895.

3) Let $f(n)$ be a real-valued additive function and let $B^2(y) - B^2(y^{2/3}) \le 10^{-4} B^2(y)$ hold for $y \ge y_0(f)$. Let $\sum\sum_{\substack{p^k \le y\\ k \ge 2}} p^{-k} f^2(p^k) \le 10^{-5} B^2(y)$ hold for $y \ge y_1(f)$. Then, there is an absolute constant C so that the inequality

$$\sum_{n \le x} |f(n) - \lambda|^2 \ge 10^{-2} \cdot x \cdot B^2(x)$$

D. Wolke. *Das Selbergsche Sieb für zahlentheoretische Funktionen* I. Arch. Math. 24 (1973), 632–639.

§ XVI. 4 Erdős-Kac theorem

1) Let $f(n)$ be a strongly additive function. Put $A(n) = \sum_{p \le n} \frac{f(p)}{p}$ and suppose that $B(n) = \left(\sum_{p \le n} \frac{f(p)^2}{p}\right)^{1/2} \to \infty$ as $n \to \infty$. Then

$$v_n\left(m; \frac{f(m) - A(n)}{B(n)} \le z\right) \Rightarrow \frac{1}{\sqrt{2\pi}} \int_{-\infty}^{z} e^{-t^2/2} dt = \Phi(z)$$

$(n \to \infty)$ (where v_n denotes the frequency and "$\Rightarrow$" weak convergence.) (Erdős-Kac theorem.)

P. Erdős and M. Kac. *On the Gaussian law of errors in the theory of additive functions.* Proc. Nat. Acad. Sci. USA 25 (1939), 206–207;

P. Erdős and M. Kac. *The Gaussian law of errors in the theory of additive number-theoretic functions.* Amer. J. Math. 62 (1940), 738–742.

2) Let $f : \mathbb{N} \to R$ be additive, not identically zero, and suppose that

(*) $$\sum_{\substack{p \le n\\ |f(p)| > \varepsilon B'(n)}} f^2(p)/p = o(B'^2)$$

$(n \to \infty)$ for every $\varepsilon > 0$. Then

$$v_n\left(m; \frac{f(m)-A'(n)}{B'(n)} \leq z\right) \Rightarrow \Phi(z)$$

$(z \in \mathbb{R})$, where $B'(n) = \left(\sum_{p^k \leq n} (f(p^k))^2/p^k\right)^{1/2}$ and $A'(n) = \sum_{p^k \leq n} f(p^k)/p^k$

J. Kubilius. *Probabilistic methods in the theory of numbers.* AMS Providence, 1964;

and

H.N. Shapiro. *Distribution functions of additive arithmetic functions.* Proc. Nat. Acad. Sci. USA 42 (1956), 426–430.

Remark. It has been conjectured by Shapiro, but still not proved, that the given condition $(*)$ is actually necessary for the result to hold.

§ XVI. 5 Erdős-Wintner theorem

1) Let $f(n)$ be an additive function.

a) In order that $f(n)$ should possess a limiting distribution, it is both necessary and sufficient that the three series

$$\sum_{|f(p)|>1} \frac{1}{p}, \sum_{|f(p)|\leq 1} \frac{f(p)}{p}, \sum_{|f(p)|\leq 1} \frac{f^2(p)}{p}$$

converge. When this condition is satisfied, the characteristic function, $v(t)$, of the limiting distribution, has the representation

$$v(t) = \prod_p \left(1-\frac{1}{p}\right)\left(1+\sum_{k=1}^{\infty} p^{-k}\exp\left(itf(p)^k\right)\right)$$

The limiting distribution is then of pure type and will be continous if and only if the series $\sum_{f(p)\neq 0} \frac{1}{p}$ diverges. (Erdős-Wintner theorem.)

P. Erdős and A. Wintner. *Additive arithmetical functions and statistical independence.* Amer. J. Math. 61 (1939), 713–721.

Note. The "sufficiency" part of the theorem was previously discovered by Erdős in 1938.

P. Erdős. *On the density of some sequences of numbers* III. J. London Math. Soc. 13 (1938), 119–127.

b) Let the series $\sum_{|f(p)|>1} \frac{1}{p}, \sum_{|f(p)|\leq 1} \frac{f^2(p)}{p}$ converge. For each integer n define $A(n) = \sum_{p\leq n, |f(p)|\leq 1} \frac{f(p)}{p}$. Then the frequencies

$$v_n(m; f(m) - A(n) < z)$$

$(n = 1, 2, \ldots)$ converge weakly. The characteristic function, $\psi(t)$, of the limiting distribution has the representation

$$\psi(t) = \prod_{|f(t)|>1} (1 + g(p)) \prod_{|f(p)|\le 1} (1 + g(p)) \cdot e^{-itf(p)/p}$$

where $g(p) = -\frac{1}{p} + \left(1 - \frac{1}{p}\right) \sum_{k=1}^{\infty} p^{-k} \cdot \exp(f(p^k))$. The limiting distribution is of pure type, and will be continuous if and only if the series

$$\sum_{f(p)\neq 0} \frac{1}{p}$$

diverges

P. Erdős. *On the smoothness of the asymptotic distribution of additive arithmetical functions.* Amer. J. Math. 61 (1939), 722–725.

2) Let (a_m) be a sequence of positive integers satisfying:

(i) $a_m \ll m$
$(m = 1, 2, \ldots)$

(ii) $\sum_{m, a_m = k} 1 = O(1)$
uniformly for all positive integers k; and $\sum_{m \le x, a_m \equiv 0 (\bmod d)} 1 = x\rho(d)/d + o(x)$ as $x \to \infty$, where $\rho(d)$ is the positive multiplicative function, and the estimate $o(x)$ may depend on a and d

Let f be a real-valued additive function, and for primes p, let us denote $f'(p) = f(p)$ if $|f(p)| \le 1$ and $f'(p) = 1$ if $|f(p)| > 1$. Then the sequence

$$(f(a_m))$$

has a limit distribution function if and only if $\sum_p f'(p)/p$, $\sum_p f'(p)\rho(p)/p$, $\sum_p f'^2(p)/p$ converge.

K.-H. Indlekofer. *Grenzverteilung additiver Funktionen.* Litovsh. Mat. Sb. 16 (1976), 81–91, 241.

3) For $x \ge y \ge 2$ and $A \subset \mathbb{N}$ let

$$v_{x,y}(A) = \frac{\text{card}\{n \in \mathbb{N} : x - y < n \le x, n \in A\}}{\text{card}\{n \in \mathbb{N} : x - y < n \le x\}}$$

Let $f(n)$ be an additive arithmetic function, and consider an increasing sequence (N_j) of natural numbers such that

$$N_j \le N_{j+1} \le N_j^c$$

for some $c > 1$, all $j \ge 1$; and a sequence (M_j) of natural numbers such that

$$M_j \le N_j$$

for all $j \geq 1$, $\log M_j / \log N_j \to 1 (j \to \infty.)$ Then the sequence of distribution functions $v_{N_j, M_j}((n : f(n) \leq z))$ converges weakly, as $j \to \infty$, if and only if the series $\sum_{|f(p)|>1} 1/p$, $\sum_{|f(p)|\leq 1} f(p)/p$ and $\sum_{|f(p)|\leq 1} f^2(p)/p$ converge.

P.D.T.A. Elliott. *A localized Erdős-Wintner theorem.* Pacific J. Math. 135 (1988), 287–297.

§ XVI. 6 Value distribution of differences of additive functions

1) Let $f(n)$ be a real-valued additive arithmetic function $\alpha(x)$, defined for $x \geq 1$, for which the frequencies

$$v_x(n; f(n) - \alpha(x) \leq z)$$

converge to a limiting distribution as $x \to \infty$, it is both necessary and sufficient that there be a constant c so that $f(n) = c \log n + h(n)$, where the series

$$\sum_{|h(p)|>1} 1/p, \sum_{|h(p)|\leq 1} h^2(p)/p$$

are convergent. Moreover, when this condition is satisfied a suitable function $\alpha(x)$ may be defined by

$$\alpha(x) = c \log x + \sum_{p \leq x, |h(p)| \leq 1} h(p)/p$$

With this choice the limiting distribution has the characteristic function

$$\frac{1}{1 + ict} \prod_{|h(p)|>1} w_p(t) \prod_{|h(p)|\leq 1} w_p(t) \cdot \exp(-ith(p)p^{-1})$$

where $w_p(t) = \left(1 - \frac{1}{p}\right) \cdot \left(1 + \sum_{k=1}^{\infty} p^{-k} \cdot \exp(ith(p^k))\right)$. It is of pure type, and is continuous if and only if the series

$$\sum_{f(p)\neq 0} 1/p$$

diverges.

P.D.T.A. Elliott and C. Ryavec. *The distribution of the values of additive arithmetical functions.* Acta Math. 126 (1971), 143–164;

and

B.V. Levin and N.M. Timofeev. *On the distributions of values of additive functions.* Acta Arith 26 (1974/75), 333–364.

2) a) Let $f(n)$ be a real-valued additive function. The distribution function

$$D_x(z) = \frac{1}{[x]}\{\text{card } 1 \leq n \leq f(x) : f(n+1) - f(n) \leq z\}$$

converges weakly to a limit function $D(z)$ as $x \to \infty$ if and only if there exists a real number λ such that the function $h(n) = f(n) - \lambda \log n$ satisfies

$$\sum_{|h(p)|\leq 1} (|h(p)|^2/p) < \infty, \sum_{|h(p)|>1} (1/p) < \infty$$

If this condition is satisfied, then the characteristic function of $D(z)$ is given by

$$\prod_p \left(1 - \frac{2}{p} + 2\left(1 - \frac{1}{p}\right) \operatorname{Re} \sum_{m \geq 1} \frac{e^{ith(p^m)}}{p^m}\right)$$

A. Hildehrand. *An Erdős-Wintner theorem for differences of additive functions.* Trans. Amer. Math. Soc. 310 (1988), no. 1, 257–276.

b) Let $a > 0, A > 0, b, B$ be integers satisfying $aB \neq Ab$, and let f_j $(j = 1, 2)$ be real-valued additive functions. There exists a function $\eta(x)$ such that the distributions

$$D_x(z) = \frac{1}{[x]} \text{card} \{n \leq x : f_1(an + b) - f_2(An + B) - \eta(x) \leq z\}$$

converge weakly to a distribution function as $x \to \infty$ if and only if there are real numbers α_j such that the series

$$\sum_{|f_j(p) - \alpha_j \log p| > 1} 1/p, \quad \sum_{|f_j(p) - \alpha_j \log p| \leq 1} (f_j(p) - \alpha_j \log p)^2/p$$

$(j = 1, 2)$ converge.

P.D.T.A. Elliott. *The value distribution of differences of additive arithmetic functions.* J. Number Theory 32 (1989), 339–370.

§ XVI. 7 Erdős-Wintner theorem for normed semigroups

Let S be a normed semigroup isomorphic to the semigroup of natural numbers and $\sum_{N(p) \leq x} l = lix + O(x \exp(-(\log x)^{\alpha}))(\alpha > 0)$. Further let f be a real additive function defined on S, and $f'(p) = f(p)$ for $|f(p)| \leq 1, f'(p) = 1$ when $|f(p)| > 1$. If $\sum_p (f'(p))^2/p < \infty$, then the sequence of distribution functions

$$F_n(x) = \frac{1}{\sum_{N(a) \leq n} 1} \cdot \sum_{N(a) \leq n, f(a) - \sum_{N(p) \leq n} f'(p)/p \leq x} 1$$

converges weakly to a proper limit distribution.

R.S. Baĭbulatov. *The Erdős-Wintner theorem for normed semigroups* (Russian). Dokl. Akad. Nauk UzSSR, 1967, no. 8, 3–6.

§ XVI. 8 Turán-Kubilius inequality and the Erdős-Wintner theorem for additive functions of a rational argument

A function f of a rational argument is called additive if it satisfies $f((a/b) \cdot (c/d)) = f(a/b)f(c/d)$ whenever a, b, c, d are pairwise coprime.

If $I = [s, t]$ and $x > 0$, let Q_x be the set of numbers $m/n \in I$ with $(m, n) = 1$, $n \leq x$.

a) Let II be the set of those prime-powers p^α for which there is a$q \in Q_x$ whose decomposition contains p^α (where α may be negative). Write $A = \sum_{p^\alpha \in \Pi} f(p^\alpha) p^{-|\alpha|}$, $B = \sum_{p^\alpha \in \Pi} f(p^\alpha)^2 \cdot p^{-|\alpha|}$. Then

$$\sum_{q \in Q_x} |f(q) - A|^2 \ll x^2(t - s)B$$

where the interval I may even depend on x, and the only constraints are that $t - s \geq x^{-\mu}$ with some $0 \leq \mu < 1, s \ll x(t - s)$. The implied constant depends on μ

J. Šiaulys and V. Stakenas. *The Kubilius inequality for additive function of a rational argument.* Lith. Math. J. 30 (1990), 72–76; translation from Litov. Mat. Sb. 30 (1990), 176–184.

b) A real-valued additive function f of a rational argument is said to have a limiting distribution if there is a distribution function F such that

$$\frac{1}{|Q_x^I|} \cdot \operatorname{card} \{q \in Q_x^I : f(q) < u\} \Rightarrow F(u)$$

(weak convergence) as $x \to \infty$ for every interval I. This is equivalent to the convergence of the series $\sum_p (f(p)^* + f(p^{-1})^*)/p$; $\sum_{|f(p^\varepsilon)|>1} 1/p$; $\sum_{|f(p^\varepsilon)| \leq 1} f(p^\varepsilon)^2/p$, where $x^* = x$ if $|x| \leq 1$ and 0 otherwise, and $\varepsilon = \pm 1$

J. Šiaulys. *The Erdős-Wintner theorem for additive functions of a rational argument* (Russian). Litov. Mat. Sb. 30 (1990), 405–415.

Remark. This is an analogue of the Erdős-Wintner theorem.

§ XVI. 9 Limit theorem for additive functions on ordered semigroups

Let G be a multiplicative semigroup with countable many generators of infinite order. Let N be a homomorphism of the semigroup G in the multiplicative semigroup of positive integers. The homomorphic image $N(m)$ is called the norm of the element m. Suppose that $\sum_{m \in G, N(m) \leq x} 1 = cx^v + O(x^{v_1})$, where c, v, v_1 are constants, $c > 0$ and $0 \leq v_1 < v$. Let $f : \mathbb{N} \to \mathbb{R}$ be an additive function such that

$$\frac{1}{D(x)} \cdot \max_{N(p^\alpha) \leq x} |f(p^\alpha)| \leq \mu(x)$$

$(D(x) \to \infty, x \to \infty)$, where $\mu(x) \searrow 0$. Further, let $v_x(f(m) - A(x))/D(x) < y)$ be the frequency of the elements $m \in G$ satisfying $N(m) \le x$ and $f(m) < A(x) + yD(x)$. Then

$$v_x\left(\frac{f(m) - A(x)}{D(x)} < y\right) = \Phi_c(y) + O\left(\mu(x)\left(c^{-y^2/2}\log\frac{1}{\mu(x)} + 1\right)\right)$$

uniformly for $x > x_0$, and all y, where x_0 is independent of y. Here

$$\Phi_c(y) = \frac{1}{\sqrt{2\pi}}\int_{-\infty}^{y} c^{-u^2/2}du$$

Z. Juškis. *Limit theorems for additive functions defined on order semigroups with regular norm.* Litov. Mat. Sb. 4 (1964), 565–603.

§ XVI.10 Laws of iterated logarithm for additive functions

1) Let $h : \mathbb{N} \to \mathbb{R}$ be additive such that $g(p^r) = O(p^{r\theta})$, where $r \ge 2$, $0 < \theta < 1/4$, p a prime.
Introduce the notations $A(k) = \sum_{p \le k} h(p)/p$, $B_b^2(k) = \sum_{p \le k}(h(p) - b\log p)^2/p$.
If there exists a real number b such that

$$\frac{1}{B_b(k)} \cdot \max_{p \le k} |h(p) - b\log p| = \mu_k \to 0$$

$(k \to \infty)$, then

$$F_k(x) = \sum_R 1 = \frac{1}{\sqrt{2\pi}}\int_{-\infty}^{x} \exp(-t^2/2)\,dt + O(\mu_k)$$

uniformly with respect to x, where the summation extends over the set $R = \left\{n \le k : \dfrac{h(n) - A(k)}{B_b(k)} \le x\right\}$

N.M. Timofccv. *An estimate of the remainder term in one dimensional asymptotic laws* (Russian). Dokl. Akad. Nauk. SSSR 300 (1971), 298–301.

Remark. A somewhat weaker result was previously obtained by Kubilius and a somewhat stronger (and more complicated) by Elliott. See

P.D.T.A. Elliott. *Probabilistic number theory*, vol. II, Springer 1979/80.

2) Let $h : \mathbb{N} \to \mathbb{R}$ be an additive function. Define $h_k(n) = \sum_{p^m \| n, n \ge p} h(p^m)$, and set $A(k) = \sum_{p \le k} h(p)/p$, $B^2(k) = \sum_{p \le k} h^2(p)/p$, and $\beta_k = B(k)\sqrt{2\log\log B(k)}$.
Suppose that $\max_{p \le n} |h(p)| \ll B(n)\sqrt{\log\log B(n)}\,(n \to \infty.)$ Then

$$\lim_{n_1\to\infty} \limsup_{n\to\infty} \frac{1}{n}\{\text{card } m \le n : \max_{n_1\le k\le n} \frac{1}{\beta_k}|h_k(m) - A(k)| \ge 1+\varepsilon\} = 0 \text{ for every}$$

$\varepsilon > 0$. The constant 1 is the best possible.

E. Manstavičius. *Laws of the iterated logarithm for additive functions.* Number Theory, vol. I. Elementary and analytic. Proc. Conf. Budapest/Hung. 1987, Colloq. Math. Soc. János Bolyai 51 (1990) 279–299.

§ XVI.11 Limit laws and moments of additive functions in short intervals

1) Given an additive arithmetic function f and $x \ge y \ge 1$, define $v_{x,y}$ by

$$v_{x,y}(n; f(n) \le z) = \frac{\text{card}\{n : x - y < n \le x,\ f(n) \le z\}}{\text{card}\{n : x - y < n \le x\}}$$

Let $H(z)$ be the distribution function whose characteristic function does not vanish on the real line. Let $y = y(x)$ satisfy $1 \le y \le x$, $\log y \sim \log x (x \to \infty)$ and let $(f_x)_{x\ge 1}$ be a family of additive functions. Then there are constants $\gamma = \gamma(x, y)$ such that $v_{x,y}(n; f_x(n) - \gamma \le z) \Rightarrow H(z)$ holds as $x \to \infty$ through some sequence of x-values, if and only if there are further constants $\lambda = \lambda(x)$ satisfying $|\lambda| \le x^{1/2}$, such that

$$F_x(\lambda^{-1}(z+\gamma)) * v_{x,x}(n; f_x(n) - \lambda \log n \le z) \Rightarrow H(z)$$

Here P_x is defined by $F_x(z) = y^{-1}(\min(x, e^z) - \min(x - y, e^z))$,"$\Rightarrow$" denotes weak convergence of distribution function in the usual probabilistic sense, and $F * G$ denotes the convolution of two distribution functions F and G.

P.D.T.A. Elliott. *Additive arithmetic functions on intervals.* Math. Proc. Cambridge Philos. Soc. 103 (1988), 163–179.

2) For $1 \le y \le x$ let $\gamma_{x,y}$ denote the uniform measure (frequency) of the integers $x - y < n \le x$. Let f be an additive function, let $y = y(x)$ satisfy $1 \le y \le x$, $(\log y)/(\log x) \to 1$ and let β be a positive-valued function satisfying $\beta(x) \to \infty$ and $\beta(x^u)/\beta(x) \to 1$ as $x \to \infty$. Then the frequency $\gamma_{x,y}(n : f(n) - \alpha(x, y) \le z\beta(y))$ possesses a limiting distribution for some $\alpha(x, y)$ as $x \to \infty$ if and only if so do the corresponding frequencies for $y = x$ (the usual limiting distribution), and then the centers and the distributions can be chosen to be the same.

K.-H. Indlekofer. *Limit laws and moments of additive functions in short intervals.* Proc. Conf. Budapest/Hung. (Number Theory, vol. I), 1987, Colloq. Math. Soc. János Bolyai 51 (1990) 193–220.

§ XVI.12 Distribution function of the sum of an additive and multiplicative function

If $f(n)$ is an additive arithmetic function and $g(n)$ a positive multiplicative arithmetic function, and both have a distribution function at least one of which is continuous, then $f(n) + g(n)$ has a continuous distribution function.

M.B. Fein and H.N. Shapiro. *Continuity of the distribution function of the sum of an additive and multiplicative arithmetic function.* Comm. Pure. Appl. Math. 60 (1987), 779–801.

§ XVI.13 Moments and concentration of additive functions

1) Let $f : \mathbb{N} \to \mathbb{R}$ be an additive function and let
$Q(x) = \sup\limits_{a} \frac{1}{x} \cdot \operatorname{card}\{n \le x : a \le f(n) < a + 1\}$.
Put $U(x, \lambda) = \sum\limits_{p \le x} \frac{1}{p} \cdot \min(1, (f(p) - \lambda \log p)^2)$ and
$W(x) = \min\limits_{\lambda} (\lambda^2 + U(x, \lambda))$.
Then

$$Q(x) \ll (W(x))^{-1/2}$$

I.Z. Ruzsa. *On the concentration of additive functions.* Acta Math. Hungar. 36 (1980), 215–232.

2) Let v_x be a relative frequency measure, which assigns the weight $1/x$ to each number $1, 2, \ldots, x$.

a) If f is a real-valued additive function, λ a real number and
$U(f, x, \lambda) = \lambda^2 + \sum\limits_{p^k \le x} p^{-k} \cdot \min(1, (f(p^k) - \lambda \log p^k)^2) = \delta$
then with $A = \lambda \log x + \sum\limits_{|f(p) - \lambda \log p| \le 1} \frac{1}{p}(f(p) - \lambda \log p)$ and with an absolute constant c_1 we have

$$v_x(|f(n) - A| \ge 2\delta^{1/3}) \le c_1 \delta^{1/3}$$

I.Z. Ruzsa. *The law of large number of additive functions.* Studia Sci. Math. Hungar. 14 (1979), 247–253.

b) Let f be an additive function, and suppose

$$v_x(f(n) \in [a, a + h]) \ge q$$

Then there is a λ, $|\lambda| \le \mathrm{ch}/q$ such that

$$\sum_{p \le x} \frac{1}{p} \min(h^2, (f(p) - \lambda \log p)^2) \le \mathrm{ch}^2/q^2$$

where c is an absolute constant.

(See I.Z. Ruzsa (1980).)

Corollary. Let the symbol (f, v_x) denote that f is regarded as a variable with respect to v_x. For a random variable ξ let $L(\xi) = \inf_{a,\varepsilon}(\varepsilon + P(|\xi - a| > \varepsilon))$. Then, for every real-valued additive function f we have

$$V(f, x) \ll L(f, v_x) \ll V(f, x)^{1/3}$$

with the implied constants absolute. Here $V(f, x) = \min(1, U(f, x))$, with $U(f, x) = \min_{\lambda} U(f, x, \lambda)$, and $U(f, x, \lambda) = \lambda^2 + \sum_{p^k \le x} p^{-k} \min(1, (f(p^k) - \lambda \log p^k)^2)$

(See I.Z. Ruzsa (1979).)

Remark. As an application we get the following result. Let (f_x) be a sequence of real-valued additive functions. A necessary and sufficient condition for the distribution of $(f_x - a_x, v_x)$ to converge to the improper law with a suitable choice of the centering constants a_x is $U(f_x, x) \to 0(*)$

If (*) holds and λ_x is a sequence of real numbers such that $U(f_x, x, \lambda_x) \to 0$, then one may choose

$$a_x = \lambda_x \log x + \sum_{|f_x(p) - \lambda_x \log p| \le 1} \frac{1}{p}(f_x(p) - \lambda_x \log p)$$

3) Let f be a complex-valued additive function and ξ_p (p prime) independent random variables with the distribution

$$P(\xi_p) = f(p^k) = (1 - 1/p)p^{-k}$$

Put $\eta_x = \sum_{p \le x} \xi_p$. Let v_x be defined as in 2). If $A \in \mathbb{C}$ and $B > 0$, then we have

$$v_x(|f(n) - A| > B) \le cP(|\eta_x - A| > B/3)$$

with an absolute constant c

I.Z. Ruzsa. *Generalized moments of additive functions.* J. Number Theory 18 (1984), 27–33.

§ XVI.14 Local theorems for additive functions

1) a) Let $f : \mathbb{N} \to \mathbb{R}$ be additive. If the series $\sum_{f(p) \ne 0} 1/p$ is divergent, then for every number a the sequence of solutions of $f(n) = a$ has density 0.

P. Erdős. *On the distribution function of additive functions.* Ann. Math. 47 (1946), 1–20.

b) For $f : \mathbb{N} \to \mathbb{R}$ additive, put

$$G(x) = \max_{a \neq 0} \sum_{n \leq x, f(n)=a} 1$$

Then, the limit $\lim_{x \to \infty} \frac{G(x)}{x}$ exists and is $\leq 1/2$.

P. Erdős, I.Z. Ruzsa and A. Sárközy. *On the number of solutions of $f(n) = a$ for additive functions.* Acta Arith. 24 (1973), 1–9.

Remarks. i) Erdős, Ruzsa and Sárközy obtain various other results, for example:

$$\log 2 < \liminf_{x \to \infty} \max_{f} \frac{G(x)}{x}$$

$$\limsup_{x \to \infty} \max_{f} \frac{G(x)}{x} < 1 - c$$

for an absolute constant $c > 0$

ii) For another method giving such results, see also

H. Delange. *Sur une inégalité remarcable, avec une application à la théorie des nombres.* Bull. Sci. Math. (2) 106 (1982), 225–234.

c) Let $f : \mathbb{N} \to \mathbb{R}$ be an additive function. Then there is a universal constant c such that

$$\sum_{n \leq x, f(n)=a} 1 \leq cx/(E(x))^{1/2}$$

where $E(x) = \sum_{p \leq x, f(p) \neq 0} 1/p (p \text{ prime})$

G. Halász. *On the distribution of additive arithmetic functions.* Acta Arith. 27 (1975), 143–152.

Corollary. Let for every N, f_N be an additive function. If $\sum_{p \leq N, f_N(p) \neq 0} 1/p \to \infty$, then card $\{n \leq N : f_N(n) = a\} = o(N)$

2) Let $f : \mathbb{N} \to \mathbb{I}$ be an integral-valued additive function, $f(n) \geq 0$ for all n. For $q \geq 0$ integer, let $S_q(n) = card\{n \in S, n \leq x$ and $f(n) = q\}$, where S is an infinite set of positive integers whose characteristic function is multiplicated. Suppose that:

(i) $\sum_{p \leq x, p \in S, f(p)=0} p \log p \sim \alpha \log x$
as $x \to \infty (\alpha > 0$, constant)

(ii) $\sum_{p \in S, f(p)=1} 1/p = +\infty$

and, for every $r > 1$, $\sum_{p\le x, p\in S, f(p)=r} 1/p = o\left(\left(\sum_{p\le x, p\in S, f(p)=1} 1/p\right)^r\right)$ as $x \to \infty$. Then:

a) $S_0(x) \sim \dfrac{e^{-\gamma\alpha}}{\Gamma(\alpha)} \cdot \dfrac{\lambda}{\log x} \prod_{p\le x}\left(1 + \sum_{p\in S, r\ge 1, f(p^r)=0} 1/p^r\right)$

b) $S_q(x) \sim S_0(x) \cdot \dfrac{1}{q!}\left(\sum_{p\le x, p\in S, f(p)=1} 1/p\right)^q$

for $q \ge 1$

H. Delange. *A theorem on integral-valued additive functions.* Illinois J. Math. 18 (1974), 357–372.

3) Let $f : \mathbb{N} \to \mathbb{I}$ be an integral-valued additive function, which satisfies

$$\sum_{f(p)\ne 1} \frac{\log p}{p} < \infty, \sum_{f(p)\ne 1} \frac{|f(p)|}{p} < \infty, \sum_p \sum_{\alpha=2}^{\infty} \frac{|f(p^\alpha)|}{p^\alpha} < \infty$$

Then

$$\text{card}\,\{1 \le m \le n : f(m) = a\} = n\varphi(y)/\lambda + Bn/\lambda^2$$

uniformly for all integers a and $n > 30$, where $\lambda = \sqrt{\log\log n}$, $y = (a - \lambda^2)/\lambda$, and $\varphi(u) = \dfrac{1}{\sqrt{2\pi}} \cdot \exp(-u^2/2)$($B$-constant.)

J. Kubilius. *On local theorems for additive number-theoretic functions.* Number theory and analysis (Papers in Honor of E. Landau), pp. 173–191, New York, 1969.

§ XVI.15 Additive functions on arithmetic progressions

1) Let f be an additive function, and write

$$E_f(x, q) = \max_{(a,q)=1} \left| \sum_{n\le x,\, n\equiv a (\operatorname{mod} q)} f(n) - \frac{1}{\varphi(q)} \sum_{n\le x, (n,q)=1} f(n) \right|$$

and

$$F_f(x) = \sum_{p^m \le x^{1/2-\varepsilon}} p^m \cdot E_f(x, p^m)^2$$

Then:

a) $F_f(x) \ll \dfrac{\log\log x}{\log x} \cdot x^2 \cdot \sum_{p^m\le x} |f(p^m)|^2/p^m$

b) $F_f(x) \ll \dfrac{\log\log x}{\log x} \cdot x \cdot \min_A \sum_{n\le x} |f(n) - A|^2$

A. Hildebrand. *Additive functions on arithmetic progressions.* J. London Math. Soc. (2) 34 (1986), 394–402.

Remark. The above estimates can be viewed as analogues of the Bombieri-Vinogradov theorem.

2) Let $f : \mathbb{N} \to \mathbb{R}$ be an additive function and define

$$E_f(y, D, r) = \sum_{\substack{n \le y, n \equiv d (\mathrm{mod}\, c) \\ n \equiv r (\mathrm{mod}\, D)}} f(n) - \frac{1}{\varphi(D)} \sum_{\substack{n \le y, n \equiv d (\mathrm{mod}\, c) \\ (n, D) = 1}} f(n)$$

for integers $d, c > 0$ and $r, D > 0$, and real y. Let $0 < \alpha < 1/2$. Then

$$\sum_{q \le x^\alpha, (q,c)=1} \varphi(q) \max_{(r,q)=1} \max_{y \le x} |E(y, q, r)|^2 \ll$$
$$\ll x^2 (\log\log x)^4 (\log x)^{-1} \sum_{m<x} |f(m)|^2 \cdot m^{-1}$$

with prime powers q, m, holds for all $x \ge 2$. The implied constant depends at most upon α, d, and c.

P.D.T.A. Elliott. *Additive arithmetic functions on arithmetic progressions.* Proc. London Math. Soc. (3) 54 (1987), 15–37.

Remark. An important feature of the inequality is that there is no restriction on f except additivity (in particular, no growth condition is required).

3) Let f be a strongly additive function. Let $r \ne 0$ be a fixed integer, $0 < \delta < 1/180$ and assume that $f(p) = 0$ whenever $p > x^\theta$. Then

$$\sum_{p \le x} p^2 \cdot \left| \sum_{\substack{n \le x \\ n \equiv r (\mathrm{mod}\, p)}} f(n) - \frac{1}{p-1} \cdot \sum_{n \le x, p \nmid n} f(n) \right|^3 \ll$$
$$\ll x^3 \cdot (\log x)^{-1/4} (\log\log x)^5 \sum_{p \le x} |f(p)|^3 / p$$

(p prime.)

P.D.T.A. Elliott. *Applications of elementary functional analysis to the study of arithmetic functions.* Number theory, vol. 1, Proc. Conf. Budapest/Hung. 1987, Colloq. Math. Soc. János Bolyai 51 (1990), 35–43.

§ XVI.16 On differences of additive functions

1) Let $f_j(n)$ be real-valued additive functions, a_j, b_j integers, $a_j > 0$, and put

$$\Lambda = \prod_{i=1}^{k} a_i \prod_{1 \le i < j \le k} (a_i b_j - a_j b_i)$$

Then the following assertions are equivalent:

(i) There exists a constant c such that $\sum_{n \le x} \left| \sum_{j=1}^{k} f_j(a_j n + b_j) \right|^2 \le cx$ and

$\sum_{n \le x} f_j^2(n) \le cx(\log x)^2 (j = 1, 2, \ldots, k)$ hold for $x \ge 2$;

(ii) With suitable constants $A_j (j = 1, 2, \ldots, k)$, whose sum is zero,
$\sum'_{p,m} p^{-m} |f_j(p^m) - A_j \log p^m|^2 < \infty \, (j = 1, 2, \ldots, n)$ and
$\sum_{j=1}^{k} \sum'_{p^m \le x} p^{-m}(f_j(p^m) - A_j \log p^m) = O(1)$ as $x \to \infty$, where the dash indicates that summation is restricted to prime powers p^m which divide $a_j n + b_j$ for some integer n.

P.D.T.A. Elliott. *Sums and differences of additive functions in mean square.* J. Reine Angew. Math. 309 (1979), 21–54.

2) a) Let $f(n)$ be a strongly additive function. Then the following assertions are equivalent:

(i) $\sum_{n \le x} |f(n+1) - f(n)|^2 \le cx$
(c-constant), for all $x \ge 1$

(ii) $\sum_{p} \frac{|f(p) - A \log p|^2}{p}$
converges for a certain constant A

P.D.T.A. Elliott. *On the differences of additive arithmetic functions.* Mathematika 24 (1977), 153–165.

b) Let $f : \mathbb{N} \to \mathbb{R}$ be an additive function. Then for all $a \in \mathbb{N}$, the following inequalities with some constants A, B, C are equivalent:

(i) $\sum_{n \le x} (f(n+a) - f(n))^2 \le Bx$

(ii) $\sum_{n \le x} \frac{1}{n} (f(n+a) - f(n))^2 \le c \log x$

(iii) $\sum_{p} \sum_{k=1}^{\infty} p^{-1} (f(p^k) - A \log p^k)^2 < \infty$

P.D.T.A. Elliott. *On the differences of additive functions.* II. Acta Arith. 37 (1980), 249–256.

§ XVI.17 Prime-independent additive functions

1) Let $f : \mathbb{N} \to \mathbb{R}$ be an additive function such that $f(p^k)$ depends only on k for all primes p. (i.e. is prime-independent.) If $f(p^m) = O(2^{m/2})$, then

$$\sum_{n \le x} f(n) = f(p)x \log\log x + Ax + O(x/\log x)$$

where A is a constant.

S.L. Segal. *On prime-independent additive functions.* Arch. Math. (Basel) 17 (1966), 329–332.

2) Let $f : \mathbb{N} \to \mathbb{R}$ be a prime-independent additive function such that $f(p) \neq 0$, $f(p^m) = O(m^{-1} \cdot 2^{m/2})$. Then

$$\frac{1}{n}\text{card}\,\{k \le n : (f(k) - f(p)\log\log n)/f(p)\sqrt{\log\log n} < x\} = \Phi(x) + O((\log\log$$

where $\Phi(x) = \frac{1}{\sqrt{2\pi}} \int_{-\infty}^{x} e^{t^{-2}/2} dt$

J. Galambos. *On the distribution of prime-independent additive number-theoretical functions.* Arch. Math. (Basel) 19 (1968), 296–299.

Remark. This generalizes a theorem of Rényi and Turán on $\Omega(n)$

A. Rényi and P. Turán. *On a theorem of Erdős-Kac.* Acta Arith. 4 (1958), 71–84.

§ XVI.18 Moments and Cesàro means of additive functions

1) Let L_α be the set of all arithmetical functions $f : \mathbb{N} \to \mathbb{C}$ with

$$\left(\limsup_{x\to\infty} \frac{1}{x} \sum_{n \le x} |f(n)|^\alpha\right)^{1/\alpha} < \infty$$

for $\alpha > 1$, and

$$\limsup_{x\to\infty} \frac{1}{x} \sum_{n \le x} |f(n)|^\alpha < \infty$$

for $0 < \alpha < 1$

a) Let $\alpha > 0$ and f be additive. Then $f \in L_\alpha$ if and only if the series

$$\sum_{|f(p^m)|>1} |f(p^m)|^\alpha/p^m, \quad \sum_{|f(p)|\le 1} |f(p)|^2/p$$

converge and $\sum_{n \le x, |f(p)| \le 1} f(p)/p = O(1)$ as $x \to \infty$

b) Let L^* be the set of (uniformly summable) functions satisfying

$$\lim_{K\to\infty} \sup_{x\geq 1} \frac{1}{x} \sum_{\substack{n\leq x \\ |f(n)|\geq K}} |f(n)| = 0$$

Let $\alpha > 0$ and $f \in L_\alpha$ be additive. Then

$$|f|^\alpha \in L^*$$

c) Let $\alpha \geq 1$ and f any real-valued additive function. Then the following three statements are equivalent:

(i) The limiting distribution F of f exists (i.e. exists the limit $\lim_{x\to\infty} \frac{1}{x} \sum_{n\leq x, f(n)\leq y}$ 1) and $\int_{\mathbb{R}} |y|^\alpha \mathrm{d}F(y) < \infty$

(ii) $f \in L_\alpha$ and the mean value $M(f)$ of f exists.

(iii) The series $\sum_p \frac{f(p)}{p}$, $\sum_{\substack{p \\ |f(p)|\leq 1}} \frac{f^2(p)}{p}$, $\sum_{\substack{p \\ |f(p^m)|\geq 1}} \sum_m \frac{|f(p^m)|^\alpha}{p^m}$ converge.

Moreover, when one of these conditions is satisfied,

$$M(f) = \int_{\mathbb{R}} y\mathrm{d}F(y),\ M(|f|^\alpha) = \int_{\mathbb{R}} |y|^\alpha \mathrm{d}F(y)$$

K.-H. Indlekofer. *Cesàro means of additive functions.* Analysis 6 (1986), 1–24.

2) Let the additive function $f : \mathbb{N} \to \mathbb{C}$ satisfy

$$(*) \qquad \sup_{x\geq 1} \frac{1}{x} \sum_{n\leq x} F(|f(n)|) < \infty$$

where $F : [0, \infty) \to [0, \infty)$ is a non-decreasing function with a property $F(x) \to \infty (x \to \infty)$

a) If satisfies (*), and if $F(x-1) \geq c \cdot F(x)$ for $x \geq 2$, then the series

$$(**) \qquad \sum_{p,|f(p)|>1} 1/p,\ \sum_{p,|f(p)|\leq 1} |f(p)|^2/p,\ \sum_{p, f(p^m)>1} \sum_{m\geq 1} F(|f(p^m)|)/p^m$$

converge and

$$(***) \qquad \sum_{p\leq x, |f(p)|\leq 1} f(p)/p = O(1)$$

b) If $F(x+y) \leq c_1 \cdot F(x) \cdot F(y)$, then the convergence of the series (**) and inequality $(***)$ imply (*)

K.-H. Indlekofer and I. Kátai. *Generalized moments of additive functions.* J. Number Theory 32 (1989), 281–288.

3) Let f be a strongly additive function, and define

$$A(x) = \sum_{p \le x} f(p)/(p-1),\ B(x) = \left(\sum_{p \le x} f^2(p)/(p-1)\right)^{1/2}$$

If $\limsup_{x\to\infty} \frac{1}{(B(x))^k} \cdot \sum_{p\le x, |f(p)|\ge B(x)} |f(p)|^k/p < \infty$ for all k, then

(i) $$\limsup_{x\to\infty} \frac{1}{\pi(x)} \cdot \sum_{p\le x} \frac{|f(p+1) - A(x)|^k}{B(x)} < \infty$$

K.-H. Indlekofer and I. Kátai. *Moments of additive functions on the sequence* $\{p+1\}$ (German). Litov. Mat. Sb. 28 (1988), 669–679.

Remark. For positive function the condition is also necessary. For functions of Kubilius' class H the existence of a limit in (i) is established.

§ XVI.19 Minimax-theorem for additive functions

1) a) Let $f(n)$ be a nonnegative additive function, which tends monotonically to zero on the sequence of primes. Let $\psi(y) = \sum_{p\le y} f(p)$, $\beta(p) = \sup_{\alpha\ge 1} f(p^\alpha)$ For $k \ge 1$, let $E_k(x) = \max_{n\le x} \min\{f(n+1), f(n+2), \ldots, f(n+k)\}$. Assume that $\psi(2y) - \psi(y) = o(1)$, $y\to\infty$, $\psi(y)\to\infty$, $\sum_p |\beta(p) - f(p)| < \infty$. Then

$$\lim_{x\to\infty}\left(E_k(x) - \frac{\psi(\log x)}{k}\right) = B_k + C_k - \frac{\psi(k)}{k}$$

where $B_k = \sup_{n\ge 1} \frac{1}{k} \sum_{j=1}^{k} \sum_{\substack{p\le k \\ p^\alpha \| (n+j)}} f(p^\alpha)$ and $C_k = \frac{1}{k}\sum_{p\le k}(\beta(p) - f(p))$

I. Kátai. *A minimax theorem for additive functions.* Publ. Math. (Debrecen) 30 (1983), 249–252.

b) Let $f(n)$ be a nonnegative, strongly additive function which tends monotonically to zero on the sequence of primes. Let $A = \sum_p f(p)/p$ and let $C > A$ determine a sequence of integers $n_1 < n_2 < \cdots$ by the condition $f(n_i) \le C$. Let $k(x) = \frac{1}{C-A} \sum_{p\le \log x} f(p)$. Then

$$\limsup_{x\to\infty} \frac{n_{i+1} - n_i}{k(x)} \le 1$$

M. van Rossum-Wijsmuller. *A variant of Kátai's minimax theorem of additive functions.* Publ. Math. (Debrecen) 34 (1987), 323–326.

Remark. A similar result, for the case where $f(p) = 1/p$ and $c = 2$ has been obtained by Galambos, who showed that $A < 1$ and $k(x) = \log\log\log x$.

J. Galambos. *On a conjecture of Kátai concerning weakly composite numbers.* Proc. Amer. Math. Soc. 96 (1986), 215–216.

2) Let $f(n)$ be an arithmetical function satisfying the conditions below
   (i) $f(n)$ is strongly additive
   (ii) $f(p)$ is decreasing, and tends to zero as $p \to +\infty$
   (iii) $\sum f(p) = +\infty$ and $\sum_{y<p\le 2y} f(p) \to 0$ as $y \to \infty$; and
   (iv) $\sum f(p)/p < +\infty$

   Then:

   a) The limiting distribution function $F(C)$ of $f(n)$ (that is the density of the set $\{n : f(n) \le C\}$) satisfies $0 < F(C) < 1$ for all $C > 0$.

   b) For a fixed number $C > 0$, let $a_1(C) < a_2(C) < \cdots$ be those integers n for which $f(n) \le C$. Then for every $C > 0$,
   $$0 < \limsup_{j\to\infty} \frac{a_{j+1}(C) - a_j(C)}{\psi(\log a_j(C))} < +\infty$$
   where $\psi(y) = \sum_{p\le y} f(p)$

   c) With the assumptions of b) let $k^*(n) = u_n\psi(\log n)$, where $u_n \to +\infty$ with n. If we denote by $N(\ldots)$ the number of integers j for which the propriety stated in the dotted space holds, then, as $n \to \infty$,
   $$F(C) = \lim N(j \le n,\ f(j) \le C)/n =$$
   $$= \lim N(j \le k^*(n),\ f(n+j)) \le C)/k^*(n)$$

J. Galambos and I. Kátai. *The gaps in a particular sequence of integers of positive density.* J. London Math. Soc. (2) 36 (1987), 429–435.

§ XVI.20 Maximal value of additive functions in short intervals

Let $g(n)$ be a non-negative strongly additive function,
$$f_k(n) = \max_{j=1,\ldots,k} g(n+j)$$

a) Let
$$\theta(k, q) = \limsup_{x\to\infty} \frac{1}{x}\text{card}\,\{n \le x : f_k(n) > f_k(0)(1+\varepsilon)\}$$
If $\theta(k, \varepsilon) \to 0 (k \to \infty)$ for all $\varepsilon > 0$, then:

(*) $$\sum_p (g(p))^r/p < \infty$$

for every $r \geq 1$ (p denotes a prime)

b) Let $F(x)$ be the limit distribution function of $g(n)$ (the existence of which is guaranteed by $\sum_p \frac{\min(1, g(p))}{p} < \infty$) and assume that $k(1 - F(f_k(0)(1+\varepsilon))) \to 0$ holds for every $\varepsilon > 0$. Then (*) holds for every $r \geq 1$

c) If for some constant $A > 0$, $k(1 - F(f_k(0) + A)) \to 0\,(k \to \infty)$, then

$$\sum_p \frac{e^{ug(p)} - 1}{p} < \infty$$

holds for every $u > 0$

d) Let $L(k)$ be a function on $[1, \infty)$ tending to infinity arbitrary slowly. Then there exists a strongly additive non-negative $g(n)$ with limsup $g(p) = \infty$, so that

$$\sup_{x \geq 1} \frac{1}{x} \operatorname{card}\{n \leq x : \exists\, k \geq k_0,\ f_k(n) > L(k)\} \to 0$$

$(k_0 \to \infty)$

e) If $g(p) = 1/p$, then

$$\sup_{x \geq 1} \frac{1}{x} \operatorname{card}\{n \leq x : \exists\, k > k_0,\ f_k(n) > f_k(0) + \lambda_k\} \to 0$$

$(k_0 \to \infty)$, where $\lambda_k = 3/(\log\log k)$

P. Erdős and I. Kátai. *On the maximal value of additive functions in short intervals and some related questions.* Acta Math. Acad. Sci. Hungar. 35 (1980), 257–278.

§ XVI.21 Normal order of additive functions on sets of shifted primes

Let $f(n) \geq 0$ be a strongly additive function. Put $A_n = \sum_{p|n} f(p)/p$ and $\Lambda_n = \max_{p<n} f(p)$; and suppose that $A_n \to \infty$ and $\Lambda_n = o(A_n)$ as $n \to \infty$. If ℓ is a fixed positive integer, and $\varepsilon > 0$, then the number of primes $p \leq n$ for which the inequality

$$(1-\varepsilon)A_n \leq f(p - \ell) \leq (1+\varepsilon)A_n$$

does not hold is $o(\pi(n))$

M.B. Barban. *The normal or additive arithmetic functions on sets of "shifted primes"* (Russian). Acta Math. Hung. 12 (1961), 409–415.

§ XVI.22 **Uniformly distributed** (mod 1) **additive functions**

Let $f : \mathbb{N} \to \mathbb{R}$ be an additive function. Then f is uniformly distributed (mod 1) if and only if for all integers $m \neq 0$ and all reals t, the series

$$\sum_{p\,(\text{prime})} \frac{1}{p} \sin^2 m\,\pi(f(p) - t \log p)$$

diverges.

H. Delange. *Quelques résultats nouveaux sur le fonctions additive.* Colloq. de Théorie des Nombres (Bordeaux, 1969), pp. 45–53, Paris, 1971.

§ XVI.23 Additive functions and almost periodicity

Let $f : \mathbb{N} \to \mathbb{C}$ be an additive function. Then $f(n)$ is almost periodic (B^2) if and only if both series

$$\sum_{p} \frac{f(p)}{p}$$

$$\sum_{k=1}^{\infty} \sum_{p} |f(p^k)|^2/p^k$$

are convergent

P. Erdős and A. Wintner. *Additive functions and almost periodicity* (B^2). Amer. J. Math. 62 (1940), No. 3, 635–644.

Remark. Conditions which *are either* necessary *or* sufficient for the almost periodicity (B^2) of a multiplicative function $f(n)$ are implied by the results of

E.R. van Kampen and A. Wintner. *On the almost periodic behavior of multiplicative number-theoretical functions.* Amer. J. Math. 62 (1940), 613–626.

§ XVI.24 Characterization of multiplicative functions

a) Let f be a complex valued multiplicative function. If $\Delta(f(n) \to 0$ $(n \to \infty)$, then

$$f(n) \to 0$$

$(n \to \infty)$ or

$$f(n) = n^{\sigma + i\tau}$$

with $\sigma, \tau \in \mathbb{R}, \sigma < 1$

E. Wirsing. See I. Kátai. *Characterization of arithmetical functions, problems and results.* Théorie des nombres (Québec, PQ, 1987), 544–555, de Gruyter, Berlin-New York, 1989.

Corollary. Let F be a real valued additive function, and assume that $\|\Delta F(n)\| \to 0$. (Here, for $z \in \mathbb{R}$, $\|z\| = \min_{k\in\mathrm{I}} |z-k|$). Then with some suitable constant $\lambda \in \mathbb{R}$ we have that

$$F(n) - \lambda \log n$$

is an integer for every $n = 1, 2, \ldots$

§ XVI.25 Multiplicative functions with small increments

a) Let f, g be complex-valued completely multiplicative functions and suppose that $\sum_{n=1}^{\infty} \frac{1}{n} |g(n+1) - f(n)| < \infty$. Then either:

(i) $\sum |f(n)|/n < \infty, \sum |g(n)|/n < \infty,$ *or*

(ii) $f(n) = g(n) = n^{\sigma+i\tau}$ with $\sigma, \tau \in \mathbb{R}, 0 \le \sigma < 1$

I. Kátai. *Multiplicative functions with regularity properties.* I–V., Acta Math. Hung. 42 (1983), 295–308; 43 (1984), 105–130, 259–272; 44 (1984), 125–132; 45 (1985), 379–380.

b) Let f, g be complex-valued completely multiplicative functions and assume that $\sum_{n\le x} |g(n+1) - f(n)| = O(x)$ holds. Then either:

(i) $\sum_{n\le x} |f(n)| = O(x)$ and $\sum_{n\le x} |g(n)| = O(x),$ *or*

(ii) $f(n) = g(n) = n^s$ with $0 \le \operatorname{Re} s \le 1$

K.-H. Indlekofer and I. Kátai. *On some pairs of multiplicative functions.* Annales Univ. Sci. Budapest, Sect. Math. (to appear).

c) Let f be multiplicative, satisfying $\sum_{n\le x} |f(n+K) - f(n)| = O(x)$ for some K Then either:

(i) $\sum_{n\le x} |f(n)| = O(x),$ *or*

(ii) $f(n) = n^{\sigma+i\tau} \cdot u(n)$, where $0 < \sigma < 1$ and u is a periodic multiplicative function with period K

K.-H. Indlekofer and I. Kátai. *Multiplicative functions with small increments.* Acta Math. Hung. 55 (1990), No. 1/2, 97–101.

§ XVI.26 Conditions on a multiplicative function to be completely multiplicative

Let $f(n)$ be a multiplicative function with $\sum f(n)$ being convergent. Suppose that there exists a positive decreasing function $g(x)$ with the following properties:

$$g(x) = o(g(2x)) \quad (x \to \infty); \quad \Delta f(n) = o(g(n)) \quad (n \to \infty);$$

$\sum\limits_{n \geq N} n|\Delta^2 f(n)| = o(g(N))(N \to \infty)$; $\limsup\limits_{x\to\infty} \dfrac{1}{g(x)} \cdot \left|\sum\limits_{n\geq x} f(n)\right| > 0$. Then $f(n)$ is completely multiplicative.

If $\Delta^2 f(n)$ is replaced by $\Delta^2|f(n)|$ and $\sum f(n)$ is absolutely convergent, then

$$|f(n)| = n^\sigma$$

for some $\sigma < 0$

X. Yu. *A note on the multiplicative functions* (Chinese). J. Shandong Univ., Nat. Sci. Ed. 22 (1987), 48–54.

§ XVI.27 Delange's theorem on mean-values of multiplicative functions

Let $f(n)$ be a multiplicative function satisfying $|f(n)| \leq 1$.

a) If the mean-value $M(f) = \lim\limits_{x\to\infty} \dfrac{1}{x} \cdot \sum\limits_{n\leq x} f(n)$ exists, and $M(f) \neq 0$, then the series

$$\sum_p \frac{1 - f(p)}{p}$$

converges (p prime.)

b) If the above series is convergent, then f has a mean-value, and

$$M(f) = \prod_p (1 - 1/p) \cdot \sum_{k=0}^{\infty} f(p^k) \cdot p^{-k}$$

$M(f) \neq 0$, excepting when $f(2^k) = -1, k = 1, 2, \ldots$ (Delange's theorem.)

H. Delange. *Sur les fonctions arithmétiques multiplicatives*. Ann. Aci. École Norm. Sup. 78 (1961), 273–304; H. Delange. *Application de la méthode du crible à l'étude des valeurs moyennes de certain fonctions arythmétiques*. Sémin. Delange-Pisot, Paris, 1962.

c) If $\sum\limits_{n\leq x} f(n) = Mx + O(x^\alpha)$, where $M \neq 0, \alpha < 1$, then

$$\sum_p \frac{|1 - f(p)|}{p} \cdot \exp(c\sqrt{\log p})$$

is convergent for a suitable $c > 0$

R.G. Kristhal. *Über den Satz von Delange* (Russian). Dokl. Akad. Nauk UzSSR, 1976, no. 5, 5–7.

d) Let $f(n)$ be a multiplicative function with $\sum_{n\le x}|f(n)|^2 \ll x$ and having a nonvanishing mean value $M(f)$. Then for each character $\chi \pmod q$ (q positive integer), the mean value of $f\chi$ exists, and

$$M(f\chi) = M(f)\prod_{p|q}\eta(f, p)$$

if $\chi = \chi_0$ and

$$M(f\chi) = 0$$

otherwise, where $\eta(f, p) = \sum_{\upsilon=0}^{\infty} f(p^\upsilon)/p^\upsilon$

L. Lucht and F. Tuttas. *Mean values of multiplicative functions and natural boundaries of power series with multiplicative coefficients*. J. London Math. Soc. (2) 19 (1979), 25–34.

Remark. Lucht and Tuttas prove also that if $\sum_{n\le x}|f(n)|^\lambda \ll x$ for some $\lambda > 1$, then for $x \ge 0$,

$$\sum_{n\le x}|f(n)| \ll x\cdot(\log x)^{-1+1/\lambda}\cdot\exp\left(\sum_{p\le x}|f(p)|/p\right)$$

e) Assume that $\sum_{n\le x} f(n) = M(f)x + O(x/\log^a x)(x\to 0)$, where $M(f)\ne 0$ and $a > 1$. Then, there exists a positive integer r such that

$$f(2^r) = -1$$

and

$$\sum_{p\le x}(1-f(p))/p = c + O(1/\log\log x)$$

as $x\to\infty$

I.V. Elistratov. *The remainder term in Delange's theorem* (Russian). *Investigations in number theory* (Russian), 38–44, Saratov. Gos. Univ. Saratov, 1987.

Remark. This theorem strengthens a result of Postnikov.

A.G. Postnikov. *On a theorem of Delange* (Russian.) *Current problems of analytical number theory* (Russian). Minsk, 1972, 168–177, Izdat. Nauka. i Tekhnika, Minsk, 1974.

2) a) Let $f(n)$ be a multiplicative function with $M(f)\ne 0$ and $\sum_{n\le x}|f(n)|^2 = O(x)$

Then the series

$$\sum_p \frac{f(p)-1}{p}, \sum_p\frac{|f(p)-1|^2}{p}, \sum_p\sum_{k\ge 8}\frac{|f(p^k)|^2}{p^k}$$

are all convergent.

b) If the above series are convergent, then $M(f)$ exists and $M(f) = \prod_p (1 - 1/p) \cdot \sum_{k=0}^{\infty} f(p^k) \cdot p^{-k}$. $M(f) = 0$ only when for some p, $\sum_{k=0}^{\infty} f(p^k) \cdot p^{-k} = 0$.

P.D.T.A. Elliott. *A mean-value theorem for multiplicative functions*. Proc. London Math. Soc. 31 (1975), 418–438.

3) a) Let f, g be multiplicative functions, and assume that $M(g)$ exists, and that for c_1, c_2, c_3, δ positive numbers ($c_2 < 2$) the following conditions are satisfied.

(i) $\max\{|f(p^r)|, |g(p^r)|\} \leq c_1 \cdot c_2^r$

(ii) $\max\{|f(p)|, |g(p)|\} \leq c_1 \cdot p^{1/2-\delta}$

(iii) $1 + g(p) \cdot p^{-s} + g(p^2) \cdot p^{-2s} + \cdots \neq 0$ in $\operatorname{Re} s \geq 1$, for all primes p and integers $r \geq 2$;

(iv) $\max\left\{\sum_{p \leq x} |f(p)|, \sum_{p \leq x} |g(p)|\right\} \leq c_3 x$ for $x \geq 1$; and

$$\sum_p \frac{|f(p) - g(p)|}{p} < \infty$$

Then $M(f)$ exists, and

$$M(f) = M(g) \cdot \prod_p \left(1 + \sum_{r=1}^{\infty} \frac{f(p^r)}{p^r} \cdot \left(1 + \sum_{r=1}^{\infty} g(p^r)/p^r\right)^{-1}\right)$$

W. Schwarz. *Eine weitere Bemerkung über multiplicative Funktionen*. Colloq. Math. 28 (1973), 81–89.

b) Let $f : \mathbb{N} \to \mathbb{C}$ be a multiplicative function such that the series $\sum_p \sum_{k \geq 2} p^{-k} |f(p^k)|$, $\sum_p (1 - f(p))/p$ and $\sum_{\{p: |f(p)-1| > c\}} (f(p) - 1)/p$ converge for some c. Then the mean value $M(f)$ exists.

E. Heppner. *Über Mittelwerte multiplikativer zahlentheoretischer Funktionen*. Ann. Univ. Sci. Budapest, Eötvös Sect. Math. 25 (1982), 85–96.

4) Let $\lambda > 1$ and f a multiplicative function with

$$\limsup_{x \to \infty} \frac{1}{x} \cdot \sum_{n \leq x} |f(n)|^{\lambda} < \infty$$

for which $M(f)$ exists and is $\neq 0$

Then the following series are convergent: $\sum_p \frac{f(p)-1}{p}$, $\sum_{|f(p)|\le 3/2} \frac{|f(p)-1|^2}{p}$, $\sum_{|f(p)|>3/2} |f(p)|^\lambda/p$, $\sum_p \sum_{k\ge 2} \frac{|f(p^k)|^\lambda}{p^k}$, and $\sum_{k=0}^{\infty} f(p^k)/p^k \neq 0$ for all primes p.

H. Daboussi. *Sur les fonctions multiplicatives ayant une valeur moyenne non nulle.* Bull. Soc. Math. France 109 (1981), 183–205.

§ XVI.28 Halász' theorem

1) Let $f(n)$ be a multiplicative function with $|f(n)| \le 1$.

a) If the series $\sum_p \frac{1 - \operatorname{Re}(p^{i\lambda} \cdot f(p))}{p}$ is divergent for all λ, then

$$F(x) = \sum_{n\le x} f(n) = o(x)$$

b) If the above series is convergent for one λ (such λ is only one), and $f(n) = n^{i\lambda} \cdot g(n)$, then

$$F(x) = \frac{x^{1+i\lambda}}{1+i\lambda} \prod_{p\le x} (1 - 1/p) \sum_{k=0}^{\infty} f(p^k) p^{-k} + o(x)$$

(Halász's theorem.)

G. Halász. *Über die Mittelwerte multiplikativer zahlentheoretischer Funktionen.* Acta Math. Acad. Sci. Hungar. 19 (1968), 365–403.

Corollary. (Wirsing.) Any *real-valued* multiplicative function of modulus ≤ 1 has a mean value.

E. Wirsing. *Das asymptotische Verhalten von Summen über multiplikative Funktionen* II. Acta Math. Sci. Hungar. 18 (1967), 411–467.

2) Let $f(n)$ be a multiplicative function, such that

a) For certain $\delta > 0$,

$$\sum_{p\le x} \frac{|f(p)|^{1+\delta} \cdot \log p}{p} = O(\log x)$$

b) $\sum_p \sum_{k\ge 2} |f(p^k)| \cdot p^{-k} < \infty$

c) For $\operatorname{Re} s = 1$,

$$\sum_{k=0}^{+\infty} f(p^k) \cdot p^{-ks} \neq 0$$

d) For the Dirichlet series D_f of the function f,

$$D_f(s) = \frac{c}{s-1} + o\left(\frac{|s|}{\operatorname{Re} s - 1}\right)$$

with a constant c, uniformly for $\operatorname{Re} s \to 1$. Then

$$M(f) = c$$

(See G. Halász.)

3) Let $f(n)$ be a multiplicative function satisfying $|f(n)| \leq 1$. Let $F(x; k, l) = \sum_{n \leq x, n \equiv l (\mathrm{mod} k)} f(n)$. Then either

$$\frac{1}{x} F(x; k, l) \to 0$$

$(x \to \infty)$ or

$$\frac{1}{x} F(x; k, l) = C_{k,l} \cdot x^{i\lambda} \cdot \exp(i A_{k,l}(x)) + o(1)$$

where λ, $c_{k,l}$ are constants, and $A_{k,l}(x)$ is a function of x.

H. Delange. *Sur les fonctions arithmétiques multiplicatives de module* ≤ 1. Acta Arith. 42 (1983), 121–151.

Remark. This papers contains many results and generalizations of Halász' theorem 1). For such generalization, see also

A. Parson and J. Tull. *Asymptotic behaviour of multiplicative functions.* J. Number Theory 10 (1978), 395–420.

§ XVI.29 Wirsing's theorem

1) a) Every real-valued multiplicative function which is bounded and nonnegative has a mean-value.

A. Wintner. *Mean-value of arithmetical representations.* Amer. J. Math. 67 (1945), 481–485.

b) Every real-valued multiplicative function of modulus ≤ 1 has a mean value.

E. Wirsing. *Das asymptotische Verhalten von Summen über multiplikative Funktionen* II. Acta Math. Sci. Hungar. 18 (1967), 411–467.

c) Let f be a nonnegative strongly multiplicative function for which $\sum_p \frac{f(p) - 1}{p}$ converges, and such that, for all $\varepsilon > 0$ there is a $\delta > 0$ and $N > 0$ such that $\sum_{n \leq p \leq n(1+\varepsilon)} f(p) \log p / p \geq \delta$ for all $n \geq N$. Then $M(f)$ exists and is positive or 0 according as the series $\sum_p g(p)^2/p^2$ converges or diverges; in either case

$$M(f) = \lim_{x\to\infty} \prod_{p\le x} \left(1 + \frac{f(p)-1}{p}\right)$$

P. Erdős and A. Rényi. *On the mean value nonnegative multiplicative number-theoretical functions.* Michigan Math. J. 12 (1965), 321–338.

Remark. Previously, Erdős proved that if $f(n)$ is a nonnegative strongly multiplicative function such that the series $\sum_p \frac{f(p)-1}{p}$, $\sum_p \frac{(f(p)-1)^2}{p}$ converge, then $M(f)$ exists and

$$M(f) = \prod_p \left(1 + \frac{f(p)-1}{p}\right)$$

P. Erdős. *Some asymptotic formulas for multiplicative functions.* Bull. Amer. Math. Soc. 53 (1947), 536–544.

d) A multiplicative function f is called exponentially multiplicative if for all primes p and all $k \ge 2$ (integer) $f(p^k) = \frac{1}{k!}(f(p))^k$.
Let $f(n)$ be nonnegative and *exponentially multiplicative* function such that the series $\sum_p \frac{f(p)-1}{p}$ converges and such that for each $\varepsilon > 0$ there exist positive constants $\delta(\varepsilon)$ and $N(\varepsilon)$ with the property: $\sum_{n\le p\le n(1+\varepsilon)} \frac{(f(p)\log p)}{p} \ge \delta(\varepsilon)$ for all $n \ge N(\varepsilon)$. Then $M(f)$ exists and $M(f) = \prod_p \left(1 - \frac{1}{p}\right) \cdot \exp \frac{f(p)}{p}$.

(See P. Erdős and A. Rényi).

2) a) Let $f(n) \ge 0$ be a multiplicative function, satisfying $f(p^v) \le \gamma_1 \cdot \gamma_2^v$ with $\gamma_2 < 2$, p prime ($v = 2, 3, \ldots$), and $\sum_{p\le x} f(p) \sim \tau x/\log x$ ($x \to \infty$). Then

$$\sum_{n\le x} f(n) \sim \frac{e^{-\gamma\tau}}{\Gamma(\tau)} \cdot \frac{x}{\log x} \cdot \prod_{p\le x} \left(1 + \frac{f(p)}{p} + \frac{f(p^2)}{p^2} + \cdots\right)$$

E. Wirsing. *Das asymptotische Verhalten von Summen über multiplikative Funktionen.* Math. Ann. 143 (1961), 75–102.

b) Let $f(n) \ge 0$ be a multiplicative function satisfying:

(i) $\sum_{p\le x} \frac{\log p}{p} \cdot f(p) \sim \tau \log x$
($\tau > 0$);

(ii) $f(p) \ll 1$

(iii) $\sum_{p,\upsilon\ge 2} \frac{f(p^\upsilon)}{p^\upsilon} < \infty$

Then

$$\sum_{n\le x} f(n) \sim \frac{e^{-\gamma\tau}}{\Gamma(\tau)} \cdot \frac{x}{\log x} \cdot \prod_{p\le x} \left(1 + \frac{f(p)}{p} + \frac{f(p^2)}{p^2} + \cdots\right)$$

(See E. Wirsing (1967).)

c) Let $f(n) \ge 0$ be multiplicative such that $f(p^k) \ll p^{\lambda k}$ for all $k \ge 1$ and some $0 < \lambda < 1/2$. Then

$$\sum_{p\le x} f(p) \sim \tau(x) \cdot x/\log x$$

implies that

$$\sum_{n\le x} f(n) \sim \frac{\tau(x)\cdot x}{\log x} \cdot \prod_{p\le x} \left(1 + \frac{f(p)}{p} + \cdots\right)$$

B.M. Shirokov. *The summation of multiplicative functions* (Russian). Zap. Nauchn. Sem. Leningrad, Otdel. Mat. Inst. Steklov (LOMI) 106 (1981), 158–169, 172.

d) Let $f(n) \ge 0$ be multiplicative and put $T(x) = \sum_{p\le x} f(p)$, $t(x) = \sum_{p\le x} f(p)/p$. If $T(x) \sim \alpha x(\log x)^{-1} t(x)$; $\sum_p f^2(p) \cdot p^{2(\delta-1)} < \infty$; $\sum_{p,\upsilon\ge 2} f(p^\upsilon) \cdot p^{\upsilon(\delta-1)} < \infty$ with $\alpha > 0$, $\delta > 0$, then

$$\sum_{n\le x} f(n) \sim \alpha x(\log x)^{-1} \cdot \sum_{n\le x} \frac{f(n)\omega(n)}{n}$$

L. Lucht. *Abschätzungen von Summen über multiplikative Funktionen*. J. Reine Angew. Math. 280 (1976), 91–97.

§ XVI.30 Mean value of fg and $f * g$

1) Let f, g be multiplicative functions. If $|f(n)| \le 1$, $|g(n)| \le 1$ and both $M(f)$ and $M(g)$ exist and are non-zero, then $M(fg)$ also exists. (Here $M(f)$ denotes the mean value of f.)

H. Delange. *A remark on multiplicative functions*. Bull. London Math. Soc. 2 (1970), 183–185.

2) Let f, g be multiplicative, $0 < \beta \le 1$, k fixed integer,

$$\sum_{n\le x, n\equiv l(\mathrm{mod} k)} \sum f(n) = AxL(\log x) + R(x)$$

for any $l \le k$, $(l,k) = 1$, where $R(x) = O(x^\beta)$, or $R(x) = o(x^\beta)$, and where L is slowly oscillating (i.e. $|L| = 1$, $L(u')/L(u) \to 1$ as $u \to \infty, u \le u' \le 2u$) and assume that $\sum_{n=1}^{\infty} |g(n)|/n^\beta \le \infty$. Then for any $l \le k$, $(l,k) = 1$ one has

$$\sum_{n\le x, n\equiv l(\mathrm{mod} k)} (f * g)(n) = Ax \cdot L(\log x) \cdot \prod_{p\nmid k} \left(1 + \sum_{r=1}^{\infty} \frac{g(p^r)}{p^r}\right) +$$

$$+ (1-\alpha)o(x) + O(R(x))$$

P. Sabirov and S.T. Tulyaganov. *Asymptotics of the mean values of related multiplicative functions* (Russian). Izv. Akad. Nauk UzSSR Ser. Fiz.-Mat. Nauk 1987, 36–40, 89.

Remark. This generalizes a result of Heppner and Schwarz.
E. Heppner and W. Schwarz. In: Studies in pure mathematics, 323–336, Birkhäuser, Basel, 1983.

§ XVI.31 Mean value of $f(P(n))$, P a polynomial

Let $f(n)$ be a strongly multiplicative function satisfying $|g(n)| \le 1$ and $f(p_k) \to 0$ $(k \to \infty)$, where p_k is the kth prime. Let $\rho(p)$ be the number of residue classes mod p for which $P(x) \equiv 0 \,(\mathrm{mod} p)$, where $F(x) > 0$ is an irreducible polynomial with integer coefficients. If $\sum_{k=1}^{\infty} (f(p_k) - 1)\rho(p_k)/p_k$ is convergent, then $f(P(n))$ has a mean value, and

$$M(f(P)) = \prod_{k=1}^{\infty} (1 + (f(p_k) - 1)\rho(p_k)/p_k)$$

J. Galambos. *A probabilistic approach to mean values of multiplicative functions.* J. London Math. Soc. (2) 2 (1970), 405–419.

§ XVI.32 Multiplicative functions $|f| \le 1$: Summation formulas

Let $f(n)$ be a complex-valued multiplicative function with $|f(n)| \le 1$. Then:

a) $$\frac{w}{x} \sum_{n \le x/w} f(n) = w^{i\tau} \frac{1}{x} \sum_{n\le x} f(n) + O\left(\left(\frac{\log 2w_0}{\log x}\right)^{1/19}\right)$$

holds uniformly for all such functions f and $1 \le w \le w_0 \le x$ with a real number τ satisfying $|\tau| \le (\log x)^{1/19}$ and depending at most on f, x and w_0, and with $\tau = 0$ if f is real-valued.

P.D.T.A. Elliott. *Multiplicative functions* $|f| \le 1$ *and their convolutions: An overwiew.* Sém. Théor. Nombres, Paris / Fr. 1987–88, Prog. Math. 81 (1990), 63–75.

b) Let D be an odd integer. Then there is a real τ, $|\tau| \le (\log x)^{1/19}$ such that

$$\sum_{n \le x, (n,D)=1} f(n) = \eta(D) \cdot \sum_{n \le x} f(n) + O\left(\frac{x(\log\log 3D)^2}{(\log x)^{1/19}}\right)$$

where

$$\eta(D) = \prod_{p|d}\left(1 + \sum_{k \le \log x/\log p} p^{-k(1+i\tau)} f(p^k)^{-1}\right)$$

Here one can take $\tau = 0$ for real functions.

(See P.D.T.A. Elliott.)

§ XVI.33 Indlekofer's theorem

a) Let $f : \mathbb{N} \to \mathbb{C}$ be a multiplicative function and let, for any real number $\alpha \ge 1$, L_α be the space of uniformly summable functions f with $\limsup_{x\to\infty}\left(\frac{1}{x}\sum_{n \le x}|f(n)|^\alpha\right)^{1/\alpha} < \infty$. Let L^* be the space of summable functions f, i.e. satisfying $\lim_{K\to\infty} \sup_{x \ge 1} \frac{1}{x} \sum_{\substack{n \le x \\ |f(n)| \ge K}} |f(n)| = 0$. Then:

(i) If $f \in L^* \cap L_\alpha$ and if the mean value $M(f)$ of f exists and is non-zero, then the series

(*) $$\sum_p \frac{f(p)-1}{p}, \quad \sum_{\substack{p \\ |f(p)| \le 3/2}} \frac{|f(p)-1|^2}{p},$$

$$\sum_{\substack{p \\ ||f(p)|-1| > 1/2}} \frac{|f(p)|^\lambda}{p}, \quad \sum_p \sum_{k \ge 2} \frac{|f(p^k)|^\lambda}{p^k},$$

converge for $\lambda = 1$ and $\lambda = \alpha$ and, for each prime p,

(**) $$\sum_{k=1}^{\infty} \frac{f(p^k)}{p^k} + 1 \ne 0,$$

(ii) If the series (*) converges then $M(f)$ exists, $f \in L^* \cap L_\alpha$ and $M(|f|^\lambda)$ exists for $\lambda = 1$ and $\lambda = \alpha$ (and is non-zero). If in addition (**) holds then $M(f) \neq 0$.

K.-H. Indlekofer. *A mean-value theorem for multiplicative functions.* Math. Z. 172 (1980), 255–271.

§ XVI.34 Cesàro means of additive functions

Let $L_\alpha = \{f : \mathbb{N} \to \mathbb{C}, \|f\|_\alpha < \infty\}$, where

$$\|f\|_\alpha = \left(\limsup_{x\to\infty} \frac{1}{x} \sum_{n\le x} |f(n)|^\alpha \right)^{1/\alpha}$$

for $\alpha \ge 1$, and

$$\|f\|_\alpha = \limsup_{x\to\infty} \frac{1}{x} \sum_{n\le x} |f(n)|^\alpha$$

for $0 < \alpha < 1$. Let $L^* = \left\{ f : \mathbb{N} \to \mathbb{C},\ \lim_{K\to\infty} \sup_{x\ge 1} \frac{1}{x} \sum_{\substack{n\le x \\ |f(n)|\ge K}} |f(n)| = 0 \right\}$

a) Let $f : \mathbb{N} \to \mathbb{C}$ be multiplicative and $\alpha \ge 1$. Then, in order that $f \in L_\alpha \cap L^*$ and $\|f\|_1 > 0$, it is both necessary and sufficient that the series

$$\sum_{\||f(p^m)|-1|>1/2} \||f(p^m)| - 1|^\alpha / p^m, \quad \sum_{\||f(p)|-1|\le 1/2} (|g(p)| - 1)^2/p$$

converge, and that

$$\sum_{p\le x} \frac{|f(p) - 1|}{p} \le c_1$$

for all x and $\sum_{p\le x} \frac{|f(p)| - 1}{p} \to -\infty$ as $x \to \infty$

Corollary. Let $\alpha \ge 1$ and $f \in L_\alpha \cap L^*$ be multiplicative with $\|f\|_1 > 0$. Then the mean value of f exists and equals zero if and only if one of the following conditions hold.

(i) For each $t \in \mathbb{R}$ the series $\sum_{\||f(p)|-1|\le 1/2} \frac{1 - Ref(p)(|f(p)|p^{it})^{-1}}{p}$ diverges

(ii) There exists a $t \in \mathbb{R}$ such that the above series converges and $\sum_{m=1}^{\infty} f(p^m) \cdot p^{-m} \cdot p^{-imt} + 1 = 0$ for some prime p.

K.-H. Indlekofer. *Remark on a theorem of G. Halász*. Arch. der Math. 36 (1980), 145–151.

b) Let f be multiplicative and $\alpha > 0$. Then the following assertions hold:

(i) If $\|f - c\|_\alpha = 0$ with $c \neq 0$ then $c = 1$ and $f(n) = 1$ for all n

(ii) $\|f\|_\alpha = 0$
if and only if $|f|^\alpha \in L^*$ and one of the series

$$\sum_{\substack{p \\ ||f(p)|-1| \leq 1/2}} \frac{(|f(p)| - 1)^2}{p}, \quad \sum_{\substack{p \\ ||f(p)|-1| > 1/2}} \frac{||f(p)| - 1|^\alpha}{p}$$

diverges
or

$$\sum_{p \leq x} \frac{|f(p)| - 1}{p} \to -\infty$$

as $x \to \infty$

K.-H. Indlekofer. *Cesàro means of additive functions*. Analysis 6 (1986), 1–24.

§ XVI.35 Multiplicative functions on short intervals

For any real-valued multiplicative function $f(n)$, with $|f(n)| \leq 1$, and any function $3 \leq \lambda(x) \leq x$ satisfying $\log \lambda(x) \sim \log x (x \to \infty)$, the limit

$$\lim_{x \to \infty} \frac{1}{\lambda(x)} \sum_{x - \lambda(x) < n \leq x} f(n)$$

exists.

A. Hildebrand. *Multiplicative functions in short intervals*. Canad. J. Math. 39 (1987), 646–672.

See also

A. Hildebrand. *Recent progress in probabilistic number theory*. Soc. Math. France, Astárisque 147–148 (1987), 95–106.

Remarks: i) Hildebrand uses the method of proof of the above result (based on a large sieve inequality) to deduce the well-known Wirsing mean-value theorem.

ii) A generalization of this result to certain strongly additive functions was obtained by Indlekofer.

K.-H. Indlekofer. *Limiting distributions of additive functions in short intervals* (preprint).

2) Let $f(n)$ be a real-valued multiplicative function with $|f(n)| \leq 1$ and put

$$M(x) = \frac{1}{x} \sum_{n \le x} f(n)$$

For $3 \le x \le y \le x^{5/4}$ we have

$$|M(x') - M(x)| \ll \left(\log \frac{\log x}{\log(2y/x)} \right)^{-1/2}$$

where the implied constant is absolute.

A. Hildebrand. *On Wirsing's mean value theorem for multiplicative functions*, Bull. London Math. Soc. 18 (1986), 147–152.

3) a) Let f be a strongly multiplicative function, $\delta(p) = f(p) - 1 > 0$ for all primes p and $\delta(p) \downarrow 0$ for $p \to \infty$. Let $\lambda(x) \uparrow \infty$ and $h(x) > \lambda(x) \frac{1}{\log \log x}$ for $x \to \infty$. Then

$$\lim_{x \to \infty} x^{-h(x)} \cdot \sum_{x < n \le x + x^{h(x)}} f(n) = M(f)$$

the mean value of f

b) Let f and h be given as in a) and let $F \in \mathbb{I}[x]$ be an irreducible polynomial with integer cofficients. Then

$$\lim_{x \to \infty} x^{-h(x)} \cdot \sum_{x < n \le x + x^{h(x)}} f(|F(n)|) = M(f)$$

P. Erdős and K.-H. Indlekofer. *Multiplikative Functionen auf kurzen Intervallen.* J. Reine Angew. Math. 381 (1987), 148–160.

c) For each sequence $\varepsilon_m \downarrow 0$ there exists a function $h(x) \downarrow 0$ with $h(m) \ge \varepsilon_m$ and a strongly multiplicative function f with $f(p) = 1 + \delta(p)$, $\delta(p) \ge 0, \delta(p) \to 0\,(p \to \infty)$, such that the mean value $M(f)$ of f exists, but

$$\limsup_{x \to \infty} x^{-h(x)} \cdot \sum_{x < n \le x + n^{h(x)}} f(n) = +\infty$$

(See P. Erdős and K.-H. Indlekofer.)

§ XVI.36 Multiplicative functions on arithmetic progressions. Elliott's theorems

1) Let $f(n)$ be a complex-valued multiplicative function which satisfies $|f(n)| \le 1$ for all n

a)
$$\sum_{n\le x, n\equiv a(\mathrm{mod}\, q)} f(n) = \frac{1}{\varphi(n)} \sum_{n\le x,(n,q)=1} f(n) + O\left(x\left(\frac{\log\log X}{\log X}\right)^{1/8}\cdot\frac{\log X}{\log x}\right)$$

holds uniformly for $x \le X$, all $(a, q) = 1$, for all moduli q except possibly for those moduli q which are multiples of a certain $q_0 > 1$

P.D.T.A. Elliott. *Multiplicative functions on arithmetic progressions*. Mathematika 34 (1987), 199–206.

b) There is a positive constant c such that for each fixed β, $0 < \beta < \frac{1}{2}$,

$$\sum_{p\le x^\beta}{}' (p-1) \max_{y\le x} \max_{(r,p)=1} \left| \sum_{\substack{n\le y\\ n\equiv r(\mathrm{mod}\, p)}} f(n) - \frac{1}{p-1} \sum_{\substack{n\le y\\ (n,p)=1}} f(n)\right|^2 \ll x^2/(\log x)^c$$

Here $'$ indicates that the summation runs over all prime moduli, with the possible exception of at most one. The implied constant depends at most upon β

P.D.T.A. Elliott. *Multiplicative functions on arithmetic progressions*. II. Mathematika 35 (1988), 38–50.

c) For $x \ge 10$, $2 \le Q \le x^{1/3}$,

$$\sum_{n\le x, n\equiv a(\mathrm{mod}\, q)} f(n) = \frac{1}{\varphi(n)} \sum_{n\le x,(n,q)=1} f(n) + O\left(\frac{x}{q}\cdot\left(\log\frac{\log x}{\log Q}\right)^{-1/2}\right)$$

holds uniformly in $(a, q) = 1$ and $q \le Q$, as long as q is not a multiple of two exceptional moduli $q_1, q_2 \ge 2$

A. Hildebrand. *Multiplicative functions on arithmetic progressions*. Proc. Amer. Math. Soc. 108 (1990), 307–318.

Remark. The above theorem generalizes a result of Gallagher.

P.X. Gallagher. Invent. Math. 11 (1970), 329–339.

d) For integers $a, q > 0$ and real $x \ge 2$, let

$$E(x, q, a) = \sum_{n\le x, n\equiv a(\mathrm{mod}\, q)} f(n) - \frac{1}{\varphi(q)} \sum_{n\le x,(n,q)=1} f(n)$$

Then for each fixed $0 < \delta < 1/2$,

$$\sum_{\log^4 x < p \le x^\delta} (p-1) \max_{(a,p)=1} \max_{y\le x} |E(y, p, a)|^2 \ll \frac{x^2(\log\log x)^2}{(\log x)^2}$$

P.D.T.A. Elliott. Ibid. III.: *The large moduli*. Erdős 75th Anniv. vol., Cambridge Univ. Press (to appear).

e) Let $0 < \beta < 1$, $N \ge 2$. Then for any set of J primitive characters to moduli not exceeding Q

$$\sum_{j=1}^{J}{}' \max_{y\le x}\left|\sum_{n\le y} f(n)\chi_j(n)\right|^2 \ll x^2\left(\frac{\log(Q\log x)}{\log x}\right)^{1/3}(\log Q)^2 + xJQ\log Q$$

holds uniformly for $N^\beta \le x \le N$, with possibly one (and the same) character excluded from the summation.

Corollary. Let $0 < \beta < 1$, $N \ge 2$. Then if $Q \ge 2$

$$\sum_{q\le Q}{}' \varphi(q)\max_{y\le x}\max_{(a,q)=1}|E(y,q,a)|^2 \ll x^2\left(\frac{\log(Q\log x)}{\log x}\right)^{1/6}(\log Q)^{3/2}$$

holds uniformly for $N^\beta \le x \le N$, with possibly all multiples of one (and the same) modulus excluded from the outer sum.

P.D.T.A. Elliott. Ibid. IV.: *The middle moduli.* J. London Math. Soc. II. Ser. (to appear).

f) (i) Let $0 < \delta < 1/2$. Then

$$\sum_{q\le x_\delta}{}' \max_{(a,q)=1}\max_{y\le x}|E(y,q,a)| \ll (\log x)^{5/6}\cdot(\log\log x)^{5/3}$$

where $'$ denotes that the summation possibly excludes those moduli which are multiplies of a certain $q_0 > 1$.

(ii) If there is a primitive character $\lambda(\text{mod } q)$ and a real τ so that the series $\sum p^{-1}(1 - \operatorname{Re} f(p)\chi(p)p^{-i\tau})$ converges (p prime), then

$$\sum_{n\le x} f(n+1)d(n) = A\cdot\prod_{p\le x}\left(1-\frac{1}{p}+\frac{(p-1)^2}{p^2-p+1}\sum_{k=1}^{\infty}\frac{f(p^k\chi(p^k))}{p^k(1+i\tau)}\right)\cdot x^{1+i\tau}\log x + o(x\log x)$$

as $x\to\infty$, where $A = \dfrac{315\zeta(3)\chi(-1)}{2\pi^4\cdot(1+i\tau)q}\cdot\prod_{p|q}\dfrac{p^2}{p^2-p+1}$ (p prime)

and $d(n)$ is the divisor function.

P.D.T.A. Elliott. Ibid. V.: *Composite moduli.* J. London Math. Soc. II. Ser. (to appear).

Remark. The essential features of the above theorems is that from the condition $|f(n)| \le 1$ there is no restriction upon the multiplicative function f whatsoever.

§ XVI.37 Effective mean value estimate for complex multiplicative functions

For $0 \le \delta \le 1$, $0 \le \Phi \le \pi$, let $\varepsilon(\delta, \Phi)$ be the set of complex numbers z with

$$\operatorname{Im}(e^{-i\Phi}z)^2 \le \delta^2(1 - \operatorname{Re}(e^{-i\Phi}z)^2)$$

and we denote by $G(\delta, \Phi)$ the class of multiplicative functions g such that $|g(n)| \le 1$ for all n and $g(p) \in \varepsilon(\delta, \Phi)$ for all primes p. Let

$$W(z) = e^{i\Phi} \cdot (\mathrm{Re}\,(e^{-i\Phi}z) + i\delta\,\mathrm{Im}\,(e^{-i\Phi}z))$$

Then the integral equation

$$\frac{1}{2\pi}\int_0^{2\pi} |W(e^{i\theta} - K)| d\theta = 1 - K$$

has a unique solution $K = K(\delta, \Phi) \ge 0$ which is, for fixed Φ, a decreasing function of δ such that $K > 0$ if and only if $0 \le \delta < 1$. We have uniformly for $x \ge 1$ and $g \in G(\delta, \Phi)$

(*) $$\sum_{n\le x} g(n) \ll x \exp\left(-K(\delta, \Phi)\sum_{p\le x}\frac{1 - \mathrm{Re}\,g(p)}{p}\right)$$

Moreover the constant $K(\delta, \Phi)$ is sharp: given $(\delta, \Phi) \in [0, 1] \times [0, \pi)$ and $x \ge 3$, there is a $g \in G(\delta, \Phi)$ such that

(i) $$\sum_{p\le x}\frac{1 - \mathrm{Re}\,g(p)}{p} \to +\infty$$

$(x \to +\infty)$

(ii) $$\left|\sum_{n\le x} g(n)\right| \gg x \exp\left(-K(\delta, \Phi)\sum_{p\le x}\frac{1 - \mathrm{Re}\,g(p)}{p}\right)$$

R.R. Hall and G. Tenenbaum. *Effective mean value estimates for complex multiplicative functions.* Preprint 1991.

Remarks: (i) The function $K(\delta, \Phi)$ is a decreasing function of δ, and for all δ, Φ we have

$$K(\delta, \Phi) \le K\left(0, \frac{1}{2}\pi\right) = 1 - \frac{2}{\pi} \text{ and } \frac{1}{4}(1 - \delta^2) \le K(\delta, \Phi) \le \frac{1}{2}(1 - \delta^2)$$

(ii) When $\delta = 0$, $\Phi = \frac{1}{2}\pi$, we have $W(e^{i\theta} - K) = i\sin\theta$ for every K, hence $K\left(0, \frac{1}{2}\pi\right) = 1 - \frac{2}{\pi}$, and the theorem implies that

$$\max_{g\in G(0,\pi/2)}\left|\sum_{n\le x} g\,(n)\right| \asymp x(\log x)^{2/\pi - 1}$$

(iii) For $\delta = \Phi = 0$ one obtains the sharp form of an inequality of Tenenbaum.

G. Tenenbaum. *Introduction à la théorie analytique et probabiliste des nombres.* Institut Elie Cartan (Nancy) 1990 (Th. III. 4.7).

(iv) Quantitative estimates of the form (*) already appear in

G. Halász. *On the distribution of additive and the mean value of multiplicative arithmetic functions.* Studia Scient. Math. Hungar. 6 (1971), 211–233.

§ XVI.38 A theorem of Levin, Timofeev and Tuliagonov on the distribution of multiplicative functions. The Bakshtys-Galambos theorems

1) Let $f(n)$ be a real-valued multiplicative arithmetic function. In order that there exist functions $\alpha(x)$ and $\beta(x) \neq 0$, defined for all sufficiently large positive values of x, so that the frequencies

$$v_x\left(n; \frac{f(n)-\alpha(x)}{\beta(x)} \leq z\right)$$

possess a proper weak limiting distribution as $x \to \infty$, it is both necessary and sufficient that $f(n)$ not be identically one, that the series

$$\sum_{f(p)=0} 1/p$$

converges, and that there is a constant c so that the series

$$\sum_{f(p)\neq 0} p^{-1} \cdot \| \log |f(p)| p^{-c} \|^2$$

converges. When these three conditions are satisfied one may take $\alpha(x) = 0$, and $\beta(x) = x^c \cdot \exp\left(\sum_{p\leq x} p^{-1} \cdot \| \log f(p) | p^{-c} \|\right)$. Here

$$\|y\| = \begin{cases} y & \text{if } |y| \leq 1 \\ 1 & \text{if } |y| > 1 \end{cases}$$

The limit law will then be asymetric if and only if $f(2^k) = -2^{kc}$ for every positive integer k, or the series

$$\sum_{f(p)<0} 1/p$$

diverges.

B.V. Levin, N.M. Timofeev and S.T. Tuliagonov. *The distribution of multiplicative functions* (Russian). Liet. mat. rinkinys Lit. Mat. Sb. 13 (1973), 101–108.

Remark. The above results remain valid if "weak" is replaced by "modified-weak", i.e. if for the sequence of distribution functions $F_n(z)$ is there a distribution function $F(z)$ so that as $n \to \infty$, $F_n(z) \Rightarrow F(z)$ together with the additional requirement that $F_n(0) \to F(0)$ and $F_n(0-) \to F(0-)$. See

P.D.T.A. Elliott. *Probabilistic number theory.* I. 1979, Springer (pp. 274–276).

2) a) A strongly multiplicative real-valued function $f(n)$ has an asymptotic distribution if, and only if, each of the three series

$$\sum\nolimits^{*} \frac{\log|f(p)|}{p}, \sum\nolimits^{*} \frac{\log^2|f(p)|}{p}, \sum\nolimits^{**} 1/p$$

converge, where $\sum^*$ signifies summation over primes p such that $|\log|f(p)|| \leq 1$, while the other primes belong to $\sum^{**}$.

A. Bakshtys. *On the limiting distribution law of multiplicative arithmetical functions.* Litovsk. Mat. Sb. 8 (1968), 5–20;

and

J. Galambos. *On the distribution of strongly multiplicative functions.* Bull. London Math. Soc. 3 (1971), 307–312.

b) Let $f(n)$ as above and denote by $F(x)$ the limiting distribution function of $f(x)$. Put $K = \sum_{f(p)<0} 1/p$ and $\Delta(x) = 1 - F(x) - F(-x)$. Then

$$|\Delta(x)| \leq e^{-2K}$$

if we assume that $f(q_j) = -1$ for all j, where $q_1 < q_2 < \ldots$ denote those primes for which $f(q_j) < 0$

J. Galambos and P. Szüsz. *On the distribution of multiplicative arithmetic functions.* Acta Arith. 67 (1986), 37–62.

3) In order that the real-valued arithmetic functions $f(n)$ possess a weak limiting distribution it is both necessary and sufficient that the three series

$$\sum_{f(p)=0} 1/p, \ \sum_{f(p)\neq 0} 1/p\|\log|f(p)|\|, \ \sum_{f(p)\neq 0} 1/p\|\log|f(p)|\|^2$$

converge. When these conditions are satisfied, the limit law is symmetric if and only if $f(2^k) = -1$ for every integer k, or the series

$$\sum_{f(p)<0} 1/p$$

diverges. The limit law will be continuous if and only if $f(n)$ is never zero and the series

$$\sum_{|f(p)|\neq 1} 1/p$$

diverges.

(See B.V. Levin, N.M. Timofeev and S.T. Tuliagonov.)

§ XVI.39 Sums on multiplicative functions satisfying certain conditions

1) Let $f : \mathbb{N} \to \mathbb{R}$ be a multiplicative function satisfying

(i) $f(p) = \dfrac{1}{p+1}$ for all primes p

(ii) for all $\varepsilon > 0$, $p^j \cdot (f(p^j) - f(p^{j-1})) = O(p^{j\varepsilon})$ for all primes p and $j = 1, 2, \ldots$

Then

$$\sum_{n \leq x} \frac{f(n)}{n} = A \log x + B + O(x^{-1} \cdot (\log x)^{2/3} \cdot (\log\log x)^{4/3})$$

where A and B are constants depending only on f

V. Sita Ramaiah and D. Suryanarayana. *Sums of reciprocals of some multiplicative functions*. Math. J. Okayama Univ. 21 (1979), 155–164.

2) Let $k \geq 2$ be an integer, $\beta > \dfrac{1}{k+1}$ and $f_k : \mathbb{N} \to \mathbb{R}$ a multiplicative function with $|f_k(n)| \leq 1$ for all n, and *either* $|f_k(p^j) - 1| \leq p^{-1}(1 \leq j < k)$, $f_k(p^k) = 0$ *or* $f_k(p^j) = 1(1 \leq j < k)$, $f_k(p^k) = p^{-\beta}$ for all primes p. Then

$$\sum_{n \leq x} f_k(n) = x \cdot \sum_{n \geq 1} \frac{1}{n} \sum_{d \| n} \mu(d) f_k(n/d) +$$

$$+O(x^{1/k} \cdot \exp(-c_k(\log 2x)^{3/5} \cdot (\log\log 3x)^{-1/5}))$$

V. Sita Ramaiah and D. Suryanarayana. *On a Method of E. Cohen*. Boll. Un. Mat. Ital. B(6) 1 (1982), 1235–1251.

Remark. Using this result, Sita Ramaiah and Suryanarayana obtain several known asymptotic results, with better error terms, on unitarily k-free integers, (k, r)-integers, (k, r)-free integers, etc.

§ XVI.40 An asymptotic summation formula for multiplicative functions with $|f(n)| \leq 1$

Let $f(n)$ be a multiplicative function with $|f(n)| \leq 1$ and satisfying

$$\sum_{n \leq x} f(p) = \rho \operatorname{li} x + O(xL(x)/\log x)$$

where $\rho \neq 1$ and $L(x)$ is a monotonic slowly varying function such that $L(x) \to 0$ as $x \to \infty$. Then for $x \geq e^2$,

$$\sum_{n \leq x} f(n) \leq c_1 x L^*(x)/\log x + c_2 \rho x/\log\log x$$

where $L^*(x) = \displaystyle\int_1^x L(u)/u \, du$ and c_1, c_2 are constants.

J.P. Tull. *A theorem in asymptotic number theory*. J. Austral. Math. Soc. 5 (1965), 196–206.

§ XVI.41 An Ω-estimate for the remainder of sums of multiplicative functions

Let $f(n)$ be multiplicative function satisfying

(i) $|f(n)| \leq M$
with $M > 0$ a constant;

(ii) $\sum_{n \le x} f(n) = \alpha x + O(x)$
where $\alpha \neq 0$

(iii) there is a sequence of pairwise relatively prime numbers $m_1, \ldots, m_k, \ldots$ with $m_i > 1$ such that $|f(m_k)| \le \beta$ with $\beta < \frac{|\alpha|}{M}$, ($\beta$ is a constant)

(iv) when $k > k_0$, $m_1 \ldots m_k < a_1^{a_2 \ldots a_l^{k^v}}$, where $a_i > 1 (i = 1, 2, \ldots, l)$, $v > 0$, a constant. Then $\sum_{n \le x} f(n) = \alpha x + \Omega((\log_l x)^{1/v})$, where $\log_l x$ denotes the l-fold iterated logarithm of x

I.I. Il'jasov. *Ω-estimates for the remainder of sums of multiplicative functions* (Russian). Izv. Akad. Nauk. (1971), 31–37.

§ XVI.42 The distribution of values of some multiplicative functions

1) Let f be a multiplicative function such that $f(p^k) - p^k$ is a polynomial of degree $k - 1$ for primes p. Then:

a) $f(n) \le cn(\log \log n)^t$

b) $f(n) \ge dm / \log \log m$
where m denotes the largest odd divisor of n, and c, d, t are positive constants.

A. Ivić. *The distribution of values of some multiplicative functions.* Publ. Math. (Beograd) (N. S.) 22 (36) (1977), 89–94.

2) Let f be a multiplicative function such that
(i) $f(p^\alpha) = g(\alpha)$ (i.e. f is prime-independent)

(ii) for each positive integer α, $g(\alpha) \ge 1$, and there exists at least one value of α for which $g(\alpha) > 1$

(iii) $\limsup \frac{G(\alpha) \log \alpha}{\alpha} < t_1 \cdot \log 2$
where $G(\alpha) = \log g(\alpha)$ and $t_1 = \max_{\alpha \ge 1} \left(\frac{G(\alpha)}{\alpha} \right)$. Let A be the largest value of α with $t_1 = G(\alpha)/\alpha$. Then

$$\log f(n) \le G(A) \cdot \operatorname{li} \frac{\log n}{A} + O(\log n \exp(-c\sqrt{\log\log n}))$$

where $c > 0$ and $\operatorname{li} x = \int_2^x dt/\log t$. One has equality for infinitely many n

J.-L. Nicolas. *Grandes valeurs d'une certaine classe de fonctions arithmétiques.* Studia Sci. Math. Hung. 15 (1980), 71–77.

Remark. This improves a theorem of Heppner.

E. Heppner. Arch. Math. (Basel) 24 (1973), 63–66.

§ XVI.43 Multiplicative functions and small divisors

1) Let f be a positive multiplicative function for which $f(p^v) \ge 1$ depends only on v (p prime). Suppose $f(mn) \le f(m)f(n)$ for all m, n; $g(p^v) = (f(p)f(p^v))^k$, where $k \ge 2$ is an integer. Then

$$f(n) \le c \cdot \sum_{d|n, d \le n^{1/k}} g(d)$$

for a certain constant c

B. Landreau. *Majorations de fonctions arithmétiques en moyenne sur des ensembles de faible densité.* Séminaire de Théorie des Nombres, 1987–88 (Talence, 1987–88). Exp. No. 13, 18 pp. Univ. Bordeaux I, Talence.

Remark. This generalizes a result of van der Corput.

2) a) For each $k \ge 2$ there exists a constant $c_k > 0$ such that

$$\sum_{d|n} f(d) \ll_k \sum_{d|n, d \le n^{1/k}} f(d)$$

n squarefree, holds for all multiplicative functions f satisfying

$$0 \le f(p) \le c_k$$

for all primes p. For $k = 2$, $c_2 = 1$ is admissible.

K. Alladi, P. Erdős and J.D. Vaaler. *Multiplicative functions and small divisors. Analytic number theory and diophantine problems* (Stillwater, OK, 1984), 1–13, Prog. Math. 70, Birkhäuser Boston, Boston, MA, 1987.

b) $\sum_{d|n} f(d) \le (2k + o(1)) \sum_{d|n, d \le n^{1/k}} f(d)$ holds, where $o(1)$ tends to zero as $\omega(n)$ tends to infinity (where $\omega(n) = \sum_{p|n} 1$, and the notations are as in a)). (Thus $c_k = \dfrac{1}{k-1}$ is admissible for all $k \ge 2$)

K. Alladi, P. Erdős and J.D. Vaaler. Ibid. J. Number Theory 31 (1989), no. 2, 183–190.

§ XVI.44 An estimate for submultiplicative functions

a) Let $f : \mathbb{N} \to \mathbb{R}$ be an arithmetical function such that $f(1) = 1, 0 \leq f(n) \leq 1$ and f is submultiplicative, i.e. $f(mn) \leq f(m)f(n)$ for $(m, n) = 1$. Then

$$\sum_{n\leq x} f(n) \leq e^{\gamma} \cdot x \cdot (1 + O(\log\log x/\log x)) \cdot \prod_{p\leq x}(1 - 1/p) \cdot$$

$$\cdot (1 + f(p)/p + f(p^2)/p^2 + \cdots)$$

R.R. Hall. *Halving an estimate obtained from Selberg's upper bound method.* Acta Arith. 25 (1973/74), 347–351.

b) Let $f(1) = 1$ and f submultiplicative, and suppose that

$$\sum_{p\leq y} f(p) \log p \leq ky + O(y/\log^2 y)$$

for some constant $k > 0$; and

$$\sum_{p^r\geq y, r\geq 2} f(p^r) \cdot p^{-r} \cdot \log p^r \ll 1/\log y$$

for $y \geq 2$. Then

$$\sum_{n\leq x}{}' f(n) \leq kx(\log x)^{-1} \cdot m\left(\frac{x}{z}, z\right) + O(xm(x, z)/\log^2 x)$$

for $z \geq 2$, where $m(x, z) = \sum_{n\leq x}' f(n)/n$ and $\sum'$ indicates a summation over integers n with no prime factor less than z.

H. Halberstam and H.-E. Richert. *On a theorem of R.R. Hall.* J. Number Theory 11 (1979), 76–89.

§ XVI.45 Divisibility properties of some multiplicative functions

Let $f : \mathbb{N} \to \mathbb{I}$ be an integer-valued multiplicative function. Suppose there is a polynomial $P(x) \in \mathbb{I}[x]$ such that $f(p) = P(p)$ for all primes p. For given positive integer d, let $M(d, x) = \text{card}\,\{n \leq x : d|f(n) \text{ and } (d, f(n)/d) = 1\}$. Then

$$M(d, x) \sim cx(\log\log x)^{\alpha} \cdot (\log x)^{-\beta}$$

(where $\alpha, \beta \geq 0, c > 0$), or in some cases $M(d, x) = O(x^{1-\varepsilon}), \varepsilon > 0$.

W. Narkiewicz. *Divisibility properties of some multiplicative functions.* Bull. Acad. Polon. Sci. Sér. Sci. Math. Astronom. Phys. 16 (1968), 621–623.

Remarks: (i) For more general results, see

W. Narkiewicz. *Divisibility properties of some multiplicative functions.* Colloq. Math. Soc. János Bolyai 1974 (Debrecen). Topics in number theory.

(ii) For the arithmetical functions $\varphi(n)$ and $d(n)$ one obtains:

a) $\text{card}\,\{n \leq x : d\|\varphi(n)\} \sim c_1 \cdot x \cdot (\log\log x)^{\Omega(d)} \cdot (\log x)^{b(d)-1}$

for d odd, where $b(d)\prod_{p|d}\left(1-\frac{1}{p-1}\right)$; and

$\sim c_1' \cdot x \cdot (\log\log x)^{\Omega(d)-1} \cdot (\log x)^{-1}$

for d even;

b) card $\{n \leq x : d \| d(n)\} \sim c_2 \cdot x$

for d odd; and

$\sim c_2' \cdot x(\log\log x)^{a-1} \cdot (\log x)^{-1}$

where $2^a \| d$.

§ XVI.46 On multiplicative functions satisfying a congruence relation

1) For a fixed positive integer k let N_k be the arithmetic function defined by $N_k(n) = m$ if $n = m^k \cdot r$, where r is k-free.
Let A, B be fixed coprime positive integers. If f is an integer-valued multiplicative function with $f(B) \neq 0$ and $f(An+B) \equiv f(B) \pmod{N_k(n)}$ for every $n = 1, 2, \ldots$, then there are a non-negative integer α and a real-valued character $\chi \pmod{A}$ such that

$$f(n) = \chi(n) \cdot n^{\alpha}$$

for all $n = 1, 2, \ldots, (n, A) = 1$

I. Joó. *Note on multiplicative functions*. C. R. Acad. Sci. Paris, Sér. I (to appear).

2) Let M be a positive integer and let f be an integer-valued multiplicative function. If $f(M) \neq 0$ and $f(n+M) \equiv f(M) \pmod{n}$ holds for every positive integer n, then $f(n) = n^{\alpha}$, $(n = 1, 2, \ldots)$, where $\alpha \geq 0$ is an integer.

Bui Minh Phong and J. Féher. *Note on multiplicative functions satisfying a convergence property*. Ann. Univ. Sci. Budapest. Rolando Eötvös. Sect. Mat. 33 (1990), 261–265.

§ XVI.47 Exponential sums with multiplicative function coefficients

Suppose that for every K there is a finite set P of primes such that $\sum_{p\in P} 1/p > K$ and $\sum_{n\leq x} e(t(pn) - (qn)) = o(x)$ for all $p, q \in P$, $p \neq q$.

Then, for every uniformly summable multiplicative arithmetical function $f(n)$ and for all function t satisfying the above condition,

$$\sum_{n\leq x} f(n)\, e\,(t(n)) = o(x)$$

K.-H. Indlekofer and I. Kátay. *Exponential sums with multiplicative coefficients*. Acta Math. Hung. 54 (1989), 263–268.

§ XVI.48 Ramanujan expansions of multiplicative functions

Let $f(n)$ be a multiplicative function and $c_q(n)$ be Ramanujan's sum. If the series $\sum_p \frac{f(p)-1}{p}, \sum_p \frac{|f(p)-1|^2}{p}, \sum_p \sum_{k=2}^{\infty} |f(p^k)|^2/p^k$ converge, then

$$\sum_{q=1}^{\infty} \varphi(q) \cdot |\alpha_q(f)|^2 = \lim_{x\to\infty} \frac{1}{x} \sum_{n\le x} |f(n)|^2$$

where $\alpha_q(f) = \frac{1}{\varphi(q)} \cdot \lim_{x\to\infty} \frac{1}{x} \sum_{n\le x} f(n) c_q(n)$

R. Warlimont. *Ramanujan expansions of multiplicative functions*. Acta Arith. 42 (1983), 111–120.

§ XVI.49 Asymptotic formulae for reciprocals of quotients of additive and multiplicative functions

1) a) Let f be a multiplicative function and let $\mathscr{F}$ be the class of these functions f with the property

$$\prod_{p|n} (1 - 1/p^{\beta})^{\gamma} \le f(n)n^{-\alpha} \le \prod_{p|n} (1 - 1/p^{\beta})^{-\gamma}$$

for all positive integers n, where α, β, γ are positive numbers satisfying $0 < \beta \le 1$ and $\gamma > \alpha$

If $f \in \mathscr{F}$, then for every positive integer M

$$\sum_{\substack{n\le x \\ f(n)>1}} \frac{1}{\log f(n)} = x \cdot \sum_{m=1}^{M} \frac{(-1)^{m-1} \cdot F^{m-1}(0)}{(\alpha \log x)^m} + O(x/\log^{M+1} x)$$

where the O-constant depends on M, and for $-1/\gamma \le t \le 0$,

$$F(t) = (\alpha t + 1)^{-1} \cdot \prod_p (1 - 1/p) \cdot \left(1 + \sum_{m=1}^{\infty} p^{-m} \cdot (f(p^m)p^{-\alpha m})^t\right)$$

E. Brinitzer. *Eine asymptotische Formel für Summen über die reziproken Werte additiver Funktionen*. Acta Arith. 32 (1977), 387–391.

b) A multiplicative function f belongs to the class D if for every prime p and every positive integer k there exist numbers $a_{1,k}, a_{2,k}, \ldots, a_{k,k}$, such that

$$f(p^k) = p^k + a_{1,k} \cdot p^{k-1} + a_{2,k} \cdot p^{k-2} + \cdots + a_{k,k}$$

where $-1 \le a_{i,k} \le K$ uniformly in i and k with some $K > 0$.

If $f \in D$, and $a_{k,k} \ge -1/2$ for $k \ge k_0$, then

$$\sum_{\substack{n\le x\\ f(n)>1}} \frac{1}{\log f(n)} = \frac{x}{\log x}\cdot\left(1+O\left(\frac{\log\log\log x}{\log x}\right)\right)$$

A. Ivić. *The distribution of values of some multiplicatuve functions*, Publ. Inst. Math. (Beograd) 22 (36) (1977), 87–94.

c) Let $f(n)$ be a multiplicative function such that for all primes p, and for $v=1,2,\ldots$, we have $f(p^v)=g(v)$, where $g(1)=1, g(v)>1$ for $v\ge 2$ and $\liminf\limits_{v\to\infty} g(v)>1$. Then we have

$$\sum_{\substack{n\le x\\ f(n)>1}} \frac{1}{\log f(n)} = x\cdot\int_{-\infty}^{0}\left(C(t)-\frac{6}{\pi^2}\right)dt + O(x^{1/2}\log^{1/2} x)$$

where $C(t)=\prod_p\left(1+\sum_{k=2}^{\infty}(g^t(k)-g^t(k-1))p^{-k}\right)$

J.-M. de Koninck and A. Ivić. *An asymptotic formula for reciprocals of logarithms of certain multiplicative functions*. Canad. Math. Bull. 21 (1978), 409–413.

2) a) Let f and g be two additive functions such that for all prime p and all integers $r\ge 1$, $f(p)=g(p)=1$; $1\le f(p^r)<c_1r, 0\le g(p^r)<c_2r$, with c_1 and c_2 two positive absolute constants. Then

$$\sum_{\substack{n\le x\\ f(n)\ne 0}} \frac{g(n)}{f(n)} = x + \frac{Ax}{\log\log x} + O\left(\frac{x}{(\log\log x)^2}\right)$$

J.-M. de Koninck. *Sums of quotients of additive functions*. Proc. Amer. Math. Soc. 44 (1974), 35–38.

b) Let f and g be two additive functions for which there exist two non-zero constants a and b such that, for each prime p and each integer $r\ge 1$,

$$g(p^r)=ar\log p+R_g(p^r),\ f(p^r)=br\log p+R_f(p^r)$$

with $|R_h(p^r)-R_h(p^{r-1})|<cp^{-r\lambda}$, uniformly in $r\ge 1$, for some $\lambda>0$ and some $c>0$, whenever $h=g$ or $h=f$. Assume that $f(n)\le 0$ for all $n\ge 2$. Then

$$\sum_{2\le n\le x} \frac{g(n)}{f(n)} = \frac{ax}{b}\cdot\left(1+\sum_{i=1}^{M} A_i/(\log x)^i + O(1/\log x)^{M+1}\right)$$

where M is a fixed integer, and A_i are computable constants.

J.-M. de Koninck and A. Ivić. *Topics in arithmetical functions*, North Holland, 1980 (See p. 101).

3) a) Let $f(n)$ be a non-negative integer-valued additive arithmetical function such that for every prime p, $f(p)=1$, and $f(p^k)<Ck$ for every $k\ge 2$ and some fixed $C>0$. Then for every fixed integer $N\ge 1$ there exist computable constants $C_1,\ldots,C_N$ such that

$$\sum_{\substack{n\le x\\ f(n)\neq 0}} \frac{1}{f(n)} = C_1 x L_1(x) + \cdots + C_N x L_N(x) \log^{1-N} x + O(x \log^{-N} x)$$

where each $L_j(x)$ $(j = 1, \dots, N)$ is a slowly oscillating function asymptotic to $1/\log\log x$.

b) Let $f(n)$ be a non-negative integer-valued additive arithmetical function such that for every prime p, $f(p) = 0$, $f(p^2) = 1$ and $0 < f(p^k) < Ck$ for every $k \ge 3$ and some fixed $C > 0$. Then for every fixed integer $N \ge 1$ there exist computable constants $e_0, e_1, \dots, e_N$ such that

$$\sum_{\substack{n\le x\\ f(n)\neq 0}} \frac{1}{f(n)} = e_0 x + e_1 x^{1/2} \cdot L_1(x) \log^{-1} x + \cdots + e_n x^{1/2} \cdot L_N(x) \log^{-N} x +$$

$$+ O(x^{1/2} \cdot \log^{-N-1} x)$$

where each $L_j(x)(j = 1, \dots, N)$ is a slowly oscillating function asymptotic to $1/\log\log x$.

c) Let $f(n)$ and $g(n)$ be two non-negative integer-valued additive arithmetical functions such that for every prime p, $f(p) = g(p) = 1$, and $f(p^k) < Ck$, $g(p^k) < Ck$ for every $k \ge 2$ and some $C > 0$. Then for every fixed integer $N \ge 1$ there exist computable constants $a_j, b_j (j = 1, \dots, N)$ such that

$$\sum_{\substack{n\le x\\ f(n)\neq 0}} \frac{g(n)}{f(n)} = x \cdot \left(a_1 + b_1 L_1(x) + \frac{a_2 + b_2 L_2(x)}{\log x} + \cdots + \frac{a_N + b_N L_N(x)}{\log^{N-1} x} \right) +$$

$$+ O(x/\log^N x)$$

(See J.-M. de Koninck and A. Ivić, pp. 133–134.)

4) Let $f(n)$ be a non-negative integer-valued arithmetic function such that:

(*) $f(n)$ is multiplicative and $f(p) = 1$ for all primes p *or*

(**) $f(n)$ is additive and $f(p) = 0$ for all primes p

If $h = o(x)$ as $x \to \infty$, then, uniformly on $|z| \le 1$,

$$\sum_{x<n\le x+h} z^{f(n)} = G(z) \cdot h + O(hx^{-1/4} \log x) +$$

$$+ O(x^{1/(3-\alpha)} \log^{(4-\alpha)/(3-\alpha)} x)$$

where $G(z) = \sum_{n=1}^{\infty} g_z(n)/n$, $g_z(n) = \sum_{d|n} \mu(d) z^{f(n/d)}$ and α is a constant for which the asymptotic formula $\sum_{n \le x} d(n) = x \log x + (2\gamma - 1)x + O(x^{\alpha} \log^2 x)$ holds (e.g. $\alpha \le 346/1067$).

A. Ivić. *The distribution of values of the enumerating function of non-isomorphic abelian groups of finite order.* Arch. Math. 30 (1978), 374–379.

Remark. For an important result on $\sum_{n \le x} z^{f(n)}$, where f is a non-negative, integral-valued additive function, see

H. Delange. *Sur des formules de Atle Selberg.* Acta Arith. 19 (1977), 105–146.

For such a function, let $f(p) = 1$ for every prime p. For every $\rho \ge 0$ let $\sigma_0(\rho)$ denote the infimum of the set of real numbers $\sigma > 1/2$ for which $\sum_{p,k \ge 2} \rho^{f(p^k)} \cdot p^{-k\sigma} < +\infty$ if this set is non-empty and $\sigma_0(\rho) = +\infty$ otherwise.

Let E be the set of all $\rho \ge 0$ for which $\sigma_0(\rho) < 1$ and let $R > 1$ be the supremum of the set E. Then for every fixed integer $N \ge 0$ there exist functions $A_0(z), A_1(z), \ldots, A_N(z)$ analytic on $|z| \le 1$ such that

$$A_0(0) = A_1(0) = \ldots = A_N(0) = 0, \text{ and}$$

$$\sum_{n \le x} z^{f(n)} = x(\log x)^{z-1} \cdot \left(\sum_{j=0}^{N} A_j(z) \log^{-j} x + O(\log^{-N-1} x) \right)$$

where the O-constant is uniform for $|z| \le 1$.

§ XVI.50 Semigroup-valued multiplicative functions

1) An arithmetical function f is *stable*, if the set $f^{-1}(g)$ posseses an asymptotic density for every $g \in \mathbf{Im} f$. If these sets posses logarithmic densities, f is called *logarithmically stable*.

   a) Let G be a commutative semigroup. Then every G-multiplicative function (i.e. a G-valued multiplicative function) is logarithmically stable.

   b) We call a commutative semigroup G *stabilizing*, if every G-multiplicative function is stable. Call a *semigroup almost-group*, if $G = G_1 \cup G_2$, where G_1 is a group and G_2 is finite.
   Any semigroup which is the direct product of finitely many almost group is stabilizing.

I.Z. Ruzsa. *Semigroup-valued multiplicative functions.* Acta Arith. 62 (1982), 79–90.

2) Let G be an Abelian group and let f be G-multiplicative. Let $d(A)$ denote the (asymptotic) density of the set A. Call f *concentrated* if there is a finite subgroup G_1 of G such that $\sum_{f(p)\notin G_1} 1/p < \infty$ and *deconcentrated* otherwise.

   Then either:

   (i) f is deconcentrated, $d(f^{-1}(g)) = 0$ for all $g \in G$ *or*

   (ii) f is concentrated, $d(f^{-1}(g)) > 0$ for every $g \in \mathbf{Im} f$ and $\sum_{g\in G} d(f^{-1}(g)) = 1$

I.Z. Ruzsa. *General multiplicative functions*. Acta Arith. 32 (1977), 313–347.

3) A class $\boldsymbol{R}$ of subsets of a semigroup G is called *divisible* if $K \in \boldsymbol{R}$, $g \in G$ imply $g^{-1}\boldsymbol{K} \in \boldsymbol{R}$.

   Let G be a semigroup, f, f_0 be G-multiplicative functions and $\boldsymbol{R}$ a divisible class of subsets of G. Suppose moreover that $\sum_{f(p)\neq f_0(p)} 1/p < \infty$ and that either $f(2^k) = f_0(2^k)$ for all k or $f_0(2^k) = f_0(2)^k$ for all k. If the sequences $f_0^{-1}(\boldsymbol{K})$ have densities for all $\boldsymbol{K} \in \boldsymbol{R}$, then so do the sequences

$$f^{-1}(K)$$

   The same assertion holds for logarithmic density.

(See I.Z. Ruzsa (1977).)

Index of authors

GPSR Compliance

The European Union's (EU) General Product Safety Regulation (GPSR) is a set of rules that requires consumer products to be safe and our obligations to ensure this.

If you have any concerns about our products, you can contact us on ProductSafety@springernature.com

In case Publisher is established outside the EU, the EU authorized representative is:

Springer Nature Customer Service Center GmbH
Europaplatz 3
69115 Heidelberg, Germany

Batch number: 10739618

Printed by Printforce, the Netherlands